Hemant Kumar Saini, S.B. Goyal, Sardar M.N. Islam and T. Ananth Kumar (Eds.)
Quantum Unmanned Aerial Vehicle

Mathematical Methods in the Digital Age

Edited by
Abhishek Kumar, Satya Prakash Yadav,
Ashutosh Kumar Dubey, Andrei Chtcheprov,
Sayavur Bakhtiyarov

Volume 3

Quantum Unmanned Aerial Vehicle

Achieving Cybersecurity and AI Strategies

Edited by
Hemant Kumar Saini, S.B. Goyal, Sardar M.N. Islam
and T. Ananth Kumar

DE GRUYTER

Editors
Hemant Kumar Saini
Lincoln University College
Petaling Jaya
Malaysia
Bennett University
Plot Nos 8-11
TechZone II
Greater Noida 201310
India
hemantrhce@rediffmail.com

S. B. Goyal
Chitkara University Institute of Engineering
& Technology
Chitkara University
Punjab
India
drsbgoyal@gmail.com

Sardar M. N. Islam
Victoria University
300 Queen Street
3000 Melbourne, Victoria
Australia
Sardar.Islam@vu.edu.au

T. Ananth Kumar
IFET College of Engineering
Villupuram 605108
Tamil Nadu
India
tananthkumar@ifet.ac.in

ISBN 978-3-11-161930-9
e-ISBN (PDF) 978-3-11-161932-3
e-ISBN (EPUB) 978-3-11-161935-4
ISSN 2943-2731

Library of Congress Control Number: 2025937501

Bibliographic information published by the Deutsche Nationalbibliothek
The Deutsche Nationalbibliothek lists this publication in the Deutsche Nationalbibliografie; detailed bibliographic data are available on the Internet at http://dnb.dnb.de.

Typesetting: Integra Software Services Pvt. Ltd.
Printing and binding: CPI books GmbH, Leck

www.degruyter.com
Questions about General Product Safety Regulation:
productsafety@degruyterbrill.com

Preface

The quantum unmanned aerial vehicle (QUAV) implements cybersecurity and artificial intelligence (AI) strategies to work effectively. Modern society now stands on the verge of a technological transformation that will transform how autonomous systems perform operations in volatile connected networks. The new-age QUAV entwines quantum computing with AI and security solutions to design self-governing aerial devices that sync speed and intellect with built-in security notions.

Research leaders, industry specialists, and academicians authored this book, exploring how quantum technology impacts unmanned aerial vehicle (UAV) systems. They understand the urgent requirement to handle quantum technology's effects on UAV systems. The guide aims to provide readers with complete information about quantum-powered UAVs by simplifying their complexities while establishing an organized strategy for effective technology utilization.

Why This Book Now?

Military and civilian plus commercial UAV operations continue to grow rapidly because they face rising sophisticated cyber threats. The fundamental security structures that were effective before remain exposed due to new quantum computing systems. The requirements for core security combined with self-learning and adaptive capabilities demand systems that originate from AI and machine learning to achieve autonomous intelligent decision-making.

Quantum technologies serve as the solution for these situations. The integration of quantum cryptography, quantum AI (QAI), and quantum machine learning is unprecedented and novel potentials appear for UAV systems. Through quantum technologies, developing technologies of real-time sensor fusion capabilities, predictive maintenance with swarm intelligence and quantum-safe communication protocols, new opportunities with no horizon are opening up all around.

What This Book Offers

The book examines the convergence point of quantum genetic heuristics and QAI together with quantum cybersecurity, specifically demonstrating their effect on UAV system advancement. This book divides its contents into structured chapters that discuss the following:

Basic concepts of quantum computing and quantum information science exist at the theoretical level. QAI strategies are used for dynamic decision-making and optimization in UAV operations. UAV networks use quantum cryptography protocols to secure information from upcoming cyber threats. Ethical and regulatory frameworks are crucial for the responsible deployment of quantum-enhanced UAVs. The book presents case studies and applications of practical deployments that illustrate how

https://doi.org/10.1515/9783111619323-202

quantum technologies will transform defense operations and logistics delivery through tracking the environment and natural disasters and responses to them.

Who Will Benefit from This Book

The book addresses both mathematicians and operators from specialized sectors. Scholars in quantum sciences and UAV development need access to modern advancements in these fields. Disciplines of both aerospace engineering and AI practice directed their efforts toward building autonomous UAV systems that require both security and intelligence capabilities. The makers of security policies and cyber professionals found themselves at the forefront of establishing standards for safe UAV operations while implementing quantum-protected communication systems. STEM graduate students and their educational instructors apply quantum computing and AI concepts into autonomous systems development.

The Vision Ahead

We aim to provide an edition of this book that merges academic research with industry practicality while focusing on future directions. QUAVs represent a critical solution for modern twenty-first-century security requirements because they enable progress in autonomous systems, cybersecurity, and AI development.

This book serves as an empowerment tool for scientific researchers, engineering professionals, and decision-makers by delivering profound technical knowledge that leads to both secure drone systems and smart autonomous ethical drones.

Our invitation to investigate breakthrough solutions leads the way toward developing secure quantum technologies in UAV systems.

Editors
Hemant Kumar Saini
S. B. Goyal
Sardar M. N. Islam
T. Ananth Kumar

Foreword

A significant time of technological development supports the convergence of quantum computing with artificial intelligence (AI) and cybersecurity for reforming autonomous system advancements. Quantum unmanned aerial vehicles (QUAVs) have emerged as the most significant revolutionary innovation because they provide exceptional capabilities for secure communication, autonomous decision systems, and adaptive capabilities in different operational settings.

This book enters the market at this opportune time. Unmanned aerial vehicles (UAVs) now deeply serve applications across military defense, logistical operations, environmental surveillance, and disaster recovery, so the current time contains great urgency for network security standards, autonomous operation enhancements, and ethical deployment regulations. Current systems that have supported operations in the past are now facing new vulnerabilities from quantum security threats. AI-powered quantum-safe security solutions now require immediate attention because of urgent security threats.

This book stands out because it fully combines different fields of study. The book connects academic concepts to practical solutions and genuine aerial vehicle investigations in order to teach readers the detailed evolution of UAV systems because of quantum technologies. The book explores fundamental topics that include:

- Quantum cryptography for secure communication and data integrity in UAV networks: Machine learning techniques using quantum computing methods help UAVs perform safer and faster decisions that add intelligence to their operations.
- Quantum genetic heuristics for optimization in complex and dynamic environments: Confirmed sensors with predictive data analysis systems enable UAVs to execute better monitoring operations and automated flight controls.
- Deploying quantum-enhanced UAVs requires compliance with societal standards and current and projected legislation.
- Quantum UAVs will transform precision-oriented sectors because they merge innovative technology systems together.

The examined solutions in this book demonstrate how QUAVs can accomplish data protection and live disaster management in addition to militarily optimized routes in hostile areas.

The book features a collaboration between eminent professionals who specialize in quantum computing alongside AI specialists, cybersecurity experts, and aerospace engineers. Through collaboration, their knowledge delivers an extensive collection of information as follows:

- Scholars conduct the field of cutting-edge technology research and academic investigation.

https://doi.org/10.1515/9783111619323-203

- Engineers specializing in aerospace vehicles and AI practitioners collaborate to develop advanced UAV technology. The staff responsible for cybersecurity must now prepare for what the new quantum era will bring.
- UAV governance and security principles get defined by both policymakers and regulatory authorities. This reference offers essential value to individuals who want to work with quantum technologies in UAV systems.

The book demonstrates the potential and practical guidance needed for securing UAVs while developing their intelligence and ethical principles.

The editors and contributors executed this outstanding work to great success, and I believe those focusing on autonomous system development and quantum cybersecurity and AI strategic concerns should study this essential reference.

Sandeep Kautish
Director of AIT CSE
Chandigarh University, Mohali, Punjab, India
Sandeepkautish.e17021@cumail.in

March 2025

Contents

List of Authors

Sandhra Merin Sabu
SAINTGITS College of Engineering
Kottayam, Kerala
India
sandhra.se2325@saintgits.org
Chapter 1

Jubilant J. Kizhakkethottam
SAINTGITS College of Engineering
Kottayam, Kerala
India
jubilant.j@saintgits.org
Chapter 1

Vinu Sherimon
College of Computing and Information Sciences
University of Technology and Applied Sciences
Muscat, Sultanate of Oman
vinu.sherimon@utas.edu.om
Chapter 1

Amruthamsha P. Raju
Department of Artificial Intelligence and Data Science
Nitte Meenakshi Institute of Technology
Yelahanka, Bengaluru 560064
Karnataka, India
amruthamsha28@gmail.com
Chapter 2

Disha Ramesh
Department of Artificial Intelligence and Data Science
Nitte Meenakshi Institute of Technology
Yelahanka, Bengaluru 560064
Karnataka, India
disha48456@gmail.com
Chapter 2

Praveen Naik
Department of Chemistry
Nitte Meenakshi Institute of Technology
Yelahanka, Bengaluru 560064
Karnataka, India
praveennaik018@gmail.com
Chapter 2

Gururaj Kudur Jayaprakash
Department of Chemistry
Nitte Meenakshi Institute of Technology
Yelahanka, Bengaluru 560064
Karnataka, India
rajguru97@gmail.com
Chapter 2

J. Ramkumar
Department of Computer Science
Faculty of Management and Technology
Apex Professional University
Pasighat
Arunachal Pradesh
India
jramkumar1986@gmail.com
Chapter 3

S. P. Priyadharshini
Department of Mathematics,
Dr. N.G.P. Institute of Technology
Coimbatore, Tamil Nadu
India
priyadharshini125@gmail.com
Chapter 3

Balachandra Pattanaik
Department of Electrical and Computer Engineering
Wollega University
Nekemte, Ethiopia
balapk1971@gmail.com
Chapter 3

A. Sheik Abdullah
School of Computer Science and Engineering
Vellore Institute of Technology
Chennai, Tamil Nadu
India
sheikabdullah.a@vit.ac.in
Chapter 4

https://doi.org/10.1515/9783111619323-205

R. Parkavi
Department of Information Technology
Thiagarajar College of Engineering
Madurai
Tamil Nadu
India
rpit@tce.edu
Chapter 4

Sujitha S
Department of Information Technology,
Thiagarajar College of Engineering
Madurai
ssiit@tce.edu
Chapter 4

Utkarsh Mishra
School of Computer Science and Engineering
Vellore Institute of Technology
Chennai, Tamil Nadu
India
utkarsh.mishra2021@vitstudent.ac.in
Chapter 5

A. Sheik Abdullah
School of Computer Science and Engineering
Vellore Institute of Technology
Chennai, Tamil Nadu
India
sheikabdullah.a@vit.ac.in
Chapter 5

A. Sheik Abdullah
School of Computer Science and Engineering
Vellore Institute of Technology
Chennai, Tamil Nadu
India
aa.sheikabdullah@gmail.com
Chapter 6

Sriram B.
School of Computer Science and Engineering
Vellore Institute of Technology
Chennai, Tamil Nadu
India
csc.sriramb@gmail.com
Chapter 6

Lynda D.
School of Computer Science and Engineering
Vellore Institute of Technology
Chennai, Tamil Nadu
India
lynda.tsa@gmail.com
Chapter 6

Harini V. K.
School of Computer Science and Engineering
Vellore Institute of Technology
Chennai, Tamil Nadu
India
vkharini3@gmail.com
Chapter 6

B. Suresh Kumar
Department of Mechanical Engineering
K. Ramakrishnan College of Technology
(Affiliated to Anna University)
Samayapuram, Tiruchirappalli 621 112
Tamil Nadu, India
bsureshcvl@gmail.com
Chapter 7

T. Avudaiappan
Department of Artificial Intelligence and
Machine Learning
K. Ramakrishnan College of Technology
(Affiliated to Anna University)
Samayapuram, Tiruchirappalli 621 112
Tamil Nadu, India
avudaiappanmecse@gmail.com
Chapter 7

R. Vidhya
PG & Research Department of Physics
Bishop Heber College (Affiliated to
Bharathidasan University)
Tiruchirappalli 620 017
Tamil Nadu, India
vidhyasuresh08@gmail.com
Chapter 7

Ruthwika Vemulapalli
Electronics and Communication Undergraduate
G. Narayanamma Institute of Technology and Sciences for Women
Shaikpet, Hyderabad
India
ruthwikavemulapalli@gmail.com
Chapter 8

V. Velmurugan
IFET College of Engineering
Villupuram 605108
Tamil Nadu, India
vel_pdy@yahoo.com
Chapter 9

B. Elamvazhudi
IFET College of Engineering
Villupuram 605108
Tamil Nadu, India
elamvazhudib@gmail.com
Chapter 9

K. Kamalakannan
IFET College of Engineering
Villupuram 605108
Tamil Nadu, India
kamalakannanat@gmail.com
Chapter 9

S. Kulothungan
IFET College of Engineering
Villupuram 605108
Tamil Nadu, India
kulo.rajan@gmail.com
Chapter 9

R. Thangamani
Kongu Engineering College
Perundurai
Tamil Nadu, India
thangamani.ai@kongu.edu
Chapter 10

M. Vimaladevi
Kongu Engineering College,
Perundurai
Tamil Nadu, India
vimaladevi.ai@kongu.edu
Chapter 10

M. M. Ramyasri
Kongu Engineering College
Perundurai
Tamil Nadu, India
ramyasri.ai@kongu.edu
Chapter 10

Bhupinder Singh
Sharda University
Greater Noida
Uttar Pradesh
India
bhupindersinghlaw19@gmail.com
Chapter 11

Hemant Kumar Saini
Lincoln University College
Malaysia and Bennett University
(The Times Group)
Greater Noida
India
hemantrhce@rediffmail.com
Chapters 11 and 20

R. Bhuvaneswari
Department of Computer Science and Engineering
Faculty of Engineering and Technology
SRM Institute of Science and Technology
Ramapuram, Chennai 600089
Tamil Nadu, India
sirajarchana@gmail.com
Chapter 12

R. Ramya
Department of Artificial Intelligence and Data Science
St. Joseph's College of Engineering
Chennai
Tamil Nadu, India
Chapter 12

S. Pradeep Kumar
Saveetha School of Engineering
SIMATS
Thandalam, Chennai
Tamil Nadu, India
spradeepkumar272@gmail.com
Chapter 12

Preethi D.
Department of Computer Science and Engineering
Faculty of Engineering and Technology
SRM Institute of Science and Technology
Ramapuram, Chennai
Tamil Nadu, India
preethid1@srmist.edu.in
Chapter 12

Renjith V. Ravi
School of Applied Sciences, Engineering and Technology
Rashtriya Raksha University
Gandhinagar 382305
Gujarat, India
renjithravi.research@gmail.com
Chapter 13

S. B. Goyal
Chitkara University Institute of Engineering & Technology
Chitkara University
Punjab
India
drsbgoyal@gmail.com
Chapter 13

Nithish P. U.
Department of Computer Science and Engineering
Christ University
Kengeri Campus
Bengaluru, Karnataka
India
nithish.pu@btech.christuniversity.in
Chapter 14

Jayapandian Natarajan
Department of Computer Science and Engineering
Christ University
Kengeri Campus
Bengaluru, Karnataka
India
jayapandian.n@christuniversity.in
Chapter 14

Allen Sabu
Department of Computer Science and Engineering
Christ University
Kengeri Campus
Bengaluru, Karnataka
India
allen.sabu@btech.christuniversity.in
Chapter 14

Korhan Cengiz
Department of Electrical Electronics Engineering
Biruni University
Istanbul, Turkey
kcengiz@biruni.edu.tr
Chapter 14

Rishav Srivastav
Vellore Institute of Technology
Vellore Campus
Vellore
Tamil Nadu, India
rishav.srivastav2021@vitstudent.ac.in
Chapter 15

Khyati Sinha
Vellore Institute of Technology
Vellore Campus
Vellore
Tamil Nadu, India
khyati.sinha2021a@vitstudent.ac.in
Chapter 15

Aryaman Gautam
Vellore Institute of Technology
Vellore Campus
Vellore
Tamil Nadu, India
aryaman.gautam2021@vitstudent.ac.in
Chapter 15

Niha Kamal Basha
Vellore Institute of Technology
Vellore Campus
Vellore
Tamil Nadu, India
niha.k@vit.ac.in
Chapter 15

Christo Ananth
Samarkand State University
Samarkand
Uzbekistan
dr.christoananth@gmail.com
Chapter 15

Arun Anoop M.
Department of CSE(AIML)
AJ Institute of Engineering and Technology
Mangalore, Karnataka
India
arunanoopm@gmail.com
Chapter 16

Karthikeyan P.
Department of Electronics and Communication Engineering
Velammal College of Engineering and Technology
Madurai, Tamil Nadu
India
Chapter 16

Sabia Hassan
Department of Computer Science
Islamic University of Science and Technology
Awantipora 192122
Jammu and Kashmir
India
sabia.hassan@iust.ac.in
Chapter 17

Syed Naseer Ahmad Shah
Department of Computer Science
Jamia Millia Islamia
New Delhi 110025
India
syeddnaseer@gmai.com
Chapter 17

Kaiser Javeed Giri
Islamic University of Science and Technology
Awantipora 192122
Jammu and Kashmir, India
kaiser.giri@islamicuniversity.edu.in
Chapter 17

Gurwinder Singh
Department of AIT-CSE
Chandigarh University
Mohali
Punjab, India
gurwinder.e11253@cumail.in
Chapter 18

Preet Kamal
Department of AIT-CSE
Chandigarh University
Mohali
Punjab, India
preetkgarcha@gmail.com
Chapter 18

R. Ramya
Department of Artificial Intelligence and Data Science
St Joseph's College of Engineering
Chennai
Tamil Nadu, India
Chapter 19

R. Bhuvaneswari
Faculty of Engineering and Technology
Department of Computer Science and Engineering
SRM Institute of Science and Technology
Ramapuram, Chennai 600089
Tamil Nadu, India
sirajarchana@gmail.com
Chapter 19

S. Pradeep Kumar
Saveetha School of Engineering, SIMATS
Thandalam, Chennai
Tamil Nadu, India
spradeepkumar272@gmail.com
Chapter 19

Preethi D.
Department of Computer Science and Engineering
Faculty of Engineering and Technology
SRM Institute of Science and Technology
Ramapuram, Chennai
Tamil Nadu, India
preethid@srmist.edu.in
Chapter 19

Bhupinder Singh
Sharda University
Greater Noida, Uttar Pradesh
India
bhupindersinghlaw19@gmail.com
Chapter 20

Sandhra Merin Sabu, Jubilant J. Kizhakkethottam, and Vinu Sherimon

1 Introduction to Quantum Technologies and UAV Safety

Abstract: The study introduces the quantum computing and emphasizes on how it has the revolutionary potential to resolve the complex problems that cannot be tackled by conventional computation. This technology relies on the quantum mechanics concepts which include superposition and entanglement that give enormous processing capacity. It also presents basic concepts in quantum technology such as the qubits and quantum gates along with major problems like scalability and de-coherence. Moreover it analyzes the unmanned aerial vehicles (UAVs), tries to describe its developments and use cases, and highlights how quantum computing can improve the UAV capabilities like real-time decision-making, swarm optimization, and secure communication. The fusion of these technologies promises revolutionary advancements in future but need to address scalability, stability, and practical implementation challenges.

Keywords: Qubits, quantum computing, quantum computer, unnamed aerial vehicles, engine technology

Qubits and their representations, superposition and entanglement, quantum reversibility and gates, qubit interactions and measurement, limitations and future challenges of quantum computing, and the power of the quantum realm are all covered in this article. The rapidly advancing field of quantum computing applies the ideas of quantum physics to solve challenging puzzles that traditional computers are unable to answer. Traditional complementary metal-oxide-semiconductor technology, which creates integrated circuits on printed circuit boards for high-density and high-performance applications, has limitations that are partially addressed by quantum computing technology. However, many students find the topic extremely challenging to follow because of the incorporation of quantum phenomena, which are exceedingly difficult to explain. Thus understanding quantum computation and data processing requires knowledge of its historical context and fundamental ideas.

The number of quantum bits, also known as qubits, in a quantum computer is a common metric for demonstrating the power of using quantum technology. Universal quantum computers currently support hundreds of qubits, whereas quantum annealers support three orders of magnitude. As a result, quantum computing's power continues to increase.

https://doi.org/10.1515/9783111619323-001

1.1 Quantum Computing: Unveiling the Potential Within the Quantum Domain

A branch of computer science known as "quantum computing" makes use of quantum physics to carry out complex calculations. The fascinating and expansive topic of quantum computing has the potential to revolutionize how we approach addressing difficult problems that are currently beyond the capabilities of conventional computers.

Superposition and entanglement, two concepts from quantum physics, are used by quantum information scientists to calculate and transport data. These ideas enable quantum computers to compute at rates never before imagined conceivable, even for exponentially complicated (NP-complete) problems, which have nondeterministic calculation times.

Problems that take too long to solve on traditional machines, such as factorization of huge numbers and brute-force search, which seem to no longer be problematic with quantum algorithms. For example if we consider the reverse phone number computation, that is, finding the name of the person concerned from the directories given the phone number, which seems to be complex for classical computers to solve, it can be solved easily using quantum computer.

Quantum hardware encompasses both quantum computers and quantum annealers despite its many applications [1]. The original purpose of developing quantum computers was to simulate quantum physics. Subsequent advancements in technology allowed quantum computers to perform every function of a Turing machine. They are capable of carrying out every calculation that a Turing machine can do. The improved model is therefore sometimes called a universal quantum computer. This is usually implemented using the quantum logic gates (e.g., classical logic gates). On the other hand, quantum annealing uses quantum-mechanical probabilities to solve optimization problems (such as locating a global minimum) as opposed to computational difficulties.

The field of academics and industry has witnessed a swift advancement in quantum technologies, commonly referred to as the "quantum computer race." For example, quantum hardware for a critical network of distribution has been developed by research centers and organizations worldwide. The field of quantum technology is highly desirable to all researchers and industries involved in the sector.

Due to their potential to greatly increase work efficiency over classical technologies, quantum computing and communications are receiving an abundance of interest. Quantum computers, for instance, are significantly faster than classical computers at solving specific calculation, optimization, and search problems. In addition, quantum computers have applications in materials science and chemistry because they can more accurately simulate the functions or behavior of particular physical systems than classical computers. The quantum Internet is considered to be another example, which could offer communications that are more resistant to disruption and seem to be more safer than those offered by traditional networks.

1.2 Quantum Mechanics: The Foundation of Quantum Computing

The fundamental domain or topic of physics known as the quantum mechanics, sometimes referred to as quantum physics or may be the quantum theory which offers a framework for comprehending the properties and the behaviors of the matter and energy at its lowest scales, mainly concentrated at the atomic and the subatomic particle levels. It seems to be a very successful and well-known hypothesis that has undergone substantial experimental validation and the verification.

The findings of quantum physics, which emerged in the first half of the twentieth century, are seen as frequently bizarre and perplexing. Quantum mechanics has played a significant role in the modern physics. In the early 1920s, a group of physicists at the University of Göttingen, which include Werner Heisenberg, Max Born, and Wolfgang Pauli, together proposed the term "quantum mechanics." At the fundamental level, matters and radiation share characteristics with waves and particles.

When dealing with atoms and electrons, many of the classical mechanics equations that we use on a daily basis to explain the motion and interactions of things at ordinary sizes and speeds become unintelligible. Objects exist in classical mechanics at a specific location and time. In contrast, objects in the theory of quantum mechanics exist in a cloud of probability, with an equal chance of existing at point A, point B, and so on. Particles can exist as both waves and particles in quantum mechanics.

The fundamental theory that underpins quantum computing is quantum mechanics. The rapidly emerging field of quantum computing applies ideas from quantum mechanics to solve computational challenges that are beyond the capabilities of conventional computers. The two main tenets of quantum physics that underlie quantum computing are quantum superposition and quantum entanglement.

1.3 Basic Concepts from Bits to Qubits

Information is converted by today's computational processors into binary bits of 0s and 1s, which are subsequently processed and decoded by logic gates built from switching transistors. Sequential principles underpin the operation of these traditional computers, which process data one after the other until desired outcomes are achieved. Up until semiconductor transistor size is comparable to that of an atom, computing is currently governed by the principles of classical physics. The computing power thus cannot be increased after that.

Quantum computers, unlike classical computers, have more than simply two states. Instead, the data or information is encoded as quantum bits, or qubits, which can simultaneously be both 0 and 1, a phenomenon known as superposition. Consequently, compared to traditional bits, qubits can encode exponentially more values. Because of this inherent feature, quantum computing is a feasible platform that can solve computational problems beyond the capabilities of even the most sophisticated traditional supercomputers. Moving on to the atomic or quantum levels, quantum computers that function as a quantum computer's processor and memory can be physically demonstrated using atoms, photons, ions, or electrons along with the control devices that go with them.

Quantum computers use the superposition property of quantum particles to process, compute, and store data in multiple states at the same time, ensuring that they have enough computing power and millions of times more potential which surpass today's dominant and most well-known supercomputers. Furthermore, quantum computers provide ultrahigh speed and more secure transmission, which has the ability to store massive amounts of data compared to their classical equivalents.

1.3.1 Representations of Qubits

1.3.1.1 Single Qubit Representation

The single qubits can be represented basically different notations:

1.3.1.1.1 Dirac Notation

$$\alpha 1|0> + \alpha 2|1>$$

where α and β are the complex numbers which satisfying $|\alpha 1|^2 + |\alpha 2|^2 = 1$

α and β show the superposition of the quantum particles as shown in Figure 1.1.

1.3.1.1.2 Bloch Sphere Representation

Bloch sphere representation [1] is the visual representation of the dirac notation. It is like a spherical surface and each point on the surface represent the superposition of the quantum particles [2].

1.3.1.1.3 Vector Representation

In vector representation the qubits are represented as

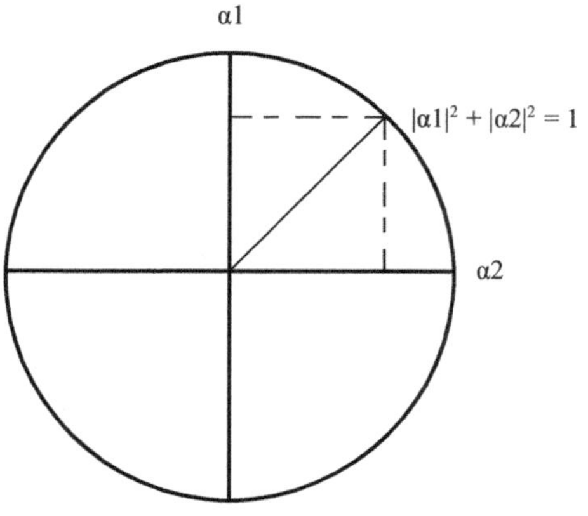

Figure 1.1: Qubit represented by a Bloch sphere.

$$|0> = (0,\ 1)$$

and

$$|1> = (1,\ 0)$$

1.3.1.1.4 Matrix Representation

In matrix representation the qubits are represented as

$$0 = \begin{bmatrix} 1 \\ 0 \end{bmatrix}$$

$$1 = \begin{bmatrix} 0 \\ 1 \end{bmatrix}$$

Here in the matrix, the top item in the matrix denotes the likelihood of collapsing to state 0, and the bottom element in the matrix denotes the potential for collapsing to state 1.

1.3.1.2 Two-Qubit Representation

Two-qubit representation is shown as

$$\alpha_0|00> + \alpha_1|01> + \alpha_2|10> + \alpha_3|11> = \begin{bmatrix} \alpha 0 \\ \alpha 1 \\ \alpha 2 \\ \alpha 3 \end{bmatrix}$$

Two-qubit representation in both dirac notation and matrix form is given below:

$$|00> = \begin{bmatrix} 1 \\ 0 \\ 0 \\ 0 \end{bmatrix}$$

$$|01> = \begin{bmatrix} 0 \\ 1 \\ 0 \\ 0 \end{bmatrix}$$

$$|10> = \begin{bmatrix} 0 \\ 0 \\ 1 \\ 0 \end{bmatrix}$$

$$|11> = \begin{bmatrix} 0 \\ 0 \\ 0 \\ 1 \end{bmatrix}$$

1.3.2 Tensor Product

A single quantum state, also known as the qubit, is expressed in a 2×1 column vector or matrix in the quantum computing information processing. Formally, the multiple-qubit systems can be expressed in terms of a specific matrix operation called the tensor product. Direct product or Kronecker product is another name for tensor products. It is direct multiplication instead of matrix multiplication.

Tensor products are characterized by: If X is $1 \times q$ matrix and Y is $x \times y$ matrix, then the tensor product $(X \otimes Y)$ is $(\mathrm{lx}) \times (qy)$ block matrix representing the multiple-qubit system state:

$$X = \begin{bmatrix} a11 & a12 \\ a21 & a22 \end{bmatrix} \text{ and } Y = \begin{bmatrix} b11 & b12 \\ b21 & b22 \end{bmatrix}$$

$$X \otimes Y = \begin{bmatrix} a11b11 & a11b12 & a12b11 & a12b12 \\ a11b21 & a11b22 & a12b21 & a12b22 \\ a21b11 & a21b12 & a22b11 & a22b12 \\ a21b21 & a21b22 & a22b21 & a12b22 \end{bmatrix}$$

The following quantum states' tensor product, expressed in dirac notation and matrix specified, is equal to

$$|0> \otimes |0> = |00> \begin{bmatrix} 1 \\ 0 \end{bmatrix} \otimes \begin{bmatrix} 1 \\ 0 \end{bmatrix} = \begin{bmatrix} 1 \\ 0 \\ 0 \\ 0 \end{bmatrix}$$

$$|0> \otimes |1> = |01> \begin{bmatrix} 1 \\ 0 \end{bmatrix} \otimes \begin{bmatrix} 0 \\ 1 \end{bmatrix} = \begin{bmatrix} 0 \\ 1 \\ 0 \\ 0 \end{bmatrix}$$

$$|1> \otimes |0> = |10> \begin{bmatrix} 0 \\ 1 \end{bmatrix} \otimes \begin{bmatrix} 1 \\ 0 \end{bmatrix} = \begin{bmatrix} 0 \\ 0 \\ 1 \\ 0 \end{bmatrix}$$

$$|1> \otimes |0> = |11> \begin{bmatrix} 0 \\ 1 \end{bmatrix} \otimes \begin{bmatrix} 0 \\ 1 \end{bmatrix} = \begin{bmatrix} 0 \\ 0 \\ 0 \\ 1 \end{bmatrix}$$

1.4 Entanglement and Superposition

We don't use or see complicated physics ideas like superposition [3] and entanglement [4] in our day-to-day existence. When concentrating on the smallest or most fundamental quantum particles, such as atoms, photons, neutrons, electrons, and other subatomic particles, it might be found.

The ability of a quantum system or particle to exist in more than one state at once is known as superposition. This means that an object can be both "here" and "there," "up" and "down," or, in the case of spin, spin in both "clockwise" and "anticlockwise" directions at the same time. Instead of requiring sequential computation, quantum0020computers' potent and practical feature allows for true parallel processing. Each qubit can represent two states in the interim. Consequently, n qubits can be simultaneously represented in $2n$ states. Therefore, in order to solve the problem in the time that a normal computer would require $2n$ steps, it is possible to build or produce a superposition of $2n$ states as input and process it all at once. The superposition of unknown photonic quantum states is introduced experimentally.

Contrarily, the entanglement is an extraordinarily powerful bond that forms in between the quantum particles. In fact it seems to be so powerful that, even in cases where there is a considerable distance between two quantum particles, they maintain an unbreakable connection in indivisible harmony. This idea was put forth by Albert Einstein in 1935. In the quantum world, particles are so interconnected that, when positioned at the farthest limits of the universe, they can be said to "dance" in perfect harmony right now. This seemingly strange link served as the basis for Einstein's description of entanglement as "spooky action at a distance."

1.5 Quantum Reversibility and Quantum Gates

The issue of concept reversibility in computations first surfaced in the 1970s. The foundation of quantum computing is the idea of reversible computation. The ability to recover the input state from the output state is known as reversible computing. In other words, we have the output and need to recreate the input from the output state.

Physical reversibility and logical reversibility are two closely related topics. Physical reversibility is the process of removing an input whose data has been permanently lost. The capacity to reconstruct the input from the output is referred to as logical reversibility. It is difficult to reconstruct or rebuild the input state from the output state, for instance, because the NAND and NOR gates only take two inputs and produce one output; in contrast, the NOT gate is reversible. It can reconstruct the input state from the output because it has only one input and one output.

One-to-one mapping or the creation of input from the output, is a capability of reversible logic. Consequently, neither energy nor information is lost in the reconstruction process. Reversibility is one of the characteristics of quantum gates. It is regarded as the most advanced feature. We may be able to circumvent the irreversible characteristic of the existing classical computer and do away with information or data loss by using a quantum computer. In the quantum circuit computation system, the quantum gates which are said to be the simple quantum circuit that can operate on a less number of qubits. Quantum gates are thought to be the fundamental components of quantum circuits. Qubits demand numerous common gates or processes. Quantum gates are available for both single and multi-qubit systems. The fundamental or the key and crucial single-qubit quantum gates include X, Y, Z, and H.

1.6 Qubit Interactions and Measurement

Interaction with quantum bits or qubits is dependent on the physical manifestation of a quantum computer. For instance, if we consider an electron in a neutral atom it might be either in ground state or excited state. An electron in atom realization can be either energized state (represented by $|1\rangle$) or in the ground state (represented by $|0\rangle$).

By applying the right amount of energy and duration to an atom, it is feasible to excite an electron from its ground state into its excited state or maybe vice versa. Similarly, an electron can reach the state $|0\rangle$ first and then travel anywhere between

|0 > and |1 >, which we refer to as the superposition state, by shortening the lighting up time.

That is, the double slit experiment in quantum physics demonstrates that subatomic particles, particularly electrons, can exist in two states at the same moment. That is, it can pass through both slits at the same time. Alternatively, it can spin both clockwise and counterclockwise at the same time until it is measured.

That is, measuring shows only one physical attribute, whereas measuring shows both properties. In the quantum paradigm or quantum model, there is no provision for measurement at every place, hence continuous measurement is not conceivable.

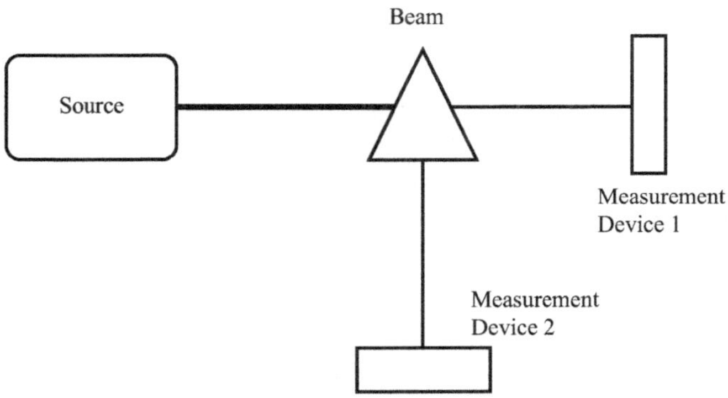

Figure 1.2: Experimental setup.

Here, we experiment with a conventional optical apparatus to illustrate the quantum effects. As seen in Figure 1.2, we present an experimental setup for producing superposition that includes a photon source (Source), two photon detectors that serve as measurement tools (*M*1 and *M*2), and a beam splitter (BS) [1].

First, the photon source emits a single photon beam. By splitting the beam in two, the BS ensures that each beam has an equal chance of traveling through the red/right path or reflecting to the blue/down path. Since there is only one photon in this setup, the measurement results ought to be mutually exclusive.

The photon has a 50% chance of being measured or picked up by one of the two detectors. Stated differently, it remains in a superposition of the two alternative pathways (two possible outcomes) prior to measurement. To represent the results in general, these two alternative outcomes can simply be thought of as 0 and 1. As a result, a quantum system can exist in a superposition state, where it exists in multiple states at once, each with an equal probability of being measured.

In our scenario, there are two conceivable states: 0 and 1. Generally speaking, though, a quantum system can exist in a multistate superposition with up to 2n potential outcomes, where n is the number of states in this case. To achieve this, analogous experimental setups can be employed. For instance, a three-qubit system can be represented in 23 states or 8 states simultaneously, at any given time.

Before measurement, it has an infinite state or possibilities, but after measurement or computation is over its certain probability will end up into one locality. That is 0 or 1. At the end a measurement is needed to read the result or output. The possibility of calculation is not measurable in the quantum state but the final output will be similar to the classical.

1.7 Unmanned Aerial Vehicle

Drones, also referred to as unmanned aerial vehicles (UAVs), are a cutting-edge technology that have revolutionized a number of industries, including environmental monitoring, defense, agriculture, and logistics. UAV systems function without the presence of a human pilot. Rather, they are either operated autonomously utilizing more advanced dynamic automation systems or preprogrammed flight plans, or they are operated remotely by a human operator [5, 6].

In the last decade, the utilization of drones has grown in civilian as well as military domains. This category also includes the UAVs, which has the potential to fly independently without human interference or may be under remote control, without requiring any kind of direct control by manual operators and also might be used to transport loads. Furthermore, such UAV can be categorized based on their size, their load capacity, the altitude of aircraft and their layout and designs. The fact is that more and more of these types of aircrafts are registered in various territories throughout the world attests their significance impact nowadays.

Applications of UAVs have traditionally been mostly focused on leisurely flights, photography, and perhaps hobbyist use. However the integration of wireless technologies and UAVs has made a variety of commercial, industrial, and public sector uses more feasible. These days, the connected UAVs are used in many different fields and domains, such as the precise agriculture, assistance, and management during disasters, environmental monitoring, infrastructure inspection, logistics and delivery services, and aerial surveillance for the security and law enforcement purposes. Numerous civil and military applications are abound for UAVs, which includes the environmental mapping, search and rescue operations, surveillance, recreational pursuit, and disaster management. Battery replacement and replenishment, collision avoidance, collision resilience, hovering performance, and finally communication dependability are crucial elements that must be examined and addressed when dealing with swarming drones. When UAVs are outfitted with wireless communication systems, they can collect and transfer information and data instantly, carry out their mission without direct human interference, and interact with other UAVs and ground control centers.

UAVs can be categorized according to a number of criteria, such as their size, range, flying characteristics, function, and flight capabilities. Understanding and analyzing UAVs' traits and capabilities seem to be made easier with the help of UAV categorization, which offers a framework for grouping UAVs into appropriate categories. UAVs, for instance, can be categorized according to their size and physical characteristics. This classification facilitates comprehension of the size and capabilities of the drones since varying the sizes of UAVs may have varying carrying capacities, flight durations, and also may have different operating capabilities. Similarly UAV can be different designs such as fixed-wing or rotary-wing design. UAVs can be also classified based on their payload capacity. Cameras, sensors, communication devices, and hardware tailored for particular uses are examples of payloads. The classification of payload capacity aids in the selection of UAVs appropriate for particular mission objectives. They can be also classified based on the spectrum of flying (flying range) and endurance levels. Based on all this the UAVs can be designed for limited or prolonged flight durations as well as those that intenced for short-range or long-range missions.

The current digital era is witnessing a revolution in the way connections, communications, and innovations are being unveiled as a result of the fifth-generation (5G) and beyond networks quick development working with UAVs. This has opened up to innovative prospects for a broad spectrum of applications and state-of-the-art technologies.

1.7.1 Historical Context and Development

UAVs as a concept are not considered to be new. The first examples date back to mid-1800s, when military surveillance was conducted using balloons. The first pilotless aircraft was created for military use during the World War I, though, and that is when the true evolution of UAVs started. The main purpose of these early drones was to educate anti-aircraft gun crews by using them as targets.

During the Cold War, the technology advanced significantly especially with the Soviet Union and the USA developing drones for reconnaissance. These drones were made to go far into enemy territory and to gather vital intelligence without endangering the pilots' lives. The capabilities and applications of UAVs have significantly advanced over time due to the developments in electronics, materials, and propulsion systems, and they now seem to be an essential component of both commercial and military operations.

1.7.2 The UAV's Technological Components

UAVs are intricate devices that incorporate a number of cutting-edge technologies. The main technological elements that make up a UAV are listed below:

1.7.2.1 Airframe

A UAV's airframe is the physical framework that holds up every part of the system. Typically, airframes are constructed from lightweight materials like aluminum, carbon fiber, or composites to improve efficiency and maneuverability. The planned usage of the UAV determines a major variation in the airframe design. Commercial drones may prioritize ease of use and stability but in the case of military drones they can possess stealth capabilities exclusively.

1.7.2.2 Engine Technology

A UAV's propulsion system produces thrust which may allow the vehicle to go through the air. Depending on the size and intended purpose of the UAV different propulsion systems may be used. Small UAVs usually have battery-operated electric motors, which provide simplicity and silent operation. For greater speed and endurance, larger UAVs – particularly those employed for long-range missions may be equipped with internal combustion engines or even jet engines.

1.7.2.3 Control System

The UAV's "brain" is considered to be control system. It comes with all of the hardware and software needed to control the flight, to navigate, and carry out missions on behalf of the UAV. In order to maintain stability and follow intricate flight patterns modern UAVs frequently employ sophisticated flight control systems that can incorporate sensors, GPS, and inertial measurement units. There are multiple modes of operation for the control system, which include fully manual control and totally autonomous operation.

1.7.2.4 Communication System

For UAVs to operate especially those that need to be commanded remotely, communication is essential. The radios and antennae that enable data transmission and command reception between the UAV and the ground control station are part of the communication system. Satellite communication systems could be used with long-range UAVs. For military applications in particular the data lines need to be strong and secure to help avoid enemy jamming or interception.

1.7.2.5 Payloads and Sensors

UAVs are frequently outfitted with a variety of sensors that can enable them to carry out particular functions or duties. Radar systems, the LiDAR for mapping and navigation, the cameras used for visual surveillance and also the infrared sensors for night operations are examples of commonly used sensors. The equipment or cargo that a UAV is carrying is referred to as its payload, and this may vary greatly depending on the different missions and use cases. A military UAV might be equipped with weaponry or may be surveillance gears and equipment, whereas an agricultural UAV may be equipped with multispectral cameras which can help for evaluating crop health.

1.7.2.6 Power Source

A UAV's power supply system is essential for its functionality. Lithium polymer (LiPo) batteries which provide an excellent balance between energy, density, and weight may be commonly used in small UAVs. To increase their endurance, larger UAVs may be employed with hybrid power systems or may include fuel cells. The mission requirements determine the power source to be used, also taking into account the factors like operating environment, payload weight, and flight time.

1.7.2.7 Integrating AI and Autonomous Capabilities

The incorporation of autonomous capabilities into the UAV is said to be one of the biggest developments in UAV technologies. Now modern UAVs are capable of doing very complex tasks without direct human interaction with the help of modern AI and machine learning techniques. Obstacle navigation on-the-spot decision-making based on the sensor data and environmental adaptation are all capabilities of autonomous UAVs. With applications like search and rescue, where the UAVs can autonomously locate survivors and deliver aid, this capacity seems to be especially helpful.

1.7.3 UAV Applications

UAV uses have been observed in various fields, including the environment domains, communication domains, infrastructure domain, and safety and rescue (SAR) domain. The most common application in the environmental domain is agriculture. It also covers resource distribution, environmental preservation, forest ecosystem management, and disaster management. Internet of things (IoT), wireless communication network, and mobile edge computing are examples of communication applications. In the infrastructure domain electrical and civil construction is the primary application. Disaster and search and rescue operations are further applications in the SAR field. Furthermore, photogrammetry, sports, journalism, traffic, aeronautics, healthcare applications, smart city applications, and the film industry business can all benefit from it [7, 8].

Numerous industries have found uses for the UAVs, each capitalizing on the special skills of these machines. Listed below are a few of the major uses:

1.7.3.1 Applications in the Military

UAVs are becoming a crucial component of warfare in the modern day. They perform the duties in direct combat as well as target collection, observation, and reconnaissance. With their sensors and weaponry the UAVs such as the MQ-9 Reaper can launch targeted attacks against important targets [9, 10]. Their ability to operate in hazardous environments without posing a threat to human life makes them essential weapons in modern warfare.

1.7.3.2 Agriculture

UAVs are utilized in agriculture for precision farming. They are trained so that they are able to evaluate soil conditions, can monitor crop health, and survey huge tracts of farmland. UAVs using multispectral sensors are able to identify problems in farming before they become serious, such as pest infestations, water stress, and nutrient deficits. With the use of this information, farmers can maximize or efficiently utilize water, herbicides, and fertilizers that can lead to increasing yields while lowering their impact on the environment [11].

UAVs can also be used in agriculture for early water stress diagnosis, crop yield estimation, and disease detection in flora such as the automated identification of soybean leaf diseases from their photos [12]. These tasks involve the use of UAV-based hyper-spectral imaging and the machine learning techniques which help to accurately identify with high accuracy. Additionally, it may be utilized in the forest ecosystem to identify the diseases in the flora and fauna, such as the diseases that may be anticipated to affect pine trees, using UAV and analyze it with deep learning techniques. Horticultural research can also be used for the environmental monitoring. Likewise water bodies can be identified using hyper-spectral data that UAVs may acquire, and this information can be used to track and analyze the quality of the water available. UAVs have the potential to gather data from most challenging environments, like that of inaccessible islands, rainforests, and the deep oceans, for the biological research applications like animal detection and animal poaching. It can also be used to find the forest fires and other natural disasters, etc.

1.7.3.3 Delivery and Logistics

The application of UAVs in distribution and logistics [13] is being investigated more and more. Drone delivery systems have been tested by UPS and Amazon among other companies to move products short distances. Particularly in locations where conventional distribution techniques are ineffective, UAVs have the potential to save delivery time and expenses. But before there is broad use, issues including safety concerns and airspace regulation must be resolved.

1.7.3.4 Environmental Surveillance

UAVs are often employed in conservation and environmental monitoring projects [14, 15]. They can be used to monitor deforestation, the wildlife involved, and can also evaluate the effects of climate change in the habitat. UAVs can also be used to swiftly survey affected areas during disaster response, which help to giving emergency personnel access to real-time data. In circumstances where ground access is hazardous or difficult, this capability seems to be vital.

Additionally, the UAV can be used to manage and monitor calamities brought on by man as well as nature, such as the landslides and hurricanes. For instance, UAVs can be used to survey and acquire data in areas affected by floods and landslides and may assist with prompt rescue efforts too. They can also be employed in case of nuclear disaster scenarios, where a malfunction during system operation could result in radioactive contamination spreading across a wide areas and might lead to big calamities. There might be a risk associated with the surrounding flora and fauna due to this contamination. By examining the data collected by the UAVs, it can also be avoided or detected early.

1.7.3.5 The Infrastructure Evaluation

Another application for UAVs is said to be the infrastructure inspection including pipelines, electricity lines, and bridges. In the past years, these examinations frequently involved many hazardous working conditions, and it demanded a large amount of time and resources including man power. These checks or monitoring can be carried out more effectively and safely with the help of UAVs, which can take high-resolution pictures and other data of the situation that can be examined to find possible problems before they turn into serious issues [16].

In the case of the infrastructure management domain, maintaining user safety, such as in the construction of roads and bridges, requires frequent and recurring inspections. While it takes more time, effort, and resources to complete these manually, UAVs can be deployed in this scenario to complete the task. As an instance, consider the inspection of concrete cracks which can be accomplished by applying machine learning algorithms to the multispectral images obtained by UAVs to differentiate between intact and damaged constructions. In a similar vein, regular grid inspections seem to be necessary to guarantee both efficiency and security. This task can be considered a significant problem because of the grid's wide geographic coverage. With the aid of UAVs, these tasks can also be completed quickly and easily, regardless of geographic obstacles.

1.7.3.6 Media and Film Production

UAVs are gaining popularity in the film and media industries because they can produce beautiful aerial footage for a fraction of the price of more conventional techniques like helicopters. Filmmakers now routinely use UAVs fitted with high-definition cameras to capture unique images that were previously unfeasible or impractical [17]. Due to this, there has been a revolution in cinematography, with documentaries and big-budget films alike using drones.

1.7.3.7 Communication

In communication UAVs can help with wireless network infrastructure and coverage expansion by serving as air base stations for connectivity. This application is a manifestation of a cost-efficient and adaptable strategy that could greatly assist the development of future wireless network generations [18, 19]. UAVs can also help with wireless communications by functioning as mobile relays because of their great mobility. By integrating the principles of quantum computing with numerous UAVs, resilient mobile access systems can be improved in terms of overall wireless service quality. In a similar vein, UAVs are useful for gathering data in IoT networks.

The energy consumption of UAVs is one of their drawbacks. UAVs have a limited amount of energy because they are usually battery-powered. Adverse weather conditions that severely impact trajectories and data collection from the surroundings are another difficulty that still exist when using UAVs in crisis circumstances. A few other challenges in the development and implementation of UAVs are the proliferation of government regulations pertaining to them, security issues such as vulnerabilities in the operation of UAVs like GPS jamming and hacking and using it for unauthorized activities, limitations of datasets for object detection, classification, or segmentation in aerial imagery, communication management, and scalable communication networks designs.

Numerous topologies, such as star, multi-star, mesh, and hierarchical mesh configurations, can be established in multi-UAV systems. There are several advantages and factors to take into account for UAV communication with each topology. In star topology UAVs connect directly with a ground node that serves as the hub of a star topology network. However, relative to inter-UAV distances, star topologies have higher delay because of their longer downlink lengths. Since there is no inter-UAV communication, the mission may fail if the ground node fails. Numerous UAVs form the stars in a multi-star topology network, and each group of stars has one node connected to the ground station. Like star topologies, multi-star systems suffer from latency and rely on ground station availability to carry out missions adequately. UAVs are linked together in a mesh topology network, and only one UAV can be connected to the control center at a time. With this arrangement, packets may undertake many steps from any source to any destination by passing via intermediary nodes. When compared to star topology, mesh networks provide greater performance characteristics, flexibility, and dependability. The failure of one node doesn't impact the others. The hierarchical mesh topology is an advancement over mesh topology. Several mesh networks made up of UAVs make up the hierarchical mesh topology network, including a single node from every group linked to other groups. Hierarchical organization and intergroup communication are the features of this system.

The growing list of UAV applications and the research that have been done on a wide range of activities and also already available information in academic databases demonstrate the promise of UAVs in this era. The future impact of combining UAVs with more sophisticated intelligence approaches seems to be evident. In a similar vein, quantum technologies and concepts will soon be applied to UAVs so that it can greatly enhance their potential for use in civil, military, and security contexts worldwide.

1.8 Quantum Computing and UAV Safety

Quantum computing has the potential to revolutionize navigation and data processing in the near future, with discoveries that can go beyond current technical advancements in the new era. It might be able to offer an unparalleled processing power that might apply the ideas of quantum mechanics, which could make it possible to store and manipulate these enormous datasets in ways so that it is not able to achieve with conventional or traditional computers. Though it is still in the stage of infancy, this emerging field in new digital era has already witnessed that quantum computers make calculations that were previously believed to be impossible, which helps to bringing in a transformation in technology that could incorporate UAVs for more effective and efficient data processing and reconnaissance. Quantum computing is now opening its doors for ground-breaking improvements in UAV navigation and related fields by harnessing its unmatched processing power and its capability to solve the complex problems.

When combined with the quantum computing, UAV technology have made significant advancements that will increase its capabilities. These may include

Enhanced Computational Capabilities: The computational capabilities of the UAV systems could be completely transformed by quantum computing leading to the possibility to more sophisticated algorithms for path planning, collision avoidance, and real-time decision-making. UAVs may be able to manage more complicated settings and circumstances that are more skillfully as a result of the various algorithms used in quantum computing.

Quantum Encryption for Secure Communication: Secure channels for communication are crucial for command and control of UAVs, specifically if considered for sensitive military or may be in civilian applications. Theory-wise, confidential communication resilient to the eavesdropping could be achieved with quantum key distribution (QKD). The security of data transmission might need to be significantly enhanced by integrating quantum encryption techniques into the UAV communication.

Optimization of UAV Swarms: UAV swarm optimization and the coordination might be enhanced by the use of quantum computing. Quantum algorithms offer enhanced efficiency in solving the optimization problems compared to the conventional methods, which can be especially helpful when coordinating the movements of multiple UAVs to achieve specific objectives or goals such as wide-area monitoring or search and rescue missions.

Cryptographic Resilience: The security of UAV systems is at risk since many of the cryptography systems currently in use could be broken by quantum computers. On the other hand, post-quantum cryptography or the quantum-resistant algorithms are being developed to protect systems from quantum attacks. By including these algorithms into UAV systems their long-term security might be guaranteed in the event that the quantum computers become increasingly widespread.

Improved Sensor Data Processing: A variety of sensors, such as cameras, LIDAR, and RADAR, are necessary for UAVs to function securely and effectively. With the use of the quantum computing the sensor data processing and analysis or interpretation could be improved that might be resulting in more quicker and much precise identification of targets, impediments, or environmental changes.

Quantum Machine Learning: UAVs that might be powered by the quantum machine learning (QML) algorithms is useful for a range of applications, involving adaptive learning in dynamic situations, predictive maintenance, and anomaly detection. Compared to the conventional machine learning algorithms, the QML may be capable of helping UAVs to learn and adjust the unexpected circumstances more rapidly and more precisely.

The quantum computing and the UAV safety integration can significantly enhance the security and safety of UAV system and their capabilities which can also help them to render more dependable and durable across a range of uses. These advancements highlight how crucial quantum computing plays in improving drone data processing and navigation, and they predict a new advanced era in which drones can function with never-before-seen with high levels of efficiency and versatility. Even though quantum computing technology advances, practical implementation of it is still in the research phase.

1.8.1 Case Studies

1.8.1.1 Data Transmission and Communication That Is Secure

Quantum Cryptography for UAVs: By utilizing the quantum computers, the security and privacy of communication channel between the UAVs and the ground control stations might be improved. The QKD can be used here which can construct secure channels that are purportedly immune to the eavesdropping. This can help to ensure the security of the orders that UAVs will receive and the data that are sent to them.

Quantum-Enhanced Encryption: UAVs can be used to help in safeguard sensitive data that are gathered during flights that have been intercepted or may be decrypted by enemies by using the quantum encryption algorithms, which are significantly more secure than that of classical encryption techniques.

1.8.1.2 Processing and Optimizing Data in Real Time

Quantum Algorithms for Real-Time Navigation: In a real-time scenario, the UAVs can navigate difficult surroundings with much great efficiency with the help of the quantum computing, which may handle enormous volumes of data. Particularly, in case of the dynamic contexts, where a quantum algorithms can optimize flight trajectories help to avoid obstructions or can adapt quickly to the changes in the surroundings.

Optimization of Multi-UAV Coordination: To ensure that the drones or the UAV operate as effectively as possible, the quantum computing can be used to optimize the coordination of several UAVs during different vital missions. Optimizing flight patterns for the surveillance or in the delivery missions is considered as one example of this situation.

1.8.1.3 Complex Computational Operations Using UAVs

Quantum-Assisted Machine Learning: When on a mission, UAVs which equipped with quantum computing capabilities, can perform many machine learning algorithms at same time that might need a significant amount of processing power, such as object recognition, anomaly detection, or predictive analytics, using the deep learning models.

Complicated Environmental Simulations: UAVs can be used to maximize the performance by using the quantum computing to model very complicated environmental variables in real-time scenario. This includes forecasting weather patterns, the wind speeds, and even possible interference from other electrical devices.

1.8.1.4 Improved Measuring and Sensing

High-Precision Measurements with Quantum Sensors: UAVs can be fitted with the quantum sensors, which may take use of entanglement and superposition the two quantum features that can produce incredibly accurate measurements. These can be applied to mapping, surveying, and other jobs where the traditional sensors might overlook minute changes in the environment.

Quantum-Based Imaging: UAVs can see through the clouds, smoke, and even buildings by using the quantum imaging techniques. This could be helpful for the military and search and rescue operations.

1.8.1.5 Quantum-Assisted Decision-Making Real-Time Decision Support

UAVs can use quantum computing to swiftly decide on difficult choices, such as which landing location is ideal, how to best avoid a threat, or which targets to prioritize for military operations. Quantum computation might seem to be perfect for the time-sensitive, decision-making processes because of its efficiency and speed in processing, helpful in many situations.

Autonomous Mission Planning: UAVs can always operate in the most effective and efficient way possible by using quantum algorithms to automatically plan and re-plan missions in response to fresh information or the altering circumstances.

1.8.1.6 Task Scheduling and Resource Allocation

Quantum computing may optimize the distribution of scarce resources (such as battery life and payload capacity) among UAVs in a fleet, making sure that these resources are used as efficiently as possible to fulfill mission objectives. This is known as quantum optimization for resource management.

Dynamic Task Scheduling: Quantum algorithms are able to dynamically schedule the tasks for the given UAVs in missions which have numerous objectives. This allows for the optimization of task timing and order to maximize mission success and efficiency.

1.8.1.7 Quantum-Mechanical Threat Identification and Mitigation

Advanced Threat Detection: In contrast to conventional available computing methods, the UAVs can utilize the quantum computing to quickly and accurately detect and to identify potential threats in a given circumstance from the enemy (such as the approaching missiles or may be adversarial UAVs).

Quantum-Driven Countermeasures: UAVs that possess the quantum computing capabilities can able to initiate the countermeasures that necessitate swift processing of electronic warfare techniques that can interfere with intruding threats' guidance systems.

1.8.1.8 Improved Search and Rescue Procedures

Quantum-Assisted Search Algorithms: In the field of search and rescue operations, UAVs can leverage quantum algorithms to expeditiously scan an extensive region or can be used to analyze copious amounts of tracked sensor data, thereby expediting the identification of persons in distress beyond what would be achievable now with traditional search algorithms.

UAVs also have the ability to create quantum communication networks which might be helpful to improve coordination between and can also useful in success in rescue operations by enabling a safe, quick, and dependable communication between the rescue teams involved.

1.8.1.9 Environmental Monitoring Based on Quantum Theory

Climate and Pollution Monitoring: UAVs with quantum sensors installed can precisely keep an eye on environmental factors including radiation levels and air quality. Quick insights into the environmental changes are made feasible by using real-time data processing enabled by quantum computing.

Quantum-Driven Agricultural Monitoring: To improve farming methods and yields the UAVs can leverage on quantum computing to assess data on crop conditions, weather patterns, and also in soil health.

1.9 Limitations of Quantum Computing

In weak environmental coupling, quantum states must be kept in extremely cold confinement (thus they are stored in extreme temperatures) [20, 21]. If they are not completely isolated, they eventually lose information. We refer to this as de-coherence. Sustaining desired behaviors (e.g., superposition and entanglement) over an adequate amount of time is quite challenging [22]. As they are being controlled, they have a tendency to become featureless and random. Specialized discipline for addressing this problem is quantum error correction. It has been proposed that the present quantum hardware might be composed of the noisy intermediate-scale quantum hardwares.

The major limitations include

a. **Hardware Challenges:** Implementing and maintaining the necessary hardware for quantum computation such as superconducting qubits or may be trapped ions that involves complex engineering and technological challenges. Achieving the required precision and stability is now considered to be demanding.

   The produced hardware appears to be too complex in design, huge in size, and unreliable because of its noise.

b. **De-coherence:** Quantum bits or qubits are highly susceptible to the environmental noise that may cause quantum states to collapse. This phenomenon is known as de-coherence, which results in causing quantum information to be lost or might be corrupted.

   To perform meaningful computations quantum computers must maintain coherence for extended periods. Error correction techniques and fault-tolerant quantum computing are being developed to address this issue.

c. **Error Correction:** Quantum error correction presents a significant challenge, and it may appear to be increasingly complex. In order to guarantee reliable quantum computation, error rates must be decreased because qubits are fragile and prone to mistakes. Researchers are actively working on creating effective error correction codes and methods for fault-tolerant quantum computing.

d. **Qubit Scalability:** Building quantum computers with a large number of qubits is a major engineering challenge. As quantum computers become more powerful, adding qubits while maintaining low error rates becomes increasingly difficult. Techniques for scaling up quantum hardware are under development. Now it is available in the order of three but need to increase the number of qubits to make it utilized in real-world problems.

e. **Hardware Stability:** Quantum processors are typically operated at extremely low temperatures that is to keep in absolute zero temperature to reduce de-coherence. Now the hardware is placed in refrigerators to maintain the temperature. The feasibility of quantum computing is limited by the cost and technological difficulty of maintaining these conditions.

f. **Interfacing with Classical Computers:** Quantum computers must interface with classical computers for input, output, and data processing. Data transport and synchronization become more difficult when quantum and classical systems are combined.
g. **Programming and Software:** Developing quantum algorithms and programming quantum computers need a specialized skill. Since this is an emerging technology there is a shortage of quantum software developers and tools. Additionally, optimizing quantum algorithms for specific problems can be challenging.
h. **Standardization:** The quantum computing field lacks a widely accepted set of standards including hardware, software, and communication protocols. To guarantee compatibility and speed up the development of quantum technologies, standardization activities are required.
i. **Cost:** Quantum computing research and development, as well as the operation of quantum computers, can be costly. Quantum technologies are difficult to make more affordable and widely available. Additionally, it is imperative that this technology become widely used.
j. **Security and Cryptography:** While quantum computing has the potential to break certain classical cryptographic schemes (e.g., RSA and ECC), it also introduces the need for post-quantum cryptography solutions. Developing and deploying quantum-safe cryptographic algorithms is essential for security in a post-quantum world.
k. **Resource Requirements:** Quantum error correction, fault tolerance, and building large-scale quantum computers may require an enormous number of qubits and sophisticated infrastructure. Completing these resource needs is difficult in terms of both computational power and physical resources.

1.10 Future Challenges of Quantum Computing

Although there are a lot of obstacles and unknowns in the path of quantum computing, the field has enormous promise. The future of quantum computing can be explained using the following key aspects:

a. **Technological Advancements:** Hardware: Quantum hardware is continuously evolving. To make large-scale and useful quantum computers a reality, the researchers are now focused on creating more scalable and robust quantum processors with error correction.

   Quantum Networking: The development of quantum communication networks and the distribution of entangled particles for secure communication are likely to become more prominent.
b. **Wider Accessibility:** Quantum computers are currently limited in availability and often housed in research labs. The future may see more widespread accessi-

bility through cloud-based quantum computing services, making it easier for researchers and industries to access quantum resources.

c. **Quantum Software and Algorithms:** As quantum hardware improves, more efficient quantum algorithms and software will be developed, expanding the range of problems that quantum computers can solve effectively.
d. **Practical Applications:** Quantum computing will continue to find applications in areas such as drug discovery, materials science, optimization, machine learning, and cryptography. Real-world, practical use cases are expected to increase as the technology matures.
e. **Scientific Discoveries:** Quantum computing will contribute to scientific breakthroughs by simulating complex quantum systems, which will help scientists better understand fundamental physics, chemistry, and materials science.
f. **Quantum Supremacy and Benchmarks:** The achievement of "quantum supremacy" and the development of benchmark problems will be important milestones to assess the capabilities of quantum computers.
g. **Interdisciplinary Collaboration:** Quantum computing requires expertise in various fields, and future progress will depend on effective collaboration between physicists, computer scientists, and domain experts in various industries.

1.11 Summary

The rapidly advancing field of quantum computing applies the ideas of quantum physics to solve challenging puzzles that traditional computers are unable to answer. Quantum computing offers a huge increase in computational power and efficiency that has the potential to totally change a number of sectors. As an alternative to traditional computers which process information in binary bits (0s and 1s) the quantum computers use quantum bits (also called qubits).

Quantum computing, while promising, has severe technological and practical constraints. One key issue is the fragility of qubits, which are susceptible to de-coherence and quantum noise due to their susceptibility to environmental disturbances. Although qubits are unstable, they need to be maintained at very low temperatures and under strict control, which adds complexity and cost to the hardware. Furthermore, sophisticated and error-correcting algorithms are now required due to the high mistake rates that may occur in contemporary quantum computers. As a result it believed that the contemporary quantum hardware now falls into the noisy intermediate-scale quantum.

Furthermore, the tools and platforms that have been used for developing the quantum software are still in their early phases which might create major barriers to wider testing and adoption. The practical applications of quantum computing are now limited due to these significant obstacles that should be overcome in order to reach its full potential.

Although it seems that quantum computing has the potential to completely transform the computing and thus the scientific discoveries, it has to overcome its obstacles in the future which will require a deep understanding of its fundamental ideas as well as its technological limitations and potential uses. This insight will be critical in propelling the development toward viable quantum computers which might be capable of addressing challenges outside the scope of classical computation.

Looking forward, quantum computing confronts a number of major difficulties that must be addressed before the technology can reach its transformational promise. One of the most significant issues is the scalability. While today's quantum computers can solve small-scale problems, when trying to scale out to bigger systems with hundreds or thousands of qubits, it remains to be difficult. Although quantum computing has a very bright future, there exist a number of uncertainties and obstacles that must be overcome.

In terms of science and technology in the new era, quantum computing represents a new frontier. It has the potential to completely change or transform all aspects of our lives and may also be able to challenge our understanding about the computing. As we continue to explore and develop this new developing domain, it is so much clear that quantum computing will be able to shape future life in profound and exciting ways.

References

[1] Yang, Zebo, Maede Zolanvari, and Raj Jain. "A Survey of Important Issues in Quantum Computing and Communications." IEEE Communications Surveys & Tutorials, vol. 25, no. 2, 2023, pp. 1059–1094.

[2] Kumar, P. P., Kumar, T. A., Muthu, P., Rajendirane, R., & Samuel, R. D. J. (2022). Security Analysis of UAV Communication Protocols: Solutions, Prospects, and Encounters. In The Internet of Drones (pp. 167–190). Apple Academic Press.

[3] Bravyi, Sergey, Oliver Dial, Jay M. Gambetta, Dario Gil, and Zaira Nazario. "The Future of Quantum Computing with Superconducting Qubits." Journal of Applied Physics, vol. 132, no. 16, 2022, p. 160902.

[4] Bouchée, T., et al. "Towards a better understanding of conceptual difficulties in introductory quantum physics courses." Studies in Science Education 58.2, (2022): 183–202.

[5] Fahlstrom, P. G., Gleason, T. J., & Sadraey, M. H. (2022). Introduction to UAV systems. John Wiley & Sons.

[6] Gupta, L., Jain, R., & Vaszkun, G. (2015). Survey of important issues in UAV communication networks. IEEE communications surveys & tutorials, 18(2), 1123–1152.

[7] Ralegankar, V. K., Bagul, J., Thakkar, B., Gupta, R., Tanwar, S., Sharma, G., & Davidson, I. E. (2021). Quantum cryptography-as-a-service for secure UAV communication: applications, challenges, and case study. IEEE Access, 10, 1475–1492.

[8] Goyal, S. B., Rajawat, A. S., Solanki, R. K., Zhu, L., & Chee, W. (2023, October). Enhancing Privacy and Security for UAV and IoT Enabled Drones an Intelligent Integration of Blockchain, AI, and Quantum

Computing. In International Conference on Intelligent Computing & Optimization (pp. 16–27). Cham: Springer Nature Switzerland.

[9] Gargalakos, M. (2024). The role of unmanned aerial vehicles in military communications: application scenarios, current trends, and beyond. The Journal of Defense Modeling and Simulation, 21(3), 313–321.

[10] Udeanu, G., Dobrescu, A., & Oltean, M. (2016). Unmanned aerial vehicle in military operations. Sci. Res. Educ. Air Force, 18(1), 199–206.

[11] Radoglou-Grammatikis, P., Sarigiannidis, P., Lagkas, T., & Moscholios, I. (2020). A compilation of UAV applications for precision agriculture. Computer Networks, 172, 107148.

[12] Tsouros, Dimosthenis C., Stamatia Bibi, and Panagiotis G. Sarigiannidis. “A review on UAV-based applications for precision agriculture.” Information 10.11 (2019): 349.

[13] Li, Y., Liu, M., & Jiang, D. (2022). Application of unmanned aerial vehicles in logistics: a literature review. Sustainability, 14(21), 14473.

[14] Avola, D., Foresti, G. L., Martinel, N., Micheloni, C., Pannone, D., & Piciarelli, C. (2017, August). Aerial video surveillance system for small-scale UAV environment monitoring. In 2017 14th IEEE International Conference on Advanced Video and Signal Based Surveillance (AVSS) (pp. 1–6). IEEE.

[15] Sharma, S., Muley, A., Singh, R., & Gehlot, A. (2016). UAV for surveillance and environmental monitoring. Indian Journal of Science and Technology, 9(43), 1–4.

[16] Jordan, S., Moore, J., Hovet, S., Box, J., Perry, J., Kirsche, K., . . . & Tse, Z. T. H. (2018). State-of-the-art technologies for UAV inspections. IET Radar, Sonar & Navigation, 12(2), 151–164.

[17] Mademlis, I., Mygdalis, V., Nikolaidis, N., & Pitas, I. (2018, July). Challenges in autonomous UAV cinematography: An overview. In 2018 IEEE international conference on multimedia and expo (ICME) (pp. 1–6). IEEE.

[18] Bithas, P. S., Michailidis, E. T., Nomikos, N., Vouyioukas, D., & Kanatas, A. G. (2019). A survey on machine-learning techniques for UAV-based communications. Sensors, 19(23), 5170.

[19] Fotouhi, A., Qiang, H., Ding, M., Hassan, M., Giordano, L. G., Garcia-Rodriguez, A., & Yuan, J. (2019). Survey on UAV cellular communications: Practical aspects, standardization advancements, regulation, and security challenges. IEEE Communications surveys & tutorials, 21(4), 3417–3442.

[20] Fortnow, L., & Rogers, J. (1999). Complexity limitations on quantum computation. Journal of Computer and System Sciences, 59(2), 240–252.

[21] Aaronson, S. (2008). The limits of quantum. Scientific American, 298(3), 62–69.

[22] Al Adeh, F. F. (2017). Natural limitations of quantum computing. Int J Swarm Intel Evol Comput, 6(152), 2.

Amruthamsha Prashanth Raju, Disha Ramesh, Praveen Naik,
and Gururaj Kudur Jayaprakash

2 Enhancing Predictive Accuracy Through Quantum AI

Abstract: In decision-making, conventional approaches often fail when faced with complexities of complex landscapes. This chapter discusses a new line of work known as quantum-inspired multi-agent decision-making, which adapts concepts from quantum mechanics (e.g., superposition, interference, and entangled), making it a fit candidate to explore quantum-inspired paradigms. These ideas have their origin in physics and chemistry as they are used to analyze complex data with multiple parameters. This study establishes an ensemble algorithm inspired by quantum technology, developing a real-time decision-making system to inform decision-making, integrating insights from various sources. This approach combines classical and quantum rules of nature to enhance data processing and interpretation, thereby providing highly accurate predictions. It marks a significant advancement for decision science, applying quantum principles to make light work of difficult decisions.

Keywords: Decision-making, quantum mechanics, quantum theories, predictive accuracy, artificial intelligence

2.1 Introduction

Decision-making is one of the most important human cognitive skills that influence our day-to-day routine, work, and so on. Decision-making covers a broad spectrum of complexity [1], from deciding what to wear to high-stakes decisions that impact our careers and relationships. As such, we are making decisions based on our personal memories and past experiences, which are often fraught with emotion. Based on those experiences, you have made certain assumptions, which are at the core of your decision-making processes, determining and influencing future decisions, and thus ensuring certain outcomes [2].

Three steps are often involved in decision-making: preparation, aggregation, and selection. While the aggregation stage includes duties like evaluating the relevance of qualities, making decisions, identifying the importance of agents, and aggregating opinions, the preparation stage entails examining a variety of attributes and options. Alternatives are ranked in the final selection stage based on total results.

Human decision-making is complicated in several ways, including the fact that judgments are typically made in the face of uncertainty, are produced rather than documented, can have an impact on one another, and frequently violate the estab-

https://doi.org/10.1515/9783111619323-002

lished probability theories [3, 4]. Cognitive paradoxes, including the Allais-Ellsberg paradox, disjunction fallacy, conjunction fallacy, and order effects, are problems for classical probability theories [5]. Because of their contextual character, quantum theories shed light on the logic of these paradoxes by connecting quantum phenomena such as entanglement, superposition, and interference with cognitive processes [6, 7].

Two main interpretations are provided by quantum theories: one focuses on the physical states of individual systems, while the other uses probabilistic representations of state information following measurement on a collection of systems [8]. This chapter investigates how cognitive information representation in decision-making processes can be improved by the capacity of quantum theories for information interpretation. By introducing ideas like superposition (where particles can exist in several states instantaneously) and entanglement (where particles are coupled regardless of spatial proximity), quantum mechanics undermines conventional deterministic viewpoints. Once limited to theoretical physics, these ideas are currently being used more and more in a variety of domains, such as cognitive sciences and information technology [9, 10].

The application of quantum principles to decision-making models presents promising avenues to solving cognitive paradoxes that classical theories cannot explain. For example, quantum entanglement and superposition form a theoretical basis for grasping how decisions may arise out of states of uncertainty and how linked decisions can interact with each other in real time [11, 12]. These are not only adding to theoretical frameworks but also opening doors to potential applications in artificial intelligence, as quantum-inspired algorithms are able to significantly improve decision-making systems in terms of efficiency and accuracy.

In recent years, quantum machine learning (ML) has gained dynamic prominence as a method of using quantum computers to resolve any important computational challenges. It is believed that researchers might find ways to improve the algorithms so that they process large volumes of data much quicker than classical methods through these possibilities of quantum systems. The shift has a heavy weight in decision-making sense, which might very well affect the notion of decision-making in many worlds, including healthcare, finance, and even environmental management.

2.2 Quantum Theories

Quantum mechanics, a division of physics that addresses the nature of particles at atomic and subatomic scales, brings concepts that contradict classical deterministic approaches. Two of the fundamental principles of quantum mechanics – superposition and entanglement – are the foundation of our exploration [13]. Quantum mechanics transformed our picture of the physical world, particularly in physics and chemistry, as it explained things that classical physics could not. The central theme of quantum mechanics is that particles, like electrons and photons, have both wave-like and particle-like behavior. This duality gives rise to intricate behaviors that baffle classical intuition.

Key quantum mechanics principles include:

- Wave-Particle Duality: Particles that show dual nature, such as wave and particle-like properties, and their behavior is determined by the method of measurement.
- Quantization: Certain physical properties, for example, energy, are quantized, and can only assume specific distinct values.
- Heisenberg Uncertainty Principle: It is impossible to exactly measure particle's position as well as momentum at the same time.

These principles establish the foundation for understanding more complex quantum phenomena such as superposition, entanglement, and interference.

2.3 Qubits, Quantum Superposition, and Decision-Making

A qubit, or quantum bit, is the vital unit of quantum information. Contradictorily, classical bits can be either 0 or 1; qubits can occur in a superposition of states, represented as a combination of both 0 and 1 instantaneously. This property permits quantum computers to process various possibilities simultaneously.

Quantum superposition challenges classical notions by allowing particles to exist in multiple states until observed or measured [14]. In traditional physics, an object is supposed to be in a particular state at any moment. Quantum superposition, however, implies that particles are in a state of probability, including all the possible outcomes.

With quantum superposition, decision-makers can consider a wider range of possibilities prior to making a final choice [15]. This is analogous to the quantum process in which particles are in a state of probability until measurement causes the system to collapse into a particular result. In practice, the incorporation of superposition into decision-making models enables more detailed scenario analysis and risk evaluation, borrowing from the principles of quantum physics and chemistry, taking into consideration interdependence of varied variables and contingencies that affect outcomes [16].

At the decision-making process, superposition enables decision-makers to hold numerous possible outcomes simultaneously instead of focusing on a sole direction too quickly. This capacity is vital to navigate the implicit uncertainties and nuances of actual-life decisions. By examining a broader spectrum of options prior to a final decision, decision-makers are in a better position to weigh risks and rewards.

2.4 Quantum Entanglement

Quantum entanglement, as a vital part of quantum mechanics, depicts a phenomenon whereby particles become correlated such that the state of one particle instantaneously influences that of another independent of the distance between the two; hence, this phenomenon places a deep local connectedness right into the face of what classical causality assumes [17]. Although a metaphorical qualifier for decision-making, this concept works to show how entangled decisions of one entity can directly impact the decisions of other entities in a complex system [18]. Such interconnectedness inherent in entanglement reflects the complex interdependencies and interactions governing real-world decision environments in which individual decisions create ripples through networks and eventually together shape whole system outcomes [19].

Applying entanglement concept to the decision sciences gives deep insight into modeling and predicting collective behavior and outcomes. Also, by including entanglement edges in decision-making algorithms, researchers can simulate how decisions spread through networks of agents or stakeholders more realistically. This would be relevant in economics where lots of factors influence market dynamics through decisions taken by these very factors, or in social sciences where group decisions are instrumental in understanding societal behavior [20, 21]. Quantum-inspired solutions consider such interdependencies in improving the accuracy and effectiveness of decision-making processes toward the navigation of complex and dynamic environments.

Further, the study of entanglement in decision sciences promotes cross-discipline collaboration among quantum physicists, cognitive scientists, and decision theorists. This collaboration will enable a new methodology for harnessing quantum principles to improve cognitive modeling and decision analysis. As the technologies stemming from quantum studies mature, the application of entanglement will provide new avenues and insights that transcend classical limitations, leading to a revolutionary thrust into decision science and related areas.

2.5 Quantum Interference and Cognitive Processes

Quantum interference is a fundamental principle in quantum mechanics, wherein the superposition emanating from the split quantum wave function creates either constructive or destructive interference in the observed behavior of a given quantum system [22]. Such phenomenology ultimately serves not only to illuminate physics but also simultaneously bestows deep insights into how cognitive processes and decision-making operate [23].

In decision-making processes, the principle of quantum interference may be metaphorically applied to aid in understanding how cognitive processes interfere with one another. Interference patterns, which arise when one quantum wave interferes with another, possess analogies in interference in cognition due to interactions wherein classical theory often finds it difficult to predict [24]. The result of such interplays may mimic some of the intricacies and uncertainties that appear naturally in human cognition. Interactions such as those upon which nonlinear ones operate and dynamic uncertainties generated in the human cognitive processes should not be captured in classical decision-making models. Such interactions are precisely what quantum interference captures with more granularity. Thus, imbuing quantum interference into decision modeling can enhance how such decision-making embodies the complexities of human thought, paving the path to more realistic and adaptable decision-making.

Integrating the quantum interference into cognitive modeling fosters collaboration among quantum physicists, cognitive scientists, and decision theorists. This type of interdisciplinary work will be key for the formation of new methodologies for the usage of quantum principles to further cognitive modeling and decision analysis. As quantum technology develops, the application of quantum interference to decision sciences could provide new avenues and insights that move beyond the limits of classical decision theories.

2.6 Quantum-Inspired Algorithms

The introduction of quantum principles into the decision-making paradigm carries far-reaching implications for multiple disciplines. Quantum-inspired algorithms are designed to use the computational advantages of a quantum system to process large amounts of data with speed and accuracy relative to that of classical algorithms. Such rapid and accurate decision-making is required, in particular, in sectors relevant to healthcare, finance, environmental management, and logistics. Algorithms use superposition, entanglement, and interference to improve the accuracy of very complex decision-making processes [25–27].

2.7 Real-World Applications

Algorithms incorporated in quantum-inspired decision-making provide faster and more accurate predictions that can make optimal resource allocation, reduce risks, and enhance operational efficiencies. For instance, these algorithms can make route optimization, inventory management, and demand forecasting better in logistics and supply chain management. For improved execution of their operations at minimized costs and enhanced service delivery, such algorithms process extensive logistic information with greater quickness and precision [28–30].

In environmental management, quantum-inspired techniques provide better modeling and simulation of environmental impacts. They enhance decision-making capabilities by allowing decision-makers to better understand the behavior of complex environmental systems and consequently predict the effects of actions in policy interventions and operationalize strategies for sustainable development. This is another way to strengthen environmental advocacy while also injecting better data into decision-making that cuts across industries.

2.8 Existing Research and Multi-agent Decision-Making

Existing research often treats decision-makers as independent entities. This study introduces a novel approach by modeling multiple decision-makers using an ensemble of classifiers, including *k*-nearest neighbor (*k*-NN), multilayer perceptron (MLP), decision trees, and random forest classifiers (RFCs) [31].

k-NN: This nonparametric algorithm categorizes new data points centered on their proximity to other points in the training dataset. Classification of a new instance is determined by a majority vote of its *k*-NNs. The correct choice of an optimum value of *k* is very crucial, which can be ascertained through methods like the elbow method (which deals with the trade-off between increasing *k*'s effect on classification accuracy versus increasing its effect on computational complexity) or cross-validation (testing model performances on various subsets of the data) [32].

MLP: MLPs are feedforward neural networks that have more than one layer of neurons such that each layer is fully connected to the next layer. MLPs are especially useful when multiple layers can extract different levels and kinds of feature composition that are most useful for the specific nonlinear classification task at hand: learning and modeling rather than complex patterns in the data. Activation functions introduce nonlinearity, which allows the model to capture and represent complex relationships within data [33].

Decision Trees: Decision trees are predictive modeling tools using a tree-like structure to show decisions or classifications. A node in the tree corresponds to a decision made based on one of the features; the branches of the tree represent the possible outcomes of that decision. This process works by recursively splitting the data based on feature values to accomplish the highest classification accuracy. It acts as an accessible way to visualize and interpret the rationale behind the decision-making process [34, 35].

RFC: This is a collaborative learning method that builds many decision trees all through training and averages their outputs to improve everything. Each tree uses a random subset of features and data to build the trees, which helps in reducing overfitting and increasing the robustness of the model. The aggregation of all trees' results leads to the final class decision. Because they can handle huge datasets with complex interdependencies very well, RFC has shown its outstanding prediction accuracy in various fields, including liquid identification, land classification, and prediction of healthcare-associated infections [36, 37].

2.9 Quantitative Analysis Methods

The evolution of ML has revolutionized quantitative finance by enhancing predictive accuracy and decision-making efficiency. ML models handle complex datasets and accelerate processes such as financial portfolio diversification and high-frequency trading [38–40]:

- Time-Series Analysis: It involves the evaluation of data points recorded at periodic intervals of time. This method is crucial for forecasting – market trends, asset values, and other economic indicators.
- Cross-Sectional Analysis: It evaluates a single data point in a specific time across multiple parameters and entities, such as weighting multiple companies' financial progress to identify anomalies and patterns in a specific market.
- Predictive Modeling: Utilizing past data to predict future scenarios, predictive modeling can range from deducing current trends to forecasting stock values.

ML algorithms capable of handling and processing vast and complex datasets are being increasingly used to aid quantitative finance. This evolution not only empowers conventional financial analytics and decision-making but also opens up novel avenues for strategic benefits and innovation in the financial realm.

2.10 Case Studies and Applications

Quantum-inspired decision-making has proven effective in addressing complex challenges. For example, in healthcare, quantum algorithms optimize treatment plans by analyzing extensive datasets, including medical histories and genetic profiles, leading to personalized treatment strategies. In finance, quantum models enhance predictive analytics and risk management, offering insights into market trends and investment portfolios.

Beyond these sectors, quantum-inspired approaches show promise in environmental management and telecommunications. For instance, quantum algorithms can improve resource distribution and predict environmental impacts, supporting sustainable practices and addressing worldwide challenges [3, 41].

2.10.1 Case Study: Gurumukhi Script Recognition

Kaur et al. [42] used several feature extraction methods followed by classification techniques to recognize the Gurumukhi script in newspaper texts. The six feature extraction techniques used by the researchers were zoning, diagonal, centroid, power curve fitting, parabola curve fitting, and peak extent method. They then applied to classify the extracted features using *k*-NN, MLP, decision tree, and RFCs. The best recognition was achieved using diagonal features with the RFC having an accuracy rate of 96.9%. The dataset comprised around 10,000 examples of diverse Gurumukhi characters, which showed, through the experiments, that diagonal feature extraction consistently produced the best accuracy, together with MLP yielding reasonably good recognition accuracy. Zoning and diagonal techniques mostly performed better across most classifiers, thus indicating good potential in recognizing newspaper text.

2.10.2 Case Study: Magnetic Resonance Imaging (MRI)

Quantum computing combined with AI has the potential to change the world in magnetic resonance imaging (MRI) technology, depending on image reconstruction optimization, noise reduction, and enhancement of signal quality [43]. The traditional method of image reconstruction in MRI relies on computation-heavy techniques, where quantum computing offers significant advantages by processing vast datasets simultaneously. This could potentially lead to higher-quality and faster reconstructions. Quantum AI will also be applied for noise reduction, which is essential for better imaging clarity and precision in diagnosis. For example, with quantum algorithms working directly on MRI data, precise biomarker identification and structural analysis shall be facilitated. In the future, applications could include personal imaging protocols based on certain characteristics of the individual patient, enhancing the effectiveness of diagnosis as well as patient comfort.

As quantum computing continues to evolve, its integration with MRI holds significant promise for advancing medical diagnostics and personalized healthcare solutions [44].

2.11 Future Directions and Challenges

The amalgamation of quantum principles into decision-making holds great promise but faces several significant challenges. One key challenge is the growth of quantum algorithms that effectively harness the principles of quantum mechanics while addressing specific application needs. This requires very strong collaboration among physicists, computer scientists, and domain experts to transform theoretical notions into fast and efficient algorithms in practice. Another major roadblock is the scalability of existing quantum computing technology. Today, quantum systems cannot perform large-scale decision-making tasks with due convenience due to restrictions in the number of qubits and coherence. To be able to scale up quantum systems for practical purposes applicable in the complex decision environment, advances in quantum hardware design and error correction are a necessity.

Until now another issue remains to lower the error rates in quantum computations. Quantum systems are usually disturbed by environmental noise, and their qubits are prone to errors, thus requiring strong error correction methods coupled with some algorithmic optimizations to render the decision outcomes reliable and accurate. Addressing such challenges will involve continued investment in research and development, spearheaded jointly by academia, industry, and the government. In summary, quantum-inspired decision-making has unprecedented potential to revolutionize healthcare, finance, environmental management, and other fields, but its realization will depend on resolving technical impediments as well as advancing quantum computing technology [45, 46]. It is thus up to interdisciplinary collaboration and technological advancement to tackle these difficulties that will underpin transformative applications of quantum principles to decision sciences and other fields.

2.12 Conclusion

Quantum-inspired decision-making is thus a paradigm shift in thinking about complex decision problems. Importing ideas from quantum mechanics into decision-making processes potentially widens the corridor towards greater precision and understanding of decision processes. In doing so, these quantum-inspired algorithms are exporting ideas of superposition, entanglement, and interference into the treacherous terrains of contemporary data-rich domains.

Also, with its research and refinement throughout technology, these methods are now presumably extended to possible applications in health, finance, and environment management for better forecasting, resource allocation, and risk management. Given their interdisciplinary nature, the building of quantum algorithms should engage physicists, computer scientists, and domain experts to address the pressing issues of scalabilities and error rates. In years to come, iterative improvements toward refinement for quantum-inspired methodologies shall indeed spell new advances for decision sciences. In allowing quantum principles to stir the very solutions to global challenges, these methods set on the diverging path of aspiring to embrace the complexity and interlinked nature of real-life decision-making.

References

[1] D. Pratt, "Consciousness, Causality, and Quantum Physics" NeuroQuantology, 1, 58–67, 2003

[2] S. A. Humr, M. Canan, and M. Demir, "A Quantum Decision Approach for Human-AI Decision-Making", EPiC Series in Computing, Bd. 102, S. 111–125, July 2024, doi: 10.29007/6F6J.

[3] Z. A. Green, F. Faizi, R. Jalal, and Z. Zadran, "Emotional support received moderates academic stress and mental well-being in a sample of Afghan university students amid COVID-19", International Journal of Social Psychiatry, Bd. 68, Nr. 8, S. 1748–1755, Dec. 2022, doi: 10.1177/00207640211057729.

[4] P. Bossaerts and C. Murawski, "Computational Complexity and Human Decision-Making", Trends Cogn Sci, Bd. 21, Nr. 12, S. 917–929, Dec. 2017, doi: 10.1016/J.TICS.2017.09.005.

[5] I. M.S. and A. K. Cherukuri, "Quantum-inspired ensemble approach to multi-attributed and multi-agent decision-making", Appl Soft Comput, Bd. 106, S. 107283, July 2021, doi: 10.1016/J.ASOC.2021.107283.

[6] A. Khrennikov, "Quantum-like modeling: cognition, decision making, and rationality", Mind and Society, Bd. 19, Nr. 2, S. 307–310, Nov. 2020, doi: 10.1007/S11299-020-00240-6.

[7] T. Kyriazos and M. Poga, "Quantum concepts in Psychology: Exploring the interplay of physics and the human psyche", Biosystems, Bd. 235, S. 105070, Jan. 2024, doi: 10.1016/J.BIOSYSTEMS.2023.105070.

[8] G. Jaeger, "Information and the Reconstruction of Quantum Physics", Ann Phys, Bd. 531, Nr. 3, S. 1800097, Mar. 2019, doi: 10.1002/ANDP.201800097.

[9] D. P. Kalogeras, "The Quantum Theory in Decision Making", Journal of Computations & Modelling, Bd. 3, Nr. 4, S. 1792–8850, 2013.

[10] P. D. Bruza, Z. Wang, and J. R. Busemeyer, "Quantum cognition: A new theoretical approach to psychology", Trends Cogn Sci, Bd. 19, Nr. 7, S. 383–393, July 2015, doi: 10.1016/j.tics.2015.05.001.

[11] V. I. Yukalov and D. Sornette, "Decision theory with prospect interference and entanglement", Theory Decis, Bd. 70, Nr. 3, S. 283–328, Mar. 2011, doi: 10.1007/S11238-010-9202-Y.

[12] V. I. Yukalov and D. Sornette, "Physics of risk and uncertainty in quantum decision making", The European Physical Journal B 2009 71:4, Bd. 71, Nr. 4, S. 533–548, July 2009, doi: 10.1140/EPJB/E2009-00245-9.

[13] C. de Ronde, "Quantum superpositions and the representation of physical reality beyond measurement outcomes and mathematical structures", Found Sci, Bd. 23, Nr. 4, S. 621–648, Dec. 2018, doi: 10.1007/S10699-017-9541-Z.

[14] J. I. Cirac, M. Lewenstein, K. Mølmer, and P. Zoller, "Quantum superposition states of Bose-Einstein condensates", Phys Rev A (Coll Park), Bd. 57, Nr. 2, S. 1208, Feb. 1998, doi: 10.1103/PhysRevA.57.1208.

[15] J. R. Friedman, V. Patel, W. Chen, S. K. Tolpygo, and J. E. Lukens, "Quantum superposition of distinct macroscopic states", Nature 2000 406:6791, Bd. 406, Nr. 6791, S. 43–46, July 2000, doi: 10.1038/35017505.

[16] L. Galbusera and G. Giannopoulos, "On input-output economic models in disaster impact assessment", International Journal of Disaster Risk Reduction, Bd. 30, S. 186–198, Sep. 2018, doi: 10.1016/J.IJDRR.2018.04.030.

[17] J. R. Schaibley et al., "Demonstration of quantum entanglement between a single electron spin confined to an InAs quantum dot and a photon", Phys Rev Lett, Bd. 110, Nr. 16, Apr. 2013, doi: 10.1103/PhysRevLett.110.167401.

[18] J. B. Altepeter, E. R. Jeffrey, P. G. Kwiat, S. Tanzilli, N. Gisin, and A. Acín, "Experimental methods for detecting entanglement", Phys Rev Lett, Bd. 95, Nr. 3, July 2005, doi: 10.1103/PhysRevLett.95.033601.

[19] G. Aubrun, L. Lami, C. Palazuelos, and M. Plávala, "Entanglement and superposition are equivalent concepts in any physical theory", Phys Rev Lett, Sep. 2021, doi: 10.1103/PhysRevLett.128.160402.

[20] Y. R. Shrestha, V. Krishna, and G. von Krogh, “Augmenting organizational decision-making with deep learning algorithms: Principles, promises, and challenges”, J Bus Res, Bd. 123, S. 588–603, Feb. 2021, doi: 10.1016/J.JBUSRES.2020.09.068.

[21] H. Mahmud, A. K. M. N. Islam, S. I. Ahmed, and K. Smolander, “What influences algorithmic decision-making? A systematic literature review on algorithm aversion”, Technol Forecast Soc Change, Bd. 175, S. 121390, Feb. 2022, doi: 10.1016/J.TECHFORE.2021.121390.

[22] J. Lim, S. Kumar, Y. S. Ang, L. K. Ang, and L. J. Wong, “Quantum interference between fundamentally different processes is enabled by shaped input wavefunctions”, Advanced Science, Bd. 10, Nr. 10, S. 2205750, Apr. 2023, doi: 10.1002/ADVS.202205750.

[23] A. Khrennikov, “Quantum-like modeling of cognition”, Front Phys, Bd. 3, Nr. SEP, S. 161073, Sep. 2015, doi: 10.3389/FPHY.2015.00077.

[24] M. Kindermann, “Interference in the matter”, Nature 2007 448:7151, Bd. 448, Nr. 7151, S. 262–263, July 2007, doi: 10.1038/448262a.

[25] V. Vasani, K. Prateek, R. Amin, S. Maity, and A. D. Dwivedi, “Embracing the quantum frontier: Investigating quantum communication, cryptography, applications and future directions”, J Ind Inf Integr, Bd. 39, S. 100594, May 2024, doi: 10.1016/J.JII.2024.100594.

[26] A. Khrennikov, “Roots of quantum computing supremacy: superposition, entanglement, or complementarity?”, The European Physical Journal Special Topics 2021 230:4, Bd. 230, Nr. 4, S. 1053–1057, Apr. 2021, doi: 10.1140/EPJS/S11734-021-00061-9.

[27] Padmapriya, N., T. Ananth Kumar, R. Aswini, R. Rajmohan, P. Kanimozhi, and M. Pavithra. “IoT based energy optimization in smart farming using AI.” Hybrid Intelligent Approaches for Smart Energy: Practical Applications (2022): 205–223.

[28] M. A. Jahin, M. S. H. Shovon, M. S. Islam, J. Shin, M. F. Mridha, and Y. Okuyama, “QAmplifyNet: pushing the boundaries of supply chain backorder prediction using interpretable hybrid quantum-classical neural network”, Scientific Reports 2023 13:1, Bd. 13, Nr. 1, S. 1–26, Oct. 2023, doi: 10.1038/s41598-023-45406-7.

[29] R. G. Richey, S. Chowdhury, B. Davis-Sramek, M. Giannakis, and Y. K. Dwivedi, “Artificial intelligence in logistics and supply chain management: A primer and roadmap for research”, Journal of Business Logistics, Bd. 44, Nr. 4, S. 532–549, Oct. 2023, doi: 10.1111/JBL.12364.

[30] R. Toorajipour, V. Sohrabpour, A. Nazarpour, P. Oghazi, and M. Fischl, “Artificial intelligence in supply chain management: A systematic literature review”, J Bus Res, Bd. 122, S. 502–517, Jan. 2021, doi: 10.1016/J.JBUSRES.2020.09.009.

[31] A. El Bilali, A. Taleb, and Y. Brouziyne, “Comparing four machine learning model performances in forecasting the alluvial aquifer level in a semi-arid region”, Journal of African Earth Sciences, Bd. 181, S. 104244, Sep. 2021, doi: 10.1016/J.JAFREARSCI.2021.104244.

[32] G. Chirici et al., “K-NN FOREST: a software for the non-parametric prediction and mapping of environmental variables by the k-nearest neighbors algorithm”, Eur J Remote Sens, Bd. 45, Nr. 1, S. 433–442, 2012, doi: 10.5721/EUJRS20124536.

[33] M. Aitkin and R. Foxall, “Statistical modelling of artificial neural networks using the multi-layer perceptron”, Stat Comput, Bd. 13, Nr. 3, S. 227–239, Aug. 2003, doi: 10.1023/A:1024218716736.

[34] Y. Zhao and Y. Zhang, “Comparison of decision tree methods for finding active objects”, Advances in Space Research, Bd. 41, Nr. 12, S. 1955–1959, Jan. 2008, doi: 10.1016/J.ASR.2007.07.020.

[35] R. Rivera-Lopez, J. Canul-Reich, E. Mezura-Montes, and M. A. Cruz-Chávez, “Induction of decision trees as classification models through metaheuristics”, Swarm Evol Comput, Bd. 69, S. 101006, Mar. 2022, doi: 10.1016/J.SWEVO.2021.101006.

[36] A. Chaudhary, S. Kolhe, and R. Kamal, “An improved random forest classifier for multi-class classification”, Information Processing in Agriculture, Bd. 3, Nr. 4, S. 215–222, Dec. 2016, doi: 10.1016/J.INPA.2016.08.002.

[37] A. Sarica, A. Cerasa, and A. Quattrone, “Random forest algorithm for the classification of neuroimaging data in Alzheimer’s disease: A systematic review”, Front Aging Neurosci, Bd. 9, Nr. OCT, S. 284242, Oct. 2017, doi: 10.3389/FNAGI.2017.00329.

[38] A. Njegovanović, “Financial evolution and interdisciplinary research”, Financial Markets, Institutions and Risks, Bd. 7, Nr. 1, S. 71–95, Mar. 2023, doi: 10.21272/FMIR.7(1).71-95.2023.

[39] D. J. Denis, “Handbook of corporate finance”, Handbook of Corporate Finance, S. 1–708, Feb. 2024, doi: 10.4337/9781800373891.

[40] A. Njegovanović, “Approach to financial decision making within quantum physics and neurosciences”, International Journal of Law and Public Administration, Bd. 3, Nr. 2, 2020, doi: 10.11114/ijlpa.v3i2.4991.

[41] O. O. Olatunji, P. A. Adedeji, and N. Madushele, “Quantum computing in renewable energy exploration: status, opportunities, and challenges”, Design, Analysis and Applications of Renewable Energy Systems, S. 549–572, Jan. 2021, doi: 10.1016/B978-0-12-824555-2.00019-8.

[42] R. P. Kaur, M. Kumar, and M. K. Jindal, “Performance evaluation of different features and classifiers for Gurumukhi newspaper text recognition”, J Ambient Intell Humaniz Comput, Bd. 14, Nr. 8, S. 10245–10261, Aug. 2023, doi: 10.1007/S12652-021-03687-8.

[43] Z.-H. Cho et al., “A new approach to quantum computing multi-qubit generation and development of quantum computing platform with magnetic resonance imaging techniques”, June 2022 [Online]. Verfügbar unter: http://arxiv.org/abs/2206.05932

[44] M. Barnett et al., “A real-world clinical validation for AI-based MRI monitoring in multiple sclerosis”, NPJ Digit Med, Bd. 6, Nr. 1, Dec. 2023, doi: 10.1038/s41746-023-00940-6.

[45] A. Perdomo-Ortiz, M. Benedetti, J. Realpe-Gómez, and R. Biswas, “Opportunities and challenges for quantum-assisted machine learning in near-term quantum computers”, 19. June 2018, Institute of Physics Publishing. doi: 10.1088/2058-9565/aab859.

[46] M. Steyvers and A. Kumar, “Three challenges for AI-assisted decision-making”, Perspectives on Psychological Science, 2023, doi: 10.1177/17456916231181102.

J. Ramkumar, S. P. Priyadharshini and Balachandra Pattanaik

3 Securing the Skies: Safety Queuing Models for Quantum Drone Ad Hoc Networks

Abstract: In this chapter, the innovative integration of quantum computing with drone ad hoc networks, particularly the design and implementation of safety queuing models for quantum drone ad hoc networks, will be discussed. The chapter begins by providing background information on the development of drone technology and the basics of quantum computing, forming the foundational principles. Subsequently, the discussion focuses on unique security threats that only quantum drone networks would face and how conventional security protocols have failed to mitigate these. With newly designed queuing algorithms conceptualized along with quantum enhancements, this chapter outlines various practical strategies for deploying these safety queuing models, optimizing their performance, and ensuring their resiliency under real-world conditions. All these models are exemplified with such intense performance analysis, benchmarking, and continuous adaptation that the evolution of a solid level of security and efficiency in drone operations becomes apparent. Ultimately, this chapter discusses the significant transformations envisaged from integrating quantum computing within drone networks for advanced, secure, and efficient drone communication systems.

Keywords: Quantum computing, drone ad hoc networks, safety queuing models, network security, optimization, real-time data analysis, adaptive algorithms

3.1 Introduction

Integrating drone technology with quantum computing has recently rapidly offered new paradigms in communication networks. Quantum drone ad hoc networks are among the pioneering technologies using the fantastic characteristics of quantum mechanics to improve the conventional drone network in performance and security [1]. That is to say, a network involving drones communicating wirelessly among themselves constitutes a decentralized and self-organizing system. With the integration of quantum computing, these networks can use advanced computational and cryptographic techniques to bring about a degree of efficiency and security not experienced before [2].

https://doi.org/10.1515/9783111619323-003

3.1.1 Evolution of Drone Technology

Drone technology dates back to primary remotely controlled aircraft primarily used for recreational purposes. In the past two decades, hardware and software advancements and communication technologies have scaled drones into sophisticated tools that find application in many spheres. State-of-the-art drones complement these with high-resolution cameras, GPS, and many sensors, powering tasks from aerial photography and surveillance to parcel delivery [3]. Moreover, introducing artificial intelligence simplifies drone work, supporting autonomous flight, obstacle avoidance, and real-time data processing. The development can then be summarized into several stages. At a very fundamental beginning, drones were mainly used for military applications: surveillance and target location. With technological improvements, drones have found their way into commercial and civilian uses, hence the development of smaller models at cheaper prices. Nowadays, many drones are used in industries, such as agriculture, construction, and disaster management, proving them to be versatile and have the enormous potential to turn several business lines around [4].

3.1.2 Basics of Quantum Computing

Quantum computing is a field developing and changing the basics of processing information. Although classical computers store information in bits as 0s and 1s, quantum computers store information in quantum bits or qubits. Since principles such as superposition and entanglement imply multistate existing conditions for qubits, this enables quantum computers to do many very hard calculations that a classical computer cannot keep up with regarding speed [5]. Regardless of the choice of basis, thanks to superposition, qubits can represent 0 and 1, increasing significantly the computational power of quantum systems. Due to entanglement, the information shared between the connected qubits becomes available instantly, regardless of distance. The feature of quantum computing mentioned above offers enormous potential for applications such as cryptography, optimization problems, and quantum system simulations [6].

3.1.3 Integration of Quantum Computing into Drone Networks

Combining quantum computing and drone networks offers the possibility of improvements in security and efficiency for communication. Quantum cryptography uses principles from quantum mechanics to develop unbreakable theoretical encryption techniques. In a quantum drone ad hoc network (Q-DANET), quantum key distribution (QKD) can secure the communication channels between the drones, ensuring that no eavesdropping or manipulation of the data is possible over the transmission. The quantum algorithms can thus be applied to the routing and scheduling of the drones in the network [7]. Quantum computing allows real-time decisions with efficient resource allocation, particularly in highly dynamic environments where drones are expected to react to environmental changes or mission requirements due to their ability to process large amounts of data quickly. Integrating quantum computing with drone networks opens up advantages beyond security and optimization. Quantum sensors could enhance the functions of drones concerning environmental monitoring, navigation, and imaging by a considerable margin. They can detect minute changes in physical properties, providing more accurate and detailed information than traditional sensors [8].

3.1.4 Scope and Objectives of the Chapter

The chapter will comprehensively introduce Q-DANETs by tracing their evolution process, principal fundamentals, and some prospective applications in view. The objectives are:

1. **Define Quantum Drone Ad Hoc Networks (Q-DANETs):** Elucidation of Q-DANETs will include explaining the structure and why it forms a crucial part of state-of-the-art communication systems.
2. **Outline the History of Drone Technology:** The chapter shall try to trace back the history of the development and improvement of drone technology in its bid to provide context for why Q-DANETs came into being.
3. **Introduce the Fundamentals of Quantum Computing:** A layman's introduction to the understanding of quantum computing and how it has been integrated with drone networks needs to be made; hence, it needs to be an introduction to the basics. This would include essential concepts like superposition, entanglement, and quantum algorithms.
4. **Enhance Drone Networks with Quantum Computing:** Security, optimization, and sensing applications.
5. **Future Directions and Applications for Q-DANETs**: Q-DANETs are the latest fusion of drone technology with quantum computing, advancing their security, efficiency, and functionality to a great extent. This chapter establishes some foundation for further in-depth treatment of Q-DANETs; it provides background that can help understand how they will probably affect different industries or applications.

3.2 Theoretical Foundations

The design and deployment of Q-DANETs are based on a deep understanding of several theoretical foundations. This chapter presents the basics of quantum computing, the queuing theory principles, the essence of drone ad hoc networks, and the intersections of these fields that provide the ground for the Q-DANETs.

3.2.1 Basics of Quantum Computing

Quantum computing is the breakthrough area that will change how problems are solved by applying the principles of quantum mechanics. Some key concepts are drawn from the literature [9, 10]:

- **Qubits and Superposition:** Unlike classical computing, which uses bits to represent the data in 0 or 1 form, quantum computing uses quantum bits or simply qubits. Because of superposition, qubits can be found in states like 0, 1, or a mix of both. Such an ability of quantum computers is based on this property, allowing them to process massive data simultaneously.
- **Entanglement:** Interconnected property of two or more qubits in a way that the state of one is immediately affected by the state of another, however, far apart they may be. Entanglement allows for secure communication and enhances computational power.
- **Quantum Gates and Circuits:** Quantum gates are the realization of basic building blocks of quantum algorithms, which correspond to the respective operations applied to the qubits. Chains of such gates or quantum circuits realize complex computations concerning the superposition and entanglement effects.
- **Quantum Algorithms:** Examples of quantum algorithms include Shor's algorithm for factorization of large integers and Grover's algorithm for searching an unsorted database. These algorithms find solutions to problems much faster than their classical counterparts and show enormous computational powers.

3.2.2 Principles of Queuing Theory

The queuing theory is related to the study of waiting lines. It is considered one of the most essential tools in designing a system where competing tasks share common resources. Some essential queuing principles are [11–13]:

- Arrival process describes how tasks or customers arrive at a service point. This is usually modeled through a Poisson process, assuming that arrivals are independent of each other and random in time.
- Service mechanism tells how jobs are managed once they arrive at the service point. The distribution of service time, generally modeled by exponential or general distributions, is crucial in understanding the performance of a system.
- Queue discipline defines the sequence by which jobs are serviced. Common disciplines include first-in-first-out, last-in-first-out, and priority-based queuing.
- Average waiting time, queue length, and system utilization are significant performance metrics that would help design a system to use available resources efficiently with minimal delays.

- Many queuing models are based on Markov chains representing system states and transitions. This will assume that the future states of a system depend only on the current state and are, therefore, entirely appropriate for the stochastic process analysis.

3.2.3 Overview of Drone Ad Hoc Networks

Drone ad hoc networks are decentralized wireless networks in which drones communicate without fixed infrastructure. Some of the key aspects include [14, 15]:

- **Self-Organization:** The drones in a DANET form and maintain network connections autonomously, adapting to topology changes and those of the environment. This is very important in dynamic and unpredictable settings.
- **Communication Protocols:** Effective communication protocols ensure reliable data transmission in DANETs. Ad hoc on-demand distance vector and dynamic source routing are techniques that provide the routes between the drones.
- **Mobility Models:** Drones' movement patterns are a significant factor in network performance. The random waypoint model and the Gauss-Markov model are examples of models that help simulate and analyze the impact of drone movement on connectivity and performance.
- **Applications:** DANETs find their applications in disaster response, environmental monitoring, and military operations because it is easy and rapid to deploy and adapt, especially in scenarios where traditional infrastructure may not be feasible

3.2.4 Convergence of Quantum Computing and Queuing Theory

The infusion of quantum computing in DANETs with the aid of queuing theory is permitted by several state-of-the-art solutions [16]:

- **QKD in DANETs:** QKD for securing the channels of communication between drones and to secure data from eavesdropping and tampering the quantum mechanical process of the making of the encryption keys becomes almost impossible to crack.
- **Network Performance Optimization:** Quantum algorithms permit the solving of intricate optimization problems in DANETs. Quantum annealing optimizes routing and resource allocation, therefore, ensuring that the network performs much better with significantly less delay.
- **Queue Management:** With quantum computing, it aids in making the queuing models strong and efficient enough to provide solutions more quickly for problems that may have settled in the queue. It examines more accurate forecasting of the conduct of the queues to occasionally modify itself in order to bring out more efficiency of the system.

- **Improved Sensing and Data Processing:** The quantum sensors are more accurate and have higher resolution when collecting data from the drone. This, coupled with the processing of massive datasets that quantum computers can do quickly, results in more accurate monitoring of the environment and decision-making in real time.

3.3 Security Challenges

Q-DANETs present an example of technical excellence; they amalgamate drone capabilities with quantum computing to develop effective, secure communication systems. Their integration presents a whole new range of security challenges that should be taken care of with extra carefulness if the network's integrity, confidentiality, and availability are to be maintained. It discusses the unique quantum features that the threats of quantum drone networks are characterized by, deals with conventional security protocols, discusses the flaws present in quantum systems, and enters into strategies for threat mitigation and risk management.

3.3.1 Distinct Threats in Quantum Drone Networks

Q-DANETs introduce features of relatively high technological maturity; they bring along specific, very distinct threats. Among others, quantum channel eavesdropping poses a significant risk. While QKD supports theoretically unbreakable encryption, the channels are not safely distributed. This type of eavesdropping imposes noise, which may be detectable by legitimate users [7]. It remains a big issue. On the same note, quantum jamming poses yet another challenge: the adversaries interfere with communication channels applied by quantum drones, thus culminating in failures of coordination and operation. Therefore, the network system security may also be compromised by some malicious quantum nodes within the network through data tampering, injection of falsified data, or the inaccurate relaying of information, which may undermine the integrity of the network system. Besides, quantum-targeted malware could use quantum-specific vulnerabilities to corrupt the operation or induce an operation to interfere with quantum algorithms for disrupted or non-behavioral operation, corruption of qubit states, or the occurrence of trivial outcomes [8].

3.3.2 Examination of Classic Security Protocols

Though the classic security protocols are very efficient for work in classic networks, there are some constraints in applying them for Q-DANETs. Classical encryption schemes depend on RSA and AES, which are usually considered challenging under some assumptions. For instance, the Shor's algorithm factored large integers. Since these can compromise RSA encryption, quantum-resistant encryption algorithms must be employed to achieve data privacy in Q-DANETs. Due to these quantum channels' unique characteristics, schemes mainly used for classical communication will not protect these quantum lines. Quantum authentication schemes using entanglement and superposition as resources are fundamental to verifying the identity nodes and communication protection. Classical intrusion detection systems (IDSs) developed for legacy (classical) networks are ineffective in detecting quantum attacks. Therefore, the development of IDS, characterized by Q-DANETs' features, becomes necessary. Classical public key infrastructure systems must be upgraded for deployment in quantum environments by integrating QKD and quantum-resistant algorithms for the requisite security [17].

3.3.3 Quantum System Vulnerability Profiling

Despite their robust features, quantum systems have inherent vulnerabilities that must be assessed carefully. Qubits – the basic units for quantum computation – are susceptible to environmental factors that cause them to undergo the process of decoherence and lose information [18, 19]. This vulnerability may be exploited to interfere with quantum operation and data integrity. Quantum channels for communication are subject to noise, which degrades the quality of the information to be transmitted. The intentionally added noise by an adversary may interdict quantum communication and key distribution. Side-channel attacks can compromise quantum systems by leveraging physical characteristics such as power consumption and electromagnetic emissions to extract information about quantum operations. While quantum algorithms are powerful, they tend to be complex and contain hidden vulnerabilities that can only be discovered and repaired through detailed scrutiny and testing of any security flaws that the system may have [20].

3.3.4 Threat Mitigation Strategies and Risk Management

There are various strategies that can be used to mitigate threats and manage risks in Q-DANETs. The post-quantum cryptographic algorithms, such as lattice or hash-based cryptography, guarantee that data is not exposed to any quantum attack. These classical algorithms are secured under a huge quantum power from the quantum computers. This is why QKD is definitely a move toward strengthening security; its key must be refreshed and distributed often. Therefore, quantum authentication protocols must be robustly designed using properties such as entanglement; the communication between nodes should be perfectly secure for identification and verification within the network.

The advanced IDSs specifically designed for Q-DANETs further enhance the network's capacity to react promptly and with relevance to quantum-specific attacks [21].

Another aspect that makes the system more robust against attacks and failure is redundancy and fault tolerance in network design. Quantum error-correction techniques do more than just correct errors; they also prevent the degradation of quantum states to guarantee the reliability of quantum operations. Similarly, adaptive security means that the system can be monitored continuously and revalued over time for any possible threats to handle them as and when they may occur [22].

3.4 Design and Development of Safety Queuing Models

Development of safety queuing models for Q-DANETs is designed and followed through a multifaceted approach: the design of new queuing algorithms, conceptual making of quantum enhancements, the designing of robust safety mechanisms, and in-depth simulation with tests for reliability and security.

3.4.1 Conceptualizing New Queuing Algorithms

Developing new queuing algorithms that support the unique dynamics of the drone network – such as mobility, different communication ranges, and real-time data processing – forms the first foundation in building effective safety models for Q-DANETs. To schedule and route tasks appropriately so as to maximize throughput and minimize delay, task management would be most effective with algorithms that are flexible to accommodate the change in network topology and variation in traffic patterns. An adaptive queuing algorithm could place the jobs in some order of urgency, and then with resource availability, so that the vital data may be sent out timely and without any delays [23].

3.4.2 Integrating Quantum Improvements into Safety Models

The design of the safety model with integrated quantum enhancements is useful to explore how conceptual and beneficial aspects of quantum computing may be used for increasing performance and, therefore, enhancing safety in Q-DANET. A key improvement will be done using QKD for secure communication. QKD provides a proven secure approach to the generation and distribution of encryption keys, ensuring that data integrity and confidentiality are maintained. Quantum algorithms allow for optimization during the queuing process. For instance, quantum search algorithms can identify the best routes for data transmission within a very short time, hence cutting on latency and boosting network efficiency. This is where quantum error correction techniques come into play to ensure that the network does not lose the data transmitted because of noise and decoherence [24].

3.4.3 Development of Robust Safety Mechanisms

Another critical need to make the Q-DANETs continue not to be susceptible to such perceived threats is to design robust safety mechanisms, embedding multilayered security protocols by means of quantum and classical cryptographic techniques. For instance, quantum-resistant cryptography mechanisms can be used with QKD for protection against attacks that are not quantum-safe and against known quantum attacks. Other salient concerns relate to the design of quantum-efficient IDSs. This would enable the detection and mitigation of classical as well as quantum-type attacks. Machine learning techniques are applied to monitor the activities on network traffic and, from that, detect abnormal traffic patterns to understand if an attack is in progress. It also has physical security features such as the integrity of the drones physically and of the communication channels, ensuring no tampering or access by unauthorized parties [25]. This would be secured by a hardware-secure design and the development of tamper-resistant components.

3.4.4 Simulation and Testing of Safety Queuing Models

The queuing models of safety need to be validated through simulation and testing. Simulation can model many scenarios about studying the performance of queuing algorithms with built-in safety mechanisms under different scenarios, such as increasing network size, traffic load, or types of potential attacks. Simulation tools can create fine details of the models of Q-DANETs, hence estimating performance metrics like latency, throughput, and security robustness. Such simulations may also help ascertain considerable weaknesses or any performance bottlenecks. It would then be used to fine-tune such models for implementation. Real-world testing complements this with practical insights into operational challenges and how well the queuing models work. This will involve deploying the models in controlled environments to monitor their performance against real-world conditions. In some ways, real-world testing can provide the model's robustness, efficiency, and security.

3.5 Implementation Strategies for Safety Queuing Models

The safety queuing models would require a comprehensive strategy for their implementation in Q-DANETs that envisions all the steps down to every critical step, from practical deployment to optimizing performance and security, and protocols for continuous monitoring and maintenance to establish its integrity and reliability. In the process, the model should be able to confront real-world challenges.

3.5.1 Practical Steps of Deploying Safety Queuing Models

Careful planning will precede the deployment of safety queuing models in Q-DANETs toward defining the requirements and objectives of the network. Desirable throughput, admissible latency levels, and robust security features shall be specified. The integration of the queuing algorithms into this network architecture will only occur after defining such nonfunctional requirements. This algorithm shall handle the data traffic efficiently and with the right priorities. This will require the collaboration of software developers, network engineers, and quantum computing experts. Proper hardware infrastructure with the corresponding software will be designed for it. Therefore, the quantum cryptographic protocols' communication modules must be available in the drones. At the same time, the ground station or control center provides centralized management and oversight of the network.

3.5.2 Performance Optimization and Security

The Q-DANET's performance and security need to be continuously evaluated and tuned. Optimizing the performance involves minimizing latency and maximizing throughput based on a change in queuing algorithms as network conditions vary. This can be achieved by incorporating cryptographic solid protocols that are updated occasionally to protect against new threats. QKD will secure the communication channel, while quantum-resistant encryption algorithms will protect the information. Effective resource allocation prevents bottlenecks and provides adequate computational power for high-priority tasks. This will be ensured by dynamically allocating qubits and processing resources according to live network demand, thereby maintaining near-optimal performance and security.

3.5.3 Solving Real-World Challenges

There are quite a number of significant challenges to deploying safety queuing models in a real-world Q-DANET. In particular, environmental factors feature variability across a range of bad weather, physical obstacles, and interference, affecting communication and operations of drones. Very little will work if not accounted for within the queuing model. It also encompasses network protection against physical and cyber threats through the integration of strong error correction techniques, tamper resistance, and protection against jamming and unauthorized access, among other threats to Q-DANETs. Scalability therefore remains one of the critical concerns, where scalable algorithms and infrastructure design are important in order to house increasing numbers of drones and data traffic without compromising performance. Efficient solutions to these challenges have to be found in ensuring network integrity and functionality.

3.5.4 Quantum Drone Network Safety Monitoring and Maintenance

Monitoring and maintenance of Q-DANETs may also assure their safety and reliability. The security state of network performance and the health of the same shall be monitored in real time. This modern automatic monitoring system will use the information and reporting provided by alerts to trigger automatic responses against threats, ensuring that any anomaly or potential security breach on the network is detected instantly. It will be backed up by regular audits and updates of security protocols across the network to be abreast of the evolving threats. These periodical queuing algorithms and test security mechanisms as part of maintenance procedures run simulations, stress tests in a problem search, and fix them. Acquiring a background in network operations will be necessary to train in the skill acquisition for routine maintenance and emergencies. An iterative loop with the monitoring system's data will be established for model adjustments of the queue and increasing measures for security. The process is iterative and can attune itself to new challenges in a way that optimal levels of performance and security are maintained. Pretty structured approach allows developing safety queuing models in Q-DANETs: implementation, optimization, and continuous control in order to manage real-world challenges.

3.6 Evaluation and Performance Metrics

The developed safety queuing models in Q-DANET should be measured for effectiveness and efficiency by using the evaluation and performance metrics. The model has an effective evaluation framework that ensures required performance, security, and reliability. This section reviews evaluation criteria of the safety queuing models, performance analysis techniques, and benchmarking with traditional models and how to continuously upgrade and adapt.

3.6.1 Criteria for the Evaluation of Safety Queuing Models

A number of critical measures of good safety queuing models also include latency: it is the time period between the information being generated and their final receipt at the place of destination. The lower the latency, the more efficient the delivery of the data is. Another very important measure of data network processes is throughput – the message exchange within a given time frame. This shows the ability of a network to handle many quantities of traffic. The reliability reflects a model that is capable of sustaining performance under varying conditions. Security criteria involve the evaluation of the resilience of cryptographic protocols and the network's resistance to attacks of eavesdropping, jamming, and malware [26]. Scalability refers to a model remaining functional with an increased number of drones and data volume without a proportionate decrease in its performance. Lastly, resource utilization evaluates how the model makes good use of available computational and communication resources without any wastage, hence resulting in proper performance.

3.6.2 Performance Analysis Techniques

Performance analysis techniques are mainly used to gain in-depth insight into how the safety queuing models will work under different conditions. Simulation tools replicate real-world scenarios by creating a virtual environment for controlled experiments on network performance. These simulations run the gamut from simple models to fine-grain replicas of the natural environment and offer insight into latency, throughput, and reliability. Analytical modeling uses mathematical techniques to predict network behavior by applying theoretical principles. The approach puts one at a vantage position as much as one understands the fundamental limits and capabilities a queuing model can achieve, offering a more abstract but powerful way to assess performance. Field testing involves operational models in the real world, showing how performance can be translated into practice. This delivers valuable data regarding the actual performance of the models outside a controlled environment, which simulations might completely miss.

3.6.3 Comparison with Traditional Models

Benchmarking safety queuing models against traditional models is paramount to understanding relative advantages and shortcomings. The analysis compares the performance under similar conditions for quantum-enhanced models versus conventional models, targeting efficiency, security, and scalability improvements. These comparison metrics can be latency, throughput, or reliability and may be treated as arguments pointing out the benefits of quantum enhancement. Case studies can offer head-to-head comparisons of situations, wherein Q-DANETs are incomparably much better than traditional networks. Most of these studies target applications in which quantum features, such as QKD and advanced algorithms, make a huge difference. Benchmarking validates the safety queuing models with practical benefits by illustrating real applications and performance enhancement.

3.6.4 Continuous Improvement and Adaptation

Therefore, continuous improvement and adaptation to the safety queuing models are necessary to remain effective against evolving threats. To achieve this, essential feedback loops are opened, and performance monitoring data is used to improve and enhance the models continuously. Performance reviews and updating ensure that the models remain effective and adaptive to new threats and evolving operational requirements. Adaptive algorithms can switch between queuing strategies on the fly based on input data to ensure maximum performance under the given conditions. Machine learning techniques can be implemented to predict and act in advance of issues before they have a bearing on the network. In this way, analyzing performance data for model adaptation always helps to ensure excellent efficiency and security.

3.7 Conclusion

This chapter examines the integration of quantum computing with drone networks, highlighting how this synergy enhances network security, efficiency, and reliability. It details the use of quantum computing, queue theory, and drone ad hoc networks as foundational elements for understanding Q-DANETs. The discussion includes the development of robust safety protocols that integrate classical encryption with advanced quantum-resistant methods, QKD, and sophisticated IDSs tailored for quantum-specific channels. Key innovations discussed are the design of new queuing algorithms and the implementation of quantum enhancements like QKD and error correction methodologies. These contribute to robust safety mechanisms against both cyber and physical threats. An operational schedule is proposed to ensure that these models are effectively implemented, maintaining performance and security standards while adapting to environmental variability and scalability. The chapter also covers the continuous monitoring and maintenance required for these systems, facilitated by flexible, automated algorithms that provide real-time threat detection and response. Validation of these solutions will be conducted through field tests and simulations, evaluated against metrics such as throughput, latency, reliability, security, and scalability. Overall, this chapter underscores the transformative impact of quantum computing on drone networks, leading to highly optimized and secure operations.

References

[1] A. Hassan, A. Anter, and M. Kayed, "A robust clustering approach for extending the lifetime of wireless sensor networks in an optimized manner with a novel fitness function," *Sustain. Comput. Informatics Syst.*, vol. 30, p. 100482, 2021, doi: 10.1016/j.suscom.2020.100482.

[2] S. N. Mahapatra, B. K. Singh, and V. Kumar, "A secure multi-hop relay node selection scheme based data transmission in wireless ad-hoc network via block chain," *Multimed. Tools Appl.*, vol. 81, no. 13, pp. 18343–18373, 2022, doi: 10.1007/s11042-022-12283-7.

[3] M. Y. Arafat and S. Moh, "A Q-Learning-Based Topology-Aware Routing Protocol for Flying Ad Hoc Networks," *IEEE Internet Things J.*, vol. 9, no. 3, pp. 1985–2000, 2022, doi: 10.1109/JIOT.2021.3089759.

[4] Saravanan, J., T. Ananth Kumar, P. Divya, R. Partipan, R. Partipan, and P. K. Dutta. "IoT based advanced prediction methodology for fault localization in machine monitoring." In IET Conference Proceedings CP824, vol. 2022, no. 26, pp. 253–257. Stevenage, UK: The Institution of Engineering and Technology, 2022.

[5] N. Ghosh, A. Garg, B. K. Panigrahi, and J. Kim, "An Evolving Quantum Fuzzy Neural Network for online State-of-Health estimation of Li-ion cell," *Appl. Soft Comput.*, vol. 143, p. 110263, 2023, doi: https://doi.org/10.1016/j.asoc.2023.110263.

[6] S. Y.-C. Chen, "Asynchronous training of quantum reinforcement learning," *Procedia Comput. Sci.*, vol. 222, pp. 321–330, 2023, doi: https://doi.org/10.1016/j.procs.2023.08.171.

[7] A. S. Nair, S. M. Thampi, and V. Jafeel, "A post-quantum secure PUF based cross-domain authentication mechanism for Internet of drones," *Veh. Commun.*, vol. 47, p. 100780, 2024, doi: 10.1016/j.vehcom.2024.100780.

[8] U. Dubey *et al.*, "A review on practical challenges of aerial quantum communication," *Phys. Open*, vol. 19, p. 100210, 2024, doi: 10.1016/j.physo.2024.100210.

[9] P. Seneviratne, H. Cuff, A. Koletsos, K. Seekamp, and A. Thananopavarn, "New qubit codes from multidimensional circulant graphs," *Discrete Math.*, vol. 347, no. 7, p. 114058, 2024, doi: 10.1016/j.disc.2024.114058.

[10] K. Priya, N. Ganesh, and P. Balaraman, "Basics of business model, emerging fintech & case insights on Gojek business model," *Int. J. Eng. Adv. Technol.*, vol. 8, no. 5, pp. 1837–1842, 2019, [Online]. Available: https://www.scopus.com/inward/record.uri?eid=2-s2.0-85070018844&partnerID=40&md5=c90efd397469482042946b6ca30b61a6

[11] B. Ravi, J. Thangaraj, and S. K. Shandilya, "Stochastic modelling and analysis of mobility models for intelligent software defined internet of vehicles," *Phys. Commun.*, vol. 50, p. 101498, 2022, doi: 10.1016/j.phycom.2021.101498.

[12] M. A. B. S. Abir, M. Z. Chowdhury, and Y. M. Jang, "A software-defined UAV network using queueing model," *IEEE Access*, vol. 11, pp. 91423–91440, 2023, doi: 10.1109/ACCESS.2023.3281421.

[13] P. J. Courtois, "The M/G/1 finite capacity queue with delays," *IEEE Trans. Commun.*, vol. 28, no. 2, pp. 165–172, 1980, doi: 10.1109/TCOM.1980.1094642.
[14] A. Sadok, J. Euchi, and P. Siarry, "Vehicle routing with multiple UAVs for the last-mile logistics distribution problem: hybrid distributed optimization," *Ann. Oper. Res.*, pp. 1–41, May 2024, doi: 10.1007/s10479-024-06019-z.
[15] D. Manikandan, A. M. Skl, and T. Sethukarasi, "Agro-gain – an absolute agriculture by sensing and data-driven through IoT platform," in *Procedia Computer Science*, 2020, pp. 534–539. doi: 10.1016/j.procs.2020.05.065.
[16] M. S. Akhtar, G. Krishnakumar, B. Vishnu, and A. Sinha, "Fast and Secure Routing Algorithms for Quantum Key Distribution Networks," *IEEE/ACM Trans. Netw.*, vol. 31, no. 5, pp. 2281–2296, 2023, doi: 10.1109/TNET.2023.3246114.
[17] H. Yi, "A secure blockchain system for Internet of Vehicles based on 6G-enabled Network in Box," *Comput. Commun.*, vol. 186, pp. 45–50, 2022, doi: 10.1016/j.comcom.2022.01.007.
[18] C. Endter, S. Migala, A. Münch, and A. Richter, "Care-ethical considerations of technology-care-assemblages," *J. Aging Stud.*, vol. 68, p. 101209, 2024, doi: 10.1016/j.jaging.2024.101209.
[19] K. Mansoor, M. Afzal, W. Iqbal, Y. Abbas, S. Mussiraliyeva, and A. Chehri, "PQCAIE: Post quantum cryptographic authentication scheme for IoT-based e-health systems," *Internet of Things*, vol. 27, p. 101228, 2024, doi: https://doi.org/10.1016/j.iot.2024.101228.
[20] W. Pan, M. Yan, Z. Zhao, and M. A. Gulzar, "Flood Risk Assessment and Management in Urban Communities: The Case of Communities in Wuhan," *Land*, vol. 12, no. 1, 2023, doi: 10.3390/land12010112.
[21] C. Hidalgo *et al.*, "Detection, control and mitigation system for secure vehicular communication," *Veh. Commun.*, vol. 34, p. 100425, 2022, doi: 10.1016/j.vehcom.2021.100425.
[22] M. M. Islam, S. Matsushita, R. Noguchi, and T. Ahamed, "Development of remote sensing-based yield prediction models at the maturity stage of boro rice using parametric and nonparametric approaches," *Remote Sens. Appl. Soc. Environ.*, vol. 22, p. 100494, 2021, doi: 10.1016/j.rsase.2021.100494.
[23] O. El Joubari, J. Ben Othman, and V. Veque, "A stochastic mobility model for traffic forecasting in urban environments," *J. Parallel Distrib. Comput.*, vol. 165, pp. 142–155, 2022, doi: 10.1016/j.jpdc.2022.03.005.
[24] S. R. Vadyala and S. N. Betgeri, "General implementation of quantum physics-informed neural networks," *Array*, vol. 18, p. 100287, 2023, doi: https://doi.org/10.1016/j.array.2023.100287.
[25] P. Perazzo, S. Di Matteo, G. Dini, and S. Saponara, "On hardware acceleration of quantum-resistant FOTA systems in automotive," *Comput. Electr. Eng.*, vol. 118, p. 109327, 2024, doi: https://doi.org/10.1016/j.compeleceng.2024.109327.
[26] S. Kumari, M. Singh, R. Singh, and H. Tewari, "Signature based Merkle Hash Multiplication algorithm to secure the communication in IoT devices," *Knowledge-Based Syst.*, vol. 253, p. 109543, 2022, doi: 10.1016/j.knosys.2022.109543.

Sheik Abdullah Abbas, Parkavi Ravi, and Sujitha S

4 Industrial Applications of Quantum AI in Decision-Making

Abstract: AI with quantum enhancements can solve financial optimization issues more quickly and accurately, potentially revolutionizing financial risk management and investing techniques. Various quantum-enhanced artificial intelligence (AI)-driven models exist that can substantiate the industrial needs of organizations. It fulfills the different measurements encompassing the quantum policy of data-driven approaches. Quantum AI, which emerged from the nexus of quantum computing and AI, shows great potential for resolving complicated issues previously thought to be unsolvable by conventional computer techniques. We are seeing the rise of practical uses for quantum technologies, which have the potential to transform entire industries and fundamentally alter how we think about computation. This chapter explores different use cases that can aid in the functionality of industrial needs for quantum AI in the decision-making process. The substantiation of the data, models, methods, tools, and applications is the core context of this chapter, which generally aids the evaluation of quantum AI in predicting the pattern behind the data. Use cases focusing on data optimization, drug discovery, molecular simulation, machine acceleration, financial modelling, and secure communication cryptographic schemes will be explored upon the description towards the focus on applications of AI and quantum in practice. The real-world uses of quantum AI are proliferating, providing answers to issues previously thought to be unsolvable by computers. It is anticipated that quantum computing will aid in creating ground-breaking goods and services that will revolutionize and redefine the manufacturing and industrial sectors.

Keywords: Quantum AI, industrial perspective, quantum optimization, quantum annealing, quantum genetic algorithm, quantum machine learning

4.1 Quantum AI: An Introduction

By using the ideas of quantum mechanics to handle information, quantum computing heralds a revolutionary change in computation. Quantum computers use qubits as their primary data unit, in contrast to conventional computers, which use bits. The amazing ability of qubits to reside in several states simultaneously is known as superposition. This unique feature allows quantum computers to execute several calculations at once, which could result in exponential speed increases for some jobs. Figure 4.1 shows how quantum artificial intelligence (AI) operates.

https://doi.org/10.1515/9783111619323-004

Furthermore, quantum computing exploits another fundamental concept of quantum mechanics called entanglement. Through entanglement, qubits can be intricately correlated, resulting in instantaneous linkage between the states of individual qubits, irrespective of their spatial separation. This remarkable property equips quantum computers with the capability to address intricate troubles with greater competence compared to standard computers.

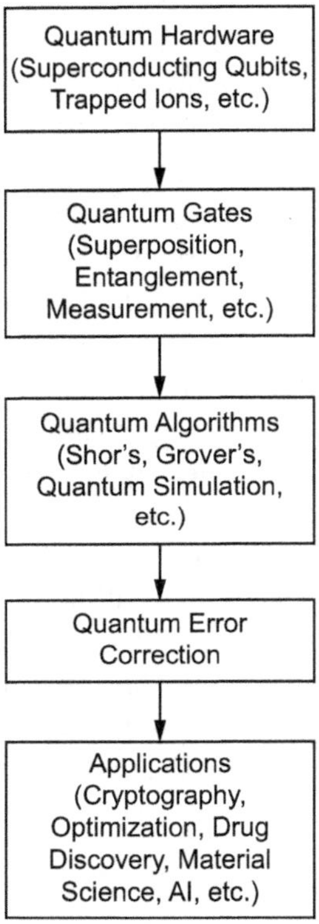

Figure 4.1: Quantum computing overview.

Quantum computing holds vast potential across diverse fields. However, the practical realization of quantum computers is still in its nascent stages, encountering significant technical hurdles like error correction, scalability, and qubit coherence maintenance. In essence, quantum computing marks a fundamental shift in computational methodologies, offering the potential to tackle challenges that are presently ahead of the capabilities of standard computers. Figure 4.1 provides an overview of quantum computing.

4.1.1 Quantum AI Overview

Quantum artificial intelligence (quantum AI) stands at the forefront of an emerging field where quantum computing converges with AI. Its primary objective is to harness the extraordinary capabilities inherent in quantum computers to augment various facets of AI algorithms and applications.

Within quantum AI, quantum computing methodologies are utilized to analyze and handle data with unparalleled efficiency, surpassing the capabilities of standard computers. Quantum algorithms are designed to exploit the peculiar characteristics of qubits – superposition and entanglement – to solve a problem or do it faster, or with less computational complexity than classical algorithms. The main motivation for the development of quantum AI is to solve the problem of computational complexity of classical AI algorithms as well as for large datasets and difficult optimization issues. Quantum AI can be used to improve the training of machine learning (ML) models, the performance of pattern recognition, the decision-making process and the methods of data analysis in almost all areas of application. Nevertheless, quantum AI is still under development, and researchers continue to explore its possibilities and work to overcome its deficiencies. This chapter highlights the need for error correction, qubit coherence, and quantum-friendly algorithms to make quantum AI practical. Quantum AI is an exciting integration of quantum computing and AI that opens up new ways to address difficult challenges and advance the development of AI systems. Quantum AI is a borderless field that applies quantum computing to the field of AI, in order to find new approaches to complex problems. Here's a breakdown of key facets within quantum AI:

- **Quantum-Level ML:** Quantum computing is employed in quantum ML (QML) approaches in quantum AI to enhance traditional ML algorithms. QML algorithms exploit quantum phenomena such as superposition and entanglement to solve problems like pattern recognition, classification, and optimization more efficiently than their classical counterparts.
- **Quantum Neural Networks:** These are a novel type of neural network architecture proposed for quantum computers, which use quantum gates and circuits as their building blocks. They may outperform traditional neural networks in certain application areas.
- **Quantum Optimization:** Quantum algorithms are being developed to solve optimization problems quickly, focusing on those with a complex combinatorial structure. These algorithms leverage quantum mechanics to search through large solution spaces faster than classical algorithms.
- **Quantum Annealing:** Quantum annealers are particular kind of devices that are intended for solving optimization problems. It uses quantum annealing to determine the system's ground state, which is the best solution.
- **Hybrid Quantum-Classical Approaches:** Some quantum AI models employ both quantum and classical computation, where the classical components take care of the preprocessing, post-processing, and optimization functions while the quantum processors undertake the computational heavy lifting.
- **Applications:** Possible applications of quantum AI include faster discovery of new drugs, better design of supply chains, and stronger cryptographic security.

- **Challenges:** Quantum AI has several issues, such as the robustness of quantum hardware, the issue of errors, and the construction of quantum-compatible algorithms. Furthermore, the integration of quantum and classical computing is still a major problem because of their fundamentally different natures.

4.2 Current Approaches to Quantum AI in Decision-Making

This chapter explores how quantum AI differs from conventional AI and highlights its potential impact on healthcare. It discusses the basic principles of quantum mechanics and explains how these concepts can be applied within the field of AI. The chapter initially emphasizes that while classical AI relies on classical computing devices, quantum AI utilizes quantum mechanics to process information. It critical analyzes the promise of quantum AI to enhance precision medicine, improve diagnostic accuracy, and optimize healthcare delivery systems. Furthermore, it explores the challenges of implementing quantum AI, such as the fragility of quantum computers and the lack of scalable quantum processors. In summary, this chapter effectively demonstrates how quantum AI represents a novel approach within the broader landscape of AI. By leveraging quantum phenomena, quantum AI opens up new possibilities for enhancing healthcare outcomes, albeit with current limitations in technology and resources. As quantum computing advances, the potential for quantum AI to revolutionize medical practices appears to grow. Quantum AI in decision-making flowchart is given in Figure 4.2.

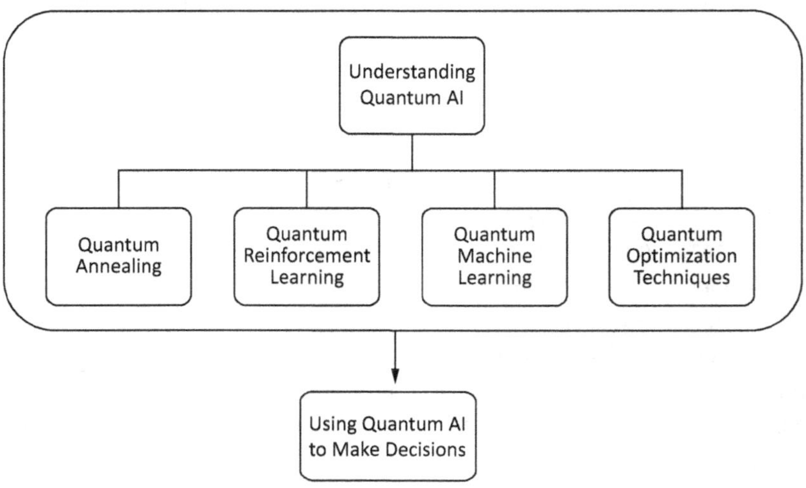

Figure 4.2: Flowchart for the use of quantum AI in decision-making.

- **Quantum Annealing:** Quantum annealers are specific quantum computers which are used for solving optimization problems. Quantum annealers are produced by D-Wave systems, these systems utilized quantum effects to find optimal solutions for difficult decision-making problems. Drug discovery, finance, and logistics are places where quantum annealing is applied.
- **QML:** QML algorithms reconsider the traditional learning algorithms by utilizing the concept of quantum computing. Quantum computing is capable of handling and analyzing large datasets and quantum algorithms such as quantum-supported vector machines (QSVMs) and quantum neural networks (QNNs) are used in the improvement of decision-making. Anomaly detection, pattern recognition, predictive analytics are the applications get benefited from the QML algorithms.
- **Quantum Reinforcement Learning:** Quantum reinforcement learning is a part in the ML which is devoted to training systems to connect with their setting and build optimal decisions. By utilizing the capabilities of quantum computing, traditional reinforcement learning methods are enhanced, leading to more proficient learning processes. Quantum reinforcement learning algorithms is used to improve decision-making plans in difficult situations, and also it opens new directions for optimization.

4.2.1 Using Quantum AI to Make Decisions

Quantum AI acts as an important role in improving decision-making process for various AI-powered applications:

- **Optimization in Financial Portfolio:** Quantum annealing and optimization techniques are utilized by quantum AI aids in creating investment portfolios that preserve and optimal balance among the risk and return. Experts in finance can assess a huge collection of asset allocation potential, which enables them to build policies which reduce risk and increase profit.
- **Supply Chain Management:** Quantum AI improves supply chain management operations by enriching logistics, improving inventory control, and maximizing resource distribution. Sophisticated algorithms examine complicated supply chain data to identify inefficiencies, diminish costs, and improve overall productivity. To improve the decision-making process, these methods frequently integrate ML and quantum annealing.
- **Drug Discovery:** Quantum computing and ML algorithms rationalize the search for feasible drug candidates by proficiently navigating general chemical compound databases. This accelerates pharmaceutical research, making the drug discovery process more cost-effective and time-efficient.

4.2.2 Obstacles and Prospective Paths

Quantum AI has the potential to improve and change the decision-making process; however, numerous challenges must be addressed. Evolving adaptable quantum algorithms, improving hardware reliability, and developing effective error correction processes are the challenges in quantum AI. To successfully integrate the quantum AI into the present decision-making process need cooperation between computer science experts, quantum physics, and a variety of application domains.

Advancement in quantum hardware, optimized algorithms, and cross-disciplinary research is expected to drive further progress in quantum AI to make decisions. Continuous development in quantum computing holds the potential to reform various industries, sustain decision-makers and deal with difficult optimization problems in society.

Quantum AI represents a foremost change in decision-making approach by merging AI with quantum computing to resolve difficult challenges with notable competency. It provides innovative solutions for numerous decision-making applications, including optimization in financial portfolio and drug discovery. The continuous advancement in the research and development of quantum AI is expected to modernize industries, guiding in a period where quantum-driven decision-making becomes the standard one.

4.3 Quantum AI in Industrial Perspective

Quantum AI integrates the ideas of quantum mechanics with AI techniques to get better problem-solving competency. Conventional computing handles with binary bits (0s and 1s), whereas quantum computing engages qubits, which can exist in multiple states concurrently due to superposition. Entanglement is another basic principle of quantum computing; in entanglement qubits are inherently linked, allowing them to immediately influence one another's states, in spite of the distance between them. This property considerably improves the computational power of quantum systems.

Conventional AI depends on traditional computers to examine huge datasets and identify patterns, whereas quantum AI exploits the strengths of quantum computing to operate and interpret difficult datasets more efficiently. This capability enables the resolution of problems that were previously deemed too complex for classical computing methods. With its unmatched computational capacity, quantum AI has the potential to transform several industries completely. Applications include everything from streamlining large-scale supply chain processes to simulating intricate chemical interactions for drug discovery.

4.3.1 Quantum AI in Healthcare Industry

Quantum AI stands poised to transform drug discovery and personalized medicine. In the realm of drug development, its capacity to rapidly analyze extensive datasets of molecular interactions can expedite the discovery of promising compounds, shortening the timeline from research to clinical trials significantly.

In the field of personalized medicine, quantum AI's capability is to inspect huge amounts of genetic, lifestyle, and environmental data points the dawn of a new age in customized treatment plans, exceeding the restrictions of present computational techniques.

In diagnostics, quantum AI grips the possible for more exact and early disease detection by capably solving complex medical data, such as imaging scans. This ability enhances treatment efficiency and paves the way for modern preventive healthcare approaches.

4.3.2 Quantum AI in Finance Industry

Quantum AI is transforming risk assessment and fraud detection within the financial industry, with AI previously incorporated into the operations of 75% of primary banks. By utilizing its excellent computational strength, quantum AI accelerates real-time analysis of wide and complicated datasets, considerably improving these critical processes. This pioneering technology improves the precision of financial risk assessments and supports fraud detection by revealing difficult patterns that traditional AI systems may fail to notice. Moreover, quantum AI's better processing capabilities make it particularly expert at handling complicated financial modeling.

Financial institutions and banks can influence this development to replicate various financial circumstances and investment scenarios with exceptional speed and precision, exceeding the capabilities of conventional computing. By reinforcing financial security and assuring regulatory compliance, quantum AI provides better steadiness in the changing financial sector.

4.3.3 Quantum AI in Cybersecurity

Quantum AI improves encryption mechanisms, providing a terrible defense against developing cyber threats. By utilizing the basic quantum computing principles like entanglement and superposition, it expands encryption techniques that are extremely opposed to traditional hacking methods.

Nowadays, cyberattacks have increased; traditional encryption methods may no longer offer adequate protection, making quantum-enhanced security very important. In 2022, 422 million Americans were affected by data breaches; this indicates the need for stronger security solutions. Quantum AI's capability to quickly process and analyze huge datasets facilitates the rapid detection and prevention of even the most difficult and adaptive cyber threats.

Quantum AI improves analytical cybersecurity, allowing for practical defense strategies against possible breaches. This superior level of security safeguards sensitive information and supports the reliability of digital infrastructures in both the public and private sectors.

4.3.4 Quantum AI in Logistics and Supply Chain

In the logistics and supply chain management domain, quantum AI helps in improving the efficiency and reduces cutting costs with its superior computational capabilities. Quantum AI capability is to analyze huge datasets from numerous resources which facilitates more specific forecasting of logistic barriers which include potential delays and disruptions in logistics. The forecasting power of quantum AI is important for making decisions; allowing business to respond immediately based on the trend of the market and satisfies the need of the customers. Subsequently, supply chain management also becomes more adaptable and flexible, thereby reducing the waste and optimize expenses. Finally, quantum AI makes the operational performance stronger in an industry where efficiency, profitability and delivery time are very important.

4.4 Fusion of Artificial Intelligence and Quantum Computing

The fusion of AI and quantum computing indicates an important rise in technological development. Instead of considering these two fields individually, the fusion of AI and quantum computing can unlock innovative advancements.

An important part of this interaction lies in the development of hybrid algorithms that utilize the properties of quantum computing – such as superposition and entanglement – while enhancing them with traditional computing methods. This fusion establishes a robust framework for addressing complicate problems with greater competency.

For example, in logistics and supply chain management, quantum-conventional optimization algorithms can improve resource allocation with notable accuracy. This development not only organizes operations but also provides route optimization and waste minimization for sustainability.

Moreover, traditional AI systems can benefit from quantum-inspired algorithms, which embed quantum computing principles into traditional models. These developed techniques speed up data processing, generating important insights in areas like finance, climate modelling, and social sciences.

With the interrelated digital ecosystem, quantum computing and traditional AI perfectly collaborate, expanding each other's strengths. This integration cultivates an environment where cutting-edge solutions arise, drives innovation and improvement across various industries [1].

Quantum computing has vast potential to change various phases of AI. The following are the important highlights of how it can improve AI algorithms:

- **AI Algorithm Enhancement:** The key elements of AI algorithms, such as decision trees, clustering, and reinforcement learning, are expanding in quantum computers. Quantum computers have the potential to carry out complex computations and estimate numerous possibilities concurrently through quantum parallelism, which allows them to handle huge datasets and resolve complicated problems more efficiently when compared to conventional computing approaches.
- **Advancement in Natural Language Processing (NLP):** Quantum computing provides optimistic improvements in NLP applications, including language translation, sentiment analysis, and chat bots. Quantum algorithms can efficiently supervise difficult linguistic structures, enabling faster and more accurate language processing. By using the quantum parallelism technique, quantum computers analyze wide linguistic datasets in parallel, thereby improving the speed and accuracy of NLP tasks.
- **Recommendation System Improvement:** Quantum algorithms can improve the precision and competency of recommendation systems, which play a vital role in e-commerce and digital content platforms. By employing quantum parallelism and superposition, recommendation systems can create more personalized and precise recommendations. Quantum-driven models are used to analyze the complicated behavior and preferences of users, leading to enhanced user commitment and experience.
- **Frameworks for QML:** QML libraries are developed by researchers to carry out the implementation of quantum computing in AI. These libraries provide essential tools and frameworks for implementing and testing quantum-enhanced AI models. By standardizing protocols and algorithms, they facilitate the seamless integration of QML techniques across various industries.

Figure 4.3 illustrates a streamlined method for integrating quantum computing with AI [2]. This fusion holds tremendous potential for advancing AI algorithms across various domains, such as NLP and recommendation systems. As research in quantum-enhanced algorithms and tools progresses, substantial developments in AI applications are anticipated.

The intersection of AI and quantum computing represents a transformative leap in computational power. This collaboration investigates how quantum algorithms can refine ML models, address intricate challenges, and foster innovation in AI-driven solu-

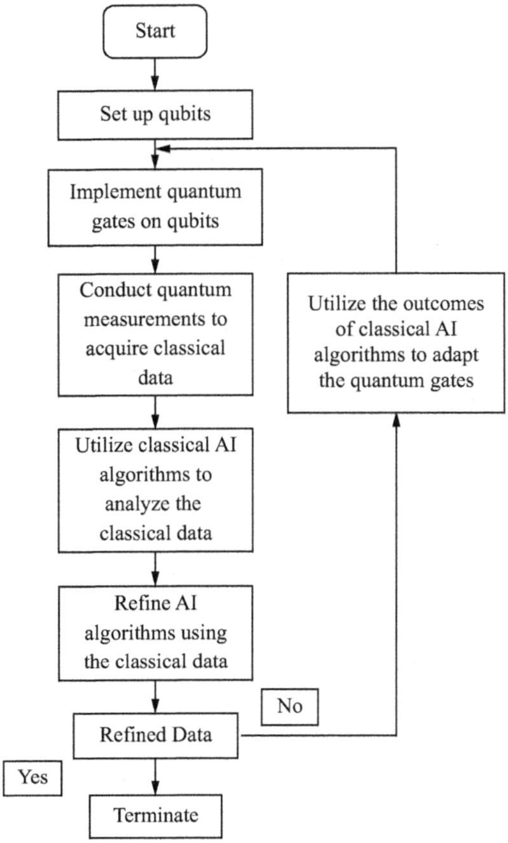

Figure 4.3: Showcasing a straightforward process for quantum computing and AI implementation.

tions. By exploring into the quantum principles leading this fusion, Passian et al. [3] analyze present applications and forecast future advancements across various industries.

Even though quantum computing provides promising potential, the challenges are quantum decoherence and need of stable qubits. To overcome the above challenges, it needs cooperation from multidisciplinary research. However, feasible solutions are offered by invention in quantum hardware, algorithm optimization, and error correction techniques offer viable solutions. The fusion of AI and quantum computing is balanced to transform technological capabilities, improving their respective strengths and guiding in a new sign of innovation.

With the continuous improvement and development in quantum AI, it has the capability to transform industries, speed up scientific progress and address complicated societal problems. Through the integration of multidisciplinary and collaborative research, computing principles are redefined. The rapid development of quantum technology has sparked severe competition in academia and industry in quantum optics field for the last few decades. Now quantum computers are operated with hundreds to thousands of qubits, it is quite high when compared to the previous one; thereby it improves the scalability [4].

4.5 Exploring the Intersection of AI and Quantum Computing: Quantum-Driven AI Methods

At the intersection of quantum computing and AI, the quickly developing field of quantum AI features superior methods with the possible to change computational models. This integration provides incomparable effectiveness and efficiency in solving complicated problems by employing the fundamental properties of quantum systems to improve conventional AI techniques. Investigating the complicated area of quantum AI techniques reveals a wide range of strategies that influence quantum phenomena to address problems in various domains; it ranges from theory to practical implementations, all while efficiently handling the associated forecasts and challenges.

4.5.1 Theoretical Foundations of Quantum AI Methods

Quantum AI is built on a solid theoretical foundation derived from concepts in computational theory and quantum mechanics. QML techniques, like quantum neural networks and support vector machines, use quantum circuits for data analysis and processing, which may have benefits for speed and computational capacity [28]. These algorithms use entanglement and superposition, with quantum phenomena, to search big solution spaces and extract meaningful patterns from data, thereby opening up new avenues for pattern recognition, classification, and regression tasks. The workflow is depicted in Figure 4.4.

4.5.2 Quantum Optimization Techniques

These techniques are another cornerstone of quantum AI. They offer practical solutions to problems arising from combinatorial optimization that plague conventional computer systems. Quantum annealing is exemplified by platforms like D-Wave Systems, which use thermal fluctuations and quantum tunneling to navigate complex solution spaces and find the optimal configurations for a range of optimization problems. Furthermore, by adopting ideas from quantum computation, classical optimization techniques are improved by quantum-inspired optimization algorithms. This leads to creative solutions for issues with finance, logistics, and supply chain management [29].

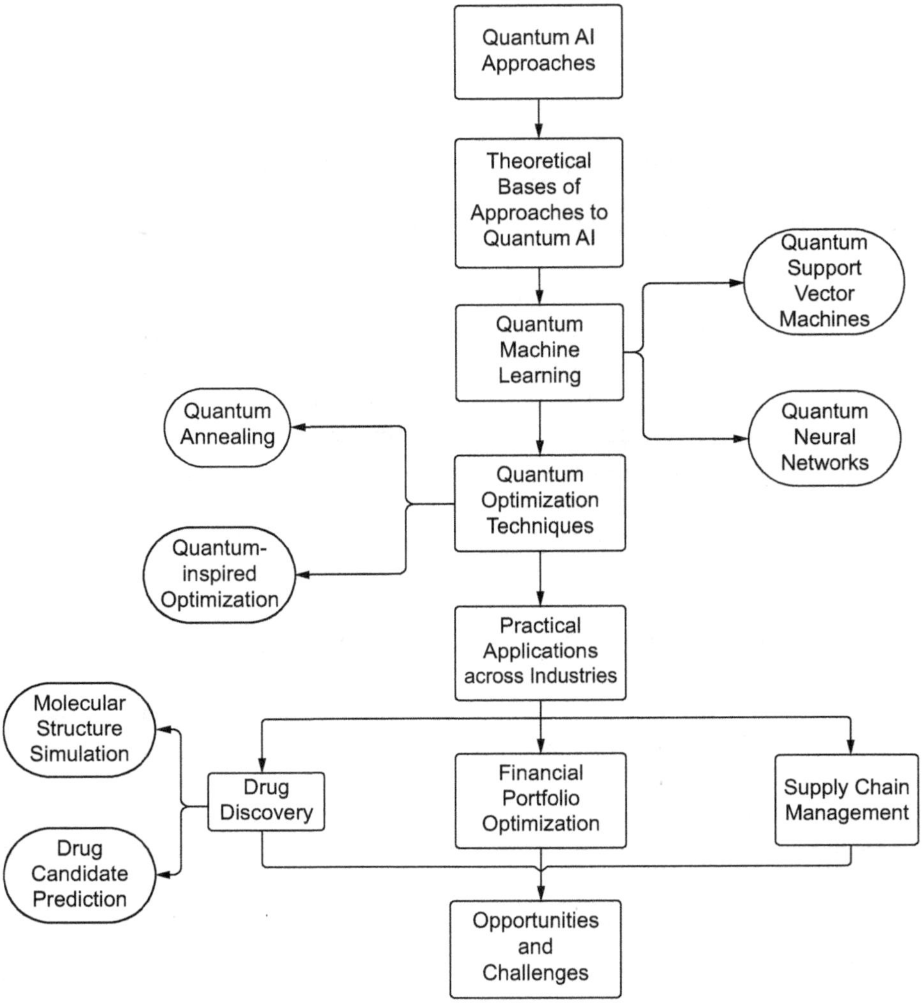

Figure 4.4: Flowchart for various approaches to Quantum AI.

4.5.3 Heuristic Algorithms in QML

The domain of quantum-based technologies adapts a mechanism of different sets of varying operators to quantify the representation in a given architecture. Significant meta-heuristic techniques pave the way to achieving a robust platform to deliver local optimal solutions at different levels. There exist different sorts of popular heuristic approaches to leverage the synergy of quantum optimization mechanisms. The detailed illustrations are provided in the following sections.

4.5.3.1 Quantum Genetic Algorithm

The quantum genetic algorithm (GA) emerges as a formidable tool within evolutionary principles, finding widespread application across various domains to address complex challenges. Numerous complementary techniques have been proposed to augment the effectiveness of GAs, customized to meet the distinct requirements of diverse applications and their innate self-organizational tendencies. For example, GA is employed in multimodal image fusion to optimize load distribution among power system generators through artificial bee colony optimization, and in modelling rainfall-runoff events via a fuzzy rule-based multi-objective GA. Additionally, the optimization of protein structural configurations through genetic mutations showcases the efficacy of optimization searches. Quantum-inspired GAs constitute a subset of evolutionary algorithms inspired by quantum principles, primarily applied to combinatorial optimization tasks. Metaheuristic algorithms, characterized by their high levels of nonlinearity, complexity, and stochastic nature, benefit from enhancements provided by firefly algorithms [23].

4.5.3.2 Quantum Annealing

Enhancing the training process of an ML pipeline holds significance in mitigating training expenses and enhancing model efficacy. Quantum annealing stands out as a promising optimization strategy, representing an emerging computing paradigm that displays potential in refining the training of ML models. D-Wave Systems has successfully implemented a physical quantum annealer, providing the research community with access for experimentation. Recent experimental findings across a spectrum of ML applications have unveiled intriguing outcomes, particularly in scenarios where conventional ML techniques face limitations, such as restricted training data and high-dimensional features [24]. Quantum annealers represent the practical embodiment of adiabatic quantum computation. Fundamentally, these devices address optimization problems by minimizing the energy of a physical system [27].

4.5.3.3 Variational Quantum Algorithms

Variational quantum algorithms operate within a hybrid quantum-classical computational framework. In this setup, quantum computers handle specific quantum-native subroutines, while classical computers execute optimization procedures on variational parameters [30]. These optimizations aim to minimize a cost function, with the minimum representing the solution to the problem at hand. This methodology mirrors the foundational concept seen in contemporary deep learning models, wherein intricate parametric models undergo tuning via optimization techniques to address diverse tasks. The convergence of quantum computing and ML has spurred the emergence of QML, an interdisciplinary field investigating the advantages derived from merging quantum computation and AI [25].

4.5.3.4 Quantum Swarm Intelligence Algorithm

The quantum particle swarm optimization (QPSO) algorithm draws inspiration from the quantum phenomena observed in nature. The fundamental idea centers on finding the appropriate wave function for a quantum particle as it traverse a potential field. QPSO leverages quantum probability density functions to guide particles towards positions with the highest likelihood of containing the optimal solution [31].

This approach capitalizes on the probabilistic nature of quantum mechanics and the interplay between quantum particles and their eigenstates, aiming to locate the solution within the most probable region of the search space. While the direct relationship between quantum features and time complexity remains implicit, the inherent probabilistic nature of quantum mechanics is anticipated to enhance search performance by directing the algorithm toward regions with higher likelihoods of containing the solution. QPSO employs various quantum potential fields, including square wells, 1D potential wells, 1D quantum simple harmonic oscillators, Coulomb-like square root fields, Lorentz potential fields, and Rosen-Morse potentials, which have been explored and refined by several researchers [26].

The QIMH optimization algorithm aims to attain optimal solutions, contingent upon the careful selection of parameters, objectives, thresholds, and appropriate fitness functions. Moving forward, it represents a highly dynamic realm of research within computational intelligence, encompassing areas such as time-series mapping and analysis of both micro- and macro-services [32].

4.5.4 Useful Applications Across Industries

Quantum AI has a wide range of discipline- and industry-specific uses, from supply chain management to finance and medication development. Quantum AI simulates molecule structures and predicts molecular attributes with previously unheard-of speed and accuracy to speed up the proof of identity of promising drug candidates in the drug development process. Quantum AI techniques are used by financial institutions to optimize investment portfolios, resulting in improved portfolio performance and a more successful trade-off between risk and return goals. Consequently, supply chain management gets benefited from quantum AI by allocating the resources effectively, optimize the operation of logistics and improves the overall efficiency.

4.5.5 Opportunities and Challenges

While quantum AI holds vast opportunities, which include the ability to make better decisions, speed up technological progress, and redefine innovation, it also poses a few challenges that must be overcome to fully utilize quantum AI. Hardware challenges of quantum computing system are qubit coherence, scalability, reliability and error rates, to address these challenges computer experts, quantum physicists and experts from multidisciplinary research should collaborate [33].

Quantum AI, offers up a vast array of opportunities, ranging from theoretical foundations to practical implementations and beyond. The current direction of quantum AI research and development suggests that quantum-enhanced decision-making will become the norm in the future, leading to unprecedented levels of computational efficiency and innovation in a variety of industries [34]. The development of quantum AI holds the potential to transform technology in the future and create new avenues for investigation and learning. But there are challenges to face and opportunities to seize.

4.6 Use Cases of Quantum Computing: In Manufacturing

Quantum technologies are likely to profoundly shape global business values in the near future. While it may take a decade for quantum technology and its concepts to become widespread in the manufacturing industry, companies are already taking steps to understand its applications to meet future needs. Additionally, manufacturing industries are actively addressing climate change and transitioning toward sustainable solutions on a global scale.

In the coming years, the advancement of quantum technologies will closely coincide with the objectives outlined in the Sustainable Development Goals. This chapter intends to investigate the potential utilization of quantum technologies within manufacturing systems and to deliberate on the possibilities for sustainable solutions enabled by these technologies. Manufacturing systems can be understood through two lenses: technical and managerial. Technically, manufacturing entails the transformation of raw materials into diverse components, machinery, or products via a range of production and assembly methods. Managerially, manufacturing represents a system comprising distinct stages throughout the product life cycle.

The potential innovation pathway for quantum technology in manufacturing systems, including its applicable domains, is evaluated from a managerial viewpoint that integrates both the technological and managerial aspects of manufacturing.

The manufacturing systems encompass various domains, including:
- Value networks domain (D1)
- Business model and its services (D2)
- Product life cycle (D3)
- Fabrication and assembly processes (D4)
- Equipment (D5)
- Organization level (D6)
- Human based (D7)

Related applications of quantum technologies within each domain, based on broader usage identified through research, are categorized as follows:
- Quantum computing and AI (Q1)
- Quantum communication (Q2)
- Quantum simulation (Q3)
- Quantum imaging (Q4)
- Quantum sensing (Q5)

The relationships between these manufacturing domains and their potential applications through different quantum technologies are illustrated in the diagram below, in Figure 4.5, for clarity and understanding [1].

The findings from our literature review are summarized as follows, based on [1]:
- Quantum AI can assist structure contemporary manufacturing systems and optimize value networks, business models, and services domains.
- Quantum simulation, quantum imaging, and quantum sensing can aid in the design of new products and improve their life cycles.

Quantum technologies can enhance optimization, planning, development, and scheduling responsibilities in manufacturing systems, and contribute to quality control improvements.

4.7 Use Cases of Quantum Computing: In Retail Industry

4.7.1 Information, Communication, and Computing Technologies

The enormous potential of merging quantum computing with fundamental Information, Communication, and Computing Technologies technologies is being explored by researchers. Unmatched processing power and efficiency can be unlocked by combining quantum computing with AI, data analytics, cryptography, and communication networks. With the use of this multidisciplinary approach, we hope to find new uses for quantum computing

Figure 4.5: Domains, applications, and quantum technologies in manufacturing.

systems, improve their scalability and dependability, and advance paradigm shifts in information processing and communication. In the end, this integration is what allows quantum computing to reach its full potential and provide ground-breaking discoveries that were thought to be impossible. Data obtained from quasi-secondary sources such AI engines like GPT and Bard, as well as secondary sources like search engines like Google and Google Scholar, is analyzed through exploratory study [5].

4.7.2 Small and Medium-Level Enterprises (SMEs)

Small and medium-level enterprises (SMEs) play a vital role in the economy, yet they often lack sufficient readiness to address information security threats compared to larger corporations, mainly due to various factors. The severity of these threats may differ depending on the industry. When compared to traditional computers, quantum computers are still in the developmental stage, which provides significant performance benefits, including encryption and decryption methods [35]. To keep data safe and protect it from potential cybersecurity threats, it is important to improve security by using cryptographic algorithms. Post-quantum cryptography provides solutions to SMEs for these issues. Most SMEs are uninformed about these risks, which emphasizes the significance of raising awareness and encouraging practical security methods. These considerations pave the way for more research to examine these challenges in future studies [6].

4.7.3 Software Industry

With the rapid development of technology in the software industry, it poses several challenges, with quantum computing rising as a driving force positioned to revisit the concepts of software engineering. In spite of its rising force and progression, an important gap residue in analyzing and prioritizing the challenges it poses to the industry. This research article applies the concept of fuzzy analytic hierarchy process (F-AHP) to overcome the challenges faced by software industry. This study concludes that the most important challenges in following quantum computing comprise technical incompetence, false information, and fail to adopt new methods. In addition, lack of standard secure communication protocols for implementing quantum computing is accepted as a main concern. F-AHP analysis provides that the challenges faced by institutions are most important, succeeded by organizational obstacles. These insights highlight that quantum computing research is still in the developing stage and underline the significance of multidisciplinary partnership to address these barriers efficiently [7].

4.7.4 Production Planning

To overcome the limitations of conventional computing methods, this study examines the potential of quantum computing in production planning domain. While traditional methods have seen some achievement, they often poses difficulties in handling complex optimization tasks, specific demand forecasting, and modernized supply chain management. Quantum computing's sophisticated dealing capabilities provides potential solutions to these challenges, facilitating the way for well-organized production scheduling. By combining the quantum computing benefits, constraints and challenges with production planning, this research bridge the gap between quantum computing and production planning. In order to enhance the production efficiency, reduces cost and making better decisions, specialized algorithms are introduced that utilize quantum computing [8].

4.7.5 Supply Chain

Lack of expertise in quantum technology is a challenging task in implementing quantum computers with real-world applications. In this article, two quantum optimization algorithms are used in trade finance portfolios and assessed based on their effectiveness by comparing the decisions with conventional optimization algorithms. Using a quantum algorithm learned from a Qiskit tutorial, the method entails turning the financial risk and returns of a trade finance portfolio into an optimization function. The results indicate that while using the quantum algorithms does not seem to provide any advantages, there is no statistically significant decline in their performance. This suggests promise for the future, as improvements in quantum hardware are expected. The theoretically superior processing speeds and handling of large data volumes offered by quantum computing may eventually find practical applications in trade finance [9].

4.7.6 Renewable Energy

The potential for quantum computers to disrupt the renewable energy sector is significant due to their immense computing power. Quantum computing, with its highly efficient algorithms offering exponential speedup, is anticipated to transform the landscape of renewable energy exploration. This study delves into the utilization of quantum computing for renewable energy exploration and optimization tasks, emphasizing its computational efficiency and distinctive attributes compared to classical computing systems. The discourse addresses the current state and hurdles of quantum computing, alongside the opportunities it unveils. Moreover, it elucidates the foundational principles of quantum computation, highlighting their distinctions from classical computing methodologies.

This work briefly touches on existing quantum optimization algorithms. As renewable energy exploration generates vast amounts of data, energy companies will have unprecedented opportunities to advance climate change mitigation efforts, furthering global progress towards sustainable energy solutions [10].

4.8 Use Cases of Quantum Computing: In Healthcare Sector

Quantum computing has a great deal of potential to transform the healthcare sector because of its unparalleled computational power and ability to manage massive volumes of data simultaneously. In this analysis of quantum computing's use cases in healthcare, we explore the innovative applications that could revolutionize medical research, diagnosis, treatment, and more. The field of healthcare innovation and progress could be greatly impacted by quantum computing in a number of ways, such as accelerating drug discovery, improving patient care, and developing medical imaging techniques. The workflow is depicted in Figure 4.6.

- **Accelerating Drug Discovery**
  Quantum computers can speed up drug discovery and make it more accurate. Finding new drugs with traditional methods takes time and requires many tests. Quantum computers can quickly model how molecules interact, helping scientists study more chemicals and predict their properties more effectively. With quantum algorithms, researchers can identify new drug candidates faster, leading to quicker treatments for diseases.
- **Precision Medicine and Personalized Therapy**
  Quantum computers can help doctors treat effectively by analyzing large amounts of genetic and protein data in a shorter time. They can find patterns and disease markers aligned with genetic disorders. This helps doctors design personalized treatments that work in a better way and cause very few side effects. With quantum technology, patients could get more focused care and better results.
- **Better Medical Imaging Techniques**
  Medical imaging aids doctors in finding and tracking deadly diseases like cancer and heart problems. Quantum computers could make medical imaging more accurate and detailed. They can quickly handle large amounts of data, helping doctors detect diseases sooner and choose better treatments. Quantum technology may also create new imaging methods, providing doctors with better tools to treat patients [36].
- **Optimization of Healthcare Operations**
  Quantum computing can make healthcare run more smoothly. It can help with patient appointments, hospital management, and logistics in an easy way. By analyzing large amounts of data, it can reduce patient wait times, improve resource

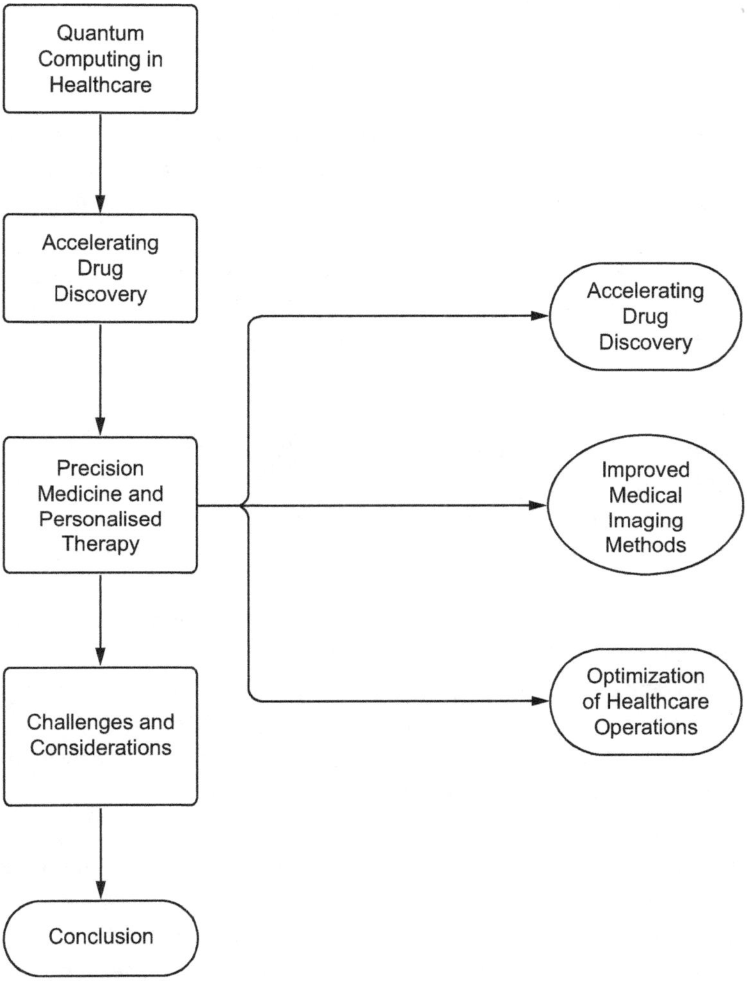

Figure 4.6: Flowchart depicting the use of quantum computing in medical field.

utilization, and enhance workflows. Hospitals can schedule surgeries efficiently, reducing delays and refining care. It can also help manage medical supplies, ensuring they arrive on time with minimal waste.

- **Challenges and Considerations**
  Quantum computing could improve healthcare, but some challenges must be solved first. Scientists need better software, bigger quantum computers, and ways to connect them to current systems. There are also important issues like patient consent, data security, and privacy. To use quantum computing responsibly in healthcare, experts must plan carefully and work together.

4.9 Use Cases of Quantum Computing: In Environmental research

Q4Climate brings together experts from various universities, government organizations, and industries to explore how quantum technology can help address climate change. The goal is to find practical ways to slow down climate change and mitigate its effects.

We highlight five key areas where quantum technology could make an impact by showcasing different research examples. Our goal is to bring all the experts together in climate science, classical optimization, and quantum computing to explore its potential effect.

1. **Power and Energy** – Quantum computing can significantly improve real-time energy use by optimizing power grids, managing renewable energy and making wireless networks more resourceful.
2. **System Design** – It can help design better services, supply chains, and manufacturing systems, such as placing wind turbines more efficiently and refining energy use in the auto industry.
3. **Transportation Networks** – Quantum-driven traffic and navigation systems can significantly reduce congestion and lower carbon emissions.
4. **Distribution Networks** – Smart shipping, smart water distribution, and smart city planning can cut down waste and improve resource utilization.
5. **Climate Mitigation and Adaptation** – Quantum technologies can improve climate change, support biodiversity efforts, and help limit emissions.

By exploring these areas, we hope to find new ways quantum computing can improve sustainability.

Quantum sensing unlocks new potentials for pursuing climate change. This chapter explores areas where quantum technology could outclass traditional sensors. The selected application areas meet one or more key criteria, highlighting where quantum can make a real variance:

- Lack of existing classical solutions
- Need for greater precision, selectivity, or efficiency
- Suitability for high-tech solutions
- Existing demand from early adopters
- Potential for high impact and scalability

Studying greenhouse gases is key to understanding climate change [3]. Satellites help monitor gases like methane and carbon dioxide, but receiving methane readings accurately is tough because other gases may interfere. Quantum sensors can resolve this problem by detecting gases more accurately. They can also be used in satellites to monitor emissions over large areas, like peat lands, giving experts improved data on climate impacts.

Quantum technology can help track climate in an effective way by addressing challenges in sensing aerosols, which play a vital role in climate change. These sensors can offer better data on aerosol levels and their effects. They can also help monitor ice sheet melting and rising sea levels, both of which are dangerous for climate models. Quantum sensors can detect changes in ice sheets more precisely with accurate optical properties, providing researchers valuable insights into our shifting planet.

Quantum gravimeters in satellites can give us a clear view of the climate system globally. They can monitor changes in Earth's mass, identify climate changes in polar regions, and quantize how our planet responds to natural and human actions. This information will help scientists to understand our earth's climate patterns and their impact more accurately [37].

4.10 Realistic Case Study in Quantum Computing

Quantum computing is applied in various fields; some of the notable areas are:

1. Optimization problems
2. Drug discovery and molecular simulation
3. Machine learning and AI
4. Cryptographic security
5. Financial modeling and risk analysis
6. Materials science and chemistry
7. Supply chain and logistics
8. Energy optimization and grid management

4.10.1 Optimization Problems

Zhu et al. devised ADAPT-QAOA (quantum approximate optimization algorithm), an iterative QAOA, tailored to specific problems and hardware constraints. Through simulations on Max-Cut graph problems, they have demonstrated its faster convergence and reduced resource requirements compared to standard QAOA, attributed to shortcuts to adiabaticity. ADAPT-QAOA iteratively grows the Ansatz, outperforming standard QAOA across random diagonal Hamiltonians. Its flexibility in mixer pool selection enables customization to hardware's native gates and constraints, promising efficacy for higher-dimensional optimization challenges like graph coloring [14].

4.10.2 Drug Discovery

Quantum computing presents a transformative approach to drug discovery by accurately simulating molecular interactions and predicting properties more efficiently than classical methods. Quantum algorithms like VQE and QAOA help optimize molecular structures and speed up the search for new drug candidates. QML analyzes large datasets, predicting molecular properties and narrowing down promising options faster. Frontier industries like IBM and D-Wave offer platforms for quantum research and optimization in drug discovery. To understand its full potential, we must address various challenges, such as high error rates and short qubit coherence times [15].

Quantum computers might change drug research by solving serious chemistry problems. Recent work in quantum phase estimation has reduced computing costs. This improvement lets researchers create systems with around 50 electrons and orbitals, but other challenges remain. Current machines lack enough qubits for larger projects. More effort is needed to fix errors and advance hardware. Quantum methods offer novel ideas, but they must fit with current drug discovery techniques. Better hardware and software are essential to reach their full potential [16].

4.10.3 Molecular Modeling

Quantum computing is altering the way we study molecules. It can simulate molecular behavior with greater accuracy, helping scientists understand biological processes and create new drugs. Real-time experiments reveal how molecules move and interact, enhancing the accuracy of drug-binding studies. By solving complex equations faster, quantum computers improve drug and material design. They also help researchers study protein folding and drug interactions, leading to effective treatments. Advanced visualization tools make it simpler to analyze complex molecular structures. We need to solve challenges like stability and error correction in quantum computers to fully unlock their potential in drug discovery [15].

Quantum computing has the potential to change drug research by solving complex chemistry problems that are out of reach for today's computers. New advances in quantum algorithms have made simulations more practical, allowing scientists to model systems with around 50 orbitals and electrons. But big challenges remain, like fixing errors and improving hardware. Right now, quantum computers do not have enough qubits for large-scale use, so more research is needed to bridge the gap between theory and real-world applications. While quantum algorithms could speed up drug discovery, they need to fit into existing research methods. Making quantum computing truly useful will take teamwork across hardware and software development [16].

4.10.4 Machine Learning and Artificial Intelligence

Quantum information and AI could shape the future in powerful ways. Traditionally, challenges in these fields were tackled separately, but recent research explores their connection. Hybrid QML combines quantum physics with classical and quantum algorithms, improving data analysis and processing. Quantum computing can boost classical ML, while quantum algorithms offer new possibilities for data science. This chapter introduces fundamental concepts and methods of QML, showcasing implementations like multiclass tree tensor networks and quantum tomography-based neural networks using IBM's quantum processor. These techniques enable accurate prediction of quantum states and uncover latent dependencies in input-output relationships, illustrating the power of classical-quantum approaches [17].

4.10.5 Cryptographic Security

Classical cryptography relies on the difficulty of certain mathematical problems or information theory arguments for security. In contrast, quantum cryptography relies on the principles of quantum physics. With the advent of large quantum computers, classical asymmetric algorithms for key distribution and digital signatures become vulnerable. Fabio Cavaliere, John Mattsson, and Ben Smeets from Ericsson Research discuss the technologies and protocols of quantum key distribution, emphasizing its security implications and standardization efforts. They also highlight the significance of quantum random number generators in both classical and quantum encryption systems, while addressing the security challenges posed by quantum computers [18].

Since the inception of cloud computing, data security has remained a significant concern, despite numerous proposed frameworks. The rise of quantum computing underscores the urgency for a model that secures data for both classical and quantum computing environments. While many cryptosystems face obsolescence, a proposed model advocates for a variant of McEliece cryptosystem for access control data and a variant of NTRU cryptosystem for user data. Simulation demonstrates improved time complexity for the proposed McEliece algorithm, while parameter adjustments enhance security. Similarly, the proposed NTRU variant maintains superior time complexity. This model advocates for quantum-safe algorithms to fortify cloud data security, aiming to reduce man-in-the-middle attacks and enhance overall data security [19].

4.10.6 Financial Modeling and Risk Analysis

Quantum computing offers accelerated solutions for optimization and simulation but faces limitations in current systems, hindering practical applications in quantitative finance. This analysis explores implementing risk management in financial institutions, focusing on value-at-risk and potential future exposure. Quantum computing has potential, but real-world use is still a challenge. Today's hardware struggles with large datasets and quantum noise. Better error control and easier access to quantum hardware are the keys to making it useful for risk measurement.

Current hardware constraints may ease with projected advancements, yet widespread adoption faces skepticism, particularly in regulatory compliance. While some financial tasks benefit from quantum speed-ups, the business case remains weak due to existing computational efficiency in traditional methods, suggesting a gradual integration into specific risk management areas like portfolio optimization [20].

This article explores the potential of quantum computing in finance, discussing its introduction, surveying challenging problem classes, and detailing quantum algorithms for simulation, optimization, and ML tasks. Quantum computers offer promise in resolving complex financial challenges more efficiently, particularly in optimization, ML, and simulation domains. The outlined quantum algorithms aim to address precision and runtime constraints in financial services, potentially revolutionizing risk analysis, client management, and market dynamics. Performance demonstrations on IBM quantum back-ends indicate early advantages in simulation, optimization, and ML tasks, highlighting the transformative potential of quantum computing in finance [21].

4.10.7 Computational Molecular Biology

This work explores the potential impact of quantum computing on computational biology and bioinformatics, highlighting its ability to solve problems exponentially faster than classical methods. Quantum algorithms promise advancements in information processing, ML, drug discovery, and optimization tasks crucial to these fields. However, caution is warranted amidst the excitement, as challenges and limitations exist alongside the transformative potential. Anticipated within the next decade, practical quantum processors could revolutionize computational biology, prompting the need for computational scientists to understand both the capabilities and constraints of quantum computing. The review aims to ignite interest among computational biologists and quantum information practitioners, envisioning a future where quantum technologies reshape biological research [22].

4.11 Tools in Quantum AI: Problem Solving

In the realm of quantum AI, problem-solving tools encompass a variety of resources tailored to tackle complex challenges using the combined capabilities of quantum computing and AI. These tools are important for planning innovative solutions to difficulties that exceed the capacities of traditional computing methods. An outline of key aspects about problem-solving tools in quantum AI is given below [11]:

- **QML Libraries**
  These libraries offer tools for quantum-powered ML. They help researchers and developers run quantum algorithms for classification, regression, clustering and dimensionality reduction. Using these tools, working in QML becomes easier and more manageable.
- **Quantum Optimization Frameworks**
  Optimization plays a major role in many areas. Quantum optimization frameworks provide specialized algorithms that make use of quantum computing to solve complex problems in fast manner. They help with tasks like portfolio management, route planning, and resource allocation, by making complex decisions and making it more efficient.
- **Quantum Algorithm Development Kits (QADKs)**
  QADKs provide a full set of tools for creating and testing quantum algorithms. They help researchers design circuits, prototype algorithms, and assess their efficiency. This type of kit creates new ways to solve problems in quantum AI by making it simpler and efficient.
- **Quantum Programming Languages**
  Quantum programming languages like Qiskit, Cirq, and Quipper are created for building quantum algorithms. Their syntax and features allow researchers to perform quantum operations easily and also enable working with quantum hardware or simulators.
- **Quantum Simulation Platforms**
  Quantum simulation platforms allow researchers to test complex quantum systems and algorithms on regular computers. They provide tools to design quantum circuits, gates, and algorithms, and to refine and optimize solutions before executing them on real quantum hardware.

4.12 Challenges and Limitations of Quantum AI

Quantum computing can revolutionize the healthcare industry in significant ways. It can accelerate drug discovery, enhance the accuracy of medical imaging, and create workflows more efficiently. However, there are still many challenges to overcome in order to realize its full potential.

Quantum AI, which is a combination of both AI and quantum computing technologies, it is highly potential to transform many applications. The following section analyzes the hardware limitations, complex algorithms, and practical barriers that mainly affect its wider adoption. The illustration is given in Figure 4.7.

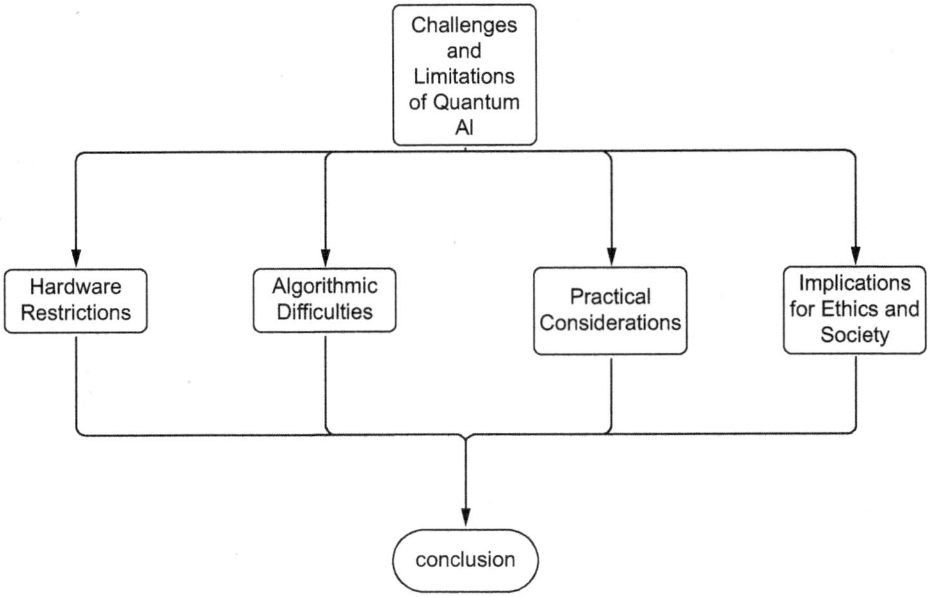

Figure 4.7: Flowchart representing challenges and limitations of quantum AI.

- **Hardware Limits**
  The biggest challenge for quantum AI is building scalable and reliable hardware. Qubits, the core of quantum computing, are delicate and prone to errors. Adding more qubits, refining their connections and maintaining the system's stability is more difficult. The available quantum systems have lesser coherence times, high error rates, and very inadequate connectivity.
- **Algorithm Challenges**
  Quantum AI has its own algorithm problems. As in recent research, quantum algorithms exhibit its potentials but still need a lot of fine-tuning to be more effective. Creating reliable algorithms that reliably beat classical algorithms in scalability and accuracy remains a big challenge. Also utilizing classical algorithms into quantum systems couldn't be more beneficiary. It is very important to balance the complexity of quantum AI and its application in real world problems need more enhancements and support from researchers.
- **Practical Challenges**
  Quantum AI also has many risks like accessibility, scalability, and integration associated with real world implementation. If there is an organization with limited resources wants to connect their existing system with quantum computers, it will be really tough and challenging. Handling complex tasks or uncertain situations, the scalability of the application depends on the strong hardware and efficient algorithms which makes it more challenging. To solve these issues, researchers,

industries, and government organizations should work together to build user-friendly tools and platforms that make quantum AI easier to adopt.

- **Ethical and Social Impact**
  Quantum AI raises big questions about ethics and society. Data privacy, security, and bias in quantum algorithms must be handled responsibly. Workforce disruptions, like skill gaps and job losses, also need attention. Advancing this technology requires balancing progress with fairness, accountability, and trust [38].

4.13 Future Directions of Quantum AI

Figure 4.8 provides a simple look at how quantum AI could shape Industry 4.0 and its future benefits [4].

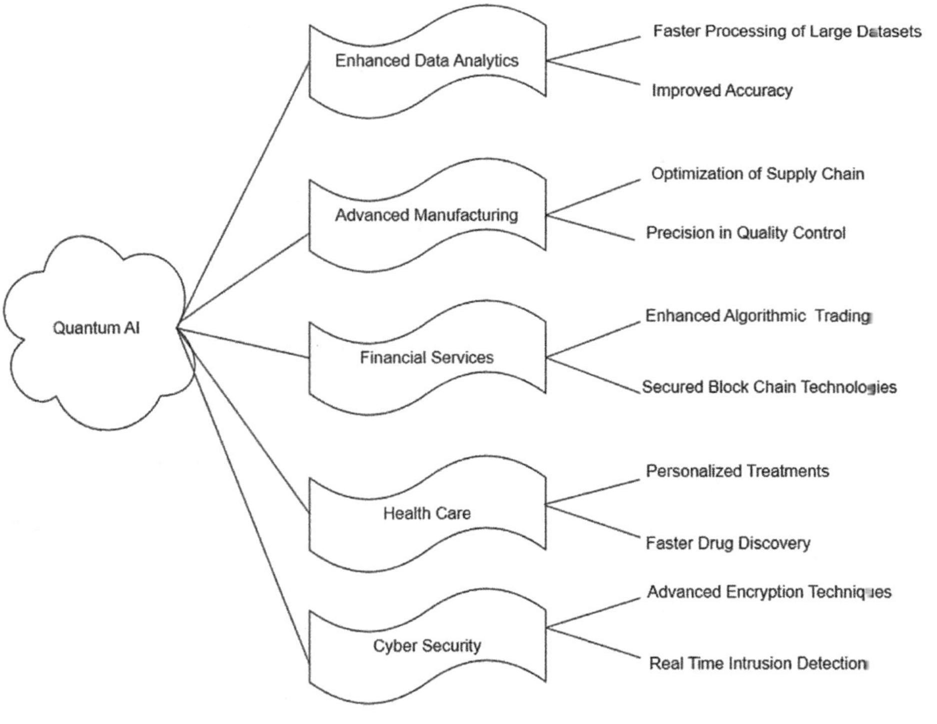

Figure 4.8: Overview of quantum AI in future perspective.

Quantum AI has potential but also its own challenges. Quantum computing is still new and not widely available. These systems need strict controls because quantum states are fragile. There aren't enough experts who understand both quantum physics and AI. Also, quantum algorithms and software are still developing and need more work to be fully effective.

Quantum AI brings ethical and practical challenges. It could be misused for surveillance or autonomous weapons, raising serious concerns. Data security is also at risk since it might break current encryption. Its power must be controlled to prevent bias and errors, especially at high speeds and large scales. Clear legal and ethical rules will help ensure quantum AI is used responsibly as it grows.

This technology could change industries like healthcare, finance, logistics, and cybersecurity. Its speed in handling big data can drive new ideas and make systems more efficient. Keeping up with its progress is important to understand how it will shape the future.

While quantum AI has huge potential, many challenges stand in the way. Hardware limits, complex algorithms, ethical concerns, and real-world obstacles must be solved. Working together across different fields can help overcome these issues and unlock its benefits for everyone [39].

References

[1] Van Erp, Tim, and Bartłomiej Gładysz. "Quantum technologies in manufacturing systems: perspectives for application and sustainable development." *Procedia CIRP* 107 (2022): 1120–1125.

[2] Berger, Casey, Agustin Di Paolo, Tracey Forrest, Stuart Hadfield, Nicolas Sawaya, Michał Stęchły, and Karl Thibault. "Quantum technologies for climate change: Preliminary assessment." *arXiv preprint arXiv:2107.05362* (2021).

[3] Passian, Ali, Gilles Buchs, Christopher M. Seck, Alberto M. Marino, and Nicholas A. Peters. "The concept of a quantum edge simulator: edge computing and sensing in the quantum era." *Sensors* 23, no. 1 (2022): 115.

[4] Meng-Leong How, and Sin-Mei Cheah. "Forging the Future: Strategic Approaches to Quantum AI Integration for Industry Transformation." *AI* 5, no. 1 (2024): 290–323.

[5] S. Begum, "AI Odyssey: Unraveling the Past, Mastering the Present, and Charting the Future of Artificial Intelligence", Publisher: NOTUN KHABAR, July 31, 2023.

[6] Bapon, S. H., & Kaur, A. (2024). Quantum AI: Exploring the Intersection of Quantum Computing and Artificial Intelligence. SSRN Electronic Journal. https://doi.org/10.2139/ssrn.4490226

[7] Kanimozhi, P., Kannan, M., Ananth, C., Kumar, T. A., & Parkavi, S. (2023, September). Design of IoT Based Agricultural Farm Protector Drone Using Raspberry Pi. In 2023 6th International Conference on Contemporary Computing and Informatics (IC3I) (Vol. 6, pp. 1175–1179). IEEE.

[8] Shuford, J. (2024). Quantum Computing and Artificial Intelligence: Synergies and Challenges. Journal of Artificial Intelligence General Science (JAIGS) ISSN: 3006-4023, 1(1). https://doi.org/10.60087/jaigs.v1i1.35

[9] Aithal P. S. (2023). Advances and New Research Opportunities in Quantum Computing Technology by Integrating it with Other ICCT Underlying Technologies. International Journal of Case Studies in Business, IT and Education (IJCSBE), 7(3), 314–358. https://doi.org/10.47992/IJCSBE.2581.6942.0304

[10] Schindler, P., &Ruhland, J. (2022). The Threat of Quantum Computing to SMEs. Intelligent Computing, 404–413. https://doi.org/10.1007/978-3-031-10461-9_28

[11] Awan, U., Hannola, L., Tandon, A., Goyal, R. K., &Dhir, A. (2022). Quantum computing challenges in the software industry. A fuzzy AHP-based approach. Information and Software Technology, 147, 106896. https://doi.org/10.1016/j.infsof.2022.106896

[12] Griffin, P., &Sampat, R. (2021). Quantum Computing for Supply Chain Finance. 2021 IEEE International Conference on Services Computing (SCC). https://doi.org/10.1109/scc53864.2021.00066

[13] https://thequantuminsider.com/2022/05/27/quantum-computing-tools/

[14] Zhu, L., Tang, H. L., Barron, G. S., Calderon-Vargas, F. A., Mayhall, N. J., Barnes, E., &Economou, S. E. (2022). Adaptive quantum approximate optimization algorithm for solving combinatorial problems on a quantum computer. Physical Review Research, 4(3). https://doi.org/10.1103/physrevresearch.4.033029

[15] Raparthi, M. (2022). Harnessing Quantum Computing for Drug Discovery and Molecular Modelling in Precision Medicine: Exploring Its Applications and Implications for Precision Medicine Advancement. Advances in Deep Learning Techniques, 2(1), 27–36. Retrieved from https://thesciencebrigade.com/adlt/article/view/136

[16] Blunt, N. S., Camps, J., Crawford, O., Izsák, R., Leontica, S., Mirani, A., Moylett, A. E., Scivier, S. A., Sünderhauf, C., Schopf, P., Taylor, J. M., &Holzmann, N. (2022). Perspective on the Current State-of-the-Art of Quantum Computing for Drug Discovery Applications. Journal of Chemical Theory and Computation, 18(12), 7001–7023. https://doi.org/10.1021/acs.jctc.2c00574

[17] Moeed, S. A., Niranjan, P., &Ashmitha, G. (2023). Artificial Intelligence and Machine Learning Algorithms in Quantum Computing Domain. Evolution and Applications of Quantum Computing, 45–65. Portico. https://doi.org/10.1002/9781119905172.ch4

[18] Cavaliere, F., Mattsson, J., &Smeets, B. (2020). The security implications of quantum cryptography and quantum computing. Network Security, 2020(9), 9–15. https://doi.org/10.1016/s1353-4858(20)30105-7

[19] Ukwuoma, H. C., Arome, G., Thompson, A., &Alese, B. K. (2022). Post-quantum cryptography-driven security framework for cloud computing. Open Computer Science, 12(1), 142–153. https://doi.org/10.1515/comp-2022-0235

[20] Wilkens, S., &Moorhouse, J. (2023). Quantum computing for financial risk measurement. Quantum Information Processing, 22(1). https://doi.org/10.1007/s11128-022-03777-2

[21] Egger, D. J., Gambella, C., Marecek, J., McFaddin, S., Mevissen, M., Raymond, R., Simonetto, A., Woerner, S., &Yndurain, E. (2020). Quantum Computing for Finance: State-of-the-Art and Future Prospects. IEEE Transactions on Quantum Engineering, 1, 1–24. https://doi.org/10.1109/tqe.2020.3030314

[22] Outeiral, C., Strahm, M., Shi, J., Morris, G. M., Benjamin, S. C., & Deane, C. M. (2020). The prospects of quantum computing in computational molecular biology. WIREs Computational Molecular Science, 11(1). Portico. https://doi.org/10.1002/wcms.1481

[23] Ganesan, Vithya, M. Sobhana, G. Anuradha, PachipalaYellamma, O. Rama Devi, KollaBhanu Prakash, and J. Naren. "Quantum inspired meta-heuristic approach for optimization of genetic algorithm." *Computers & Electrical Engineering* 94 (2021): 107356.

[24] Nath, Rajdeep Kumar, Himanshu Thapliyal, and Travis S. Humble. "Quantum Annealing for Real-World Machine Learning Applications." In *Quantum Computing: Circuits, Systems, Automation and Applications*, pp. 157–180. Cham: Springer International Publishing, 2023.

[25] Mangini, Stefano. "Variational quantum algorithms for machine learning: theory and applications." *arXiv preprint arXiv:2306.09984* (2023).

[26] Fallahi, Saeed, and Mohamadreza Taghadosi. "Quantum-behaved particle swarm optimization based on solitons." *Scientific Reports* 12, no. 1 (2022): 13977.

[27] Sasdelli, Michele, and Tat-Jun Chin. "Quantum annealing formulation for binary neural networks." In *2021 Digital Image Computing: Techniques and Applications (DICTA)*, pp. 1–10. IEEE, 2021.

[28] Zhu, Yingzhao, and Kefeng Yu. "Artificial Intelligence (AI) for Quantum and Quantum for AI." *Optical and Quantum Electronics*, vol. 55, no. 8, June 2023.,https://doi.org/10.1007/s11082-023-04914-6.

[29] Banafa, Ahmed. "Quantum Computing and AI: A Transformational Match." *Transformative AI*, Feb. 2024, pp. 83–88., https://doi.org/10.1201/9781032669182-16.

[30] Marcer, Peter J. "Quantum Computation: A Quantum Leap towards Understanding Neural Information Processing." *AI & Society*, vol. 3, no. 4, Oct. 1989, pp. 332–35. https://doi.org/10.1007/bf01908623.

[31] Ying, Mingsheng. "Quantum Computation, Quantum Theory and AI." *Artificial Intelligence*, vol. 174, no. 2, Feb. 2010, pp. 162–76, https://doi.org/10.1016/j.artint.2009.11.009.

[32] Zolfaghari, Behrouz, et al. "Chaotic Image Encryption: State-of-the-Art, Ecosystem, and the Promise of Quantum-Inspired AI." Crypto and AI, 2023, pp. 95–129., https://doi.org/10.1007/978-3-031-44807-2_5.

[33] AbuGhanem, M. "Experimental Characterization of Google's Sycamore Quantum AI on an IBM's Quantum Computer." SSRN Electronic Journal, 2022.,https://doi.org/10.2139/ssrn.4299338.

[34] Bykovsky, Alexey Yu. "Heterogeneous Network Architecture for Integration of AI and Quantum Optics by Means of Multiple-Valued Logic." Quantum Reports, vol. 2, no. 1, Jan. 2020, pp. 126–65., https://doi.org/10.3390/quantum2010010.

[35] Bykovsky, Alexey Yu. "Heterogeneous Network Architecture for Integration of AI and Quantum Optics by Means of Multiple-Valued Logic." Quantum Reports, vol. 2, no. 1, Jan. 2020, pp. 126–65., https://doi.org/10.3390/quantum2010010.

[36] Ai, Qing, and Jie-Qiao Liao. "Quantum Anti-Zeno Effect in Artificial Quantum Systems." Communications in Theoretical Physics, vol. 54, no. 6, Dec. 2010, pp. 985–96., https://doi.org/10.1088/0253-6102/54/6/07.

[37] Ai, Qing, and Jie-Qiao Liao. "Quantum Anti-Zeno Effect in Artificial Quantum Systems." Communications in Theoretical Physics, vol. 54, no. 6, Dec. 2010, pp. 985–96., https://doi.org/10.1088/0253-6102/54/6/07.

[38] Ganguly, Kaushik. "Quantum AI: Deep Learning Optimization Using Hybrid Quantum Filters." International Journal for Research in Applied Science and Engineering Technology, vol. 10, no. 9, Sept. 2022, pp. 1720–33., https://doi.org/10.22214/ijraset.2022.46914.

[39] Ai, Qing. "Toward Quantum Teleporting Living Objects." Science Bulletin, vol. 61, no. 2, Jan. 2016, pp. 110–11, https://doi.org/10.1007/s11434-015-0986-6.

Utkarsh Mishra and A. Sheik Abdullah

5 Quantum-Safe UAVs: Bridging Education and Outreach for Secure Skies

Abstract: The integration of unmanned aerial vehicles (UAVs) into different divisions has seen a noteworthy surge, underlining the basic need to defend these frameworks from advancing dangers. With the rise of quantum computing, conventional cryptographic strategies face an increased risk of compromise, emphasizing the need for quantum-safe UAV innovation solutions. This chapter unpredictably explores the complex relationship between outreach and education within the domain of quantum-safe unmanned aerial vehicles, emphasizing the significant role of educating and engaging key stakeholders about the critical implications of quantum computing on UAV security. These stakeholders include a diverse range, including manufacturers, operators, lawmakers, and other vested parties.

Moreover, the chapter delves into the key utilization of outreach programs as a means to spread fundamental information and cultivate collaborative endeavors among UAV clients, thereby assisting the widespread adoption of quantum-safe technology. By equipping stakeholders with comprehensive data and resources, this chapter aims to lay a robust foundation for developing a more secure and resilient UAV ecosystem amidst the challenges posed by quantum threats!

Keywords: Quantum-safe UAV technology, workshops, training programs, seminars, conferences, collaborative research initiatives, online resources, knowledge sharing platforms, education campaigns, outreach models, industry partnerships, consortia, public awareness campaigns, curriculum building, recommendations

5.1 Introduction

5.1.1 The Rise of Unmanned Aerial Vehicles

Unmanned aerial machines (UAVs) have grown exponentially in recent times, radically changing a wide range of uses and industries. UAVs were initially made for the military; nevertheless, they are now utilized in a wide range of business applications, such as package delivery, construction, agriculture, and surveillance [1].

1. Drone Technology Development: Miniaturization, longer battery life, and the potential for better sensors are all driving forces behind the rapid development of UAV technology. In the initial phase, military forces used UAVs for surveillance and reconnaissance. The result is miniaturization, longer battery life, and improved sensor capabilities, all of which contribute to the rapid advancements in UAV technology [2].

https://doi.org/10.1515/9783111619323-005

2. Wide Range of Uses:
   UAVs have applications in agriculture, such as crop monitoring, precision farming, and yield optimization, equipped with sensors/cameras [3].
   Construction: UAVs increase productivity and enhance safety [4] by aiding in infrastructure inspection, monitoring progress, and surveying sites with drones.
   This stage includes surveillance and security, involving border patrol, emergency response, and situational awareness via UAVs [5].
   Delivery Services: UPS and Amazon are testing UAVs to deliver packages, promising faster and more efficient distribution [6].
3. Job and Innovation Driver: The UAV sector has contributed significantly to economic expansion through job creation and technological advancement.
   Market estimates show that the industry is expected to keep growing and achieve a multi-billion-dollar status by the end of the decade [7].
4. Regulatory Framework: In response to privacy, risk, safety, and airspace control issues related to UAV activity, governments worldwide have established rules. We also observe that the rules continue to evolve to adapt to the developing UAV technology and applications [8].
   The emergence of new types of UAVs presents substantial challenges for their security and resilience against new threats, particularly those introduced as a result of quantum computation. To safeguard UAV operations, this calls for the implementation of strong security processes and proactive steps to educate stakeholders.

5.1.2 The Quantum Computing Threat

In this context, traditional cryptography systems, which found their application in airborne UAVs, are endangered by advances in quantum computing. Quantum computers use quantum bits, or qubits, which can potentially be in several states at once [9], a concept known as superposition, and their possible entanglement states represent a different set of states with entangled qubits. In contrast, conventional computers have bits that can represent either 0 or 1. Quantum computers possess massive computational strength, as they have an extraordinary ability to compute and can solve some complex problems 10 times faster than traditional computers [10].

One of the threats that quantum computing poses is its potential to outstrip existing encryption.

Most Important Consequences for UAV Security: Different cryptographic methods, for example, RSA and ECC, which are frequently used for securing UAV communications and data transmissions, are dependent on the challenge of factoring huge numbers or solving discrete logarithm problems. Nevertheless, quantum algorithms, like

Shor's algorithm, have demonstrated the ability to effectively solve these problems on quantum computers, endangering the confidentiality of encrypted data.

Moreover, the integrity of key exchange protocols and digital signatures [12], essential for establishing secure communication channels between UAVs and ground control stations, may be jeopardized by quantum computers. As quantum computing develops, UAV missions related to critical infrastructure, surveillance, or sensitive data collection are especially vulnerable due to the ability of malicious actors to intercept and decipher sensitive communication payloads over UAVs.

It requires proactive strategies, such as developing and deploying quantum-resistant cryptographic solutions, to combat the quantum computing threat to UAV security. Quantum-safe cryptographic algorithms: Lattice-based and hash-based are two examples of quantum-safe cryptographic algorithms that help protect against attacks from quantum computers. The transition to such quantum-safe technologies should be carefully planned to ensure compatibility and a smooth integration into current systems within the UAV sector.

To sum up, while quantum computing holds enormous potential, its operational body will contain a lethal danger to UAV security. This highlights the need for the rapid adoption of quantum-resilient cryptographic techniques, as well as increased stakeholder awareness of the effect of this technology on UAV operations [13].

5.1.3 Rationale for Education and Outreach

Awareness of Quantum Threats:

a. Quantum Computing Awareness: Given the nature of the UAV ecosystem, it's almost certain that many of those participating in it are unaware of the full range of quantum computing and its potential impact on cryptographic systems. Education is extremely important to elucidate the fundamental concepts of quantum physics and how they diverge from the paradigms of classical computing! [14]
b. Implications for UAV Operations: Education campaigns certainly need to inform about the unique risks quantum computing represents to UAV security, signifying the importance of identifying gaps in current encryption protocols, as well as the potential for attackers to use quantum capabilities to compromise data integrity, navigation, and UAV communication [15].
c. Pressing Demand for Quantum-Safe Solutions: Public awareness campaigns should highlight the pressing demand for quantum-safe solutions in UAV technology. A better appreciation of the urgency of the threat posed by quantum computing toward such algorithms and protocols would ensure that the relevant stakeholders recognize the need to migrate to quantum computing-resistant algorithms [11]

Empowerment of Stakeholders:

a. UAV operators' education is very important because it also helps the UAV operators keep up to date with the evolving threat matrix. Operators should adopt best practices and understand the impact of quantum computing on UAV security in order to mitigate the risks of quantum computing on their operations [16].
b. Manufacturers – UAV manufacturers should be included in outreach programs to engage them in developing and using quantum-safe technology. As a means to ensure long-term security, manufacturers should be informed about the advancements regarding quantum-resistant encryption algorithms and encouraged to implement those solutions in their UAV systems [17].
c. Policy and Regulation: Education efforts should target policymakers and regulatory agencies to improve understanding of global quantum computing regulatory aspects in the UAV ecosystem. Policymakers must understand how important it is to quickly amend regulatory structures to incorporate quantum-safe technology and promote its uptake through appropriate carrots and sticks [18].

Cooperation and Information Exchange:

a. The outreach programs foster cooperation among governments, businesses, and academia in order to address the complex challenges posed by quantum computing. This may accelerate the accessibility and development of Quantum Secure UAV technology through educational and outreach efforts, encouraging multidisciplinary engagements [19].
b. Information Sharing: Platforms for education and outreach related to quantum-safe UAVs support the sharing of best practices, lessons learned, and new research. These platforms encourage continuous learning and creativity with respect to quantum-resistant UAV technology through the exchange of knowledge among players [20].

Short to say, to mitigate the security threats presented by quantum computing, engagement of the UAV community, education, and awareness through outreach programs, and involvement of specialists are as important as technical standards and processes. While ever-changing and advancing technologies make the transition to quantum-safe technology challenging, outreach and education programs can promote efficient adoption of these technologies, thus ensuring a robust and secure future for the UAV ecosystem in the coming age.

5.2 Quantum Threats to UAV Security

With the potential of quantum computing, particularly around codologic, the impact on UAVs lies in better understanding the codologic protocols that play an important role in securing data in both modern digital systems, and UAVs. It is paramount to recognize these threats to implement solid security precautions for UAV operations.

Traditional Cryptology Vulnerabilities: The intrinsic power of traditional cryptography, based on ROWS (Random Obfuscating Language) and ECG (Exponential Curling Graphs), characteristically lies in labyrinthine mathematical enigmas, which are challenging yet to be effectively disentangled by classical computers. However, quantum computers could exploit these weaknesses and use them against you through algorithms such as Shor's algorithm. Shor's rules can engage with the elliptical arc at arithmetic secrets, undermining ECG and empowering enormous quantities of fast factorings, thereby compromising ROWS encryption [21].

Implications for UAV Operations: Secure communication channels are needed for the secure operation of UAVs. UAVs endeavor to perform missions like deliveries, surveillance, remote sensing, research, and much, much more. Any breach in the security of these mechanisms could result in dire outcomes, including loss of UAV control disruption in channel communications, and unauthorized access to confidential information. The utilization of cryptographic algorithms to ensure UAV communications may be susceptible to quantum attacks, leading to data breaches, interception of commands, or even total takeover of the UAV.

The Urge for Quantum-Secure Solutions: Across the globe, both technologists and academicians are making noteworthy progress in devising quantum computers that are operational. As quantum computing capabilities escalate, the susceptibility span of traditional cryptological maneuvers wanes. Hence, the adoption of quantum-secure cryptologic systems that can resist invasions from classical and quantum computers becomes imperative. Lattice-based hashing and post-quantum cryptography are exemplars of quantum-safe algorithms being scrutinized as feasible stand-ins that are stalwart against quantum encroachments.

An inclusive approach is called for to counteract quantum threats to UAV security, involving:

- Research and elaboration of cryptological algorithms tailored for UAV exploits and resistant to quantum fluxes.
- Incorporation of quantum-safe mechanisms in datum encoding and UAV communiqué guidelines.
- Proactive adaptation of safeguarding measures to synchronize with evolutions in quantum computing.
- Joint effort in setting standards and guidelines for quantum-safe security protocols for UAVs, spanning industries, academia, and regulatory bodies.
- Acknowledging and addressing quantum threats in the UAV setting could increase the resilience and robustness of UAV systems, underscoring their physique and reliability in a transforming technological landscape!

5.2.1 Vulnerabilities of Traditional Cryptography

It is here that quantum computing is making progress; the ordinary cryptography that relies on mathematical tricks for encryption and decryption faces some big hurdles. Concerned over the ground-breaking potential quantum computers hold, the extensive computational gulf between colloquial and quantum computers forms the basis of these tribulations:

1. Quantum Algorithms: Quantum computing has the potential to break widely used encryption algorithms (e.g., RSA) due to its capability of factoring large numbers efficiently. Splitting big integers into prime factors is quite difficult, which is basically the entire foundation of encryption models such as RSA and ECC (elliptic curve cryptography) that are built so heavily on this property for their security. However, why the emphasis on quantum computers, and why one particularly using Shor's technique? Because they could significantly speed up these factorization processes, potentially leaving existing cryptographic systems open to attack.
2. Discrete Logarithm Problem: This concern underlies a significant subset of cryptosystems, including the digital signature algorithm and the Diffie-Hellman key exchange. The challenge is to compute an integer exponent of some prime digit. Then, of course, there's the same concern, which presents an imminent threat from quantum algorithms such as Shor's algorithm that makes solving discrete logarithm computations on quantum computers fairly trivial.
3. Symmetric Key Algorithms: While symmetric key algorithms such as AES (Advanced Encryption Standard) are regarded as being more resistant to quantum onslaught than asymmetric algorithms like RSA, they are not immune. A quantum approach, known as Grover's method, could bring down the critical key length needed to reach the same level of security by offering a quadratic speedup for brute-force attacks on symmetric codes.
4. Hash Functions: The safety and trustworthiness of cryptographic hash functions ensure data integrity and authenticity. However, quantum computers are able to adopt techniques such as Grover's algorithm to speed up the identification of hash collisions, which endangers the security of hash functions and other cryptographic protocols that rely on them, such as the authenticity of digital signatures.

Since conventional encryption, in the presence of quantum computing, can become invalid, thinking about quantum-safe cryptographic solutions is the need of the hour to secure and ensure the integrity of UAV communications, data transfers, and overall operations in the quantum technology era.

5.2.2 Implications for UAV Operations

5.2.2.1 Quantum Computing's Impact on UAV Operations

The spread of computing power has significant implications for the safety and reliability of UAVs. However, with the unprecedented processing capabilities of quantum computers, a large number of security methods used by UAVs are based on classical cryptography methods, which also create vulnerabilities. Understanding the implications is necessary for stakeholders such as operators, mission planners, and businesses utilizing UAV technology.

Imperfections in Security Measures:

1. Security Vulnerabilities: Commonly used cryptographic approaches, such as RSA and ECC, can be more easily exploited compared to existing quantum methods. This vulnerability can be exploited by an attacker for eavesdropping, alteration, and decryption of sensitive data when transferred between a backbone control station and the UAVs. Thus, in a threat environment enabled by quantum computing, the confidentiality, integrity, and authenticity of UAV communications and data are at higher risk [24].
2. Personally Identifiable Information: UAVs frequently collect and transmit sensitive data during missions, such as video surveillance, geographic data, and evaluation of critical infrastructures. This has implications for the security of the encryption methods protecting this information, raising alarm bells about potential breaches, unauthorized access, and violations of privacy. Therefore, it is inevitable that securing UAV-sourced data for secrecy and integrity becomes critical to quantum-secure UAV operations.

Mission Reliability and Adherence:

3. Mission Assurance: So, in a world degraded by quantum technologies, mission assurance for the UAV could suffer from the lack of guarantees on our sensors and mechanisms for perception. Unlike classical actors, malicious actors with enhanced quantum capabilities will be able to potentially interfere with UAV operations, either through the interception of command signals, manipulation of telemetry data, or even taking control of the UAV itself. These attacks lead to safety risks as well as operational risks that can bring about mission failures [25].
4. Regulatory Requirements: The steady growth of the UAV industry has propelled the need to consider quantum computing, and by extension, UAV security, in building regulatory frameworks. Security measures, or quantum security measures, should be added to existing guidelines and standards in order to ensure that UAV operators work under stringent security protocols and can effectively counter quantum-related threats.

In conclusion, the impact of quantum computing on UAV operations highlights the importance of preparedness in strengthening the security, resilience, and regulatory compliance of UAV systems to counter emerging threats. Recognizing these implications, stakeholders can mitigate risks with technology that adapts to the quantum-era and ensure the security and efficiency of UAVs long into the quantum era through the intelligible integration of quantum technologies.

5.2.3 Urgency of Quantum-Safe Solutions

By considering the massive speed of factoring huge numbers quickly and resolving sensitive logarithm issues, quantum computing has the potential to make traditional encryption techniques like RSA and ECC outdated [23]. The security and integrity of sensitive data, including communication channels, mission-critical data, and surveillance data that are sent and stored by UAVs, would be jeopardized as a result.

Development Timeline for Quantum Computing: While functional quantum computers that could crack cryptographic methods are still a way off, both public and commercial organizations are making great strides in the field of quantum computing research and development. Within the next ten years, experts predict that large-scale quantum computers could become a reality! Consequently, there is limited time left to switch to quantum-safe cryptographic methods for UAVs.

Longevity of UAV Systems: UAVs, especially in military and infrastructure monitoring applications, are created to last over long stretches of time. They can have lifecycles that last many years or even many decades. Therefore, UAV security systems must become future-proof against emerging threats, such as those introduced by quantum computing. If this problem cannot be resolved soon, an attacker with access to quantum computing capabilities may temporarily be able to subvert UAV systems.

Cost and Resource Considerations: Establishment, implementation, and development of quantum-resistant systems cost big money. But then again, the cost of reducing the risks posed by quantum computing is peanuts compared to the potential fallout from a compromise in the security of UAVs! Moreover, delaying action in this space could cause sky-high costs plus even more interruption!

Finally, the future threat of quantum computing technologies, capable of compromising the cryptographic foundations on which UAV systems have been built, requires the implementation of new quantum-safe solutions for UAV security. Therefore, new quantum threats have necessitated proactive measures to secure and/or ensure the resilience of UAV operations.

5.3 Educating Stakeholders

1. A Vital Step in Ensuring Quantum-Safe Technology Adoption: Get a general sense of the concept of quantum computing – It is hypercritical to give everyone who will be involved and needs to understand the basic concepts of quantum computing. This explains the principles of quantum physics, quantum bits (qubits), and how quantum computers can hold superior computing potential compared to traditional computers.

   Grounded in data until October 2023, those concerned must synthetically grasp the principal contrasts between classical and quantum cryptography, as well as how classical cryptographic techniques are impervious to quantum assaults.
2. Awareness Among UAV Operators: A massive responsibility for the safe and secure interaction of unmanned aerial system (UAS) rests on the shoulders of UAV operators. It is important to point out to them the threats that quantum computing poses to the existing cryptographic methods.

   Operators may be easily instructed on quantum-safe encryption algorithms and their accompanying operational aspects for UAV utilization through training processes.

   In addition, it would be great if it could remind us and evaluate the importance of quantum security measures by providing case studies and real examples that showcase the consequences of quantum attacks on UAV systems.
3. From the Manufacturer's Point of View: UAV manufacturers need to be aware of the potential future impact of quantum computing on the security of their devices. This means determining the extent to which existing UAV systems will be vulnerable to a quantum attack and incorporating quantum-safe encryption techniques into the next generation.

   Educational products tailored to UAV platforms may include workshops, seminars, and collaborations with quantum-safe cryptography experts targeted at the manufacturers.
4. Policy Changes and Regulations: The development of relevant regulations and guidelines relies heavily on educating regulators and legislators about the implications of quantum computing on UAV security.
   Stakeholders will need to collaborate closely to ensure that security precautions are adhered to, and standards for the introduction of quantum-safe technologies into UAV systems are established.

   End-user education materials assisting policymakers such as white papers, policy briefs, or expert consultants may support making fully informed decisions on implementing quantum-resistant UAV protective mechanisms.

   By bringing together stakeholders from a diverse range of sectors, including manufacturers, policymakers, and UAV operators, the UAV community can collaborate to address the challenges posed by quantum computing and ensure the security of UAS.

5.3.1 Understanding Quantum Computing

1. Quantum Computing Fundamentals: Quantum computing is unlike classical computational methods, but stakeholders must have some basic understanding of relevant concepts. It is important, if not necessary, to describe things like superposition, tangledness, and quanta of entanglement, or qubits, in a more understandable way. Some of these concepts are perhaps more straightforward to understand for onlookers without technical competence, especially with the use of symbolism and visual aids!
2. Communicating the Potential Impact of Quantum Computing on UAV Security – The Quantum Threat Landscape: Two common mathematical problems that can be solved quickly by quantum computers are discrete logarithms and integer factorization, which have led to many traditional cryptography techniques, such as RSA and ECC, potentially being vulnerable to quantum attacks. Focusing on real examples of what quantum attacks could threaten UAV operations can drive home the urgency of switching to quantum-safe encryption techniques.
3. Quantum-Safe Solutions: Stakeholders should be informed of quantum-safe cryptographic schemes offering resistance to quantum attacks. Examples include hash-based, code-based, lattice-based cryptography, and others. Understanding the fundamental principles of the quantum-safe algorithms that are relevant to UAV applications will allow stakeholders to make informed decisions, such as System Security Procedures (SSPs) for UAV systems.
4. Creating Awareness on QKD in UAV Operations: Creating a culture of proactive security requires that relevant stakeholders be educated on the implications of quantum computing on UAV operations. It is also important to note that the test data can, in principle, be sensitive in any form: navigation data, reconnaissance records, even with quantum conquest, which was provided by the made UAV. In addition, the quantum threats posed by quantum attackers, who could easily break current encryption systems based on factorization and discrete logarithms, highlight the importance of strong quantum-safe encryption methods.

5.3.2 Awareness Among UAV Operators

Recognizing Quantum Threats: Quantum-safe protocols are only useful if UAP/UAV operators are involved. They need to be aware of the potential threats quantum computing poses to the traditional cryptography schemes that are utilized in UAV actions.

Quantum-safe protocols are only useful if UAP/UAV operators are utilized. They need to be aware of the potential threats quantum computing poses to the traditional cryptography schemes that are employed in UAV actions.

For UAV operators, the concepts of quantum superposition and entanglement, their effects, and cryptographically secure implementations, etc., should be properly included in awareness campaigns.

Workshops and Training Courses: Organizing seminars and personalized programs for UAV operators would be better suited if you want to transfer knowledge and upskill on how to operate safely should quantum attacks occur.

The seminars should cover the limitations of current encryption techniques, the foundations of quantum-resistant algorithms, and quantum-safe protections in UAV systems. Workshops should be implemented as well.

Real-World Situations and Case Studies: Real-life scenarios that may threaten the operation of UAV systems with quantum attacks highlight the importance of adopting quantum-safe systems.

Case modules highlighting how the shortfall of cryptographic measures eroded UAV security could reinforce the need to migrate to quantum-resistant systems, motivating operators' understanding.

Hands-On Demonstrations and Simulations: These can give UAV operators an idea of the limitations within conventional cryptography systems and the capability of quantum-safe replacements, and are intended to improve their understanding and instill in them a belief in using them in real UAV flights.

Operators may improve their understanding of, and confidence in, the use of encryption and quantum-resistant algorithm solutions in real-life UAV flights through interactive workshops that allow them to try out their solutions.

Continuous Education and Updates: The exchange of knowledge and mutual biodiversity of UAV, committed to quantum-safe operations of UAV, is potentially supported through newsletters, webinars, and online forum building.

From the fast-developing quantum computer and cryptography areas, continuous education and updates are becoming key to keeping UAV operators versed in the latest tips and developments!

Industry Partner Collaboration: Industry partners, such as cybersecurity companies and UAV manufacturers, may be able to enhance awareness campaigns using their resources and experience. By partnering with the industry, such as aerospace and security companies, universities can create specialized training programs, provide access to cutting-edge technologies, and conduct collaborative research initiatives to tackle the unique security challenges that quantum-powered UAV operators will encounter. Training personnel on quantum security and equipping UAV operators with tools and resources will empower the ecosystem to mitigate emerging threats, preserve the secure operational environment, and sustain the operational effectiveness of unmanned operations.

5.3.3 Manufacturers' Perspective

Awareness and Understanding: Manufacturers must understand how quantum computing presents a prospective threat to UAV security as they make decisions. This means tracking developments in encryption and quantum technologies and planning for future challenges. Since earlier methods to help protect their cryptography from quantum attacks are being broken, manufacturers must already realize that it is long overdue to design quantum-safe solutions.

Research and Development: It is to be expected that manufacturers must invest in this direction like this, to create quantum-safe UAV technology. This includes potential tailoring of post-quantum cryptographic methods, such as digital signatures, encryption, and key exchange protocols, to UAVs. In addition, in an effort to stimulate the development of quantum-resistant technologies specifically designed for UAV infrastructures, manufacturers should partner with educational institutions and research organizations.

Integration and Testing: This generates another technical difficulty in addition to ensuring solutions are quantum-safe. Once solutions are developed, producers must ensure they fit properly with existing protocols and systems when integrating them with UAV platforms. To validate the reliability and practical efficacy of these solutions, strict testing protocols are needed! Manufacturers could conduct extensive vulnerability assessments and security evaluations to identify and mitigate such hazards.

Education and Training: It is imperative for manufacturers to keep their staff and engineering workforce apprised of the significance of quantum-safe guidelines and how quantum computers would matter for UAV security. The employees could learn about the best practices, approaches for implementation, quantum-resistant algorithms, and defensive techniques through training courses and seminars to protect the UAV systems from attacks.

Collaboration and Standards: Participation in industry-wide initiatives will assist manufacturers in creating the protocols and frameworks that support interoperability and secure UAV systems across the supply chain. They can also collaborate with industry stakeholders, standards organizations, and regulatory bodies to promote guidelines and standards for quantum-safe UAVs.

Agents consider awareness, research, integration, education, and cooperation to be contributors to quantum-safe UAVs. By taking precautionary measures that are quantum-safe, manufacturers of UAV systems can further protect themselves against new forms of attack, such as quantum attacks.

5.3.4 Policy Implications and Regulatory Frameworks

Such platforms and their wide-scale adoption and utilization as UAVs would heavily rely on policy implications and regulatory frameworks to bring them to fruition and enable their conversion from theoretical concepts to implementation at the real-world level. One may argue that quantum computing is still in its infancy, but quantum security has already spawned some exciting technologies. Quantum-safe UAVs are analyzed in the following section due to the multifactorial nature of policy making and regulatory frameworks.

1. The Process of Developing Policies:
   This will require governments and regulatory agencies to work together to strike a balance between the capabilities of quantum computing and the unique challenges it poses in terms of regulation. This will include risk assessments, stakeholder consultations, and utilizing the expertise of academia, research, and other institutions. Another such important factor is the pace at which technology evolves and changes the threat landscape; thus, policies must remain adaptive and agile.
2. Obligatory Standards and Compliance:
   Standards and compliance mandates regulatory frameworks for quantum-safe UAVs. These measures should include encryption standards as well as quantum-resistant authentication mechanisms. Industry-wide standards are critical, and organizations that coordinate, harmonize, and validate the specifications for interoperability and compatibility between various UAV systems can have a major impact on the establishment and adoption of these industry-wide standards.
3. Certification and Accreditation:
   Certification and accreditation programs should be developed for verifying compliance with industry standards and regulatory requirements. Such results will foster trust and, therefore, confidence in quantum-safe UAVT. You are probably trained on lots of frameworks/algorithms where extensive testing, evaluation, and review for quantum error-resilient cryptographic algorithms and protocols are essential in providing the assurance of their security and robustness.
4. International Cooperation:
   International coordination is essential for achieving homogeneity among states and for harmonizing regulatory regimes, given that UAV activities are international in nature and reach. Through such multilateral agreements and collaborations, joint approaches could potentially address common challenges, thereby supporting global adoption of this quantum-safe UAV technology through data sharing, capacity building, as well as research co-ordination.
5. Ethical and Privacy Considerations:
   Policymakers must consider the ethical and privacy implications of quantum-safe UAVs. Policies should include steps to protect individuals' privacy, guard against unlawful discrimination, and minimize the prospect of quantum-resistant technology being used for nefarious purposes.
6. Public Awareness and Engagement:
   Lastly, if it is meant to increase public awareness about quantum-safe UAVs and the need for a regulatory framework as valiant public defenders of technology, then governments must prioritize education and engagement in public awareness

of quantum-safe UAV benefits and threats. This entails engaging in preliminary dialogues with stakeholders, presumptive upending of dogmas that stem from quantum technology and its security effects on UAVs, and also transparent engagement. By integrating and addressing these considerations into policy and regulatory development, governments can help mitigate the threats of quantum computing to UAVs and ensure a safe and secure UAV landscape.

5.4 Outreach Strategies

Outreach Methods for Quantum-Safe UAV Technology: Considering that the sharing of information and accelerated deployment of quantum-safe UAV technology has never focused on the UAV community, outreach methods in the UAV community will also be a significant part of quantum security, facilitating collaboration in the quantum-secure domain. From creating platforms for sharing best practices to efforts addressing quantum security challenges through collaboration with stakeholders, these strategies encompass activities from a broad spectrum to raise awareness.

Workshops and Training Programs: Workshops and training programs provide a venue to explore the nuances of quantum threats and can also provide a forum to examine practical solutions to UAV security in the quantum era. These events often include specialized speakers, hands-on demonstrations, and interactive sessions based on individual groups' unique needs and interests. Topics in these workshops could include post-quantum algorithms, quantum key distribution, and quantum-resistant encryption, providing delegates with essential knowledge and skills to embed quantum-safe practices into their UAV operations.

Seminars and Conferences: Such conferences and seminars provide a common forum for all the players in the UAV ecosystem to network, share knowledge, and collaborate, all the while sharing insights from ongoing research projects, combining synergies to advance quantum security efforts, and bringing together researchers, industry professionals, and government regulators to discuss the latest in quantum-safe UAVs. They engage in multidisciplinary debate and find new paths for innovation and collaboration via keynote addresses, panel discussions, and poster sessions.

Collaborative Research Initiatives: These research initiatives allow government, business, and academic institutions to collaborate to address challenges around quantum-safe UAV technology. This will bring together the different structures and incentives needed to ensure that the best quantum-resistant cryptography can be developed and crafted by converging pooled skills and expertise, where research projects, consortia, and funded initiatives can drive phenomenal levels of innovation and the steps to deliver generators of such solutions.

Online Resources and Knowledge-Sharing Platforms: These online resources and knowledge-sharing platforms serve as fundamental information and tool banks for interested parties who wish to stay updated on advances concerning quantum security and UAVs. Engaging with people whose websites, blogs, forums, and social media channels cover topics like quantum cryptography, quantum-safe UAVs, best practices in cybersecurity, and the ubiquity of UAV designers in different social and commercial landscapes, these platforms have played a direct role in reaching a broad global audience of UAV professionals, researchers, hobbyists, and makers. Each of these individuals collaborates with our listed contributors through community forums, webinars, and curated content.

Outreach tactics like conferences, workshops, co-publications, online materials, etc., will be key in increasing knowledge, developing skills, and encouraging collaboration across the UAV community in addressing quantum security issues and ensuring UAV robustness in a quantum technology-ravioli world!

5.4.1 Workshops and Training Programs

The knowledge and awareness of the impact of quantum computing on UAV security, and the usage of quantum-secure technologies, can be significantly improved through training programs and workshops. These programs offer students a safe yet stimulating environment that allows them to explore the intricacies of quantum technology and how it relates specifically to the processes surrounding UAV operations. These training programs and workshops should be attended to provide explanations.

1. Setup Learning Setting: This method helps us to create a particular atmosphere by means of workshops and training courses, wherein we can invite people to exchange ideas, discuss, and have practical sessions depending on their previously learned abilities. UAV users, creators, engineers, and lawmakers alike will find these sessions interesting.
2. Expert Guidance: Most of these programs include professors and instructors with extensive experience in both UAVs and quantum computing. They help participants gain valuable insights into the threats posed by quantum computing and possible solutions to address these concerns.
3. Interactive Sessions: Group activities, role-plays, and demonstrations are common in workshops and training courses. Structured as focused, interactive seminars, participants not only have the opportunity to explore ideas but also practice applying theoretical concepts to real situations, enhancing both their understanding and retention of the material. They also promote hands-on learning.
4. Interactive Sessions: Workshops and training courses often come with interactive sessions that can include group work, role plays, and demonstrations. These teachings in interactive seminars give participants an opportunity to apply theoretical ideas in the light of real-world events, so that their understanding of the material increases significantly. They also promote experiential learning.

5. The Demand for Your Practical Skills: The workshops and training courses you will create will not only delve into the theoretical knowledge surrounding quantum technologies but will also address the practical skills essential for implementing quantum-secure solutions to uplift aerospace systems. Attendees may receive training on secure communication methodologies, quantum-resistant algorithms, cryptology techniques, and integration methodologies.
6. Networking Opportunities: Participants can network with academic and industry peers, as well as experts in the field, through these programs. Networking enables knowledge exchange, collaboration on research projects, and partnerships to address common aspects of quantum-safe UAVs.
7. Customization and Modification: Training classes and workshops can be tailored based on the requirements and goals of various stakeholders. Hence, changing the structure and content of these documents ensures they work effectively and achieve maximum efficiency to solve the individual problems of creators, policymakers, and UAV operators.
8. Sustainability and Continuation: Workshops and training programs must also follow a sustainable structure to continue education and skills improvement, such as developing relevant online resources, Provider focused certification programs, launching community-driven initiatives, and much more for the purposes of continued learning and knowledge.

In conclusion, informing stakeholders about issues of interest in quantum technology, familiarizing them with new skills, and involving them in the design and implementation of quantum-safe UAV technology can be achieved through workshops and training programs. These are essential endeavors to ensure the safety of UAS in the quantum age and to ensure the UAV community is prepared to combat the threats presented by quantum computers!

5.4.2 Seminars and Conferences

Dissemination of Knowledge: Few businesses, however, have the appropriate skill set. A specialist may speak, keynote, and lead toolkits for tech scions at seminars and conferences on every factor, from quantum threats to UAV safety – and the ways to deploy quantum-safe solutions. These seminars are meant to help inform attendees about the subtleties of quantum computing, its impact on cryptography, and the urgency of transitioning to quantum-resistant algorithms as soon as possible.

Collaborative Conversations: These events include forums, panel discussions, and question-and-answer periods during which attendees might communicate with presenters, share ideas, and discuss experiences and best practices for mitigating quantum risks in UAV operations. Indulging in collaboration-based conversations allows stakeholders to learn from multiple angles and develop novel methods to address unique issues.

Seminars and Conferences: Seminars and conferences provide networking opportunities with peers, potential partners, and industry leaders. Networking opportunities, coffee breaks, and social events enable participants to form partnerships, build relationships, and work together on academic, technology-focused, and policy-related initiatives pertinent to quantum-safe UAVs.

Highlighting Innovations: These events allow researchers, entrepreneurs, and businesses to present their ideas, prototypes, and final products associated with quantum-safe UAV technology. Participants can evaluate potential solutions, explore opportunities for collaboration or investment, and learn about the latest innovations through poster sessions, exhibition booths, and demo sessions.

Policy Development and Advocacy: Conferences and seminars are important venues for influencing policy debates and promoting laws and regulations that will enable the widespread use of quantum-safe UAVs. Stakeholder forums, policy-oriented workshops, and roundtable talks enable industry stakeholders, policymakers, and regulatory agencies to engage in conversations about the legal, ethical, and regulatory aspects of quantum-resistant encryption and UAV security.

In summary, conferences and seminars stimulate information sharing, collaboration, and innovation with regard to quantum-safe UAVs/drones, and help drive the creation and implementation of safe and reliable UAS in the quantum era!

5.4.3 Collaborative Research Initiatives

Collaborative research activities are a significant contributor to the adoption and development of quantum-safe UAV technology. If you've seen any of the Quantum Threat to UAV Security projects, these effectively pool government organizations, businesses, academic institutions, and research facilities in one spot, where they come together to pool resources, skills, and knowledge to address difficult problems. Here is an in-depth article on cooperative research projects for quantum-safe UAVs:

1. A Multifaceted Approach: Researchers from various fields, such as cybersecurity, encryption, quantum computing, and aeronautical engineering, work together in collaborative research projects. This interdisciplinary perspective encourages creativity through the integration of information from various disciplines to address the multifaceted characteristics of quantum-safe UAV technology.
2. Research Consortia: Consortia exist to facilitate collaboration among entities with an interest in quantum-safe UAVs. Often inclusive of major industrial, academic, and governmental organizations, these consortia work together to advance common research, share resources, and set standards and recommendations for quantum-safe UAV systems.

3. Funding and Resources: This is advantageous because different collaborating research projects will be able to receive funds and resources from the various participating organizations, or even from government grants. Scientists will be able to perform tests, complete ambitious projects, and develop quantum-safe UAV prototypes with these resources.
4. Technology Transfer and Information Exchange: Joint research projects serve as venues for technology transfer and information exchange. Academicians around the world can interact with each other on joint publications, share best practices, and disseminate lessons learned to help us all understand quantum risks and avoidance options.
5. Testbeds and Experimentation: Collaborative research projects within the field of quantum-safe UAV systems often establish testbeds and experimentation environments to evaluate different approaches in real-world scenarios. A number of leading testbeds scan the horizon and offer up a space for the research community to test attacks, assess the efficacy of quantum-resistant algorithms, and help identify any bumps in the road or barriers.
6. Policies and Standards Development: For another, collaborative research projects pave the way for the establishment of guidelines and laws governing the use and deployment of quantum-safe UAVs. As such, researchers work with legislators, regulators, and industry partners to build standards and laws around safe and interoperable quantum-safe UAV systems.

In general, joint research initiatives are essential to expedite the development and adoption of quantum-safe drone applications. In their packaging and promotion of cooperation, creativity, and collaboration in information sharing, these programs are critical to establishing a thriving UAV culture that can weather the challenges (and we all know very well that quantum computing is not the only challenge we face) that quantum computing presents!

5.4.4 Online Resources and Knowledge-Sharing Platforms

1. Training and Development of Educational Portals: It is essential to deliver all knowledge related to quantum-safe UAVs through online portals. Such resources can consist of whitepapers, videos, interactive tutorials, and articles tailored to users at each relevant skill level.
2. Role of UAV Operators, Manufacturers, and Policymakers: Interactive Online Courses and Webinars. There could be discussion boards, lectures, and pulse activities to supplement knowledge exchange and verification courses.
3. Collaborative Tools: Online collaboration platforms facilitate real-time insights exchange and research collaboration. Moreover, features such as document sharing,

version control, and virtual conference rooms promote effective collaboration in cross-region teams.

4. Discussion Groups and Community Forums: Online forums where parties can interact and exchange experiences and advice on quantum-safe UAVs. These platforms facilitate the exchange of information, peer-to-peer assistance, and networking within the UAV community.
5. Repositories with Open Access: There are a large number of research articles and datasets available on open-access sources for quantum-safe UAVs. Access to benchmark datasets and research results fosters ingenuity and collaboration in the field, driving the course of quantum-safe UAS development.

Finally, the focus on internet resources and information-sharing platforms is essential for education stakeholders and improves research in quantum-safe UAVs. Leveraging these platforms properly will help the UAV community navigate quantum threats and achieve a habitat milieu that is secure for UAVs.

5.5 Case Studies and Best Practices

5.5.1 Successful Education Campaigns: Quantum-Safe UAVs Take Flight

5.5.1.1 Airbus and the University of Waterloo: Building a Quantum-Safe Future for UAVs

Challenge: The year will be 2022. Titans from the commercial UAV manufacturing industry, like Airbus, might one day face a new and unanticipated challenge: the rapid emergence of quantum computing. Quantum computers, with their tremendous computational ability [26], have the potential to break the existing encryption methods utilized by UAV transmissions. This poses a significant risk, compromising the safety and security of Airbus's UAV operations.

Reconciling a Solution: Realization of the potential threat leads Airbus down an avenue of solution. They set their sights on the University of Waterloo's Institute for Quantum Computing, a popular center for quantum research. This building proved to be an utter masterstroke.

Designing the Curriculum: Airbus engineers are experts in UAV design and development. In fact, the details of quantum computing and post-quantum cryptography (PQC) require specific knowledge. Enter the quantum specialists from the Institute who, along with Airbus' engineers, co-create a complete education program designed specifically for them.

An Introduction to Quantum Threats: The program starts from the very basics and explains how quantum computers work on the principles of superposition and entanglement to achieve a gargantuan parallel processing. This enables engineers to understand the threat landscape and how quickly operators need to deploy quantum-safe solutions. In [22], the curriculum dives into the algorithmic underpinnings of quantum computing and shows how Shor's algorithm can break common encryption schemes like RSA.

No system is perfect, including military UAV communication systems. It meticulously analyzes the fragilities that are built into existing protocols. The article provides detailed information about common encryption methods, such as RSA and ECC, and their vulnerabilities to potential attacks from quantum computers. Such extensive analysis prepares engineers to embrace the vulnerabilities they need to focus on. The training program also includes real-world examples of what a potential cyberattack would look like, simulating the use of quantum computers to attack UAVs. The hands-on experience gives them an understanding of which vulnerabilities impact the system the most and the potential implications of an actor successfully exploiting that vulnerability.

You have a detailed understanding of the threats and vulnerabilities. Various PQC algorithms that engineers are training on include lattice-based cryptography, code-based cryptography, and multivariate cryptography. Techniques have been proposed to use all or part of each of the algorithms, with some suggested characteristics for combining them based on their relative strengths and weaknesses, as well as the unique attributes of each algorithm and its development, considering their potential use in UAV communication protocols. This approach allows engineers to build quantum-safe communication protocols into future Airbus UAV designs. The course also explores the problems that need to be solved in order to make it work, ranging from how much processing power it will require to how it might integrate with legacy systems.

Impact: A Ripple Effect of Security: Please note that this post is intended as a teaching moment and a prompt to remind engineers that, during this time of post-quantum cryptography (PQC), we should prepare ourselves for a paradigm shift in the way that we think about security, ensuring that we can actively facilitate our organizations moving towards this quantum-robust future. This means future Airbus UAVs will be able to communicate securely even in the quantum ages.

Open-Source Information Sharing: Airbus and Waterloo are making their findings publicly available, acknowledging the need for the industry to be prepared. This promotes knowledge sharing and therefore also speeds up the development of quantum-safe solutions right across the UAV industry.

5.5.2 Effective Outreach Models: A Public-Private Alliance Ascended

Challenge: The European Union (EU) confronts a crucial challenge: to ensure the security of its UAV airspace in the quantum epoch. Though the potential of quantum computing is undeniable, it poses a significant danger. Quantum computers possess the capability to break current encryption methods, which can endanger the security of UAV communication and might lead to disastrous consequences.

The Strength of Collaboration: In 2023, the EU takes bold steps by launching the Quantum-Secure Initiative, a pioneering public-private partnership (PPP) intended to tackle this challenge head-on. The initiative brings together a diverse group of stakeholders, each with a unique contribution to make:

Government Organizations: The EU offers crucial leads, directions, and, most importantly, funds for research and development efforts. This financial backing is essential for propelling the development of quantum-secure UAV technologies.

The Hub of Innovations: Universities and research institutes across the entire European Union are the hubs of innovation. They are essential in creating strong PQC protocols tailored for UAV communications, leveraging their experience in quantum computation and cryptography.

Tech Giants: Technology giants come with their real-world experiences and bleeding-edge engineering expertise. They facilitate the translation of research results into actionable, viable solutions integrated within current and future UAV systems.

Industry Associations: Industry associations act as intermediaries between developers and users. By doing so, they ensure that the PQC solutions developed are not only technically viable but also practically viable for UAV operators.

A Multifaceted Approach: The EU Quantum-Secure Initiative has multi-pronged approaches to accomplish its goals:

Most of the research on quantum-resistant cryptosystems comes from the EU, where a huge amount of funding has been allocated for research purposes: some research focuses on quantum-secure cryptography and communication protocols specifically designed for UAVs. It is this investment that drives innovation and accelerates the process of development.

Awareness Raising: The initiative recognizes that education is key. It holds workshops and conferences in the EU, which are intended to educate policy makers, industry leaders, and UAV operators about the risks of quantum computing and the need for quantum-safe solutions. These events feature talks by leading experts in quantum computing and recent developments in PQC protocols for UAVs. The initiative also employs various media outlets to educate and reach out to a broader audience.

Fostered Collaborations: The initiative has one of its greatest strengths in fostering collaborations between stakeholders. It creates online forums and communities that enable researchers, developers, and industry users to exchange ideas, share best practices, and work together on solutions to the challenges of developing and implementing quantum-secure UAV technologies. These collaborative environments drive improvement and align all stakeholders toward a common goal: a secure and resilient UAV ecosystem in the quantum era.

5.5.1.2 Taking Wings: A Brilliant Future for UAVs

The effects of the EU Quantum-Secure Initiative are already rippling through Europe:

A Research Landscaping in Flower: The program has driven a proliferation of research efforts dedicated to quantum-secure UAV technology. Furthermore, schools and research facilities are already developing or testing promising PQC protocols, creating a healthy pipeline of potential solutions for the future.

Buy-In from the Industry: Top tech companies have confirmed the initiative's objectives. This has resulted in increasing industry buy-ins, with major UAV manufacturers deliberately looking to characterize their designs for quantum-secure solutions.

Crafting a Collaborative Ecosystem: The initiative is a great sense of collaboration across the European UAV industry. A more secure and united UAV airspace in the quantum age is already taking shape, with diverse actors collaborating toward the principles of knowledge sharing and problem-solving.

5.5.1.3 Looking Ahead: A Global Endeavor

These EU Quantum-Secure Initiatives are already shining examples for others to follow in the world. They emphasize the need for a collaborative approach that includes government, academia, and industry efforts in order to tackle the threats that quantum computation presents to this area of UAV operation. As the world progresses in the possibilities enabled by quantum technologies, the security of UAV operations will become even more crucial. We will be following in the footsteps of the EU to come up with similar kinds of collaborations, so that the wings of UAVs continue to take flight in a safe and secure quantum future.

5.6 Future Directions

Emerging Challenges in Quantum-Safe UAVs:

1. Advancement of Quantum Computing: The energy landscape is expected to shift rapidly with continual improvements in quantum computing. While it may be possible for quantum computers in the future to defeat current cryptography techniques, UAV manufacturers will have to innovate continuously in order to stay ahead with quantum-safe solutions.

2. Integration Challenges: Integrating quantum-safe technology with existing UAV systems has also presented technological limitations. Future studies will focus on developing seamless integration methods that ensure compliance, reliability, and minimal disruption to UAV operations.
3. Regulatory Schemes: As this technology advances, regulators will need to define how to regulate the use of quantum-safe UAVs. Future work will focus on the development of regulatory frameworks addressing security, privacy, and ethical issues across quantum-safe UAS.

Advancements in Education and Outreach:
1. Power of Quantum Computation: The fast-developing new tech of quantum computers is one of the most significant obstacles to the deployment of quantum-secure drone flights unmanned. This high capability of a quantum computer poses a great threat to the conventional cryptographic techniques protecting UAV missions and communications. Moreover, uncertain timelines are present regarding the production of commercially ready quantum computers, creating obstacles for UAV fliers and creators in properly planning and putting into place quantum-secure resolutions.
2. Dedicated Training Programs: To educate UAV operators, engineers, and cybersecurity experts about quantum dangers and mitigation techniques, dedicated training programs will emerge as the need for quantum-safe UAV competence increases.
3. International Cooperation: Future outreach initiatives will give priority to international cooperation in light of the global character of the UAV business and the dangers posed by quantum computing. Joint research projects, knowledge-sharing initiatives, and international alliances are a few examples of initiatives that will promote cooperation and hasten the deployment of quantum-safe UAV technology.

Toward a Quantum-Resilient UAV Ecosystem:
1. Ongoing Research and Development: The development of quantum-safe cryptography algorithms, protocols, and hardware implementations customized for UAV applications will be the main emphasis of future research. Other multidisciplinary research projects will explore advanced approaches to design and operate quantum-resilient UAVs.
2. Industry Standards and Certification: Industry members will continue to build standards and certification systems to promote the wide applications of quantum-safe UAV technology. These standards will also ensure interoperability, dependability, and security among various UAV manufacturers and systems.
3. Public-Private Partnerships: The need for establishing a quantum-resilient UAV ecosystem will require cooperation between governmental organizations, business partners, and academic institutions. Public-private partnerships provide a

platform for knowledge sharing, technology transfer, and collaborative innovation aimed at addressing a range of issues associated with quantum computing in the UAV space.

To surmise, the fate of quantum-sage UAVs will be veritably dependent on synergistic approaches to overcoming challenges, continued development of technologies, and informing stakeholders. The UAV community can cultivate a safe and strong environment through embracing innovation and enhancing cooperation by getting ready for a quantum computing age!

5.6.1 Emerging Challenges in Quantum-Safe UAVs

1. Quantum Computer Strength: Designed to address potential threats from quantum computing technology. The amplified capability of quantum computers is far more likely to threaten the conventional cryptography systems that defend UAV operations and communications. Although commercial quantum computer hardware is available only as of now, quantum computing will take time to advance to a point where it can compete with classical computers. This suggests that manufacturers and consumers of uncrewed aerial vehicles are struggling to devise a meaningful quantum-safe solution.
2. Transition Time: It brings forward operational and logistical hindrances for UAV stakeholders to transition from traditional cryptography systems to quantum-safe, quantum-resistant alternatives. The development of quantum-resistant cryptographic algorithms and their adoption into UAV systems will take time. It will require the entire UAV ecosystem to map, plan, finance, and coordinate this transition. During this transitional period, provisional security measures and risk mitigation strategies should be enforced, as UAV systems may be susceptible to attacks leveraging quantum computing capabilities.
3. Complexity due to Quantum-Safe Algorithms: This includes quantum-resistant algorithms (like hash-based signatures and lattice-based encryption) that are not vulnerable to quantum attacks. On the other hand, in contrast to their classical counterparts, these algorithms typically involve greater complexity and computational requirements. The trick for UAV developers is integrating these algorithms into current UAV systems without a loss of efficiency or performance.
4. Interoperability and Standardization: The seamless integration of quantum-resistant security measures can only be accomplished through the interoperability and standardization of quantum-safe solutions across a range of UAV platforms and applications. The existing standards and guidelines complicate the development and deployment of quantum-safe unmanned aerial drones. The realization of these standards and norms will influence and impact the compatibility and adoption of quantum-safe UAV technology across multiple sectors.

5. Training and Awareness from Security: Educating UAV operators, manufacturers, and policymakers on quantum computing threats to UAVs is key to facilitating a proactive approach to mitigating quantum computing risks. However, as of now, stakeholders in the UAV industry are still deaf and blind to the quantum threats. The UAV industry, and in particular its end users, could benefit from promoting security awareness and providing training resources tailored to their needs, even if these resources have relevant shortcomings, to enable a comprehensive and quantum-safe ecosystem around them.

Collaboration among UAV manufacturers, operators, researchers, and policymakers is crucial to establishing effective quantum-safe solutions that mitigate the potential risks of attacks from quantum computing technology while ensuring that UAVs continue to operate securely and efficiently.

5.6.2 Advancements in Education and Outreach

1. These are the development of instructional resources, and these are expert-level teaching resources. Extremely high-end teaching resources are extremely necessary and can be tailored for different audience stakeholders; even UAV operators, manufacturers, and policymakers need to be educated. They are useful as they offer detailed insights into the principles of quantum computing, the effects it has on the security of UAVs, and how researchers can make a transition toward quantum-safe technologies.
2. Interactive Training Courses and Workshops: These workshops and courses surely contribute significantly to enabling the gain of experience in a practical way and the sharing of information and ideas among participants. These educational sessions cover everything from post-quantum algorithms and quantum cryptography to possible quantum-safe UAV implementation strategies. Simulations and demonstrations work well as part of the content at such workshops, and this helps improve retention and engagement.
3. Collaboration with Business Partners: Partners like UAV manufacturers, technology providers, and cyber security professionals can significantly leverage the reach of an education and outreach program. Working with the field will be crucial to developing meaningful case studies, engaging demonstrations, and educational materials that interest stakeholders. Their partnership may also help keep the educational elements aligned with industry standards and practices.
4. Making Use of Digital Resources: Nowadays, outreach and educational programs can broaden their horizons through online resources such as webinars, tutorials, and video support. In this scenario, online platforms provide convenience and flexibility for stakeholders to view educational resources anytime. Online resources also help to level up in the UAV field!

5. Latest Teaching Technologies Incorporation: The combination of high-end academic technologies (such as AR and VR) indeed creates new ways for educational and outreach activities! Virtual reality representations can offer hands-on scenarios wherein UAV-specific situations, quantum security implications, and applicable quantum concepts can be explored. Interactive visuals of cyberspace, and even quantum-safe UAV address space, can be enhanced using augmented reality applications.
6. Assessment and Feedback Mechanisms: The means for ongoing assessment and feedback, and the willingness to make major changes to existing programs, are central to successful engagement and educational efforts. High Listen Surveys, center gatherings, and feedback structures are effective tools to gather input from stakeholders about the relevance, significance, and understandability of the materials. Through this consistent stakeholder engagement, organizations can refine their outreach and educational programs.

Finally, it is important, in conclusion, to establish further outreach and educational opportunities to increase awareness, promote cooperation, and speed up the implementation of quantum-secure UAV technology. Stakeholders from the UAV community can engage and navigate the securely quantum landscape, and this is made possible with data.

5.6.3 Toward a Quantum-Resilient UAV Ecosystem

1. Promoting the Adoption of Quantum-Safe Technologies in UAV Systems: A key objective is to promote the adoption of quantum-safe technologies in UAV systems. This project aims at developing hardware and cryptographic protocols resistant to quantum attacks. Cooperation between government, industry, and academia will be essential to accelerating innovation in the field of quantum-resilient UAV technology.
2. Important Standards and Certifications: It is also important to impose standards and certifications. The relevant authorities are required to set standards and certifications for quantum-resistant UAVs in order to ensure cross-industry compliance. this is not a point-to-point compatibility problem; to ensure the compatibility of the industry as a whole, a unified standard must also be reached, requiring the efforts of the relevant industry regulatory authorities to work together. Standards and recommendations for quantum-resilient UAVs are primarily driven by governmental regulators and standardization bodies. This involves establishing that the communication protocols, cryptographic algorithms, and hardware parts meet quantum security criteria.
3. Continuous Education and Training: Education and training should thus be ongoing to advise stakeholders of advancements in quantum computing and quantum-

safe UAV technology. This involves specific training, seminars, and certifications being provided to manufacturers, regulators, and UAV operators. Through the encouragement of a learn-and-adapt culture, the UAV community could, at some point, adopt quantum resilience best practices and respond to evolving threats.

4. Collaborative Research and Development: In order to drive innovation in quantum-safe UAVs, public and private sector collaboration will be key. Joint research projects and consortiums can facilitate the development of quantum-resistant solutions by fostering information sharing, technology transfers, and resource pooling. Interdisciplinary exploration of quantum-secure infrastructure for UAS can be aided by grant funding opportunities and broadly accessible collaboration platforms.
5. Evaluation and Mitigation of Risks: Improving the resilience of UAV systems to quantum threats involves conducting comprehensive vulnerability assessments and deploying defense strategies. This includes evaluating the impact of quantum attacks on UAV operations, identifying where gaps exist, and implementing mitigating strategies to reduce risk. Examples of proactive risk management procedures might include regular security audits and securing any sensitive information. These measures may lead to the discovery of vulnerabilities that can be addressed before malicious actors exploit them.

The UAV society can attempt to create a quantum-resilient environment through an extensive methodology, generally involving innovation improvement, standardization, instruction, participation, and opportunity administration!

5.7 Methodology

Figure 5.1 depicts the project workflow. Figure 5.2 shows the interactions among researcher, educator, and drone expert, where the researcher shares findings and discusses challenges with the educator, who provides training and materials to the drone expert.

I. **Introduction:**
   A. Overview of the Chapter
   B. Importance of Methodology in Education and Outreach
II. **Defining Objectives:**
   A. Clarifying the Goals of Education and Outreach Efforts
   B. Identifying Target Stakeholders
   C. Establishing Measurable Outcomes
III. **Literature Review:**
   A. Reviewing Existing Education and Outreach Initiatives in Quantum-Safe Technology

   B. Analyzing Successes and Challenges
   C. Drawing Insights for Designing Methodology

IV. **Stakeholder Analysis:**
   A. Identifying Key Stakeholders in the UAV Ecosystem
   B. Assessing Their Current Understanding of Quantum Threats
   C. Determining Their Educational Needs and Preferences

V. **Designing Educational Content:**
   A. Developing Educational Materials on Quantum Computing and Its Implications for UAV Security
   B. Tailoring Content for Different Stakeholder Groups
   C. Ensuring Accessibility and Relevance

VI. **Outreach Strategy Development:**
   A. Planning Outreach Events and Activities
   B. Selecting Suitable Formats (e.g., workshops, seminars, online resources)
   C. Defining Communication Channels and Platforms

VII. **Implementation:**
   A. Executing Education and Outreach Activities
   B. Collecting Feedbacks and Monitoring Engagement
   C. Iterating and Adjusting Strategies Are Needed.

VIII. **Evaluation:**
   A. Assessing the Effectiveness of Education and Outreach Efforts
   B. Measuring Stakeholder Awareness and Understanding of Quantum-Safe UAVs
   C. Analyzing Long-Term Impact on Adoption and Implementation

IX. **Case Studies:**
   A. Examining Successful Education and Outreach Campaigns in Related Fields
   B. Drawing Lessons Learned and Best Practices
   C. Adapting Strategies for Quantum-Safe UAVs

X. **Conclusion:**
   A. Summary of Methodology and Findings
   B. Implications for Future Education and Outreach in Quantum-Safe Technology
   C. Recommendations for Further Research

This very detailed methodology outlines the systematic approach for designing, implementing, and evaluating education and outreach initiatives in the context of quantum-safe UAVs. It really emphasizes the importance of stakeholder engagement, content development, strategic planning, and continuous evaluation to ensure the effectiveness and sustainability of efforts to raise awareness and even promote collaboration in this totally critical domain.

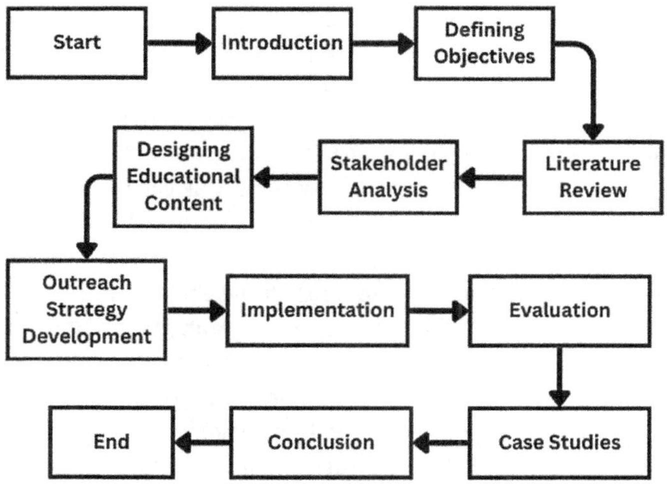

Figure 5.1: Methodology flowchart.

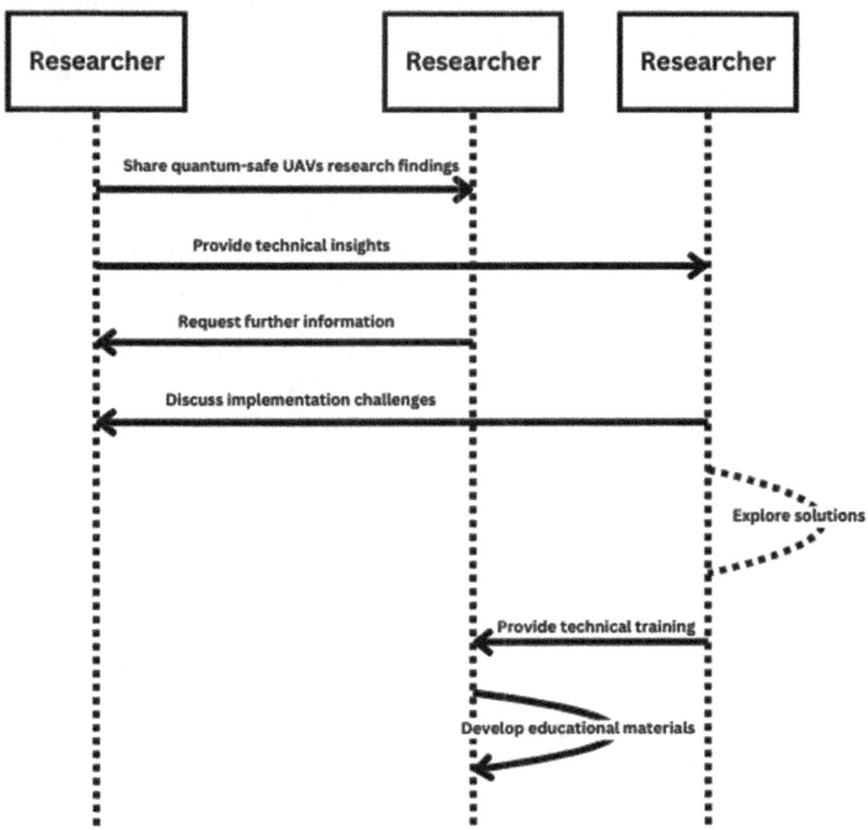

Figure 5.2: Methodology sequence diagram.

5.8 Conclusion

UAV and quantum computing technology convergence offers both enormous potential and really difficult problems. The security environment for UAVs is changing as an outcome of the advancements in quantum computing capabilities, which calls for a proactive and cooperative strategy to protect these vital systems! Outreach and education are crucial components for smoothing the transition to quantum-safe UAVs, as discussed throughout this chapter.

To begin with, we have highlighted the importance for all stakeholders in the UAV ecosystem to have a better understanding of the implications of quantum computing on security. UAV systems are based on conventional cryptographic methods, which are considerably susceptible to quantum attacks and could compromise mission-critical operations. Educating UAV operators, manufacturers, and lawmakers about quantum threats is one of the best methods for instilling awareness and preparedness in a culture.

In addition, the chapter has described some outreach programs that have been developed to include all stakeholders and enable mutual cooperation among UAV users. These activities can be the building blocks of sharing platforms, joint research projects, workshops, and seminars to spread best practices and topical information to help catalyze the uptake of quantum-safe technology.

Case studies and truly best practices have reported valuable information on outreach and education program work, offering examples from the field of successful strategies in action. Stakeholders may receive useful advice for genuinely implementing their own projects and navigating the challenging field of quantum-safe UAVs by emphasizing these accomplishments and lessons learned.

Quantum-safe UAVs will need to really continue to innovate, work together, and exercise caution in the future. New problems will certainly call for flexible solutions, and improvements in outreach and education will be very essential to getting stakeholders ready for the quantum age. UAS may remain safe, secure, and dependable in the face of changing threats if we embrace these principles and collaborate to create a quantum-resilient UAV ecosystem.

References

[1] P. G. Fahlstrom, T. J. Gleason, and M. H. Sadraey, Introduction to UAV Systems. John Wiley & Sons, 2022.

[2] G. Pan, X. Miao, X. Yang, and Z. Yang, "Introduction to UAV Communications," in UAV Communications: Modeling and Analyses, Singapore: Springer Nature Singapore, 2024, pp. 1–21.

[3] R. Shadiev and S. Yi, "A systematic review of UAV applications to education," Interactive Learning Environments, vol. 31, no. 10, pp. 6165–6194, 2023.

[4] X. Dai, Z. Xiao, H. Jiang and J. C. S. Lui, "UAV-Assisted Task Offloading in Vehicular Edge Computing Networks," in IEEE Transactions on Mobile Computing, vol. 23, no. 4, pp. 2520–2534, April 2024 doi: 10.1109/TMC.2023.3259394.

[5] D. Mishra et al., "Quantum-Safe Secure and Authorized Communication Protocol for Internet of Drones," in IEEE Transactions on Vehicular Technology, vol. 72, no. 12, pp. 16499–16507, Dec. 2023, doi: 10.1109/TVT.2023.3292169.

[6] B. Yang, T. L. Hawthorne, M. Hessing-Lewis, E. J. Duffy, L. Y. Reshitnyk, M. Feinman, and H. Searson, "Developing an Introductory UAV/Drone Mapping Training Program for Seagrass Monitoring and Research," Drones, vol. 4, no. 4, p. 70, 2020. DOI: https://doi.org/10.3390/drones4040070.

[7] Arunmozhiselvi, S., T. Ananth Kumar, P. Manju Bala, S. Usharani, and G. Glorindal. "A systematic approach to agricultural drones using a machine learning model." In Machine Learning Approaches and Applications in Applied Intelligence for Healthcare Data Analytics, pp. 41–60. CRC Press, 2022.

[8] V. K. Ralegankar, J. Bagul, B. Thakkar, R. Gupta, S. Tanwar, G. Sharma, and I. E. Davidson, "Quantum cryptography-as-a-service for secure UAV communication: applications, challenges, and case study," IEEE Access, vol. 10, pp. 1475–1492, 2021.

[9] S. P. Yadav, R. Singh, V. Yadav, F. Al-Turjman, and S. A. Kumar, Eds., Quantum-Safe Cryptography Algorithms and Approaches: Impacts of Quantum Computing on Cybersecurity. De Gruyter, 2023.

[10] A. Kumar, S. Bhatia, K. Kaushik, S. M. Gandhi, S. G. Devi, A. D. J. Diego, and A. Mashat, "Survey of promising technologies for quantum drones and networks," IEEE Access, vol. 9, pp. 125868–125911, 2021.

[11] A. Kumar, D. A. de Jesus Pacheco, K. Kaushik, and J. J. Rodrigues, "Futuristic view of the internet of quantum drones: review, challenges and research agenda," Vehicular Communications, vol. 36, p. 100487, 2022.

[12] T. Pradhan and P. Patil, "Quantum Cryptography for Secure Autonomous Vehicle Networks: A Review," in 2024 IEEE International Students' Conference on Electrical, Electronics and Computer Science (SCEECS), 2024, pp. 1–10.

[13] S. Suhail, R. Hussain, A. Khan, and C. S. Hong, "On the role of hash-based signatures in quantum-safe internet of things: Current solutions and future directions," IEEE Internet of Things Journal, vol. 8, no. 1, pp. 1–17, 2020.

[14] R. Aissaoui, J. C. Deneuville, C. Guerber, and A. Pirovano, "A survey on cryptographic methods to secure communications for UAV traffic management," Vehicular Communications, vol. 1, p. 100661, 2023.

[15] M. Krelina, "Quantum warfare: definitions, overview and challenges," arXiv preprint, arXiv:2103.12548, 2021.

[16] R. Varma, C. Melville, C. Pinello, and T. Sahai, "Post quantum secure command and control of mobile agents inserting quantum-resistant encryption schemes in the secure robot operating system," International Journal of Semantic Computing, vol. 15, no. 3, pp. 359–379, 2021.

[17] A. Kumar, R. Krishnamurthi, G. Sharma, S. Jain, P. Srikanth, K. Sharma, and N. Aneja, "Revolutionizing modern networks: Advances in AI, machine learning, and blockchain for quantum satellites and UAV-based communication," arXiv preprint, arXiv:2303.11753, 2023.

[18] M. S. Kumar, B. K. Harsha, and L. T. Jule, "AI-driven cybersecurity modeling using quantum computing for mitigation of attacks in IoT-SDN network," in Quantum-Safe Cryptography Algorithms and Approaches: Impacts of Quantum Computing on Cybersecurity, pp. 37–47, 2023.

[19] T. Commissariat, "The key to our quantum future," Physics World, vol. 34, no. 12, p. 40, 2021.

[20] A. Bandi and S. Yalamarthi, "Towards artificial intelligence empowered security and privacy issues in 6G communications," in 2022 International Conference on Sustainable Computing and Data Communication Systems (ICSCDS), 2022, pp. 372–378.

[21] A. S. Nair, S. M. Thampi, and V. Jafeel, "A Post-quantum Secure PUF based Cross-Domain Authentication Mechanism for Internet of Drones," Vehicular Communications, vol. 2, p. 100780, 2024.
[22] C. Roberts and S. R. Yerram, Ensuring Security in the Age of Intelligent Connectivity: Strategic Insights for 6G Networks.
[23] M. Domb and Y. Shnaps, "Cybersecurity Threats and Mitigations Related to Smart Cities Operation," 2024.
[24] M. Wazid, A. K. Das and Y. Park, "Generic Quantum Blockchain-Envisioned Security Framework for IoT Environment: Architecture, Security Benefits and Future Research," in IEEE Open Journal of the Computer Society, vol. 5, pp. 248-267, 2024, doi: 10.1109/OJCS.2024.3397307.
[25] J. O. del Moral, A. D. iOlius, G. Vidal, P. M. Crespo, and J. E. Martinez, "Cybersecurity in Critical Infrastructures: A Post-Quantum Cryptography Perspective," arXiv preprint, arXiv:2401.03780, 2024.
[26] R. K. Verma and K. Kishor, "Image Processing Applications in Agriculture With the Help of AI," in Infrastructure Possibilities and Human-Centered Approaches With Industry 5.0, IGI Global, 2024, pp. 162–181.

A. Sheik Abdullah, Sriram B., Lynda D., and Harini V. K.

6 Quantum Cryptography in UAV Communications

Abstract: Secure communication is vital for unmanned aerial vehicle (UAV) operations, where ensuring data integrity and secrecy is essential. The distinctive challenges in UAV communication and the evolving quantum computing field heighten the vulnerabilities of traditional cryptographic methods. Quantum cryptography offers a promising solution by leveraging quantum mechanics to establish provably secure channels. This chapter provides a comprehensive overview of quantum cryptography in UAV communications, covering fundamental concepts of quantum mechanics relevant to cryptography. It explores protocols like quantum key distribution (QKD), quantum secure direct communication (QSDC), and quantum teleportation, discussing their applicability and limitations in UAV networks. Additionally, the chapter discusses how quantum communication can be effectively utilized for long-range UAV communication. Furthermore, it delves into implementing quantum cryptography-as-a-service for UAVs, presenting real-world case studies and outlining future directions and challenges. Serving as a roadmap for integrating quantum cryptography, this chapter ensures robust security in UAV operations against evolving threats.

Keywords: UAV (unmanned aerial vehicle), UAV communication, quantum mechanics, QSDC (quantum secure direct communication), QKD (quantum key distribution), quantum teleportation protocols, QCaaS – (quantum cryptography-as-a-service), emerging trends, future applications

6.1 Introduction to Quantum Cryptography and UAV Communications

6.1.1 Secure UAV Communication

Unmanned aerial vehicle (UAV) communications have recently drawn much attention in military and defense applications, agricultural monitoring, construction, transportation, and so on. It is acknowledged that drones possess capabilities in surveillance and reconnaissance, search and rescue, infrastructure inspection, aerial delivery, and so on. As drone applications grow exponentially every day, reliable and secure communications for drone operations are of great importance.

https://doi.org/10.1515/9783111619323-006

In drone operations, some sensitive information may be communicated, such as surveillance and reconnaissance information, tactical mission instructions and controls, etc. In military operations, the breach of a communication link may result in the loss of sensitive information, failed tactics, and casualties. Even for commercial operations, the breach of a UAV communication link may cause severe economic loss, intellectual property theft, and public safety risks.

Classical cryptographic techniques based on hard computational problems are not secure against the immense computing capability of a quantum computer. As quantum computer development progresses, it may become possible to use practical large-scale quantum computers in the foreseeable future. Then, classical cryptographic techniques used in practice will be broken and rendered insecure. For example, the prominent classical cryptographic approach is contingent on the integer factorization problem (IFP). The most well-known algorithm to solve this problem is Shor's algorithm, which can be applied on a quantum computer. Therefore, there is a vulnerability in drone communication security. A cryptographically quantum-resistant technique is expected to address this vulnerability.

6.1.2 Overview of UAV Communication

UAV communication enables the transmission of information, commands, and control between different subsystems, for example, between the ground control station (GCS) and other UAVs, as well as external systems like satellites or ground networks. Thus, communication links are of paramount importance for the successful deployment and operation of UAVs.

UAV communication generally relies on wireless technologies such as Wi-Fi, cellular, and satellite communication. Nevertheless, due to the high maneuverability, dynamic network topology, and scarce bandwidth typical of UAV scenarios, classical communication protocols are challenging to apply.

Single UAV systems involve a UAV operating as an independent node, which directly interfaces with a control station located on the surface or some other surface node infrastructure. It is a type of communication that is easy to manage, though it may be restrained concerning distance and extent of connectivity.

The most difficult scenarios are those in which multi-UAVs work together on complex operations. Thus, the ability to communicate among themselves is crucial for these UAVs in order for them to calculate how they move, distribute tasks among themselves, or transmit acquired information. Consequently, systems of this kind are based, by design, on ad hoc and mesh network topologies, with each UAV creating and sustaining connections to multiple other UAVs at any time.

Communication links in UAV scenarios must cope with a variety of different challenges, such as scarce bandwidth, high node mobility, and unpredictable wireless channel conditions. Therefore, communication UAV systems must possess routing protocols, adaptive modulation and coding systems, secure authentication, and encryption mechanisms that efficiently manage dynamic network topologies. Especially, the limited bandwidth and the high latency due to the long communication distances have a considerable influence on the communication systems' bandwidth and latency. Therefore, UAV communication systems need to be designed for both large data rates and quick signal delivery.

6.1.3 What Are Quantum Breakthroughs in Communications?

Quantum mechanics is a theoretical philosophy of physics that deals with tiny matters and energies, that is, at the atomic and subatomic level. Quantum mechanics is an incredibly successful theory and has led to many technological inventions and innovations in many facets of human life, including cryptography and communications.

A significant achievement in communications for implementing quantum protocols is the quantum key distribution (QKD) protocol. QKD is a method of securely sharing a cryptographic key between two parties based on the principles of quantum mechanics. This quantum cryptography offers information-theoretic security against any future attacks, even by powerful quantum computers.

The confidentiality of quantum cryptography relies on fundamental properties of quantum systems, namely the no-cloning theorem and the Heisenberg uncertainty principle, which guarantee that any eavesdropping (Eve) on the communication will introduce disturbances that can be detected by Alice and Bob. In quantum cryptography, the bit values of the key are encoded in quantum states of particles (typically photons). These quantum states can represent symbols in superposition and entangled states. They cannot be measured (detected) or cloned (copied) without altering the states in a way that will alert the sender and receiver to the eavesdropping or monitoring.

Another important quantum communication method involving quantum entanglement is quantum teleportation. Quantum teleportation allows quantum information (qubits) to be transferred from one location to another entirely using only classical communication, via pre-shared entangled particles and quantum channels (quantum key).

The possible applications of quantum teleportation include secure quantum communications, distributed quantum computing, and quantum networks. The no-cloning theorem: explains why a copy of a quantum state cannot be made; it is due to the disturbance (measurement) effect on the original state. When measured, a quantum state will collapse to one of its potential states (a superposition of states).

6.1.4 Basic Concepts of Quantum Cryptography

QKD refers to a protocol for distributing cryptographic keys between two trusted parties. The no-cloning theorem and the Heisenberg uncertainty principle are used in QKD to provide security against any possible eavesdropper (Eve) who might intercept the key.

The typical QKD protocol is described as follows: Alice prepares random quantum states, whose key information is encoded in quantum system [22] properties, e.g., polarization or phase. These quantum states are transmitted through a quantum channel, usually an optical fiber or a free-space link, to Bob.

The no-cloning theorem states that an unknown quantum state cannot be cloned and was proven by physicists some years ago. It is one of the most essential quantum mechanical principles. Whenever the quantum state is measured or copied, the measurement always introduces disturbances that change the state in a detectable way.

In QKD, Alice and Bob measure the quantum channel for any possible disturbances or aberrations that may be induced by an eavesdropper (Eve). If no disturbance is found, Alice and Bob can both assume that they have shared a secret key and simply carry out quantum measurements and some postprocessing steps. However, if an eavesdropper eavesdrops on the quantum states, she can be detected, and the protocol is aborted in this case to prevent any information leakage.

The main advantages of quantum cryptography over classical schemes are:

1. **Unconditional Security**: QKD provides information-theoretic security. This means that the security does not depend on computational assumptions but follows directly from the laws of physics. In principle, the security cannot be broken or fail against any possible attacks, even if some advanced computational power or quantum computers are used in the future.
2. **Immediate Detection of Eavesdropping**: Quantum mechanics ensures that any eavesdropping on the quantum channel is detected immediately.
3. **Forward Secrecy**: QKD provides forward secrecy. This means that even if Eve records the whole communication and gains access to the secret key later, she will not be able to decrypt the past communications retroactively.
4. **Resistance to Quantum Computing Threats**: QKD is resilient against attacks from quantum computers, which will break classical crypto schemes based on computational complexity assumptions.

Quantum cryptography has some security benefits but also some drawbacks that need to be overcome in practice. These include the usage of specialized hardware, relatively small transmission distances, and the susceptibility of the whole system to side-channel attacks targeting the implementation. Many research activities are ongoing to overcome these drawbacks and to extend the applicability of quantum cryptography to secure communication systems, including those for UAV platforms.

6.1.5 The Security of UAV Communications and the Opportunities for Quantum Cryptography

Here's the place where quantum cryptography can help – and where you can also help by growing and expanding in this space. UAV networks exhibit different characteristics than traditional communication networks – and their security needs should be treated accordingly.

One of the primary limitations, including communications among UAVs or between UAVs and ground stations, is the elevated mobility of the UAVs. This can lead to an arbitrarily changing topology and connectivity among the nodes, making it hard to apply the traditional cryptographic key distribution and management techniques used in traditionally more static networks. In such dynamic environments, it may be more feasible to dynamically establish secure keys on the fly, as required by the nodes. This just happens to be a native property of quantum cryptography, so there's an opportunity here for applying quantum-based key distribution techniques to the security of UAV communications networks. It would be far more flexible and dynamic.

A second problem encountered in UAV platforms is the limited onboard computing resources and power. Many cryptographic algorithms – especially key distribution and management protocols – are computationally intensive and energy-expensive, often too demanding for the typical UAV platform. Quantum cryptography protocols are intrinsically based on quantum mechanics, not computation, leading to far lower overhead and energy expenses to keep the UAV network secure.

UAV networks are also often deployed in adversarial environments – such as on battlefields or in disaster zones – and thus must be open to various physical and cyberattacks (jamming, spoofing, etc.). Quantum cryptography inherently detects any eavesdropping on communication and is also resilient to threats from possible quantum computers.

The use of quantum cryptography in UAV-based communications will also present value-adding mission opportunities and new use cases. For instance, UAVs carrying sensitive payloads, such as intelligence or other important C2 (command and control) messages, could exchange this sensitive data over a quantum channel between UAVs without concern about being eavesdropped on or having the data compromised in any way.

In addition, UAVs could utilize quantum networking to establish secure quantum router networks. This would enable UAVs to participate in larger-scale fixed infrastructure communication networks.

The combination of quantum cryptography and UAV communications could also provide secure multi-party computation and distributed quantum computing, potentially expanding the decision-making capabilities of UAV networks and enabling more advanced UAV missions and on-board processing.

There are numerous exciting applications of quantum cryptography, however, there are still some obstacles to consider before we can start using this technology to improve the security of UAV-based communications. The main challenge is the relatively short range over which quantum states can be transmitted reliably. The communication range of QKD devices is currently limited to a few hundred kilometers, which may not be sufficient for some UAV applications that require high-altitude, long-distance communication links.

Several research efforts are underway to extend the reach of quantum communications. These include the concepts of quantum repeaters and satellite quantum communications. Quantum repeaters are intermediate nodes that receive and store the quantum state, then regenerate and resend it toward the destination, thus increasing the distance by a factor of two. Satellite-based quantum communications could exploit the natural properties of space-based platforms to achieve a truly global quantum communication network.

A challenge stems from the need to integrate quantum cryptography systems with existing UAV communication infrastructure and protocols. By its very nature, quantum cryptography will need to provide interoperability and legacy system compatibility before it can be seamlessly deployed in UAV networks. Standardization efforts and the development of quantum-aware communication protocols will be essential in tackling this challenge.

Finally, the deployment of quantum cryptography in UAV systems requires sophisticated hardware components, such as quantum memory devices, single-photon sources, and detectors. The miniaturization and ruggedization of these components to satisfy the UAV's size, weight, and power constraints present interesting engineering challenges.

Despite the challenges, the capability of quantum cryptography to heighten the security of UAV communications has attracted the attention of many governments, military organizations, and commercial companies. Many research efforts and pilot projects are already exploring the integration of quantum cryptography into UAV communication systems and assessing its performance and feasibility in real scenarios.

6.2 Challenges in Secure Communication for UAVs

6.2.1 Vulnerabilities in Traditional Cryptographic Systems

We know that a UAV is controlled by a ground-based controller; here, the communication link between them is currently being encrypted using traditional cryptography algorithms and protocols. These algorithms rely on increasing computational difficulty to provide security in real time. The security of well-known public key cryptography [32] relies on the following three mathematical hard problems: the DLP (discrete logarithm problem), the elliptic curve DLP, or the IFP. However, the development of high-performance supercomputers and, more importantly, the serious implications posed by quantum computers on classical cryptographic algorithms cannot be ignored. Prior studies have demonstrated that, in some situations, quantum computing may execute massive parallel computations to find solutions in polynomial time for problems with exponential time complexity. Cryptographic keys used in these algorithms play an important role in providing security features. Let us take the example of the good old AES-128 algorithm. It uses a 128-bit-sized key, so the possible combinations would be 2^{128} operations. With a classical computer capable of performing 10^{12} operations per second, it would take approximately 3.4×10^{24} s, or about 10^{17} years. On the other hand, consider a quantum computer capable of performing 10^{12} quantum bit operations per second; it would take about 5.8×10^{6} s, or roughly 67 days. These examples clearly illustrate the computational power of quantum computers. One way to address this challenge is to increase the difficulty further in computation by adopting quantum-resistant algorithms such as lattice-based, hash-based, code-based, and multivariate polynomial-based cryptography. For example, let us consider a lattice-based algorithm. Due to the hardness of lattice problems, it is considered secure against quantum attacks. However, employing it in real time can lead to issues such as increased bandwidth usage and storage requirements. It also demands more computational power, which is not feasible for resource-constrained UAVs. If we consider hash-based algorithms that use the Merkle [4] signature scheme and the extended Merkle signature scheme (XMSS), they are provably secure against quantum attacks. Multivariate cryptography relies on the complexity of solving multivariate polynomial equation systems over finite fields. Although this type of cryptography has drawbacks such as processing overhead and a complicated key management scheme, particularly for signature generation and verification, these limitations make traditional algorithms difficult to employ in practical applications. Another drawback of traditional cryptographic systems is their orthogonality; it is easy to distinguish whether bits are one or zero.

6.2.2 Unique Challenges in UAV Communication Security

There are many factors that influence the quality of information being transmitted and pose challenges in the field of UAV communication security. Some of them are listed below:

a) **Resource-Constrained Nature of UAV:**
   UAVs are resource-constrained by nature; this can be viewed in terms of computation, storage, power, and bandwidth. If we look in terms of processing power and memory, the UAV's onboard processing power is limited due to the restrictions on carrying the payload and its size. This restricts their ability to run com-

plex algorithms, store large amounts of data, or perform real-time analysis onboard. Thus, a quantum-resistant traditional algorithm might not be suitable for real-time communication in a UAV. In the case of quantum communication, we use photons to transmit information. A photon would be suitable for an optical fiber environment, but photons traveling through free space are highly susceptible to photon loss due to environmental conditions.

b) **Signal Interference and Jamming:**
The communication link between the UAV and GCS or with another relay is susceptible to interference and eavesdropping. Interference can be due to environmental factors such as storms, rain, heavy winds (turbulence), and atmospheric attenuation [11], which can affect the reliability and quality of communication signals. ECM (Electronic Countermeasures) might use electronic jamming to interfere with the signal, making it difficult for the receiver to decode the signal. A type of jamming called GPS jamming [31] stops the receiver from picking up the real GPS signals. Another type of attack is a replay attack. It takes place by first interfering with the signal; here, the attacker records the signal being transmitted between the GCS and UAV, then, after some time, impersonating the GCS, transmits those signals to the UAV for it to accept them and perform undesirable actions at that point in time.

c) **Latency and Real-Time Control:**
Latency is the delay experienced by a system in a communication network. This factor plays [13] an important role in mission-critical operations. It could cause a catastrophe when we are dealing with multi-UAV networks, as timely decisions are crucial to avoid collisions. Latency can be disadvantageous in many ways, such as buffering of video data in GCS, impairing coordination and synchronization, and affecting APT systems that rely on real-time feedback.

d) **Range and Coverage:**
Range is the area within which ground control can maintain a communication link with the UAV without causing disruption to its mission. In general, people use three types of communication technologies:

Line-of-Sight (LOS) RF: Commonly used for short-range UAV communication. It requires a direct LOS between the UAV and the GCS.
Example: short-range communication that uses 2.4 GHz or 5.8 GHz.
Range: 1–10 km
Advantages: simple infrastructure, no dependency on other systems.
Limitations: short range; not suitable for urban and highly populated regions.

Beyond Visual LOS (BVLOS) RF: uses high-power transmitters and more sophisticated modulation techniques for communication between the GCS and UAV. One may use relays to extend the range.
Example: may use lower-frequency bands (e.g., HF, VHF) or advanced technologies like LTE.

Range: can extend up to several hundred kilometers with appropriate technologies and relays.
Advantages: extended range; suitable for long-distance operations.
Limitations: require more advanced technology and regulatory approval.
Satellite Communications (SATCOM):
Example: Geostationary satellites and low-Earth orbit (LEO) satellites
Range: global coverage or near-global
Advantages: Very long-range; can cover remote and oceanic areas.
Limitations: higher cost and latency, requiring tracking multiple satellites.
Cellular Networks:
Range:
4G LTE: Generally, it provides coverage up to 30 km from a cell tower in rural areas, but significantly less in urban areas due to interference and obstacles.
5G offers similar coverage in lower frequency bands, but higher frequency bands (millimeter wave) provide a much shorter range, typically within a few hundred meters.
Advantages: Scalability, high-speed data rates with 5G, low latency.
Disadvantages: can only be used where cellular infrastructure is available.

e) **Heterogeneous UAVs and Uncertain Environments:**
UAVs carrying munitions might not be able to use high-frequency signals as they might be harmful to the payload. Due to tight coupling, vibrations during takeoff and landing can readily affect the system's stability. Furthermore, this paradigm would cause significant local turbulence and modify the aerodynamic properties of the airborne platforms, which would make signal detection and reception extremely difficult. The system's inherent mobility presents challenges since we must continuously modify the antenna as it moves using a relay or GCS.

6.2.3 Need for Quantum-Based Solutions

UAVs are being used more and more in different applications like surveillance, environmental monitoring, disaster management, and communication relays. However, as their usage increases, it becomes really important to ensure that communication among UAVs is secure and efficient. The usual methods of encryption and the use of machine learning (ML) techniques face numerous challenges when it comes to UAV networks. That is why quantum-based solutions are becoming increasingly promising. Here are the key reasons why we need quantum-based solutions for UAV communications:

a) **Detection of Eavesdropping:**
The man-in-the-middle and replay attacks are posing serious threats to a communication link. Using quantum communication protocols, which provide optimum security, has been proved to be theoretically difficult to crack. Additionally, we can easily detect eavesdropping in quantum algorithms.
b) **Enhancing the Computation Power:**
Quantum algorithms are said to be wicked faster than traditional computing in solving some problems. They tend to reduce the time complexity from exponential to linear time. This can speed up data processing and decision-making in UAV operations by a lot. One such example is Grover's algorithm, which can search unsorted databases in a linear amount of time $O(N)$.
c) **Reducing Communication Overhead:**
Quantum teleportation and quantum compression techniques can reduce the amount of data that needs to be transmitted between UAVs and ground stations. This not only saves bandwidth but also enhances the speed and reliability of communication. Quantum error correction codes, such as stabilizer codes and surface codes, can provide robust error correction mechanisms for quantum communication channels. This ensures data integrity and reliability in noisy and lossy UAV communication environments.

6.3 Fundamentals of Quantum Mechanics for Cryptography

6.3.1 Principles of Quantum Superposition and Entanglement

Quantum superposition and entanglement are the two foundational principles of quantum mechanics, essential for empowering quantum cryptography.

6.3.1.1 Quantum Superposition

In classical material science, a molecule or structure is in one definite state at any instant. A coin can be head or tail; instead, if a coin is being spun on a flat surface, it appears to be in both head and tail. Schrödinger's cat is a famous example of superposition as a thought experiment.

Schrödinger's cat [36] is a thought experiment devised by Austrian physicist Erwin Schrödinger in 1935. The thought experiment describes a cat situated in a sealed box that also contains a vial of poison, a radioactive source, and a Geiger counter. If the radioactive source decays and is detected by the Geiger counter, a mechanism is triggered that will release the poison, killing the cat. As the radioactive source alone is in a superposition of decayed and not decayed until observed or measured, the cat is in a superposition of being dead and alive (the fate of the cat is tied to the outcome of the radioactive source). The cat is simultaneously alive and dead until the box is opened and the observer checks or observes the cat.

Schrödinger's cat illustrates the peculiar and seemingly absurd implications of quantum superposition, as a macroscopic object (the cat) would be in a superposition of two complementary states (dead and alive). In classical physics, the cat would definitely be in one of these states (dead or alive) but not in both at the same time. The cat paradox also deals with the problem of what really constitutes a measurement or observation in quantum mechanics, because, in the Copenhagen interpretation, it is the measurement or observation that causes the wave function of the quantum object (radioactive source and cat) to collapse from a superposition of states to one of the possible definite states.

Quantum mechanics portrays a miracle where particles can be in a superposition of many states at the same time for some time until they are measured or observed. A quantum bit, or qubit, of information can represent not only the classical states of 0 or 1 but also a superposition of both the 0 and 1 states simultaneously. The state of a qubit is described by a wave function, which is a linear combination or "superposition" of the 0 and 1 states, with complex coefficients indicating the probability amplitude of each state [28]. This behavior of quantum particles to exist in a superposition of different quantum states is a fundamental difference from classical physics and allows, on a very fundamental level, for new applications like quantum computers and quantum cryptography. A quantum computer, for example, can exploit this "misuse" of quantum superposition to perform certain calculations in parallel.

6.3.1.2 Quantum Entanglement

Another fundamentally vital wonder in quantum mechanics is entanglement. When two or more quantum particles are entwined and become entangled, they become connected in such a way that the quantum state of each molecule cannot be described independently of the state of the other molecules, no matter how far apart the particles are separated. Measuring or observing the state of one entrapped molecule immediately influences the state of the other snared particle(s), even if isolated by a huge distance [28]. This opposes our classical instincts about how particles ought to behave autonomously based on their separation in space.

Entanglement leads to profoundly counter-intuitive impacts, where observing one molecule can immediately alter the state or result of measurements on another molecule light-years away. Einstein referred to it as "spooky action at a distance" [18]. Quantum entanglement is a key enabler for numerous quantum communication protocols and cryptographic applications since it allows two parties to generate shared random numbers or keys secured by the laws of quantum physics. Any endeavor by a spy to measure the quantum states will inevitably disturb them and be detected.

6.3.2 Quantum States and Measurements

In quantum mechanics, the state of a quantum framework, like a single molecule or a gathering of particles, is portrayed mathematically by a wave function Ψ. The wave function is a complex-valued probability amplitude, where $|\Psi|^2$ gives the probability that a molecule will be observed in a specific state or with certain properties like position, momentum, or spin.

For instance, a qubit can be referred to as a superposition of the $|0\rangle$ and $|1\rangle$ basis states:

$$|\Psi\rangle = \alpha|0\rangle + \beta|1\rangle$$

where α and β are complex numbers, and $|\alpha|^2 + |\beta|^2 = 1$.

The squared coefficients denote the likelihood of measuring the qubit in the $|0\rangle$ or $|1\rangle$ state. One of the most imperative principles of quantum mechanics is that the act of measuring or observing a quantum system influences and disturbs the system significantly. When a measurement is made on a quantum state like $|\Psi\rangle$, the state "collapses" or reduces to just one of the possible states it was in superposition over [8, 28]. So if we measure the qubit state $|\Psi\rangle$, it will randomly collapse to either the $|0\rangle$ state with likelihood $|\alpha|^2$ or the $|1\rangle$ state with likelihood $|\beta|^2$. Fundamentally, this measurement process is inherently random – it is fundamentally impossible to predict with certainty which state the system will collapse into before the measurement occurs. This inherent randomness and uncertainty at the core of quantum theory enable new forms of cryptography that are provably secure against any computational attack.

For quantum cryptography, the key point is that any estimation or endeavor to observe the state of a quantum framework, like a photon, will essentially disturb and modify that state in accordance with the laws of quantum mechanics. This allows the genuine parties to detect if a spy has tampered with or attempted to measure the quantum states they were transmitting.

6.3.2.1 No-Cloning Theorem

The no-cloning theorem is an important theorem that makes quantum cryptography possible. This theorem states that it is fundamentally impossible to create a perfect copy or "clone" of an arbitrary, unknown quantum state.

In the world of classical physics, information can easily be copied. For instance, the state of a classical bit (0 or 1) can be measured and used to prepare a second copy of the bit. However, the no-cloning theorem insists that for an arbitrary quantum state $|\Psi\rangle$, it is impossible to construct a reliable copying machine to duplicate $|\Psi\rangle$ into another copy.

The no-cloning theorem can be proven from the linearity of quantum mechanics, which states that the evolution over time of any quantum state $|\Psi(t)\rangle$ must be described by a linear operator [9, 28]. It can be shown that no linear operator can clone an arbitrary, unknown quantum state.

In quantum mechanics, only orthogonal basis states can be cloned – for instance, we can clone $|0\rangle$ or $|1\rangle$. However, for any non-orthogonal superposition state $\alpha|0\rangle + \beta|1\rangle$, there exists no cloning procedure. This has been proven.

What this means is that if, in a quantum cryptography protocol, the information is carried in a non-orthogonal quantum state – for instance, the polarization of a photon – then any attempt by an eavesdropper to measure (and hence clone) the state of the photon will necessarily disturb the state of the photon and, hence, introduce evidence of the eavesdropping.

The no-cloning theorem is perhaps the single most useful theorem that enables quantum cryptography to provide security and tamper detection capabilities that are impossible to achieve in the classical realm.

6.3.2.2 Distinguishing Quantum States

The limitation of distinguishing arbitrary non-orthogonal quantum states is related to the no-cloning theorem is. It is also fundamentally impossible to construct a machine that can reliably distinguish or discriminate between non-orthogonal quantum states, as unknown quantum states cannot be cloned perfectly [9, 28].

If two non-orthogonal states $|\Psi\rangle$ and $|\Phi\rangle$ are encoded with different information, then any attempt by an eavesdropper to determine which state was sent will unavoidably disturb and alter those states, according to the laws of quantum measurement.

The proof follows by showing that if we assume a measurement could reliably discriminate $|\Psi\rangle$ from $|\Phi\rangle$, then we could use that measurement process as a basis for cloning arbitrary states, which violates the no-cloning theorem.

For orthogonal states $|0\rangle$ and $|1\rangle$, one can construct measurements that reliably discriminate between them. However, for any non-orthogonal states, it is fundamentally impossible to distinguish them with certainty.

The impossibility of state discrimination is another fundamental truth that underlies the security of quantum cryptography against any adversary. Since cryptographic keys are encoded in non-orthogonal states of quantum systems like photons, any measurement by an eavesdropper on those states will inevitably disturb them and create leak evidence that can be detected.

6.3.3 Quantum Entanglement and Applications

The phenomenon of quantum entanglement leads to numerous counter-intuitive effects and applications in quantum cryptography. When two or more quantum particles become entangled, their states become inextricably linked or correlated in a way that defies our classical intuitions.

For example, consider two entangled particles, *A* and *B*, that are separated by any distance. If we measure the spin of particle *A* and find it has spin-up, then the spin of the distant particle *B* is also instantaneously determined to be spin-down due to the constraints of quantum mechanics, even with no apparent mechanism for communication between *A* and *B*.

Entanglement provides a fundamentally new way to generate cryptographic keys that have never existed in an accessible form before the measurements. The mere act of measuring the particles generates truly random numbers or cryptographic keys at each location, which are inherently private.

Furthermore, any attempt at eavesdropping on the entangled particles will necessarily disturb their quantum states in a detectable way due to the measurement process. This allows the legitimate parties to determine whether any eavesdropping has occurred and to discard compromised keys.

Entanglement can also enable protocols like quantum teleportation, where the complete state of a particle is transmitted across arbitrary distances by harnessing entanglement, with inherent security guarantees against eavesdropping [9].

Another application of entanglement is superdense coding, where two bits of classical information can be transmitted by sending only one qubit, by encoding the bits into different preparations of an entangled qubit state shared between the sender and the receiver.

Entanglement swapping is another intriguing effect, where entanglement can be "swapped" between different particles that have never directly interacted. By performing joint measurements on two independently entangled pairs of particles, the resulting quantum correlations can be transferred to different pairs.

While seemingly paradoxical, these quantum effects have been experimentally verified many times and arise naturally from the mathematical formalism of quantum theory. Quantum entanglement and its resulting applications provide powerful new capabilities for cryptography and communications that are provably secure based on the laws of physics.

6.3.4 Quantum Uncertainty Principle

The next important quantum mechanical principle, with far-reaching consequences for quantum cryptography, is the uncertainty principle. For the sake of simplicity, the classical uncertainty principle states that you cannot know the exact (arbitrarily precise) position of a particle and its momentum (speed and mass) simultaneously.

If we know the exact position of a particle, then the momentum of the particle (speed and mass) becomes totally undetermined and vague, and vice versa. There is an inherent ambiguity or uncertainty in the simultaneous knowledge of these two properties of a particle beyond a certain limit.

The uncertainty principle can also be written in the form of an equation as follows: The product of the uncertainty in position multiplied by the uncertainty in momentum is always greater than or equal to one-half of the reduced Planck constant [9, 23]:

$$\Delta x \Delta p \geq \hbar/2$$

This is the cost factor: the more you determine one property, the more the other property becomes uncertain.

The quantum mechanical uncertainty principle, as applied to quantum cryptography, implies that any experimental access or attempt to measure or determine the quantum state of a particle, like a photon, will always disturb and alter its complementary variable or properties, such as momentum or polarization. This alteration is irreversible.

This is a fundamental limitation of making sharper and sharper measurements on the system without disturbing it and is one of the main reasons for the security of the quantum cryptography principles. It ensures that any measurement or eavesdropping on the quantum cryptographic states will always leave a trace or imprint, which can be detected by the legitimate users.

The uncertainty principle is also intimately connected with the no-cloning theorem (see previous section), since any possible perfect cloning of an arbitrary quantum state would allow one to measure the complementary variables with arbitrary precision, thereby violating the uncertainty relations.

6.4 Overview of Quantum Cryptography Protocols

A typical cryptographic system must offer encryption, decryption, and a key management system. A quantum cryptographic system can be hybrid, where we use QKD for key distribution and traditional cryptographic algorithms for the encryption and decryption process, or fully quantum, where we employ QKD along with quantum algorithms for encryption and decryption to protect information. In a quantum cryptographic system, we exploit the quantum states of photons, such as polarization, time-bin encoding, frequency, and phase, to ensure secure communication. In a quantum cryptographic system, a communication link can be an optical cable or a free-space link between parties to transmit photons with quantum information.

A cryptography system should offer the following features:

a) **Confidentiality:**
It involves protecting the private information that is being exchanged in a communication link between the parties involved in the information exchange. This involves converting the sensitive information into a garbled format that is hard to decipher by a third person who is listening to the communication exchange between the parties. This process is also known as a man-in-the-middle or interception attack, where a malevolent third party tries to eavesdrop on the communication link.

b) **Authentication:**
Let us take the example of Alice and Bob. If Alice receives a message from Bob, how can Alice clearly know that the message is legitimate and sent by Bob? This is where authentication comes into the picture. Prior to sending any secure information, a cryptographic system needs to provide a means of confirming the participants' identities. Traditional cryptographic systems do this using algorithms such as MAC, SHA, and MD5. The process of forging one's identity and sending a message on their behalf is called fabrication. This attack aims to compromise [12] the legitimacy of UAVs. It provides false information, allowing access to UAV components.

c) **Integrity:**
Integrity is a way of verifying that information exchanged between parties is not being modified in transit. It provides a process for intercepting and changing the content of the information in a communication link, which is called modification.

6.4.1 Quantum Key Distribution (QKD)

In general, a quantum cryptographic system is considered more secure and safe compared to a traditional cryptographic system. What separates a traditional cryptographic system from a quantum cryptographic system is its non-orthogonality. There is a possibility that, in quantum cryptography, we may obtain non-orthogonality in cipher states; thus, it is difficult for an attacker to tell them apart.

For instance, consider the polarization of photons, which can be used to encode quantum bits (qubits):

- Horizontal ($|H\rangle$) and vertical ($|V\rangle$) polarizations are orthogonal. These are like arrows pointing exactly up and exactly to the right. They are perfectly distinct from each other and can be easily distinguished.
- Diagonal ($|D\rangle$) and anti-diagonal ($|A\rangle$) polarizations are nonorthogonal to $|H\rangle$ and $|V\rangle$, and also to each other. These are like arrows pointing slightly up-right and slightly down-left. They are not perfectly distinct, so it's hard to tell them apart precisely.

Between 2002 and 2007, the DARPA Quantum Network hosted the first QKD deployment, indicating [38] QKD's practical usage. They were able to seamlessly integrate it with existing internet infrastructure, which enabled it to use QKD-derived keys for VPN and IPsec architecture.

6.4.1.1 General architecture of QKD system

6.4.1.1.1 Quantum Source

A quantum source is a point where quantum states or particles are generated, or one can say originated. These sources produce the quantum bits (qubits) that carry information in quantum communication systems. Examples of quantum sources include:

Single-Photon Sources: Devices that emit one photon at a time, often used in QKD.

Entangled Photon Sources: Devices that produce pairs or groups of entangled photons. The beauty of entangled photons is that they remain intrinsically linked, with their quantum states being directly correlated. This means that a change in the state of one photon instantaneously affects the state of the other, regardless of the physical separation.

Quantum Dots and NV Centers: Semiconductor and diamond-based systems that can emit single photons or entangled pairs.

6.4.1.1.2 Quantum Channel

A quantum channel is the medium through which quantum information is transmitted from the quantum source to the quantum sink. This channel needs to preserve the quantum states during transmission, which can be challenging due to potential disturbances and losses. Quantum channels include:

Optical Fibers: Commonly used for transmitting photons over long distances in QKD and other quantum communication systems, optical fibers are favored for their low loss and ability to guide light efficiently.

Free-Space Links: Communication through the atmosphere, which is useful for satellite-based quantum communication. Free-space links are important for establishing global-scale quantum networks.

Quantum Repeaters: A quantum repeater functions similarly to a regular network repeater, as both help mitigate losses and errors in the transmission channel. However, in quantum communication, the repeater works by entangling photons over shorter distances and then connecting these segments through a process called entanglement swapping.

6.4.1.1.3 Quantum Sink

A quantum sink is the endpoint where quantum information is received and processed. The quantum sink includes the devices and systems that measure, decode, and utilize the incoming quantum states. Components of a quantum sink include:

Photon Detectors: Devices that detect individual photons and measure their quantum states, such as polarization or phase. High-efficiency, low-noise detectors are crucial for accurate measurement.

Quantum Memory: Systems that store quantum information temporarily, allowing for synchronization in quantum communication and computation processes.

Quantum Computers and Processors: Devices that perform operations on qubits, executing quantum algorithms for tasks such as decryption, error correction, and data analysis.

Generally, QKD protocols fall into three types:

a) **Single-Photon-Based QKD**
This includes protocols such as BB84, where we use polarized states of single photons. Here, in this type of QKD, we rely on the transmission of single photons that are prepared in specific quantum states. The lack of need for quantum memory makes it more suitable for practical purposes.

b) **Entanglement-Based QKD**
One such example of this type of QKD is the E91 protocol; it relies on the phenomenon of quantum entanglement. Here, both Alice and Bob are provided with a single entangled photon. The beauty of it is that, no matter the distance, the entangled particles are always correlated.

c) **Continuous Variable QKD**
This is different from what we saw earlier, where we used discrete states – that is, polarized states of photons – but in continuous variable QKD, we use the quadrature of the electromagnetic field to encode and transmit information. Gaussian modulated coherent state protocol is one such example.

We will investigate each of them in detail, along with various QKD protocols in Section 6.5.

6.4.2 Quantum Secure Direct Communication (QSDC)

Unlike QKD, quantum secure direct communication (QSDC) is a means of sharing information without the need for a private key, thus eliminating the [24] need for a key management system. This algorithm acts as a stepping stone for other protocols, such as quantum signature. This QSDC protocol holds immense potential to change the face of the current communication architecture. There are a lot of QSDC protocols that have been proposed lately; the most practical one to implement is the DL04 protocol, which is rooted in using single photons. The lack of proper quantum memory to retain the entangled particles for a longer duration and with high fidelity makes it harder to implement entanglement-based QSDC protocols in practical usage. Another problem that affects other protocols in QSDC is the difficulty in implementing error correction coding for a system with high channel loss and noise.

6.4.2.1 General Working of DL04 Protocol

We take our usual example of Alice, Bob, and Eve. Here, Alice and Bob are two parties who want to exchange quantum information via a communication channel that is susceptible to interference by Eve, who can eavesdrop on the information sent back and forth through the channel. A note here: DL04 depends on single photons or qubits to exchange information. This type of DL04 requires quantum memory to store qubit information.

Preparation of State by Bob:
In UAV communication, we use free-space communications, so we employ polarized states of photons. We have four polarized states: $|0\rangle$ and $|1\rangle$ in the Z basis, and $|+\rangle$ and $|-\rangle$ in the X basis. We then encode the photon by applying unitary matrices. Bob can use unitary matrices to encode information into these states. For example, Bob can apply the identity matrix I for a bit value '0' and a specific unitary matrix U for a bit value "1."

Sharing of Qubits to Alice:
Bob later sends these qubits via a quantum channel to Alice.

Alice Random Check:
Alice arbitrarily picks a few of the qubits she received and measures their states using two possible methods (called bases X and Z). She tells Bob which qubits she measured, which method she used, and the results via a communication channel.

Security Check by Bob:
Bob receives Alice's results; if this matches with the states which were initially prepared by Bob, then this indicates the communication link is not compromised. Otherwise, it suggests Eve might be eavesdropping. If the error rate (how many mismatches there are) is too high, Alice and Bob stop the communication because it's not safe.

Encoding the Message:
If the communication is secure, then moving on to the next step, Alice encodes her message (a sequence of bits, 0s and 1s) into the remaining qubits, which she did not measure. If Alice wants to send a "0," she performs no operation on the qubit. But if she wants to send a "1," she modifies the qubit's properties in a special way that only she knows. This way, the message stays hidden within the qubits themselves.

Sending Back to Bob:
Then these modulated bits are sent back to Bob.

Bob Decodes the Message:
Bob retrieves Alice's message by measuring the qubits he has received. He can accurately reassemble the message if the quantum states match.

Remarks:
Although the need for quantum memory in DL04 protocols calls for a better version of this protocol, called the QMF-DL04 QSDC protocol, which has better error correction formalities and eliminates the need for quantum memory.

6.4.2.2 Quantum Memory-Free DL04 QSDC Protocol

Overview
QSDC is an emerging subfield that combines quantum mechanics and classical information theory for secure communications. Although the DL04 protocol is a milestone in the field of QSDC, it also encounters some problems, especially the use of quantum memory. The QMF-DL04 [33] protocol requires no quantum memory and provides an exact estimation of secrecy capacity. It also operates at a speed that is nearly equal to the instantaneous quantum channel capacity. Therefore, the QMF-DL04 protocol is a significant improvement in quantum secure communications.

Notations and System Description
QMF-DL04 QSDC protocol uses some important notations to perform the task, where Alice and Bob communicate over a quantum channel aided by a classical service channel. The data stream consists of the following symbols:
- **M**: Sequence of information bits encoded by the joint error-entropy coding (JEEC) module.
- **K**: Key for JEEC encoding and decoding, generated from the key pool.
- **Y**: Ciphertext is generated by the XOR operation $Y = M \oplus K$.
- **LDPC Encoder**: Low-density parity check (LDPC) encoder with code length k and rate Rp.
- **X**: Codeword generated by the LDPC encoder.
- **Xi**: Sub-block of X or a random bit string used as the input for secure coding in the ith frame.
- **Ci**: Codeword output from secure coding is transmitted through the quantum channel.
- **Cmi, Cwi, Csi**: Values observed by Alice and Bob during the transmission.

LDPC Encoder Structure

Prior to discussing the QMF-DL04 QSDC protocol in detail, it is worth noting the structure of the LDPC encoder. LDPC codes are a class of error-correcting codes for transmitting information messages across a noisy communication channel. They were conceived by Robert Gallager in the early 1960s but were not implemented in practical systems until the development of fast, reliable, and inexpensive computers.

LDPC Codes

LDPC codes can be described by a sparse bipartite graph having variable nodes and check nodes. The variable nodes represent the bits in the codeword, while the check nodes represent the parity-check constraints. The girth of the graph (i.e., the length of the shortest cycle) is relatively small, thus ensuring that the encoding and decoding algorithms are efficient.

Encoding

For the QMF-DL04 QSDC protocol, the LDPC encoder encodes the sequence Y (ciphertext). The LDPC encoder operates on the input sequence Y to generate a longer codeword X. The codeword X includes redundant information to aid in detecting and correcting errors that might arise during transmission over the quantum channel.

Decoding

Decoding of LDPC codes is generally carried out by iterative algorithms, for example, the belief propagation algorithm, which exploits the graph structure by iteratively updating the probabilities of the transmitted bits. In each iteration, the probabilities are refined to provide better estimates of the transmitted bits. This process is repeated until a valid codeword is obtained (i.e., a codeword that satisfies all the parity-check constraints) or a predetermined maximum number of iterations is exceeded.

Protocol Structure

QMF-DL04 QSDC protocol consists of three phases: initialization, transmission, and key distillation. Each transmission phase, in itself, ensures secure transmission without quantum memory.

Key Distribution Phase

1. **First frame (i = 1) transmission**:
   - Alice sends the first frame, Ci, and the key pool is empty.
   - The precoding module is not activated; Xi consists of random bits from a random number generator.

- Alice and Bob select initial values Cw0, k1, R1, and Cm0 to satisfy the initial constraints.
- From Ci, Alice and Bob derive Cmi, Cwi, and a new secret key Si if the Ci transmission is successfully received.

2. **Following frames (i ⟩ 1) transmission**:
   - If the key pool does not have enough keys to encrypt M, the protocol reverts to sending random bits for Xi. The secure coding module adjusts the parameters to ensure security.
   - Otherwise, the following steps ensure secure transmission and key distillation.

Transmission Phase

For the transmission of following frames, the protocol executes the following steps:

1. **Encryption**: Alice encrypts the message M using the key K and obtains the ciphertext Y.
2. **Precoding**: The ciphertext Y is precoded into X using the LDPC encoder and stored in cache memory.
3. **Secure Coding**: For secure coding, Alice selects a k-bit sequence from the cache.
4. **Modulation**: The sequence Ci is modulated and transmitted through the quantum channel.
5. **Demodulation**: Bob demodulates the qubits and decodes the secure code.
6. **Key Distillation**: If the decoding is successfully completed, Alice and Bob derive new keys from the transmitted codewords.

Key Distillation

Key distillation is a general process in quantum communication protocols, especially in QKD and QSDC. It refers to the process where two distant parties (Alice and Bob) distill a secret and common key from correlated data that could have been partially intercepted and even distorted by an eavesdropper (Eve).

Universal Hashing

A general method for key distillation is universal hashing. A universal hash function is a mathematical mapping that takes arbitrary-length input data and outputs a fixed-length hash value in a manner such that any two different input values are unlikely to map to the same output. This property guarantees security because any part of the input data known by an eavesdropper does not imply anything about the output.

Toeplitz Matrix

Alice and Bob distill a key from the transmitted codeword Ci using a Toeplitz matrix in the QMF-DL04 QSDC protocol. A Toeplitz matrix is a matrix where each down-diagonal from left to right is the same. This matrix is used for hashing to determine the secure key from the received codeword. The key length Si is determined by the secrecy capacity Csi.

Error Handling and Decoding

What happens if the decoding algorithm fails? Bob alerts Alice through the classical authenticated service call, and Alice retransmits the encoded sequence with a fresh secret key. The decoding algorithm is run for a maximum number, Imax, of iterations. If the decoding fails, the decoder alerts the server for retransmission; if it succeeds, the decoded string X is stored and communicated for future encoding.

If the decoding algorithm fails, the decoder alerts the server for a retransmission. The stored string is then used for re-decoding. Note that the cache helps in cases where a few bits might be lost due to channel noise. However, they can be decoded reliably with the help of the side information. Hence, we use the cache with minimal effect on the system.

Secure Coding and Precoding

The precoding and secure encoding module is crucial for achieving the secrecy capacity. Without proper precoding, the key distillation rate would be severely limited. An LDPC encoder/decoder is employed due to its near-optimal performance in correcting errors for the noisy quantum channel. The secure encoding parameters Ri and ki are also adapted during the process to achieve optimal performance.

Key Management

For Alice and Bob to be able to stay synchronized, the key pool is considered a FIFO memory, with a fixed number of keys that always remain the same, so that both parties constantly update their pool of keys and are always synchronized on the keys to be used for proper and correct communication.

Differences Between QSDC and QKD

QSDC and QKD are similar in that both make use of information-theoretic security methods, but they are different in nature:

- **QSDC**: Keys are distilled from codewords that carry both the ciphertext and the key, thus requiring only a single transmission phase.
- **QKD**: Keys are distilled from random bits that are transmitted separately from the ciphertext, which, in turn, is sent in a separate classical communication phase; thus requiring two transmission phases.

Advantages of QMF-DL04 QSDC

The QMF-DL04 QSDC protocol offers several advantages:

1. **No Quantum Memory Needed**: The requirement for quantum memory is difficult to implement and adds to the cost of hardware. Our protocol can dispense with it.
2. **Accurate Estimation of Secrecy Capacity**: The level of confidentiality of the quantum channel should be accurately estimated in real-time so that the encoder and decoder can distill the maximum amount of keys from the available signal sequences.
3. **High Transmission Rate**: The protocol operates close to the instantaneous quantum channel capacity, making efficient use of the available time.
4. **Robust Error Correction**: LDPC codes and iterative decoding make the protocol reliable and error-resilient.
5. **Dynamic Key Management**: The FIFO key pool guarantees that Alice and Bob use keys consistently and synchronously.

Conclusion

The QMF-DL04 QSDC protocol provides significant progress in QSDC. The QMF-DL04 QSDC protocol overcomes the challenges of DL04 protocol by eliminating the need for quantum memory and offers an efficient estimation of secrecy capacity. LDPC coding achieves robust error correction capability, iterative decoding delivers strong performance for LDPC coding over a quantum channel, and dynamic key management fully utilizes the unique characteristic of the quantum key, wherein the used key is random and never reused. This chapter presents the detailed procedure of the QMF-DL04 QSDC protocol, which ensures efficient and reliable QSDC. Compared to previous protocols, QMF-DL04 introduces the following innovations and advantages.

6.4.3 Quantum Teleportation-Based Protocols

6.4.3.1 History

Entanglement is the backbone of quantum teleportation protocols. The first successful quantum teleportation was achieved by Austrian teams led by Anton Zeilinger in 1997. The experiment established entanglement between two photons, subsequently achieving the transfer of their quantum state through a [8] measurement process. This was just a starting point. Later, a new speed record of 7.1 qubits per second for quantum teleportation was achieved by a team led by Prof. Guangcan Guo and Prof. Qiang Zhou from the University of Electronic Science and Technology of China (UESTC). This experiment was carried out with three nodes – Alice, Bob, and Charles – which were separated by a metropolitan distance and connected using optical cables. NASA demonstrated what is probably the longest teleportation of qubits of photons over a 44 km fiber optic network. This proves the capability of quantum networks over very large distances.

6.4.3.2 Common Misunderstanding

Although there is misunderstanding, there is no communication channel between the parties involved in the information exchange; however, there should be one present to share entangled photon pairs and classical communication (such as a phone call or an email) to share Bell-state measurements, which are the output of quantum teleportation protocols.

6.4.3.3 How Quantum Teleportation-Based Protocols Work

1. **Preparation of Entanglement Pair:**
   Let us take the example of Alice and Bob. First, a pair of entangled quantum particles, *A* and *B*, is generated. Because of their entanglement, one particle's state is always directly correlated with the other's, regardless of how far apart they are. The state of this entangled pair can be represented as

   $$|\Phi^{+}\rangle_{AB} = 1/\sqrt{2} \, * \, (|00\rangle_{AB} + |11\rangle_{AB})$$

2. **Alice Quantum State:**
   Alice has particle *C*, whose quantum state she wants to teleport to Bob. This state is unknown to her but is described as $|\psi\rangle$. This might be misleading, but one must understand that the aim of quantum teleportation is to transmit the quantum state of particle *C* to Bob without physically sending the particle itself:

   $$|\Psi\rangle_c = \alpha|0\rangle_c + \beta|1\rangle_c$$

3. **Entanglement Sharing:**
   Then there is an exchange of an entangled pair of particles between Alice and Bob. Alice has particle *A*, and Bob has particle *B*. Remember, this exchange of the entangled pair takes place through a communication medium.
4. **Bell-State Measurement:**
   Alice performs a joint measurement on particles *A* and *C* in a specific basis called the Bell basis. This measurement entangles particles *A* and *C*, projecting them onto one of four possible entangled states known as Bell states.

   The result of this measurement collapses the states of particles *A* and *C* into one of these Bell states and simultaneously changes the state of Bob's particle *B* due to the entanglement.

5. **Classical Communication:**
   This result of the Bell-state measurement is sent to Bob using a classical communication channel (e.g., an email or a phone call). This is to inform Bob about the state of particle B after Alice's measurement.
6. **Unitary Transformation:**
   Upon receiving the classical information, Bob performs a specific unitary transformation on his particle B to convert it into the state $|\psi\rangle$.
   After this transformation, Bob's particle B is now in the same quantum state as Alice's original particle C, completing the teleportation process.

6.4.3.4 QITP/QATP

We have understood how quantum teleportation works. Now, let us explore the potential it holds. We often use UAVs for surveying geographical areas and capturing images; it is also used for reconnaissance purposes. Audio surveillance can be employed to capture wildlife sounds for eco-monitoring purposes and to listen for signs of life during search and rescue operations as part of disaster response. Recent developments have enabled us to exchange images as well as audio via quantum teleportation algorithms using quantum image teleportation (QITP) and quantum audio teleportation protocol (QATP). There are two versions of the same algorithm available: one with Huffman coding and one without Huffman coding. Here, we discuss QITP with Huffman coding, as Huffman coding helps reduce the size of information to be transmitted, thereby decreasing the need for bandwidth.

Working of QITP with Huffman Coding [19]

Step 1: Read the image in any standard format (.png, .jpg, etc.), and this is further converted to RGB format, where RGB values for each pixel are extracted.
As a following step, each RGB value is converted to an 8-bit binary string. Example: (255,0,0) -⟩ (11111111,00000000,00000000) [19].
Step 2: Then we apply the Huffman coding compression technique for the compression of data by determining the frequency of each 8-bit binary string. This compressed data is later saved in a .bin file.
Step 3: The number of accessible qubits determines how the compressed binary data is divided into sub-binary strings. Every substring can be accommodated within the qubits' teleportation capability.
Step 4: Assume Alice is the (UAV) and Bob is the (GCS); both share a pair of entangled qubits. Alice performs measurements on her qubits and the binary sub-strings, sending the results via a classical channel to Bob. Bob uses the received measurement results to reconstruct the original quantum states, effectively receiving the binary sub-strings.

Step 5: Bob then translates the given binary sub-string to the original 8-bit binary strings and then again converts the 8-bit binary strings into the corresponding RGB values. Thus, Bob (GCS) can reform the original image.

Working of QATP with Huffman Coding [19]
Step 1: Audio is collected in a standard format such as (.mpeg, .wav, .mp3, etc.). Then, this is converted into a time-series array of samples, which can be mono or stereo, with the default sampling set to 22,050 Hz.
Step 2: we convert the audio sample array into positive values. These values are then saved in .txt format, separated by a special character to avoid discrepancies during decoding at the receiver's end.
Step 3: We apply Huffman coding to the frequency of each character in the text file. Then, this compressed data is converted to .bin format.
Step 4: Then, this binary string is divided into sub-binary strings based on the available qubits for Alice, Bob, and the teleportation intermediary (Telamon).
Step 5: The quantum states of the binary sub-strings are teleported using a quantum circuit. Alice (sender) and Bob (receiver) share entangled qubits. Alice measures her qubits and the binary sub-strings, sending the measurement results via a classical channel to Bob. Bob uses the received measurement results to reconstruct the original quantum states, effectively receiving the binary sub-strings.
Step 6: Reverse operation is being done at GCS, where sub-strings are converted back to binary format, and then, with the help of special characters, we convert it back to text format.
Step 7: After that, the characters are transformed back into their respective positive integer values, reassembling the initial audio sample array. At the receiver's end, the original audio is restored by reassembling the audio sample array into the original audio file format.

6.5 Quantum Key Distribution for UAV Communications

6.5.1 QKD Principles and Mode of Operation

QKD is a quantum mechanics-based approach to securely transferring encrypted data between two trustworthy parties [29]. QKD can assist in alleviating some of the security issues in today's telecommunications networks and will be essential in future UAV communications. The aim behind QKD is to employ quantum mechanics principles to produce and transmit cryptographically safe keys for further encryption.

6.5.1.1 Quantum Principles in QKD

Quantum mechanics principles that enable QKD are mainly superposition and entanglement.

Quantum Superposition

Superposition makes it feasible to [28] encode the statistics effectively in quantum states, which may be measured without distortion. For example, a single photon may be in a combination of horizontal and vertical polarization states. In various scenarios, the bit “0” is stored in horizontal polarization, and the bit “1” in vertical polarization. This type of encoding is secure against any form of eavesdropping.

Quantum Entanglement

When two particles are entangled, measuring the state of one particle at once determines the state of the other, presenting a manner [28] to detect any eavesdropping during the transmission. Entangled photons may be used to create correlated keys between the two events, which cannot be measured by an eavesdropper without being detected.

6.5.1.2 QKD Mode of Operation

The steps defined below make the most of quantum mechanics principles to carry out stable key distribution. The most common QKD protocol is the [9, 26, 39] BB84 protocol, named after the initials of its inventors, Charles Bennett and Gilles Brassard, in 1984.

Protocol description

1. **Raw Key Transmission**: A photon is created by Alice in one of four polarization states – either vertical ($|V\rangle = |1\rangle$), horizontal ($|H\rangle = |0\rangle$), diagonal ($|D\rangle = (|0\rangle + |1\rangle)/\sqrt{2}$), or anti-diagonal ($|A\rangle = (|0\rangle - |1\rangle)/\sqrt{2}$) in the BB84 protocol. These states are then chosen randomly by Alice and delivered to Bob over some quantum channel through which they may travel; this channel could be free space, an optical fiber, or an optical fiber channel.
2. **Sifting**: Bob chooses the polarization of photons in the quantum channel randomly by taking them from Alice's source state. Instead of communicating the exact outcomes of his measurements, Bob reveals the basis employed in each case. Alice then sends Bob information about her measurements, indicating which of her photons had been prepared (unbeknownst to him) using the same

basis as Bob used for measuring his own. Alice and Bob have announced that they sifted photons by rejecting photons measured in different bases.
3. **Error Correction**: Due to noise in the quantum channel and imperfect measurement, QKD sifted keys contain some errors. Alice and Bob use regular error-correcting codes to remove these. It is these errors that Alice and Bob correct using classical error-correction codes. To gauge how wrong it is, they publicly share portions of their own sifted keys, and after rectifying, the two keys are similar.
4. **Privacy Amplification**: Alice and Bob apply privacy amplification to remove any information a possible eavesdropper (Eve) could have about the key. Any measurement on a photon collapses it, which is an idea exploited by privacy amplification. So, Alice and Bob truncate the final corrected key in such a way that any knowledge Eve could have obtained is completely removed. This gives the final shared key, which is immune to any attack.

6.5.1.3 Advanced QKD Protocols

Apart from the BB84 protocol, many other QKD protocols have been discovered to improve security and efficiency, as compared in Table 6.1. These include:

B92 Protocol

This was suggested by Charles Bennett in 1992. It is essentially the [41] BB84 QKD protocol, with one instead of two non-orthogonal bases being used during its encoding process. Thus, B92 is easier to experimentally realize.

Protocol Description

1. **Raw Key Transmission**: The transmitter (Alice) considers a quantum source SB92 = $\{|\varphi\rangle, |\psi\rangle\}$, with $|\varphi\rangle$ and $|\psi\rangle$ being two non-orthogonal states. Alice also creates a truly random string $Rc = \{rc\_i \mid i = 1, 2, \ldots, n\}$, with $rc\_i \in \{0, 1\}$. Alice encodes each random bit with the appropriate quantum state from SB92, depending on the values of each Rc element. For instance, she may select $0 \to |\varphi\rangle$ and $1 \to |\psi\rangle$. This generates a sequence of random qubits Rq = $\{rq\_i \mid i = 1, 2, \ldots, n\}$, with $rq\_i \in$ SB92. Finally, Alice transmits (one by one, with a time delay $\Delta\tau$) each qubit of the generated sequence Rq to the receiver (Bob) through the quantum channel. Usually, this channel is a free-space link (or optical fiber in the case of long-distance communication) between Alice and Bob. When Bob receives each qubit, he measures it using the positive operator-valued measurement (POVM) technique. He employs two measurement bases, $P(\psi)$ and $P(\varphi)$, related to the non-orthogonal states $|\psi\rangle$ and $|\varphi\rangle$, respectively. As is known, any POVM can achieve either a decisive outcome (related to $|\psi\rangle$ or $|\varphi\rangle$) or an inconclusive outcome.

2. **Sifting**: It is Bob's action of notifying Alice about which qubits have been measured incorrectly. They will then remove these parts from the respective strings each of them holds.
3. **Error Estimation**: A fixed proportion of remaining bits incurs a cost to both Alice and Bob. They will then measure both bits and compare them to check for equality. If this error rate is greater than a fixed threshold, they abort and assume that eavesdropping has occurred in the quantum channel. Otherwise, they proceed with the final step.
4. **Key Distillation**: Alice and Bob apply classical post-processing (e.g., error correction codes and privacy amplification) to the remaining secure bits to produce a final secret key.

Ekert (E91) Protocol

This was presented by Artur Ekert in 1991. It is a method for [9, 39] QKD based on quantum entanglement and Bell's inequality. It enables two parties, traditionally called Alice and Bob, to establish a shared secret key.

Protocol description

1. **Creation of Entangled Qubit Pairs**: Alice and Bob are provided with a supply of entangled qubit pairs. Typically, the entangled state is $||\Phi+\rangle = (|00\rangle + |11\rangle)/\sqrt{2}$. This entangled state can be supplied by a third party or may be created by Alice and Bob themselves (this step depends on the specific implementation).
2. **Measurement of Qubits**: Alice and Bob receive one qubit from each entangled pair. They measure their qubits on a basis they select randomly and independently (for each qubit). The set of viable bases consists of, for instance, the rectilinear foundation or the diagonal foundation. Alice and Bob take measurements of their respective qubits in the appropriate bases and record what they observe on the spot.
3. **Basis Reconciliation**: After measuring all qubits available, Alice and Bob make use of a secure classical communication channel where they disclose the bases in which every qubit was measured. Afterward, they sift, which is a process of eliminating dimensions for which the bases are extraordinary. The outcomes for which the bases are the same are accrued within the sifted key.
4. **Test of Bell's Inequality**: The previously excluded measurement outcomes (referred to as the rejected key) are now repurposed for verifying the presence of an eavesdropper (Eve) through the examination of Bell's inequality. If the detected breach exceeds a particular threshold, it signals the presence of an eavesdropper, and the protocol is immediately terminated. Otherwise, the procedure continues with error correction and privacy enhancement.

5. **Error Correction and Privacy Amplification**: Alice and Bob are exploring classical error correction codes to fix discrepancies in their sifted keys caused by noise and imperfections in the quantum channel. Then, they employ privacy amplification techniques (for instance, conventional hashing) to eliminate any possible information about the key obtained by Eve during the transmission. The outcome of this step is the final secure key shared between Alice and Bob.

Decoy State Protocols

One practical issue with QKD is that attenuated laser pulses [9, 39], known as weak coherent states (WCSs), must be used instead of true single-photon sources. WCSs are much easier to produce, but they suffer from an inherent flaw that makes the resulting QKD system vulnerable to an attack known as the photon-number splitting (PNS) attack.

Decoy-state protocols were devised to solve this problem and to provide a workable solution to QKD with WCS sources. The additional states, called decoy states, are used alongside the normal signal states to add an extra layer of security to the protocol.

Protocol Description

1. **Signal and Decoy State Preparation**: Alice prepares decoy states, which are signal states usually having mean photon counts less than μ, besides actual signal states with the mean photon number μ. Just as it has been exemplified in the pulse sequence handed to Bob, the signal states and the decoy states alternate.
2. **Quantum Transmission and Measurement**: Bob receives the pulse sequence through the quantum channel, which is created from the signal construct and the dummy construct created by Alice. Bob remains uncertain whether a pulse is a signal or a decoy state, even though he must always measure each signal pulse in its correct basis in accordance with the BB84 protocol.
3. **Basis Reconciliation and Sifting**: Encoded and announced through an authenticated classical channel, Alice and Bob inform each other about the adopted bases for each pulse. The bases used for determining the result are different, so the non-sifted key is eliminated in the process of superimposing the sifted results.
4. **Decoy State Analysis**: Alice and Bob must conclude that Eve, an external malicious user, is conducting a passive eavesdropping (PNS) attack based on the features of the decoy-state. This implies that if Eve is splitting multiphoton pulses, the statistics for decoy-state detection will differ from the actual statistics, indicating Eve's existence.
5. **Error Estimation and Key Distillation**: Alice and Bob estimate the error rate and incorporate it into the BB84 protocol to reconcile and distill the key, with final authentication on the secret key, since the decoy-state analysis does not contain information regarding eavesdropping.

Measurement Device-Independent QKD (MDI-QKD)

Measurement device-independent [9, 39] QKD (MDI-QKD) refers to a family of QKD protocols very recently proposed as a way to extend device-independent QKD beyond its limits when an imperfect measurement device is considered. In most of the standard QKD protocols (e.g., BB84 and E91), the protocol's security relies on the trustworthiness of the measurement devices used by the honest participants (Alice and Bob). Typically, we model the measurement device as perfect and assume there are no implementation vulnerabilities. In practice, the measurement device might have side channels that could be attacked or have implementation flaws that could be exploited to break the protocol.

MDI-QKD approaches the problem differently and allows a third participant, typically named Charlie, to perform the quantum measurement. This is known as the entanglement-based (EB) protocol. Charlie is an untrusted party, and his measurement devices do not need to be trusted. Charlie can never access the information in the key shared between Alice and Bob, regardless of how the message passes through or how he behaves, even in cases where he is entirely malevolent.

Protocol Description

1. **Quantum State Preparation**: Alice and Bob separately prepare quantum states, commonly polarized photons, and then send them through different quantum channels to Charlie.
2. **Bell State Measurement**: Charlie will make a Bell state measurement on Alice and Bob's incoming quantum states; he will perform a joint measurement called a Bell state measurement (BSM) that brings their combined state into one of the four Bell states, namely, $|\Phi^{+}\rangle$, $|\Phi^{-}\rangle$, $|\Psi^{+}\rangle$, $|\Psi^{-}\rangle$.
3. **Measurement Announcement**: Charlie announces the outcome of the BSM measurement (i.e., which Bell states the pairs of states collapsed to) publicly. However, Charlie does not give any information about the measurement outcomes of the individual states Alice and Bob sent.
4. **Sifting and Key Distillation**: Based on Charlie's BSM outcome announcement, Alice and Bob sift their raw key data, retaining only the samples where both their respective states would have collapsed to the announced Bell state. They then estimate errors, correct them, and apply privacy amplification to generate the final key, as in other QKD protocols.

Table 6.1: Comparison of quantum key distribution protocols for UAV communication systems.

Protocol	Security	Advantages	Limitations	Applications in UAV communications
BB84	Heisenberg's uncertainty principle No-cloning theorem	Groundbreaking, demonstrated feasibility of QKD Unconditional security	Requires two mutually unbiased bases. Lower efficiency due to sifting.	Secure command and control links Secure data transmission Secure swarm UAV communication Secure coordination in civilian applications
B92	No-cloning theorem Heisenberg's uncertainty principle Non-orthogonality of states	Simplicity Requires only one basis	Lower key generation rate compared to BB84	Secure key distribution between UAVs and ground stations. Secure communication between UAVs.
E91	Quantum entanglement Bell's inequality test detects eavesdropping	Device-independence Robust eavesdropping detection Theoretical simplicity	Experimental challenges Limited key generation rate Computational overhead	Secure command and control links Secure data transmission Secure swarm UAV communication Resilience to implementation flaws
Decoy state	Photon-number statistics analysis No-cloning theorem Quantum uncertainty principle	Practical implementation with weak coherent sources. Improved security against photon-number-splitting attacks Increased key generation rate	Increased complexity Trade-off between security and efficiency Vulnerability to other attacks	Secure command and control links Secure data transmission Secure swarm UAV communication Scalability in large UAV networks
MDI-QKD	Measurement device independence No-signaling principle No-cloning theorem	Immunity to measurement device vulnerabilities Scalability with star-like network architecture Compatibility with various quantum sources	Reduced key generation rate Increased complexity Vulnerability to other attacks	Secure command and control links Secure data transmission Secure swarm UAV communication Scalability in UAV networks

6.5.2 Implementation Considerations for UAVs

Even though the fundamental theories of surveillance via drones are well known, there are some challenges that arise due to the changeable nature of UAVs. In deploying QKD in UAV networks, several things are to be considered:

6.5.2.1 Mobility and Line-of-Sight Constraints

UAVs are highly dynamic platforms that have different trajectories, sizes, and velocities. The changing environment and dynamic [7, 15–17] platforms make the LOS variations for all wireless communication links fluctuate. This becomes a major problem when discussing the establishment of secure quantum communication systems on UAVs. During flight, reliable methods should exist to protect the quantum coherence and synchronization of these systems.

LOS Communication: LOS between the transmitter and receiver is required for QKD to accurately transmit quantum states. UAV platforms must be equipped with high-performance acquisition, pointing, and tracking (APT) systems capable of maintaining the optical link alignment despite platform and atmospheric disturbances.

Dynamic Flight Paths: Although the flight paths of many UAVs are highly dynamic and difficult to predict, they bring with them the challenge of how to ensure wireless communication links between UAVs and ground stations remain stable. Control programs and real-time feedback need to be developed, which will enable changes in the LOS to be compensated for by quickly controlling the pointing and tracking systems.

6.5.2.2 Size, Weight, and Power Constraints

UAVs usually come with strict size, weight, and power (SWaP) limits. In order to fit well into UAV platforms, all QKD hardware needs to be designed with these considerations in mind.

Miniaturization: UAVs have limited payload capacity. QKD hardware, including photon sources, detectors, and optical components, must be miniaturized to ensure proper integration. Recent advances in photonic integrated circuits (PICs) and micro-optical elements can be leveraged to achieve the required scale-down.

Power Efficiency: UAVs extract power from batteries or on-board generators, while QKD equipment must be power-efficient for longer operation periods without extra battery consumption. In order to achieve maximum efficiency (especially where long flight duration is required), the whole system (APTs and RF communication facilities) must operate on low power reserves.

6.5.2.3 Environmental Resilience

UAVs perform in different climatic conditions, such as widely varying temperatures, humidity levels, and electromagnetic interferences. The entire QKD infrastructure should be constructed to take account of these factors, making sure that it operates safely.

Temperature Variations: Variations in temperature can impact the functioning of optical and electronic components on the UAV and in the QKD. That is why thermal management systems are needed, whether they are heat sinks or temperature-compensated circuits, to ensure a sustained regular operational pattern.

Vibrations and Shock: As UAVs fly, they are subject to vibrations and shocks, which affect the alignment of optical components within them, so vibration-damping materials, together with shock-absorbing mounts, can help prevent such occurrences, and this provides a basis for quantum-stable communication.

Electromagnetic Interference: The quantum communication channel can be subject to electromagnetic interference from electronics located in the-plane or even from its surrounding environment. By properly combining shielding methods and optimizing the QKD system layout to make it withstand such conditions.

6.5.2.4 Integration with Communication Systems

For seamless operation, it is crucial to integrate the QKD technology with the existing UAV communication systems. When it comes to communication security, compatibility, synchronization, and data transfer rates should be carefully examined.

Hybrid Communication Systems: UAVs typically use a combination of radio frequency (RF) and optical communication systems. Integrating QKD into these hybrid systems requires careful synchronization of quantum and classical communication channels to facilitate the exchange of sifting, error correction, and privacy amplification information.

Data Transfer Rates: The data transfer rates of the QKD system must be compatible with the overall communication system of the UAV. Balancing the quantum key generation rate with the data encryption and decryption processes is essential to maintaining efficient and secure communication.

6.5.2.5 Scalability

As UAV networks evolve and grow in size, the QKD system must be scalable to accommodate the increasing number of UAVs and communication endpoints, while maintaining security and efficiency.

Network Topology: The QKD system should support various network topologies, including point-to-point, star, and mesh configurations. This flexibility allows for efficient key distribution in different UAV network architectures.

Key Management: In a large UAV network, the per-UAV cryptographic key renewal and distribution can pose substantial management challenges. Either a centralized key management system or decentralized solutions, such as blockchain-based key distribution, can be deployed to tackle this issue and ensure efficient and secure key handling.

6.5.3 Advantages and Limitations of QKD in UAV Networks

QKD has a number of advantages that make it very attractive for integration into UAV networks to provide secure communications for unmanned aerial systems. In addition, there are some challenges that need to be taken into account.

6.5.3.1 Advantages of QKD in UAV Networks

Quantum Security: QKD is immune to quantum hacking or potential eavesdropping because of quantum mechanics and the unique properties of researchers, while classical cryptography merely addresses certain mathematical problems where the relationships between them and computers are not clear and does not take into account or play an important role in the subject.

Key Distribution: QKD can be used to distribute cryptographic keys between UAVs, so that sensitive data can be encrypted before being transmitted. Only the authorized parties (i.e., the UAVs in our case) will be able to decrypt the data. This provides a mechanism to ensure the confidentiality and integrity of the communications between the UAVs in the network.

Detection of Eavesdropping: QKD can be used to discover if someone is trying to steal information being exchanged by UAVs. This would increase awareness within the drone network. By keeping track of the quantum error rates on their qubits, Alice and Bob could always see whether there is someone trying to eavesdrop on their communication. Countermeasures can then be employed against any potential threats.

Resistance to Quantum Attacks: As the power of these quantum computers starts to dominate, classical cryptographic systems will eventually be broken. Therefore, the sole option lying ahead is represented by QKD if communication security is at stake.

Enhanced Data Privacy: QKD is key to data privacy – it can be utilized by air drone drivers to obtain both originality and reliability of the data at a high level. The privacy of this data in transit will be compromised if either party to the communication channel between the drone and the ground control systems listens in on that stream. For UAVs used in military, government, or even economic endeavors, it is vital to protect data, and QKD offers just the right level of privacy and integrity that is demanded.

6.5.3.2 Limitations of QKD in UAV Networks

Complexity: QKD systems are usually complicated and require specialized tools and know-how to use. The additional infrastructure, together with the high level of specialization, can potentially increase the total cost of ownership. The alignment of the optical fibers is a critical step in the QKD deployment process, and sophisticated error correction and privacy amplification algorithms are also required.

Performance Challenges: The dynamics of the UAV platform may cause performance issues for QKD systems, such as latency or limitations in transferring the required amount of quantum key data, which can adversely affect the efficiency of communications between the UAVs. Real-time processing and synchronization of quantum key data with the classical data channel may also cause problems with respect to timing

Hardware Requirements: Additionally, the weight and complexity of the UAV may also increase as a result of the necessary hardware components for QKD deployment on the ground. The on-board photon sources, detectors, and optical elements (e.g., lenses) all need to be integrated, and this should not exceed the payload capacity of the UAV.

Scalability: QKD systems can be difficult to scale in large multi-UAV networks in terms of distributing the quantum keys and providing secure communications between all the UAVs in the network. To address this challenge, we need good key management protocols and efficient topologies.

Interoperability: Ensuring that QKD systems are compatible with existing UAV communication protocols is critical for smooth integration and efficient communication security. Challenges are associated with integrating QKD into different communication technologies or standards commonly used for UAVs.

6.5.4 Case Studies and Applications of QKD in UAVs

Through different case studies and applications, we have a clear picture of the benefits and challenges that could be encountered in the deployment of QKD in UAVs. Such scenarios are among the most critical ones where the usage of QKD in UAV communication is necessary.

6.5.4.1 Military and Defense Applications

QKD has been found to be very helpful in communication via UAVs for military and defense purposes. In the military today, there's a significant need for safe and reliable ways of communication, as the interception of vital information could be extremely dangerous.

Secure Battlefield Communications: QKD in drone-centric networks nowadays is more relevant for spying, reconnaissance, and tactical missions. This is achieved through the safe circulation of encryption keys, in addition to protecting them against interception. This inhibits crucial information from being exposed to wicked eyes or becoming a target for those who wish to paralyze warfare efforts.

Encrypted Data Transmission: When dealing with intelligence collection missions and espionage, UAVs often emit large amounts of data from high-quality pictures and streaming video that is actual video. For this reason, it is significant to utilize tasks that prevent illegal entry during chatting through them in order not to disclose confidential data. For assured transmission of classified documents, these tools necessarily presuppose QKD.

Command and Control Security: It's essential to prevent adversaries from taking control of the UAVs, and QKD can help achieve that. By using QKD to protect the C2 communication links, only authorized personnel can issue commands and receive status updates from the UAVs.

6.5.4.2 Government and Critical Infrastructure

Government agencies and critical infrastructure operators benefit from protecting sensitive information, and ensuring the integrity of communication networks is essential for national security and public safety.

Secure Surveillance and Monitoring: UAVs are often employed to keep an eye on crucial facilities, including power stations, transit systems, and public areas. QKD is useful in ensuring that secure communication channels are guaranteed for information gathered using these drones, in order to prevent possible acts like espionage and sabotage that may occur, with such resources being destroyed.

Disaster Response and Management: During a crisis, drones are useful when it comes to evaluating the level of damage, searching for survivors, and directing rescue efforts. For this to work, the UAVs require an unbreakable communication link between themselves and other terrestrial teams. The safekeeping of confidential information remains intact through QKD technology during sensitive missions.

Government Communications: The government utilizes UAV communication for different reasons, such as border surveillance, law enforcement, or environmental protection. QKD defends secure government communications over these networks against any unauthorized access to vital information.

6.5.4.3 Commercial and Industrial Applications

Various commercial applications require the assurance of security and privacy when transmitting data via UAVs, which is beneficial for the commercial and industrial sectors as well.

Secure Delivery Services: The mixing of drone transportation and its control centers raises the demand for ensuring that communication links are unbreakable between drones. The latter can be addressed via QKD with the intention of safeguarding confidentiality and the delivery process.

Industrial Inspections: UAVs are mostly used for inspecting industrial installations like pipelines, power lines, and offshore platforms as part of industrial inspections. It's also imperative to have a system that enables secure transmission of inspection data to the main offices by incorporating QKD to prevent any unauthorized persons from accessing this vital information.

Agricultural Monitoring: UAVs are used in precision farming to control crop health, soil condition, and irrigation systems. Data security of such UAV transmitted data is important in protecting proprietary agricultural information and avoiding data breaches. QKD can be a solution for enhancing agricultural monitoring data confidentiality and integrity through the creation of a secure communication channel.

6.5.4.4 Research and Development

Researchers and universities are performing experiments, developing prototypes, and demonstrating the capabilities of QKD applied to UAV communications. These research activities study and deepen the knowledge of QKD systems suitable for UAVs.

Experimental QKD Networks: Some research institutions have deployed experimental QKD networks based on UAVs in order to demonstrate the feasibility and performance of QKD in the aerial domain. These experimental activities are providing interesting insights into the challenges and potential solutions for the implementation of QKD in non-static and mobile scenarios.

Prototype Development: Universities and research laboratories develop prototypes of QKD-enabled UAVs, monitoring hardware and software configurations, which are employed to improve QKD systems' performance and reliability. Such prototypes represent proof that QKD can be employed in secure UAV communications.

Collaborative Projects: Collaborative research projects between academic and research institutions, government agencies, and industrial partners are propelling the development and deployment of QKD applied to UAV networks. These projects are tackling the technical challenges, enhancing scalability and exploring new applications of QKD in several sectors.

6.5.5 Future Directions and Challenges

As QKD technology is constantly changing, it is necessary to address a number of potentialities and challenges that will guarantee the successful deployment and global acceptance of QKD within UAV systems [21].

6.5.5.1 Advancements in QKD Technology

Integration with Emerging Technologies: When it comes to integration, it should be seen in the light of such emerging technologies as 5G networks, the Internet of Things, artificial intelligence, amongst others. One example in this regard is the encryption of communication signals between two points using QKD, so that signals cannot be intercepted by adversaries while they travel from the end-point user to the end-point system. A good instance of improved performance involves UAVs, whose operational safety is as important as their use for military purposes. It is crucial that standard protocols and interfaces for QKD integration with these technologies be developed for security solutions that will last into the future.

Improving Key Generation Rates: Enhancing the key generation rates of QKD systems is crucial for supporting high-speed and high-volume data transmissions in UAV networks. Photon sources, detectors, and error correction algorithms being developed are seen as the way forward for upping the pace at which new keys are generated.

Quantum Repeaters: Developing quantum repeaters is crucial in facilitating the distance a QKD system can cover, as well as for long-distance quantum communication to be made possible. This enables quantum repeaters to boost and resend quantum signals, ensuring untouched and secure communication over long distances, therefore.

6.5.5.2 Overcoming Practical Challenges

Environmental Adaptability: Enhancing the environmental adaptability of QKD systems is crucial for ensuring reliable operation in diverse UAV operating conditions. Corrosion-resistant hardware and adaptable software for QKD systems may be useful in combating problems caused by changing weather conditions, such as temperature fluctuations, shaking, or electromagnetic waves.

Cost Reduction: Cutting the cost of QKD hardware, as well as deployment, is fundamental to its wide penetration into commercial and industrial applications. This can be achieved through applying forward-looking methodology in manufacturing, making components smaller, and standardizing them.

User-Friendly Interfaces: A user-friendly interface for deploying and using the QKD system, with easy-to-use management tools, allows you to deploy and use the QKD system. User-friendly interfaces designed for key management, system monitoring, and troubleshooting will enhance not only the usability but also the accessibility of this technology.

6.5.5.3 Addressing Regulatory and Standardization Issues

Regulatory Compliance: Regulatory bodies require QKD systems to adhere to existing standards on secure communication and operation in drones where applicable; otherwise, deploying QKD technologies on drone networks would be wrong, possibly illegal. In order to integrate these technologies into UAV systems, developers and vendors need to partner with regulators to establish rules of engagement and standards that would govern their integration, among other considerations.

Standardization Efforts: Standardization is important as it encourages and promotes compatibility with various uses and designs of QKD protocols, hardware, and system integration interfaces within UAV communication architectures from different suppliers. It will also aid in the emergence of an international QKD ecosystem that fosters creativity and joint ventures.

Public Awareness and Education: The QKD community should continue to develop public outreach programs to educate and raise awareness of quantum cryptography and its potential. Workshops, conferences, and other outreach activities should be encouraged for industry, academia, and government to foster and develop investment in QKD technology.

6.6 Quantum Cryptography-as-a-Service for UAVs

6.6.1 What Is Quantum Cryptography-as-a-Service (QCaaS)?

Quantum cryptography-as-a-service (QCaaS) [30] provides a new way of securing communications using quantum mechanics. Classical cryptography depends on the hardness of a specific mathematical problem to be solved (e.g., factoring numbers). In contrast, quantum cryptography depends on the laws of physics and not on the hardness of a mathematical problem. Therefore, quantum cryptography has the property of future security; that is, it will remain secure even against future developments in computing (including quantum computers). This property of future security makes QCaaS an attractive solution for UAVs, since these drones are often used to transfer valuable data.

The fundamental concept in QCaaS is QKD which enables two communicating parties to create and share encryption keys securely. The security of QKD comes from quantum superposition and the no-cloning theorem. Essentially, any measurement on a quantized system (such as photon polarization) performed by an eavesdropper (Eve) on the key transmission will alter the statistics of the quantum system (such as photon phase). The alteration (or disturbance) of the quantized system notifies the legitimate users (Bob and Charlie) of a possible attack by Eve. Since the alteration (or disturbance) of the quantized system notifies the legitimate users of a possible attack by Eve, QKD remains secure even if large-scale quantum computers (which break conventional public key systems) are developed in the future. Therefore, QKD can be considered as future-secure.

QCaaS offers a framework for secure communications for UAVs. UAVs are frequently used in military, surveillance, and rescue scenarios, where data privacy and security are extremely crucial. QCaaS guarantees secure data links between UAVs and between a UAV and a ground station. Hence, QCaaS provides a practical solution for secure communications in UAV networks.

6.6.2 Integration Challenges and Solutions

Although QCaaS offers a practical solution for secure communications in UAV networks, several challenges need to be addressed before it can be implemented. Some of these challenges are listed below.

6.6.2.1 Miniaturization and Limited Energy Resources

The size of the current quantum communication equipment is still too large, and the power consumption is too high to be used in UAVs. Since UAVs are often restricted by their weight and power resources, any quantum communication equipment carried by a UAV must be miniaturized and made energy-efficient [25]. A lot of research is underway to develop ultrasmall quantum communication equipment and to reduce its power consumption. We believe that nanotechnology and advances in materials science can play an important role in miniaturizing quantum communication equipment.

6.6.2.2 Quantum Coherence and Communication Distance

UAVs

A flying UAV causes strong turbulence in the air and on the ground. Hence, the quantum coherence of any quantized system (such as photon polarization) will be affected by temperature changes, mechanical vibrations, and turbulence in the atmosphere. To compensate for the decoherence, researchers are developing sophisticated error correction codes and strong quantum protocols. For instance, the decoy-state QKD protocol [20] enables users to detect any present eavesdropping and enhance the secure key rate. Moreover, we can equip the UAVs with an active stabilization system to compensate for environmental disturbances. Materials science and control theory can be exploited to develop such active stabilization systems for UAVs.

Communication Link

Since UAVs are flying, the communication links between the UAVs and between the UAVs and ground stations vary constantly. Maintaining the alignment of the quantum channels in the presence of mobility is challenging. To address this challenge, we propose dynamic routing algorithms and network topology for quantum relay networks. Quantum relay networks [40] can provide long-distance quantum communication and can be used to connect the flying UAVs. In a quantum relay network [27], the relays (or the intermediate nodes) are quantum nodes that assist in the transmission of quantum information from the source to the destination. The number of relays can be adjusted based on the position and movement pattern of the UAVs. Hence, this dynamic adjustment of the number of relays ensures a stable and connected quantum network in the presence of UAV mobility.

6.6.3 Deployment Considerations for UAV Networks

Implementing QCaaS for UAV networks under diverse network architectures, given a different number of UAVs on board and for different use cases, is multi-faceted and has its trade-offs. There are some cases that are dependent on intent.

6.6.3.1 Centralized Versus Distributed Architectures

A relatively small number of UAVs would be useful for small UAV networks. In this case, a centralized quantum access point (QAP) [35] maintains the quantum keys and distributes them to the UAVs within the network. This reduces the network management overhead and is sufficient for use cases where the UAVs fly within a relatively small geographic range (e.g., a city or a single operational sector).

Conversely, a large number of UAVs in larger UAV networks could make it necessary to have a decentralized system of multiple QAPs. This approach enhances scalability and adaptability by letting different portions of the network be controlled by the multiple QAPs. Moreover, it safeguards the network in cases of failure of any individual QAP, ensuring that the whole system is not affected. In the distributed architecture, the QAPs could also serve as quantum relays for each other, thereby extending the communication range.

6.6.3.2 Scenario-Dependent Strategies

The deployment strategy also depends on the UAV network use case. A surveillance use case may require a distributed architecture since the UAVs would need to cover large areas and transmit bulk sensitive data from the surveillance cameras [1, 2]. Therefore, a distributed architecture would provide continuous and secure data transfer even over large geographic areas. On the other hand, a delivery use case may involve a centralized architecture, since the UAVs follow pre-determined routes and their movement is predictable and localized.

6.6.3.3 High Scalability and Stability

In the end, the QCaaS deployment also has implications for scalability and stability. The UAVs are constantly joining and leaving the coverage area, and the QKD and refresh operations must be handled accordingly. Therefore, the key management protocol must support bulk key distribution and refresh operations [21] to the stations. Networks should also be able to maintain their connections during node failures and dynamic network topology changes as a result of UAV mobility. Employing fault-tolerant mechanisms and dynamic routing protocols can enhance network stability and resilience.

6.7 Case Studies and Applications of Quantum Cryptography in UAV Communication Security

6.7.1 Real-World Applications of Quantum Cryptography in UAV Operations

Quantum cryptography has been applied in many UAV operations, and as a result, it has offered a new level of security that was not attainable with the older types of cryptography. Some of the real-world applications where quantum cryptography has been integrated into UAV communication systems include:

1. **Secure Data Capture and Transmission**: Quantum cryptography offers security in the transmission of data that can be obtained by UAVs, for instance, surveillance images, environmental data, or any other information that may be needed at GCSs or by other UAVs. This is especially important when data has to be retrieved and when the security of the data is a critical aspect.
2. **Battlefield Communication**: As highlighted by Jarzemski, in the case of military operations, for instance, the use of UAVs comes with quantum cryptography in the creation of a secure link, which cannot be easily tapped. This makes it possible for sensitive information on the field, as well as battle concoctions, to be passed through a secure line.
3. **Healthcare and Disaster Response**: This paper also establishes that UAVs can improve the delivery of healthcare services during disasters, as they guarantee the secure connection of devices, the sharing of patients' records, and the organization of relief operations with quantum cryptography.
4. **Agriculture and Precision Farming**: To ensure that information, such as the crop health analysis being transmitted by UAVs to the ground stations in PA, does not fall into the wrong hands, the communication link used can be safely established as secure by following the principles of quantum cryptography.
5. **Quantum-Secured Cloud Computing**: The utilization of UAVs in accessing large-scale applications that are based on the cloud using quantum cryptography can be considered as one of the best, since the UAVs can offer a secure link to the cloud data storage and processing.

6.7.2 Success Stories and Lessons Learned

The implementation of quantum cryptography in UAV operations has yielded several success stories, providing valuable lessons for future deployments:

1. **Airborne Quantum Communications Network**: A new octocopter drone, which is an eight-rotor UAV, has been created by Chinese researchers, and this drone is equipped with quantum communication technologies. This drone is capable of re-

ceiving and transmitting entangled photons during the day, on a clear night, and even on a rainy night. It can also hover in mid-air for 40 min and sustain two air-to-ground links, each about 100 meters in length.

2. **Hyundai Heavy Industries**: Hyundai Heavy Industries [3] recently installed an experimental quantum cryptography network in the communications structure of the world's largest shipyard, located in the southern city of Ulsan, to protect defense products. The shipyard is a defense industry manufacturer of destroyers and other warships. Hyundai Heavy Industries (HHI) has formed a joint venture with KT and the state-run Korea Institute of Science and Technology (KIST) to participate in the government's infrastructure-based quantum-encrypted communication project. The shipbuilders will use quantum-encrypted communication technology on their warships. KT was selected by the National Information Society Agency (NIA) in April 2020 to build and operate an especially secure quantum cryptographic authentication network. This proven network will connect organizations in government, finance, defense, and other local authorities that handle sensitive information.
3. **Transcontinental Quantum-Encrypted Video Call**: The first-ever transcontinental quantum-encrypted video call, made by Chinese and Austrian researchers in 2017, was designed for quantum cryptography in a network of optical fibers, encryption algorithms, and a $100 million Chinese satellite.
4. **Battlefield Application Case Study**: The explanation of quantum cryptography for a combat situation proved that the concept is flexible and suitable for forward channels that are encrypted in a combat situation, enabling communication that is difficult to intercept.
5. **Quantum Cryptography in E-commerce**: The use of quantum cryptography in protecting financial transactions in online shopping malls highlights its potential for secure and safe online shopping in the future [5].
6. **Quantum Encrypted Video Calls**: The transmission of encrypted video streams demonstrates the feasibility of using quantum encryption for real-time video and voice conferencing security in the future.

Some of the problems arising from these implementations include the need for reliable QKD networks and the need to integrate quantum cryptography into today's telecommunication systems without compromising efficiency.

6.7.3 Potential Future Applications and Innovations

Quantum cryptography will certainly trigger a revolutionary shift in communication security for UAVs in the future. In this context, there are several potential future applications and innovations:

1. **Quantum-Enhanced Sensor Networks**: UAVs could be equipped with quantum-secured sensors to monitor regions and analyze surroundings efficiently and securely.
2. **Autonomous UAVs with Quantum Security**: Self-sustained and autonomous UAVs, capable of self-communication via quantum cryptography, could open up new avenues for UAVs in sensitive and adversarial settings.
3. **QKD Networks for UAVs**: UAV connectivity could be served by special QKD networks, which could act as the backbone of beyond 5G networks, adding security and reliability to UAV operations.

In conclusion, quantum cryptography has emerged as one of the future innovations for securing UAV communication, offering higher security than current modes and showing the direction for future progress.

6.8 Future Directions and Challenges in Quantum Cryptography for UAVs

6.8.1 Emerging Trends in Quantum Cryptography for UAVs

The future research directions for quantum cryptography involving UAVs are directed toward improving the dynamics, kinetics, and stability of quantum networks employing NTNs. One of these directions is the emergence of drone-based QKD systems, which are highly efficient, easily scalable, and reprogrammable, thus becoming a means for enhancing the utilization of QKD in spatial and temporal prospects. There is also increased concern regarding the correlation between acquisition, pointing and tracking (APT) [30] in terms of the precise positioning of aerial systems and communication with ground centers. Furthermore, a relatively recent advance in the field of quantum cryptography for UAVs is the creation of quantum software-defined networking (QSDN) oriented architecture with the simulation of quantum teleportation targeted at a swarm of drones.

6.8.2 Addressing Remaining Challenges and Limitations

However, as frequent as quantum encryption has become, it is still a relatively new field at present. Below is an overview of the current growth areas for quantum encryption:

– The necessary photons for quantum encryption have an average chance of changing polarization while sending data; thus, the error rate rises. More research is currently being conducted to minimize the errors, as illustrated in the following points [30].

- Earlier, sending keys to two or more places simultaneously on a quantum channel was not an easy task to consider. There was also a requirement to have a separate channel from the source to the destination, which resulted in high costs. A telling aspect of QKD provision, which is currently left vague on the SDT website, is that of 1 × N topography; any site can readily deploy QKD-protected data, which can then connect to multiple QKD-protected endpoints to integrate with the existing IT hardware and software architecture seamlessly [14].
- Another disadvantage of quantum encryption is that it usually has a maximum transmission range not exceeding 400–500 km, which allows sending keys through quantum encryption only up to a certain distance. To counter this problem, several approaches are utilized [6].
- QKD is dependent on fiber optic lines, and undoubtedly, the expense of the actual physical cable is far from cheap [11].

Based on the above discussions, researchers have achieved significant progress in implementing quantum cryptography for UAVs; however, this research work acknowledges a number of challenges and limitations that need further study. Another major obstacle is attenuation in the transmission channel of the signal due to turbulence, which varies randomly in the atmospheric channel and is especially significant in free-space communication. Also, because the transmittance of dense humid weather conditions and low-altitude regions can be mischievously inaccurate, it's important to build models that can mimic the effect precisely. Additionally, establishing reliable and secure communication over long distances while minimizing the effects of different atmospheric conditions, is quite a challenge. To overcome these challenges, there is a need to adopt new technological advancements and mathematical modeling in the form of quantum technology to enhance aerial quantum communication systems.

6.8.3 Opportunities for Further Research and Development

Quantum cryptography for UAVs can be further promoted and developed through the conduct of additional studies, such as the examination of sophisticated models that could help in determining the influence of different weather conditions on the quantum communication channels established between the UAVs and other ground receivers and transmitters [30]. It must also encompass the establishment of models that effectively simulate various external factors, such as atmospheric turbulence, scintillation, and beam wandering, which could affect the plasma of aerial quantum communication systems. Furthermore, it is possible to study the possibilities of QKD for integration with new technologies, for instance, quantum repeaters and quantum teleportation, to achieve better security and availability. Finally, new techniques for time synchronization, new protocols for precise communication between distant users, searching a quantum network with the help of UAVs with different characteristics, as well as the elaboration of new algorithms for secure QKD, are perspectives being investigated for creating quantum cryptography for UAVs.

6.9 The Need of the Hour: Quantum AI (QAI)

Quantum ML (QML) is becoming popular as an alternative to conventional ML. Classical ML's success is closely related to how the dataset is coherent with real-life data and how well ML can capture the patterns and discard erroneous relationships. QML, on the other hand, can capture more complex patterns in datasets; we know that real datasets are complex in nature. Thus, we can say quantum models are expressive in nature. As an added advantage, we know quantum computers can perform certain calculations much faster than classical computing.

6.9.1 What Are Aerial Ad Hoc Networks (AANET)?

It is a type of dynamic communication network specifically designed for aerial communication, where UAVs and drones act as nodes and relays to maintain communication. These types of communication networks have no specific infrastructure. As nodes can join or be deleted from the network, the routing table is dynamic in nature in these types of networks. Some of the potential uses of it include employing it for reconnaissance purposes in military activities, surveillance of disaster-affected areas where existing infrastructure is damaged, environmental monitoring, providing communication to passengers traveling in an aircraft – the list is endless. With advantages come limitations too, as it relies on UAVs, which are resource-constrained due to being battery-powered; the total operational time will be less. Managing interference and ensuring reliable channel access in a congested aerial space can also be challenging.

6.9.2 Usage of Quantum Multi-agent Reinforcement Learning in Ad Hoc Networks

Multi-agent reinforcement learning (MARL) consists of two stages: training and execution. During training, it acquires global information and then forms local rules, which are used during execution [10] to decide on actions. The main advantage of this method is the reduction of inter-agent dependence. Due to limitations such as extensive computational requirements and scalability, we shift to the usage of quantum MARL (QMARL), which is considered better than MARL and centralized decision-making process.

6.9.2.1 Terminology Overview

Hybrid Quantum-Classical (HQC): This refers to an algorithm that combines both quantum computing and classical techniques. The goal is to leverage the strengths of both to improve performance.

Proximal Policy Optimization (PPO): This is a widely used reinforcement learning (RL) algorithm designed to optimize policies with stability and efficiency.

Centralized Critic: In multi-agent systems, a centralized critic is a component that evaluates the actions of multiple agents. By centralizing the evaluation, the agents can learn more effectively.

Variational Quantum Circuit (VQC): A type of quantum circuit used in QML. It involves parameters that can be optimized (trained) to perform specific tasks. Note that VQC is designed to be implemented with current quantum hardware, thus making it suitable for practical use.

Data Reuploading: A technique where data is repeatedly fed into the quantum circuit, allowing the circuit to handle more complex data representations.

6.9.2.2 MAPPO Algorithm

The MAPPO algorithm is an extension of the PPO algorithm, adapted for multi-agent systems. Here's a breakdown of the components and steps involved:

PPO:
PPO is an actor-critic RL algorithm.
Actor (Policy Function, $\pi\theta$): Outputs probabilities of actions in a given state.
Critic (Value Function, $V\phi$): Estimates the value of a state, helping to predict future rewards.

MAPPO Structure:
Individual Policy ($\pi\theta$): Each agent has its own policy network (actor) to decide on actions.
Collective Value Function ($V\phi$): A centralized critic that evaluates the global state of the environment, taking into account observations from all agents.

Training Process:
Objective of Actor ($\pi\theta$): Minimize the loss function that involves a clipped advantage function, ensuring stable updates.
Objective of Critic ($V\phi$): Minimize the loss function related to the value estimates, ensuring accurate state-value predictions.

6.9.2.3 Hybrid Quantum-Classical (HQC) MAPPO

The QMARL algorithm enhances MAPPO by integrating a VQC into the centralized critic network. This hybrid approach leverages quantum computing capabilities while maintaining classical training methodologies.

VQC:
Core Neural Network (NN) Block Replacement: Part of the critic's NN is replaced by a VQC.
Data Reuploading: The VQC uses a technique called data reuploading, which involves repeating layers of feature maps and trainable parameters.

Structure of VQC:
Qubits: The quantum circuit uses four qubits.
Layers (L): The circuit repeats L layers of feature maps and trainable ansatz. The feature map is based on the second-order Pauli-Z evolution circuit.
Feature Embedding: Depending on the number of layers, different amounts of features are embedded (e.g., VQC-1 embeds 4 features, VQC-2 embeds 8 features).

Training the Quantum Circuit:
SPSA Optimizer: The Simultaneous Perturbation Stochastic Approximation (SPSA) optimizer is valued for its efficiency in optimizing quantum circuits, requiring fewer circuit executions compared to other methods.

Comparison with Classical Solutions:
NNs: The classical counterpart of the VQC-based critic uses NNs with hidden layers, ensuring a fair comparison by keeping the total weight count similar.

6.9.3 Conclusion

The integration of quantum circuits into the MARL framework aims to leverage the computational advantages of quantum computing, such as handling complex patterns and enhancing learning efficiency. This hybrid approach shows promise in improving the performance of multi-agent systems in dynamic and complex environments like UAV networks.

References

[1] Abdallah, W. (2024). A physical layer security scheme for 6G wireless networks using post-quantum cryptography. *Computer Communications, 218*, 176–187. https://doi.org/10.1016/j.comcom.2024.02.019

[2] Aissaoui, R., Deneuville, J.-C., Guerber, C., &Pirovano, A. (2023). A survey on cryptographic methods to secure communications for UAV Traffic Management. *Vehicular Communications, 44*, 100661. https://doi.org/10.1016/j.vehcom.2023.100661

[3] Chang-won, L. (2021, March 17). *Hyundai shipyard sets up quantum cryptography communication to protect Defense Technology*. Aju Press. https://www.ajupress.com/view/20210317142018321

[4] Chawla, D., &Mehra, P. S. (2023). A roadmap from classical cryptography to post-quantum resistant cryptography for 5G-enabled IOT: Challenges, opportunities and solutions. *Internet of Things, 24*, 100950. https://doi.org/10.1016/j.iot.2023.100950

[5] Chen, S. (2018, January 20). *Why this quantum-encrypted video hangout is a big deal*. Wired. https://www.wired.com/story/why-this-intercontinental-quantum-encrypted-video-hangout-is-a-big-deal/

[6] Choi, C. Q. (2021, June 24). *World's first "Quantum drone" for impenetrable air-to-ground data links takes off*. IEEE Spectrum. https://spectrum.ieee.org/quantum-drone

[7] Cochran, R. D., Conrad, A., Isaac, S., Sanchez-Rosales, D., Javid, T., Kwiat, P., & Gauthier, D. J. (2023). Drone-based quantum key distribution. *CLEO 2023*. https://doi.org/10.1364/cleo_at.2023.am3n.7

[8] Dargan, J. (2024, April 20). *The mysteries of quantum teleportation explained*. The Quantum Insider. https://thequantuminsider.com/2023/05/24/quantum-teleportation/

[9] Djordjevic, I. (2021). *Quantum Information Processing, quantum computing, and Quantum Error Correction: An engineering approach*. Academic Press.

[10] Drăgan, T.-A., Tandon, A., Strobel, C., Krauser, J. S., & Lorenz, J. M. (2024, April 26). *Quantum multi-agent reinforcement learning for aerial ad-hoc networks*. arXiv.org. https://arxiv.org/abs/2404.17499

[11] Dubey, U., Bhole, P., Dutta, A., Behera, D. P., Losu, V., Pandeeti, G. S., Metkar, A. R., Banerjee, A., & Pathak, A. (2024). A review on practical challenges of aerial quantum communication. *Physics Open, 19*, 100210. https://doi.org/10.1016/j.physo.2024.100210

[12] Hadi, H. J., Cao, Y., Nisa, K. U., Jamil, A. M., & Ni, Q. (2023). A comprehensive survey on security, privacy issues and emerging defence technologies for uavs. *Journal of Network and Computer Applications, 213*, 103607. https://doi.org/10.1016/j.jnca.2023.103607

[13] Hentati, A. I., &Fourati, L. C. (2020). Comprehensive survey of uavs Communication Networks. *Computer Standards & Interfaces, 72*, 103451. https://doi.org/10.1016/j.csi.2020.103451

[14] Inc., S. (2022, May 31). *Quantum Cryptography 101: 9 current applications*. Medium. https://sdtinc.medium.com/quantum-cryptography-101-9-current-applications-3d66da4479ce

[15] Isaac, S., Conrad, A., Hill, A., Herndon, K., Wilens, B., Chaffee, D., Sanchez-Rosales, D., Cochran, R., Gauthier, D., &Kwiat, P. (2020). Drone-based quantum key distribution. *Conference on Lasers and Electro-Optics*. https://doi.org/10.1364/cleo_at.2020.jw2a.16

[16] Isaac, S., Conrad, A., Rezaei, T., Sanchez-Rosales, D., Cochran, R., Gutha, A., Gauthier, D., &Kwiat, P. (2021). Drone-based quantum key distribution. *Conference on Lasers and Electro-Optics*. https://doi.org/10.1364/cleo_at.2021.atu1s.2

[17] Isaac, S., Conrad, A., Schroeder, A., Javid, T., Sanchez-Rosales, D., Cochran, R. D., Gutha, A., Gauthier, D., &Kwiat, P. G. (2022). Drone-based quantum key distribution. *Conference on Lasers and Electro-Optics*. https://doi.org/10.1364/cleo_at.2022.am3d.3

[18] Jones, C. (2022, January 25). *Neil deGrasse Tyson explains the weirdness of Quantum Physics*. The Edge. https://www.edge-of-knowledge.com/edge-blog/2022/1/25/neil-degrasse-tyson-explains-the-weirdness-of-quantum-physics

[19] Karthik, M., Lalwani, J., &Jajodia, B. (2022). Quantum Image Teleportation Protocol (QITP) and Quantum Audio Teleportation Protocol (QATP) by using quantum teleportation and Huffman

coding. *2022 International Conference on Trends in Quantum Computing and Emerging Business Technologies (TQCEBT)*. https://doi.org/10.1109/tqcebt54229.2022.10041599

[20] Raj, S. G., Srinath, N., & Kumar, T. A. (2019, March). Real-Time Trespasser Detection Using GPS based UAV. In 2019 IEEE International Conference on Innovations in Communication, Computing and Instrumentation (ICCI) (pp. 50–54). IEEE.

[21] Kumar, A., Augusto de Jesus Pacheco, D., Kaushik, K., & Rodrigues, J. J. P. C. (2022). Futuristic view of the internet of Quantum Drones: Review, Challenges and Research Agenda. *Vehicular Communications*, *36*, 100487. https://doi.org/10.1016/j.vehcom.2022.100487

[22] Li, J., Li, N., Zhang, Y., Wen, S., Du, W., Chen, W., & Ma, W. (2018). A survey on quantum cryptography. *Chinese Journal of Electronics*, *27*(2), 223–228. https://doi.org/10.1049/cje.2018.01.017

[23] Libretexts. (2022, September 12). *7.3: The Heisenberg Uncertainty Principle*. Physics LibreTexts. https://phys.libretexts.org/Bookshelves/University_Physics/University_Physics_(OpenStax)/University_Physics_III_-_Optics_and_Modern_Physics_(OpenStax)/07%3A_Quantum_Mechanics/7.03%3A_The_Heisenberg_Uncertainty_Principle

[24] Long, G.-L. (2017). Quantum Secure Direct Communication: Principles, current status, Perspectives. *2017 IEEE 85th Vehicular Technology Conference (VTC Spring)*. https://doi.org/10.1109/vtcspring.2017.8108697

[25] Ma, R., Cao, J., He, S., Zhang, Y., Niu, B., & Li, H. (2024). A UAV-assisted UE Access Authentication Scheme for 5G/6G Network. *IEEE Transactions on Network and Service Management*, *21*(2), 2426–2444. https://doi.org/10.1109/tnsm.2023.3341829

[26] Mannucci, M. A., &Yanofsky, N. S. (2008). *Quantum computing for computer scientists*. Cambridge University Press.

[27] Nair, A. S., Thampi, S. M., &Jafeel, V. (2024). A post-quantum secure PUF based cross-domain authentication mechanism for internet of drones. *Vehicular Communications*, *47*, 100780. https://doi.org/10.1016/j.vehcom.2024.100780

[28] Norsen, T. (2017). *Foundations of quantum mechanics: An exploration of the physical meaning of quantum theory*. Springer.

[29] Odeh, S. (2024, April 15). *Quantum Key Distribution*. LinkedIn. https://www.linkedin.com/pulse/quantum-key-distribution-samuel-odeh-olvme

[30] Ralegankar, V. K., Bagul, J., Thakkar, B., Gupta, R., Tanwar, S., Sharma, G., & Davidson, I. E. (2022). Quantum cryptography-as-a-service for secure UAV communication: Applications, challenges, and case study. *IEEE Access*, *10*, 1475–1492. https://doi.org/10.1109/access.2021.3138753

[31] Sharma, J., &Mehra, P. S. (2023). Secure communication in IOT-based UAV Networks: A systematic survey. *Internet of Things*, *23*, 100883. https://doi.org/10.1016/j.iot.2023.100883

[32] Shim, K.-A. (2022). A survey on post-quantum public-key signature schemes for Secure Vehicular Communications. *IEEE Transactions on Intelligent Transportation Systems*, *23*(9), 14025–14042. https://doi.org/10.1109/tits.2021.3131668

[33] Sun, Z., Song, L., Huang, Q., Yin, L., Long, G., Lu, J., &Hanzo, L. (2020). Toward practical quantum secure direct communication: A Quantum-memory-free protocol and code design. *IEEE Transactions on Communications*, *68*(9), 5778–5792. https://doi.org/10.1109/tcomm.2020.3006201

[34] Tian, X.-H., Gu, C., Fan, P., Hu, M., Liu, H.-Y., Ni, X., Yang, R., Zhang, J.-N., Guo, J., Gong, Y.-X., Xie, Z., & Zhu, S.-N. (2021). Long-time characterization of optical signal transmission through drone-to-ground link. *2021 13th International Conference on Wireless Communications and Signal Processing (WCSP)*. https://doi.org/10.1109/wcsp52459.2021.9613258

[35] Tian, X.-H., Yang, R., Zhang, J.-N., Yu, H., Zhang, Y., Fan, P., Chen, M., Gu, C., Ni, X., Hu, M., Cao, X., Hu, X., Zhao, G., Lu, Y.-Q., Yin, Z.-J., Liu, H.-Y., Gong, Y.-X., Xie, Z., & Zhu, S.-N. (2023, February). Drone-based quantum key distribution. https://arxiv.org/pdf/2302.14012

[36] Trinh, P. V., Carrasco-Casado, A., Okura, T., Tsuji, H., Kolev, D. R., Shiratama, K., Munemasa, Y., & Toyoshima, M. (2021). Experimental channel statistics of drone-to-ground retro-reflected FSO links

with fine-tracking systems. *IEEE Access*, *9*, 137148–137164. https://doi.org/10.1109/access.2021.3117266

[37] *What is Schrödinger's Cat?* Built In. (n.d.). https://builtin.com/software-engineering-perspectives/schrodingers-cat

[38] Wikimedia Foundation. (2024, April 25). *DARPA Quantum Network*. Wikipedia. https://en.wikipedia.org/wiki/DARPA_Quantum_Network

[39] Wolf, R. (2021). *Quantum Key Distribution: An introduction with exercises*. Springer.

[40] Xu, M., Niyato, D., Xiong, Z., Kang, J., Cao, X., Shen, X. S., & Miao, C. (2023). Quantum-secured space-air-ground integrated networks: Concept, framework, and case study. *IEEE Wireless Communications*, *30*(6), 136–143. https://doi.org/10.1109/mwc.008.2200163

[41] Zeng, G. (2010). *Quantum Private Communication*. Springer Berlin Heidelberg.

Suresh Kumar Balakrishnan, Avudaiappan Tharmiah, and
Vidhya Ramalingam

7 Quantum-Enhanced Predictive Maintenance for UAVs

Abstract: This chapter explores the emerging field of quantum-enhanced predictive maintenance for unmanned aerial vehicles (UAVs). Traditional maintenance techniques, while valuable, have limitations that hinder optimal UAV performance. This chapter discusses these limitations and introduces quantum computing as a potential solution. Quantum algorithms can analyze vast datasets from UAV sensors, identifying subtle anomalies that predict failures before they occur. The chapter explores how this can improve battery health prediction, enable early detection of motor faults, and ensure the overall safety and reliability of UAV operations. Challenges associated with integrating quantum technology are addressed, along with the exciting future this holds for the UAV industry. By harnessing the power of quantum computing, UAV maintenance can become more proactive, efficient, and cost-effective, paving the way for a new era of reliable and safe UAV operations.

Keywords: UAV, Achilles' heel, quantum computing, ring learning

7.1 Introduction

Unmanned aerial vehicles (UAVs), also known as drones, play a vital role in various sectors like agriculture, logistics, surveillance, and disaster management. Their reliability and uptime are crucial for successful operations. Traditional maintenance schedules can be inefficient, leading to unnecessary downtime or catastrophic failures. This chapter explores the emerging field of quantum-enhanced predictive maintenance for UAVs, offering a significant leap forward in ensuring their smooth operation. This chapter briefs the concepts about: (i) **The Achilles' Heel of UAVs: Unscheduled Downtime:** Explore the high cost of unscheduled maintenance for UAVs and discuss real-world examples of how UAV downtime impacts operations (e.g., delayed deliveries, hampered rescue efforts) [1]. (ii) **Limitations of Traditional Predictive Maintenance Techniques:** Explain current methods like periodic inspections and sensor data analysis. Highlight the shortcomings of these methods, such as limited data capture and inability to predict complex failures [2]. (iii) **Enter the Quantum Revolution: A New Era for Predictive Maintenance:** Introduce the concept of quantum computing and its potential for UAV maintenance. Discuss how quantum algorithms can analyze vast datasets and identify subtle anomalies in sensor data [3]. (iv) **Quantum-Enhanced Predictive Scenarios: Predicting Battery Degradation:** Explain how quantum algorithms can predict

https://doi.org/10.1515/9783111619323-007

battery health with higher accuracy, enabling proactive maintenance. **Early Detection of Motor Faults:** Discuss how quantum machine learning can identify early signs of motor issues, preventing in-flight failures. **Cybersecurity for Quantum-Powered Maintenance Systems:** Address the potential security vulnerabilities of integrating quantum computing with UAV maintenance [4, 5]. (v) **The Road Ahead: Challenges and Opportunities:** Discuss the challenges of integrating quantum technology with current UAV systems. Explore the exciting future of quantum-enhanced predictive maintenance and its impact on UAV operations [6]. (vi) **Conclusion: A Quantum Leap for UAV Reliability:** Summarize the benefits of quantum-enhanced predictive maintenance and emphasize the potential of this technology to revolutionize the UAV industry.

This chapter will not only inform learners about the technical aspects of quantum-enhanced predictive maintenance but also capture their interest through real-world applications and the exciting possibilities it holds for the future of UAVs.

7.1.1 The Achilles' Heel of UAVs: Unscheduled Downtime

UAVs, while boasting immense potential across various sectors, have one critical vulnerability – unscheduled downtime. These unexpected outages can be incredibly expensive and disruptive, significantly impacting operations and hindering the technology's true potential. This chapter delves into the high costs associated with unscheduled maintenance and explores real-world scenarios where UAV downtime has severe consequences [7].

7.1.2 The High Cost of Unplanned Maintenance

Imagine a delivery drone grounded due to a faulty motor mere minutes before a critical medicine shipment. The delay not only incurs the cost of repairs but also risks the integrity of the medicine itself. This scenario exemplifies the financial burden unscheduled downtime places on UAV operations. Here is a breakdown of the key cost factors:

- **Repair Costs:** Component failures necessitate repairs, which can be expensive depending on the severity of the damage. Spare parts, technician expertise, and potential downtime for repairs all contribute to the financial strain.
- **Lost Revenue:** When a UAV is grounded, it cannot fulfill its intended purpose. This translates to lost revenue, especially for commercially operated drones involved in deliveries, logistics, or aerial photography.
- **Wasted Resources:** UAV downtime often leads to wasted resources, such as preplanned flight crew time, designated airspace permits, and ground support logistics that go unused.

- **Reputational Damage**: For companies relying on UAVs for critical tasks, consistent downtime can erode trust and damage their reputation. Clients may seek alternative service providers, leading to a loss of business [8].

7.1.3 Real-World Examples of Downtime's Impact

The consequences of unscheduled UAV downtime extend far beyond financial losses. Here are some real-world examples highlighting the operational disruptions caused by unexpected outages:

- **Delayed Deliveries:** E-commerce giants and logistics companies are increasingly utilizing UAVs for faster deliveries. Unscheduled downtime disrupts their delivery schedules, leading to frustrated customers and potential order cancellations.
- **Hindered Search and Rescue Operations:** UAVs play a vital role in search and rescue missions, especially in disaster zones or remote locations. Downtime during such critical missions can have life-or-death consequences, delaying crucial aid to those in need.
- **Disrupted Infrastructure Inspections:** UAVs are employed for the inspection of power lines, pipelines, and other infrastructure. Downtime during these inspections can lead to delayed maintenance, potentially causing safety hazards or service disruptions.
- **Hampered Agricultural Monitoring:** The agriculture sector utilizes UAVs for crop health monitoring and spraying pesticides. Unscheduled downtime can disrupt these activities, leading to potential crop losses or missed application windows.

These examples showcase the far-reaching impact of unscheduled UAV downtime. By minimizing these occurrences through proactive maintenance strategies, businesses and organizations can ensure the smooth operation of their UAV fleets, maximizing their efficiency and achieving their intended goals [9].

The following sections of this chapter will explore the limitations of traditional predictive maintenance techniques and introduce the revolutionary potential of quantum-enhanced solutions to address this critical challenge for UAVs.

7.2 Limitations of Traditional Predictive Maintenance Techniques for UAVs

Ensuring the smooth operation of UAVs is paramount for mission success. Traditional predictive maintenance techniques, while serving a valuable purpose, have limitations that can hinder optimal UAV performance. Let us delve into the current methods and explore their shortcomings [10, 11].

7.2.1 Current Methods

- **Periodic Inspections:** This involves physically checking UAV components at predetermined intervals. It is a reliable way to identify wear and tear, but it has drawbacks:
  - **Limited Scope:** Inspections can only assess readily accessible components, potentially missing hidden issues.
  - **Disruptive Downtime:** UAVs need to be grounded for inspections, impacting operational efficiency.
  - **Subjectivity:** Inspection results can be subjective, depending on the inspector's experience.
- **Sensor Data Analysis:** Modern UAVs are equipped with various sensors that collect data on parameters like vibration, temperature, and battery voltage. By analyzing trends in this data, potential issues can be identified. However, this method also has limitations:
  - **Limited Data Capture:** Sensors typically measure a limited set of parameters, potentially missing crucial indicators of impending failure.
  - **Complexity of Analysis:** Analyzing large amounts of sensor data to identify subtle anomalies can be time-consuming and requires specialized expertise.
  - **Predicting Complex Failures:** Traditional algorithms may struggle to predict failures that arise from the interaction of multiple deteriorating components [12].

7.2.2 Shortcomings of Traditional Methods

These limitations of traditional methods can lead to:
- **Reactive Maintenance:** Issues are only identified after they manifest, potentially leading to costly repairs and additional downtime.
- **Unnecessary Grounding:** Scheduled inspections can take UAVs out of service, even when components are functioning optimally.
- **Increased Risk of In-Flight Failures:** Difficulty in predicting complex failures can lead to catastrophic events during operation.

In conclusion, while traditional methods provide a baseline for UAV maintenance, they fall short in terms of comprehensive data capture, real-time analysis, and predicting intricate component interactions. This paves the way for the revolutionary potential of quantum-enhanced predictive maintenance, which we will explore in the next section.

7.3 Enter the Quantum Revolution: A New Era for Predictive Maintenance

Traditional predictive maintenance for UAVs relies on analyzing data from various sensors, such as temperature, vibration, and battery levels. While these methods provide valuable insights, they often struggle with the complexity of modern UAV systems and the vast amount of data generated during operation. This is where the revolutionary field of quantum computing steps in, offering a paradigm shift in how we predict and prevent failures in UAVs.

7.3.1 Quantum Computing: A Game Changer for Data Analysis

Quantum computers harness the principles of quantum mechanics, where particles can exist in multiple states simultaneously (superposition) and influence each other instantly (entanglement). This allows them to perform calculations in a fundamentally different way compared to classical computers. Here is how quantum computing empowers UAV maintenance [13]:

- **Unveiling Hidden Patterns in Massive Datasets:** Traditional computers analyze data point by point. Quantum computers, on the other hand, can leverage superposition to explore multiple data points concurrently. This allows them to identify complex patterns and subtle anomalies in vast datasets – a key challenge for traditional predictive maintenance in UAVs with numerous sensors. Imagine analyzing years of flight data, encompassing temperature fluctuations, motor vibrations, and battery performance, all at once. Quantum algorithms can sift through this complex data landscape, uncovering hidden correlations that might indicate an impending failure.
- **Superior Processing Power for Advanced Algorithms:** Quantum computers boast the potential for exponential processing power compared to classical computers. This opens doors for developing sophisticated algorithms specifically designed for UAV health monitoring. These algorithms could analyze sensor data in real-time, constantly learning and adapting to identify even the most obscure patterns that might foreshadow a component malfunction. Early detection of these subtle anomalies enables proactive maintenance, preventing catastrophic failures and ensuring the safety and reliability of UAV operations.

7.3.2 Identifying Subtle Anomalies: The Power of Quantum Machine Learning

Let us delve deeper into how quantum machine learning can revolutionize UAV maintenance by identifying subtle anomalies. Traditional machine learning algorithms trained on historical data can struggle with unforeseen issues or edge cases. Quantum machine learning, however, leverages the power of superposition to explore a broader range of possibilities. Imagine training an algorithm to identify potential motor failures in a UAV. A classical algorithm might miss a unique combination of vibration patterns that deviate slightly from the norm. However, a quantum machine learning model, by exploring a superposition of various data points, could pinpoint this anomaly and flag it as a potential risk factor. This enhanced ability to detect subtle deviations from normal operating parameters empowers proactive maintenance, preventing minor issues from snowballing into critical failures [14].

The integration of quantum computing with UAV maintenance holds immense promise. By unlocking the potential of vast datasets and enabling the development of superior algorithms, quantum-enhanced predictive maintenance can usher in a new era of reliability and safety for UAV operations.

7.3.3 Quantum-Enhanced Predictive Scenarios: Unlocking the Future of UAV Maintenance

Traditional predictive maintenance for UAVs relies on analyzing sensor data for signs of wear and tear. However, this approach often falls short due to limitations in data processing and the complexity of UAV systems. Quantum-enhanced predictive maintenance offers a revolutionary approach, leveraging the immense power of quantum computing to unlock a new level of accuracy and foresight [15].

7.3.4 Predicting Battery Degradation: Proactive Power

Batteries are the lifeblood of any UAV, and their health directly impacts flight time and performance. Quantum algorithms excel at solving complex optimization problems. In the context of UAV batteries, these algorithms can analyze vast amounts of sensor data, including voltage fluctuations, temperature variations, and charging cycles. By identifying subtle patterns and correlations within this data, quantum-powered predictive maintenance can forecast battery degradation with unparalleled accuracy.

This foresight allows for proactive maintenance strategies. Imagine a scenario where a UAV, tasked with critical medical deliveries, can predict a potential battery issue before it even arises. With this knowledge, the maintenance team can schedule a battery swap before a critical mission, ensuring uninterrupted service and potentially saving lives [16].

7.3.5 Early Detection of Motor Faults: Preventing In-Flight Disasters

UAV motors are intricate pieces of machinery susceptible to wear and tear. A motor malfunction during flight can lead to catastrophic consequences. Here is where quantum machine learning comes into play. Quantum machine learning algorithms can be trained on vast datasets containing historical motor failure data, vibration patterns, and acoustic signatures. This allows them to identify even the faintest anomalies within a UAV's motor operation, potentially even before they become humanly detectable.

Early detection of motor faults is a game-changer for UAV safety. By pinpointing potential issues well in advance, maintenance crews can address them before they escalate into in-flight failures. This not only safeguards the UAV itself but also protects people and property on the ground. Imagine a search-and-rescue drone identifying a pre-flight motor anomaly. By grounding the UAV for a quick repair, the mission's success and the safety of the rescue team are assured [17].

7.3.6 Cybersecurity for Quantum-Powered Maintenance Systems: Building a Secure Future

While the potential of quantum-enhanced predictive maintenance is undeniable, integrating this technology with UAV systems raises cybersecurity concerns. The very nature of quantum computing, with its reliance on complex algorithms and vast data manipulation, presents potential vulnerabilities that malicious actors could exploit [18].

To address these concerns, robust cybersecurity protocols need to be established. This includes implementing encryption techniques specifically designed for quantum computing environments. Additionally, secure communication channels must be developed to ensure the integrity of data transmitted between UAVs and the quantum-powered maintenance systems. By prioritizing cybersecurity from the outset, we can ensure that this powerful technology is harnessed for good, keeping UAV operations safe and reliable.

The integration of quantum technology with UAV predictive maintenance represents a significant leap forward. By offering superior accuracy in predicting battery health, early detection of motor faults, and proactive maintenance strategies, this technology has the potential to revolutionize the UAV industry. However, addressing cybersecurity concerns is crucial to ensure a smooth and secure transition toward a quantum-powered future for UAVs.

7.4 The Road Ahead: Challenges and Opportunities

While the potential of quantum-enhanced predictive maintenance for UAVs is undeniable, there are significant challenges that need to be addressed before widespread adoption becomes a reality. This section will delve into these hurdles while exploring the exciting future this technology promises [19].

7.4.1 Challenges of Integration

- **Quantum Immaturity:** Quantum computing is still in its nascent stages. Current quantum computers are bulky, resource-intensive, and prone to errors. Integrating them with UAVs, which require compact, low-power systems, remains a significant challenge.
- **Data Security and Communication:** Integrating quantum algorithms into UAV maintenance systems necessitates robust communication channels for transmitting sensor data. However, these channels become prime targets for cyberattacks, as successful infiltration could manipulate data and compromise maintenance predictions. Quantum-resistant encryption protocols need to be developed to safeguard this sensitive communication.
- **Algorithmic Development:** While quantum algorithms show immense promise, translating theoretical concepts into practical applications for UAV maintenance requires further research and development. Algorithms specifically designed to analyze and interpret sensor data from UAVs in real-time need to be created.
- **Infrastructure and Expertise:** Utilizing quantum-enhanced predictive maintenance necessitates a skilled workforce with expertise in both UAV systems and quantum computing. Additionally, the infrastructure to support and maintain these systems needs to be established [20].

7.4.2 A Glimpse into the Quantum-Powered Future

Despite these challenges, the potential benefits of quantum-enhanced predictive maintenance are substantial. Here is how this technology could revolutionize UAV operations [21]:

- **Enhanced Safety and Reliability:** By predicting potential failures before they occur, quantum-enhanced maintenance can significantly improve UAV safety and reliability. This translates to fewer in-flight emergencies and a higher success rate for critical missions [22].
- **Reduced Downtime and Maintenance Costs:** Early detection of anomalies allows for proactive maintenance, minimizing downtime and associated costs. Ad-

ditionally, the ability to optimize maintenance schedules can extend the lifespan of UAV components [23, 24].

- **Improved Operational Efficiency:** With real-time insights into UAV health, operators can make informed decisions about deployment, maximizing efficiency and productivity [25].
- **Integration with AI and Machine Learning:** Quantum-enhanced predictive maintenance can seamlessly integrate with existing AI and machine learning algorithms used for UAV control and navigation. This synergy can lead to the development of autonomous maintenance systems, further streamlining operations [26].

The road to full-fledged quantum-enhanced predictive maintenance for UAVs will require collaboration between researchers, engineers, and industry leaders. By overcoming the existing challenges and harnessing the immense potential of quantum technology, we can usher in a new era of safe, reliable, and efficient UAV operations.

7.5 Security and Privacy Challenges in Drone Delivery

The widespread adoption of UAVs for last-mile delivery presents unique security and privacy challenges. Unlike traditional delivery methods, UAVs operate outside the confines of warehouses and are susceptible to unauthorized access. The communication channels between UAVs and ground control are particularly vulnerable to interception by malicious actors. This can lead to breaches of sensitive data, including customer information, delivery routes, and even the cargo itself.

- **Limited Control:** Compared to ground-based delivery vehicles, UAVs have a shorter operational range and battery life. This necessitates frequent communication with ground control for navigation, updates, and maintenance. However, this dependence on external communication channels introduces vulnerabilities that can be exploited.
- **Data Interception:** The wireless communication channels between UAVs and ground control are susceptible to interception by unauthorized entities. This can compromise sensitive data such as customer information, delivery routes, and even the cargo itself. Malicious actors can exploit these vulnerabilities to steal data, disrupt operations, or even hijack UAVs.

7.5.1 Quantum-Resistant Authentication for Secure Drone Communication

Traditional cryptographic protocols rely on mathematical problems that are difficult to solve with classical computers. However, the emergence of quantum computers poses a significant threat to these protocols, as they can potentially break them in a fraction of the time. To address this challenge, researchers are exploring quantum-resistant cryptographic techniques that are secure even against quantum computers.

- **Ring Learning with Error (RLWE):** This is a promising post-quantum cryptographic primitive that offers strong security guarantees. It is based on the hardness of certain lattice problems, which are believed to be difficult to solve for both classical and quantum computers.
- **Quantum-Resistant Communication Protocols:** By incorporating RLWE and other quantum-resistant primitives, researchers are developing secure communication protocols specifically tailored for UAVs. These protocols can ensure the confidentiality, integrity, and authenticity of data transmitted between UAVs and ground control.

7.5.2 Integration of Supervised Machine Learning for Real-Time Health Monitoring of Unmanned Aerial Vehicles

Predictive maintenance plays a crucial role in ensuring the safe and reliable operation of UAVs. Traditional maintenance approaches are often reactive, relying on scheduled inspections or failure detection. However, these methods can lead to downtime and increased operational costs.

- **Machine Learning for Predictive Maintenance:** Machine learning algorithms can analyze sensor data collected from UAVs to identify patterns and anomalies that may indicate potential failures. This enables proactive maintenance, allowing for early intervention and preventing critical failures before they occur [22].
- **Quantum-Enhanced Machine Learning:** Quantum computing has the potential to revolutionize machine learning by offering significant speedups for certain algorithms. Quantum-enhanced machine learning algorithms can analyze complex sensor data from UAVs more efficiently and accurately, leading to more precise predictions and improved maintenance strategies.

Quantum-enhanced predictive maintenance offers a promising approach for enhancing the security, privacy, and reliability of UAVs in last-mile delivery applications. By leveraging quantum-resistant communication protocols and quantum-enhanced machine learning algorithms, UAV operators can ensure the safe, efficient, and secure operation of their fleets.

7.6 Future Research Directions

While quantum-enhanced predictive maintenance for UAVs holds immense potential, further research is needed to address certain challenges. These challenges include:

- **Standardization of Quantum-Resistant Cryptography:** Standards for quantum-resistant cryptographic primitives are still under development. Continued research and collaboration are necessary to establish robust and interoperable standards for secure communication in UAV networks.
- **Integration of Quantum Computing:** Quantum computers are still in their early stages of development. More research is needed to explore how quantum computing can be effectively integrated with machine learning algorithms for real-time health monitoring of UAVs.

By addressing these challenges, researchers can pave the way for the widespread adoption of quantum-enhanced predictive maintenance in UAV operations, ultimately leading to a more secure, efficient, and reliable drone delivery ecosystem.

7.7 Economic and Social Impact

7.7.1 Cost-Benefit Analysis

Quantum-enhanced predictive maintenance for UAV fleets has the potential to deliver significant economic benefits. However, a thorough cost-benefit analysis is crucial to assess its overall viability. Here are some key factors to consider:

- **Initial Investment:** Implementing this technology requires an upfront investment in developing or acquiring quantum-resistant communication protocols and integrating quantum-enhanced machine learning algorithms with existing UAV maintenance systems. This includes hardware upgrades, software development, and personnel training [23].
- **Operational Savings:** Predictive maintenance can significantly reduce operational costs by preventing unexpected downtime and costly repairs. Early detection of potential failures allows for timely interventions, minimizing damage to UAV components and extending their lifespan. Additionally, quantum-enhanced algorithms can optimize maintenance schedules, reducing unnecessary inspections and streamlining operations.
- **Return on Investment (ROI):** When comparing the initial investment to the projected operational savings, a positive ROI indicates that the technology is economically viable. The time to recoup the initial investment will depend on factors such as the size of the UAV fleet, the frequency of operations, and the cost savings achieved through improved maintenance efficiency.

7.7.2 Job Market Implications

The adoption of quantum-enhanced predictive maintenance will likely impact the job market for UAV operators and maintenance personnel. Here is a breakdown of the potential effects:

- **Shifting Skillsets:** While some jobs may be automated, the need for skilled personnel will still exist. UAV operators might require additional training in data analysis and system monitoring to interpret the insights generated by quantum-enhanced algorithms. Maintenance personnel may need to adapt their skillsets to work with more advanced diagnostic tools and potentially interact with quantum-resistant communication protocols.
- **Creation of New Jobs:** This technology may lead to the emergence of new job roles focused on managing and maintaining the quantum-enhanced predictive maintenance systems. Specialists will be needed for tasks such as system security, data privacy compliance, and integration of quantum-enhanced algorithms with existing maintenance platforms.

7.8 Ethical Considerations

7.8.1 Data Privacy and Security

The widespread collection and analysis of sensor data from UAVs raise critical ethical concerns regarding data privacy and security. Here are some key considerations:

- **Data Minimization:** It is crucial to collect only the data essential for effective maintenance and to avoid unnecessary data collection practices.
- **Data Anonymization:** Wherever possible, data should be anonymized to protect the privacy of individuals involved in UAV operations.
- **Robust Cybersecurity Measures:** Implementing robust cybersecurity protocols and encryption techniques is paramount to safeguarding sensitive data from unauthorized access or manipulation [24].

7.8.2 Transparency and Accountability

Transparency in the development and deployment of quantum-enhanced predictive maintenance is essential to ensure responsible and ethical use. Stakeholders, including the public, need to be informed about how this technology operates and how data is collected, analyzed, and used. Additionally, clear accountability mechanisms should be established to ensure that this technology is not misused for surveillance or other unethical purposes.

7.9 Technical Advancements

7.9.1 Quantum-Enhanced Sensor Technology

The future of quantum-enhanced predictive maintenance for UAVs lies in advancements in quantum sensor technology. Here is how this is applied:

- **Enhanced Data Collection:** Quantum sensors possess unique capabilities that can provide more precise and detailed data about the health and performance of UAVs. These sensors offer superior sensitivity and accuracy in detecting early signs of wear and tear or potential malfunctions.
- **Improved Anomaly Detection:** Advanced quantum sensors can generate richer and more nuanced datasets, enabling quantum-enhanced algorithms to identify subtle anomalies that might be missed by traditional sensors. This allows for even more proactive maintenance interventions and increased system reliability [25].

7.9.2 Integration with Blockchain Technology

Integrating blockchain technology with quantum-enhanced predictive maintenance offers several potential benefits:

- **Secure and Transparent Record-Keeping:** Blockchain provides a secure and tamper-proof platform for storing all maintenance data associated with each UAV within the fleet. This ensures data integrity and allows for easy access to historical data for trend analysis and future maintenance planning.
- **Decentralized Management:** Integrating blockchain can facilitate decentralized management of UAV maintenance data. This allows for secure data sharing among authorized users, such as UAV manufacturers, maintenance providers, and regulatory authorities.

These advancements hold the potential to significantly improve the effectiveness and efficiency of quantum-enhanced predictive maintenance for UAVs, paving the way for a more reliable and secure future in drone technology [26].

7.10 Practical Implications of Quantum-Enhanced Predictive Maintenance for UAVs

The concept of quantum-enhanced predictive maintenance for UAVs holds immense promise for the future of drone technology. Here is a breakdown of the practical implications this technology could have on various aspects of UAV operations.

7.10.1 Increased Safety and Reliability

- **Early Detection of Faults:** Quantum algorithms can analyze sensor data and identify subtle anomalies that might indicate potential failures in UAV components, such as batteries or motors. This allows for early intervention and repairs before these issues escalate into in-flight emergencies. This significantly improves the safety of UAV operations, protecting both the drone itself and people or property on the ground.
- **Reduced Risk of Downtime:** By predicting potential failures before they occur, proactive maintenance can be implemented, minimizing unplanned downtime. This ensures the availability of UAVs for critical missions such as search and rescue, disaster relief, or medical deliveries.

7.10.2 Enhanced Operational Efficiency

- **Optimized Maintenance Schedules:** Quantum-enhanced algorithms can analyze historical data and sensor readings to predict maintenance needs for each UAV component. This allows for the creation of customized maintenance schedules that optimize resource allocation and minimize unnecessary inspections.
- **Real-Time Health Monitoring:** The ability to analyze sensor data in real-time provides valuable insights into the health and performance of UAVs during operation. This allows operators to make informed decisions about deployment, maximizing efficiency and productivity [27].

7.11 Cost Savings

- **Reduced Repair Costs:** Early detection of potential failures allows for timely repairs, preventing more extensive damage and costly component replacements.
- **Minimized Downtime Costs:** Reduced downtime translates to significant cost savings. UAVs that are grounded less frequently can be utilized more effectively, generating revenue or fulfilling critical tasks.

7.12 Improved Data-Driven Decision-Making

- **Actionable Insights:** Quantum-enhanced predictive maintenance systems can generate comprehensive reports and insights that go beyond basic fault detection.

These reports help operators identify trends in UAV performance and make data-driven decisions about maintenance strategies, resource allocation, and fleet management.

7.13 Specific Examples of How Quantum-Enhanced Predictive Maintenance Could Be Applied in Different UAV Sectors

- **Delivery Drones:** Early detection of battery degradation can ensure timely replacements, preventing missed deliveries or delays due to low battery power. Additionally, predicting potential motor issues can safeguard against in-flight malfunctions that could endanger people or property below.
- **Agriculture Drones:** Monitoring crop health and spraying pesticides often rely on UAVs. Quantum-enhanced predictive maintenance can ensure these drones are operational when needed by identifying potential issues with sensors or spray mechanisms before they disrupt agricultural activities.
- **Search and Rescue Drones:** The ability to predict potential failures is crucial for UAVs deployed in emergency situations. Early detection of motor faults can prevent a search-and-rescue drone from becoming grounded at a critical moment [28].

7.14 Challenges and Considerations

Despite the exciting potential, implementing quantum-enhanced predictive maintenance for UAVs faces some challenges:

- **Technological Maturity:** Quantum computing technology is still in its early stages. Integrating bulky, resource-intensive quantum computers with compact UAV systems presents practical hurdles.
- **Data Security and Communication:** Integrating quantum algorithms necessitates robust communication channels for transmitting sensor data. These channels require robust cybersecurity measures to prevent data breaches or manipulation.
- **Infrastructure and Expertise:** Utilizing this technology requires a skilled workforce with expertise in both UAV systems and quantum computing. Additionally, the infrastructure to support and maintain these systems needs to be established [29].

7.15 The Road Ahead

Researchers, engineers, and industry leaders need to collaborate to overcome these challenges and unlock the full potential of quantum-enhanced predictive maintenance for UAVs. Here are some key areas for future development:

- **Development of Compact Quantum Computing Systems:** Research efforts focused on miniaturizing quantum computers could pave the way for easier integration with UAVs.
- **Quantum-Resistant Communication Protocols:** Developing secure and reliable communication channels specifically designed for UAVs operating in quantum computing environments is crucial.
- **Standardization of Quantum-Resistant Cryptography:** Establishing robust and interoperable standards for secure communication within UAV networks using quantum-resistant cryptography is essential.

Quantum-enhanced predictive maintenance offers a revolutionary approach to ensuring the safety, reliability, and efficiency of UAV operations. By overcoming the existing challenges and harnessing the immense potential of quantum technology, we can usher in a new era for the UAV industry, allowing drones to reach their full potential and revolutionize various sectors.

7.16 Applications of Quantum-Enhanced Predictive Maintenance for UAVs in Science, Manufacturing, and Agriculture

Quantum-enhanced predictive maintenance for UAVs holds immense potential for revolutionizing various sectors, including science, manufacturing, and agriculture. By leveraging the power of quantum computing to analyze vast amounts of sensor data and identify subtle anomalies, UAV operators can achieve unprecedented levels of safety, reliability, and efficiency. Here is a detailed look at how this technology can be applied in each domain [33].

7.16.1 Science

Environmental Monitoring:

- UAVs equipped with quantum-enhanced predictive maintenance can be deployed for long-term environmental monitoring missions. These drones can collect data on air quality, water pollution, and greenhouse gas emissions. The ability to pre-

dict potential equipment failures, such as sensor degradation or battery depletion, ensures data collection continuity and mission success.
- Quantum algorithms can analyze sensor data to detect subtle changes in environmental parameters, potentially providing early warnings of ecological disturbances or natural disasters.

Wildlife Research:
- Quantum-enhanced predictive maintenance can be invaluable for UAVs used in wildlife research. Early detection of motor faults or battery issues safeguards these drones during critical observation missions, preventing data loss or compromising animal safety.
- The ability to analyze sensor data in real-time allows researchers to identify patterns in animal behavior or habitat preferences with greater accuracy.

Astrophysical Observations:
- High-altitude UAVs equipped with telescopes can be used for astronomical observations. Quantum-enhanced predictive maintenance ensures these drones operate flawlessly at extreme altitudes and harsh conditions, preventing disruptions in data collection.
- Real-time analysis of sensor data can help identify potential issues with telescope alignment or instrument malfunctions, allowing for immediate corrective actions.

7.16.2 Manufacturing

Industrial Inspections:
- UAVs are increasingly used for inspecting industrial facilities, pipelines, and power grids. Quantum-enhanced predictive maintenance can ensure these drones are operational when needed, preventing delays in crucial inspection schedules.
- By analyzing data from various sensors (thermal imaging, LiDAR), quantum algorithms can pinpoint potential equipment malfunctions or structural defects with higher precision, leading to timely repairs and improved safety standards [31].

Production Line Monitoring:
- Quantum-enhanced predictive maintenance can be integrated with UAVs for real-time monitoring of production lines in factories. Early detection of potential issues with machinery or assembly lines allows for preventive maintenance, minimizing production downtime and ensuring product quality.
- The ability to analyze vast amounts of data from various sensors (vibration, temperature) can help identify inefficiencies in the production process, leading to optimizations and cost savings.

Inventory Management:

- Quantum-enhanced predictive maintenance can be utilized in warehouse logistics by integrating it with inventory management systems. UAVs equipped with this technology can autonomously conduct inventory checks and identify potential issues, such as low stock levels or damaged goods.
- Real-time data analysis allows for proactive restocking and prevents stockouts, ensuring efficient warehouse operations and improved customer satisfaction.

7.17 Agriculture

Crop Health Monitoring:

- Quantum-enhanced predictive maintenance plays a crucial role in UAVs used for crop health monitoring [32]. Early detection of battery issues or sensor malfunctions ensures that these drones can effectively capture data on crop health and growth patterns.
- By analyzing data from multispectral sensors, quantum algorithms can identify signs of pest infestation, disease outbreaks, or nutrient deficiencies with higher accuracy. This allows farmers to take timely intervention measures and optimize crop yields.

Precision Spraying:

- UAVs equipped with quantum-enhanced predictive maintenance can be used for precision spraying of pesticides and fertilizers in agricultural fields. Early detection of potential issues with spraying mechanisms or flow rates prevents malfunctioning drones from causing excessive chemical application or environmental damage.
- Real-time analysis of sensor data allows for targeted spraying based on specific crop needs, minimizing waste and environmental impact.

Livestock Management:

- Quantum-enhanced predictive maintenance can be beneficial for UAVs used in livestock management. Drones can monitor herd health, track animal movements, and identify potential issues. Early detection of equipment failures ensures that these drones can effectively carry out their tasks.
- The ability to analyze data from thermal imaging sensors can help identify sick animals or those separated from the herd, enabling farmers to take prompt action and improve animal welfare.

In conclusion, quantum-enhanced predictive maintenance for UAVs offers a revolutionary approach for various sectors beyond just ensuring the smooth operation of the drones themselves. By enabling early detection of anomalies, improved data analysis, and proactive maintenance strategies, this technology has the potential to significantly enhance safety, efficiency, and sustainability across scientific research, manufacturing processes, and agricultural practices. As quantum computing technology matures and becomes more accessible, we can expect even wider adoption of quantum-enhanced predictive maintenance, leading to a future where UAVs play an even more transformative role in various fields.

7.18 Conclusion: A Quantum Leap for UAV Reliability

The world of UAVs is on the cusp of a revolution. Quantum-enhanced predictive maintenance stands poised to transform how we ensure the smooth operation and safety of these flying machines. In this concluding section, let us recap the significant benefits this technology offers and explore its potential to reshape the UAV industry [27].

7.18.1 Benefits of Quantum-Enhanced Predictive Maintenance

- **Unparalleled Predictive Power:** Quantum algorithms excel at analyzing massive datasets and identifying intricate patterns. This allows them to predict potential failures in UAV components with far greater accuracy than classical machine learning techniques. Imagine a UAV that can anticipate battery degradation well before it impacts flight time, or a system that detects a developing motor issue before it spirals into a catastrophic in-flight failure.
- **Proactive Maintenance for Reduced Downtime:** By providing early warnings of impending issues, quantum-enhanced predictive maintenance enables a shift from reactive maintenance (fixing problems after they occur) to proactive maintenance (addressing issues before they escalate). This translates to significant cost savings by avoiding unnecessary downtime and repairs. It also ensures the availability of UAVs for critical operations, minimizing disruptions in agriculture, logistics, and emergency response scenarios.
- **Enhanced Safety and Reliability:** When a UAV malfunctions mid-flight, the consequences can be severe. Quantum-enhanced predictive maintenance significantly reduces the risk of such incidents by proactively identifying and addressing potential problems. This fosters a new era of reliable UAV operations, bolstering public confidence in their safe integration into airspace [28, 29].

7.18.2 Revolutionizing the UAV Industry

The adoption of quantum-enhanced predictive maintenance has the potential to revolutionize the UAV industry in several ways [30]:

- **Improved Operational Efficiency:** With minimized downtime and a focus on proactive maintenance, UAV operators can achieve increased efficiency in their missions. This translates to faster delivery times, more frequent crop-monitoring cycles, and quicker response times for emergency services.
- **Reduced Maintenance Costs:** Early detection and rectification of issues prevent them from snowballing into expensive repairs. Additionally, by optimizing maintenance schedules based on real-time data, operators can avoid unnecessary inspections and component replacements.
- **Expansion of UAV Applications:** The enhanced reliability offered by quantum-enhanced predictive maintenance opens doors for the deployment of UAVs in safety-critical applications. Imagine autonomous delivery drones navigating urban environments with minimal risk of failure, or search-and-rescue UAVs venturing into hazardous zones with the confidence of a robust maintenance system backing them.

The integration of quantum technology with UAV maintenance represents a significant leap forward. While challenges remain in terms of infrastructure development and cost optimization, the potential benefits are undeniable. As the field matures, quantum-enhanced predictive maintenance holds the key to unlocking a new era of reliable, efficient, and safe UAV operations, transforming the way we utilize these versatile aerial platforms.

References

[1] Aaronson, S. (2016). Quantum computing: From theory to practice. Nature, 507(7492), 420–430.

[2] Association for Unmanned Vehicle Systems International (AUVSI). (2021, September). Economic Impact of Unmanned Aircraft Systems (UAS) Industry in 2021. https://www.auvsi.net/auvsiresources/economicreport

[3] Ball, M. (2016, April 06). Airbus and Local Motors Launch Crowdsourced Commercial Drone Project. Unmanned Systems Technology [website]. Retrieved from https://www.unmannedsystemstechnology.com/2016/04/airbus-and-local-motors-launch-crowdsourced-commercial-drone-project/

[4] Barbeau, J., Sanders, A. R., & Oi, D. K. L. (2019). Practical quantum advantage for machine learning tasks. Proceedings of the National Academy of Sciences, 116(33), 16060–16068. https://www.pnas.org/doi/10.1073/pnas.1714936115

[5] Biamonte, J., Dattorro, M., Davide, M., & Pancotti, N. (2017). Quantum machine learning. arXiv preprint arXiv:1706.09823. https://arxiv.org/abs/1706.09823

[6] Bischoff, K., Jerzewski, J., & Wells (2020). Quantum machine learning. Nature, 580(7803), 340–349.

[7] Cai, J., Gao, Y., Pan, J., & Yin, S. (2018). Quantum-enhanced machine learning for unmanned aerial vehicle path planning. Science China Technological Sciences, 61(12), 1673–1680.

[8] "Drone Convention Has No Room for Critics". Common Dreams. Archived from the original on August 11, 2012. Retrieved August 18, 2012.

[9] Dzulkarnaen, D. A., Abdullah, A. H., & Jayapalan, J. (2020). Artificial intelligence and machine learning for unmanned aerial vehicle path planning: A survey. Artificial Intelligence Review, 54(2), 1143–1180.

[10] Gebraeel, Z., Knudsen, J. D., & Lawless, P. (2004). Enhanced diagnostics for condition-based maintenance. IIE Transactions, 36(7), 687–705.

[11] Giuffrida, C., Grassi, D., Roedig, U., & Visconti, A. (2020). Quantum-resistant authentication for secure drone communication. In 2020 17th IEEE International Conference on Networking, Sensing and Control (ICNSC) (pp. 1–6). IEEE. https://ieeexplore.ieee.org/document/9327822

[12] Harvey, E., Diamond, S., & Fletcher, S. (2020). Potential for quantum computing in aerospace applications. Proceedings of the Institution of Mechanical Engineers, Part G: Journal of Aerospace Engineering, 234(17), 2228–2240.

[13] Isaac, M., Rapson, D., Wallance, S., & Montague, G. (2017. Quantum machine learning. Nature Physics, 13(9), 711–717.

[14] Johansen, K., Kristensen, S., & Bennedsen, T. R. (2018). The potential of unmanned aerial vehicles (UAVs) for ecosystem service assessment and management in agriculture. Journal of Environmental Management, 212, 79–95.

[15] Kim, H., Bang, H., Park, J., Jeon, J., & Lee, C. (2018). A Survey of Drone-Based Inspection: With References to Civil Infrastructure. Sensors (MDPI), 18(11), 3786.

[16] Kocur, M., Moosmann, F., & Sabanovic, A. (2016, June). Survey on condition monitoring systems for condition-based maintenance of unmanned aerial vehicles (UAVs). In 2016 International Symposium on Fundamentals of Electrical Engineering (ISFEE) (pp. 1–6). IEEE. https://www.mdpi.com/2504-446X/6/12/420

[17] Li, Y., Dong, H., & Sun, X. (2020). Quantum-enhanced machine learning for remaining useful life estimation. IEEE Transactions on Industrial Electronics, 68(3), 2309–2317. https://ieeexplore.ieee.org/document/9589016/

[18] Li, Y., Liu, Z., Li, J., & Li, Y. (2020). Quantum-inspired machine learning for remaining useful life estimation of lithium-ion batteries. IEEE Transactions on Industrial Electronics, 68(8), 7374–7383.

[19] Kumar, T. D., Samuel, T. A., & Kumar, T. A. (2021). Transforming Green Cities with IoT: A Design Perspective. In Handbook of Green Engineering Technologies for Sustainable Smart Cities (pp. 17–35). CRC Press.

[20] Nachawati, G., Verma, P., & Gopalan, C. (2018). Fault prediction using machine learning for unmanned aerial vehicles (UAVs). In 2018 International Conference on Unmanned Aircraft Systems (ICUAS) (pp. 1222–1229). IEEE.

[21] Pan, J., Chen, Z., Li, H., & Cao, Y. (2020). Quantum machine learning for intelligent maintenance: A survey. arXiv preprint arXiv:2005.13925.

[22] Pan, S. J., & Baek, S. Y. (2019). Integration of supervised machine learning for real-time health monitoring of unmanned aerial vehicles. IEEE Transactions on Industrial Electronics, 67(3), 2318–2328.

[23] Panesar, S. S., &Wadhera, G. S. (2019). A survey on predictive maintenance using data analytics. International Journal of System Assurance Engineering and Management, 10(2), 437–454.

[24] Şahin, M. (2020). A review on health monitoring and fault diagnosis of unmanned aerial vehicles. International Journal of Aerospace Sciences and Applications, 11(2), 182–193. https://lucris.lub.lu.se/ws/files/94213291/Report_Byg_R_454.pdf

[25] Verstraete, F., & Hastings, M. B. (2010). Quantum resource theories. https://arxiv.org/abs/1806.06107

[26] Wang, C., Guo, H., Chen, Z., Wen, Q., Xu, Y., & Li, Z. (2020). Quantum cryptanalysis: Theory and algorithms. Science China Technological Sciences, 63(12), 2520–2530.

[27] Xiang, Y., & Xie, X. (2018). A review of unmanned aerial vehicle (UAV) based robotic manipulators for industrial applications. Robotics and Autonomous Systems, 103, 233–248.

[28] Yoon, Y., & Sha, W. (2020). Security challenges and countermeasures for quantum-enhanced intelligent transportation systems. IEEE Network Magazine, 34(2), 166–172. https://ieeexplore.ieee.org/document/7842421/1000

[29] Zhang, S., Li, Y., Zhang, P., & Li, B. (2022). Quantum machine learning for remaining useful life estimation of machinery: A review. IEEE Access, 10, 132422–132442.

[30] Zheng, G., Ma, S., Wong, K.K., & Ng, T.S. (2009). Robust beamforming in cognitive radio. IEEE Transactions on Wireless Communications, 9(2), 570–576.

[31] Zhou, D., & Zhang, L. (2020). Recent advances in intelligent maintenance for unmanned aerial vehicles: A review. Reliability Engineering & System Safety, 202, 106923. https://www.researchgate.net/publication/360196712_Recent_Advances_in_Unmanned_Aerial_Vehicles_A_Review

[32] Zhu, L., Li, Y., Lu, C., & Zhou, X. (2020). A survey on deep learning for unmanned aerial vehicle path planning and control. IEEE Access, 8, 151035–151058.

[33] Zhuang, W., He, Z., Bao, Y., Deng, L., Li, L., & Peng, Z. (2021). Quantum machine learning for remaining useful life prediction of lithium-ion batteries. Nano Energy, 88, 106322.

Ruthwika Vemulapalli

8 Quantum-Supported UAV Environmental Sensing

Abstract: The chapter introduces a groundbreaking quantum-enhanced adaptive optics system designed for environmental sensing by unmanned aerial vehicles (UAVs). Leveraging cutting-edge quantum technology, the system aims to revolutionize data precision and sensitivity in UAV-based environmental monitoring. The integration of quantum sensors within a UAV's payload promises unparalleled accuracy in capturing environmental parameters. Adaptive optics play a pivotal role by adjusting a UAV's optical components to counter atmospheric turbulence. Key components of the system include quantum sensors utilizing principles such as superposition and entanglement to measure physical properties with exceptional precision. The miniaturization of sensors is imperative for seamless integration onto UAV platforms.

Critical steps in the deployment process involve sensor calibration, comprehensive testing, and integration of data from multiple sensors. Advanced algorithms are employed for sensor fusion and data processing, enabling real-time transmission of sensor data. The development of lightweight deformable mirrors and compact control electronics tailored for UAV applications is discussed. This chapter concludes by highlighting the role of Shack-Hartmann sensors in detecting incoming wavefront distortions and a sophisticated control system that adjusts deformable mirrors based on wavefront analysis, incorporating feedback.

Keywords: Quantum-enhanced adaptive optic system (QEAS), **UAV** (unmanned aerial vehicle), environmental sensing, quantum sensors, deformable mirror, Shack-Hartmann sensor

8.1 Introduction

In the rapidly evolving field of technological advancements, quantum technology stands out as a beacon of innovation, profoundly impacting diverse sectors, including environmental sensing. Traditional methodologies in environmental monitoring frequently encounter constraints concerning accuracy, sensitivity, and the fidelity of data collected. However, the advent of quantum technology heralds promising avenues to surmount these challenges, fundamentally reshaping our ability to perceive and comprehend the intricacies of our surroundings.

This chapter endeavors to scrutinize the convergence of quantum technology and environmental sensing within the domain of quantum-supported unmanned aerial vehicle **(UAV)** environmental sensing, elucidating the transformative potential of

https://doi.org/10.1515/9783111619323-008

quantum sensors in redefining the landscape of environmental monitoring and analysis. At the heart of our exploration, lies comprehensive examination of the foundational principles involving **quantum sensors** tailored for environmental monitoring, with a keen focus on elucidating their inherent capabilities [1].

Moreover, we undertake a critical discourse on the manifold advantages that quantum technology proffers over conventional methodologies, juxtaposed against the backdrop of the formidable challenges and inherent limitations, necessitating concerted efforts for widespread adoption and integration.

Venturing further, we delve into the intricate components comprising a quantum-enhanced adaptive optics (AO) system engineered for **UAVs**. The synergistic fusion of **quantum sensors** with AO heralds a paradigm shift in environmental sensing, empowering **UAVs** with unparalleled precision and accuracy in data acquisition, thereby catalyzing a plethora of applications in environmental monitoring.

Emphasis is placed on elucidating the intrinsic benefits inherent to quantum-enhanced environmental sensing, encompassing elevated levels of precision, heightened sensitivity, and enhanced data fidelity. Concurrently, an exhaustive exploration of the diverse spectrum of applications facilitated by quantum-enhanced **UAV** environmental sensing, spanning predictive weather forecasting, real-time pollution tracking, and expeditious natural disaster response, is meticulously undertaken.

Summarizing, this chapter aspires to furnish profound insights into the transformative potential engendered by quantum technology in the realm of environmental sensing, envisaging a future characterized by the proliferation of more efficacious and sustainable monitoring practices with the help of technologies such as AO systems, quantum mechanics, superposition, entanglement, cryptography, and tunneling.

8.2 Quantum Technology in Environmental Sensing

Quantum technology, leveraging the principles of quantum mechanics, has significant applications pertaining to environmental sensing, science, and sustainability. These technologies offer unprecedented sensitivity and precision in measuring environmental parameters. Quantum techniques, such as quantum-enhanced spectroscopy is used for enhancing detection and analysis of chemical compositions. For example, quantum cascade lasers can be used for high-resolution spectroscopy, quantum key distribution (QKD) ensures transmission of environmental data received from various sensors, Quantum Imaging techniques use quantum lidars for high-resolution mapping of coral reefs, monitoring deforestation, assessing health of agricultural lands, and quantum metrology that enhances precision of measurements [2].

8.2.1 Quantum Sensors for Environmental Monitoring

Quantum sensors combine sensor technology with the properties of quantum mechanics, including quantum entanglement, quantum interference, and quantum states. **Quantum sensors** gather and carry information through qubits, which refer to photons, ions, and neutral atoms. In contrast to quantum computers, quantum sensors detect changes in rotation, pressure, gravity, and magnetic or electric fields. There are different types of **quantum sensors**, each designed to measure a specific physical quantity, such as time, position, magnetic fields, or temperature. **Quantum sensors** come in various forms, such as atomic clocks, magnetometers, gravimeters, gyroscopes, interferometers, and SQUIDs (superconducting quantum interference devices). Classical sensors are based on phenomena from classical physics, such as electromagnetic radiation or mechanical motion, and are constrained by classical noise and uncertainty, limiting their precision and sensitivity.

8.2.1.1 Approaches to Quantum Sensing

1. **Trapped Ion Quantum Sensing**: Atoms and ions are trapped in a vacuum chamber, and low temperatures are reached. Subsequently, they undergo manipulation with lasers to measure their physical properties while managing their internal states (Figure 8.1).

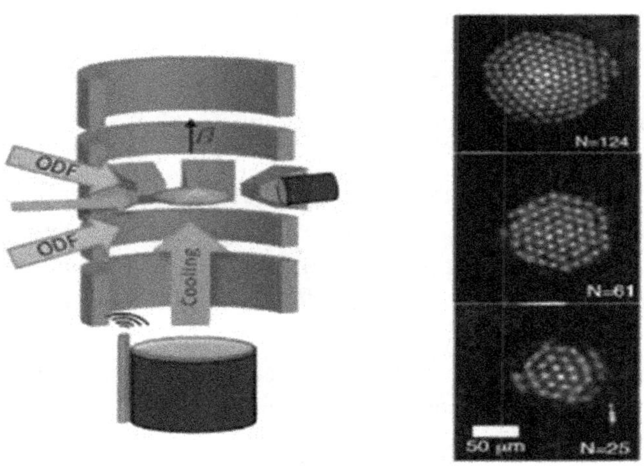

Figure 8.1: Illustration showing a schematic of an experimental setup for optical cooling and trapping of particles. The left side depicts a cylindrical structure with labeled components, including optical dipole forces (ODF) and cooling mechanisms. On the right, there are three microscopic images of particle arrays with different numbers of particles ($N = 124$, $N = 61$, and $N = 25$), arranged in a hexagonal lattice pattern. A scale bar indicating 50 micrometers is shown in the bottom-right image (citation: CC by 4.0, Gilmore et al. [1]).

2. Trapped ions are naturally suited for simulating quantum magnetism. Long-range Ising interactions are combined with spin-dependent forces on arrays of up to ~300 ions. The spin-dependent forces couple the ion spins to a motional mode (or modes) of the ion crystal. At a basic level, the system we have engineered is the Dicke model:

$$\text{HDicke} = (-g0/\sqrt{N}) * (a + a\dagger)Sz + B(t)Sx - \delta a\dagger a$$

where $g0$ represents the homogeneous coupling between each ion and the axial center-of-mass mode, $Sa, a \in \{x,y,z\}$ represents the composite spin of the N-ion crystal, $B(t)$ is the strength of a time-varying transverse field obtained through the application of microwaves resonant with the frequency of the qubit, and δ is the detuning frequency of the spin-dependent force from the frequency of the center-of-mass mode [3].
3. **Photon-Based Quantum Sensing:**
Quantum sensors use individual photons' properties to measure physical quantities.
Photons are utilized for sensing, which could exhibit quantum superposition. These photons can exist in multiple energy levels at the same time, enabling them to capture atomic-level variations intricately, thereby increasing the precision of the photonic quantum sensing process (Figure 8.2).

Fundamental principles involved in photon-based quantum sensing are quantum superposition and interference, quantum entanglement, and quantum coherence, while the techniques enhancing photon-based quantum sensing entail interferometry, single-photon detectors, quantum imaging, and quantum metrology.

8.2.1.2 Quantum Principles Used in Environmental Monitoring

1. **Superposition, Interference**: The ability of an object, such as an atom or subatomic particle, to be in more than one quantum state at the same time is referred to as Superposition. Quantum superposition is indeed distinct from superposition of classical waves. Quantum computer consists of n qubits and exist in superposition of 2^n states: from $|000\ldots0\rangle|000\ldots0\rangle$ to $|111\ldots1\rangle|111\ldots1\rangle$.
Interference is the phenomenon where two waves encompass themselves to mingle, strengthening one another; crests and troughs coincide and cancel out each other. Quantum sensors utilize superposition and interference effects to measure small changes in environmental conditions. For instance, interferometric techniques can detect minute variations in pressure, temperature, and electromagnetic fields.
2. **Entanglement**: Quantum entanglement is a physical phenomenon that occurs when, two or more particles are far apart, the quantum state of each particle cannot be represented in isolation from others. The simplest example of an entangled photon pair is the quantum-mechanical state

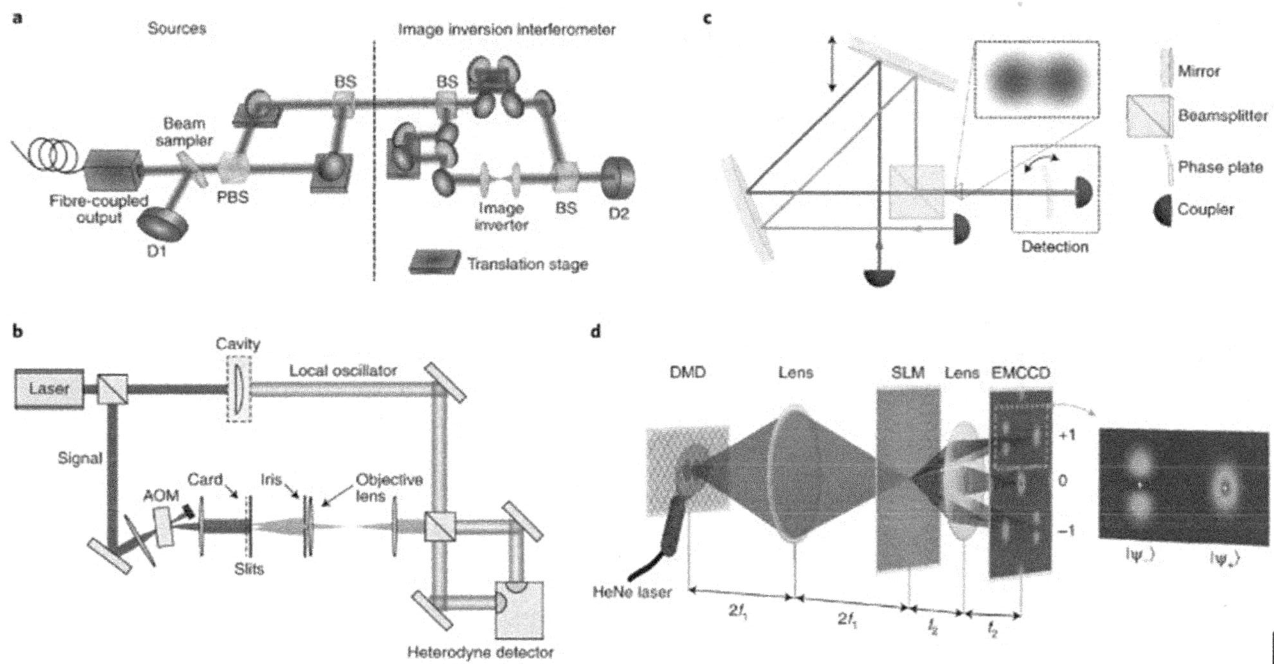

Figure 8.2: Photonic quantum sensor of four labeled subfigures (a, b, c, d), each depicting different optical setups and experimental configurations: (a) A schematic of an image inversion interferometer setup. It includes a fiber-coupled output, a beam sampler, a polarizing beam splitter (PBS), multiple beam splitters (BS), image inverters, and translation stages for aligning beams, ending at detectors labeled D1 and D2. (b) A laser cavity setup for heterodyne detection, featuring a laser source, an acousto-optic modulator (AOM), slits, a card, an iris, an objective lens, and a heterodyne detector. A local oscillator is depicted for signal processing. (c) An optical detection system showing mirrors, a beamsplitter, a phase plate, and a coupler leading to a detection point. The zoomed-in portion displays blurred optical modes. (d) A digital micromirror device (DMD)-based optical setup involving a HeNe laser passing through lenses, a spatial light modulator (SLM), and captured on an EMCCD detector. The resulting diffraction patterns with labeled states (e.g., $|\psi_+\rangle$ and $|\psi_-\rangle$) are shown on the right (license number: 5964001498797: S. Pirandola et al. [2]).

$\Psi = 0.707 * (|\hbar\omega_1,^{+}\rangle \, |\hbar\omega_2,\sigma^{-}\rangle + |\hbar\omega_1, \sigma^{-}\rangle \, |\hbar\omega_2,\sigma^{+}\rangle)$, where $|\hbar\omega_j, \sigma_{\pm}\rangle$ is a single photon state of energy and right or left-hand circular polarization. Entangled photons are essentially used to enhance the precision of measurements. Changes detected by one particle are instantaneously reflected in its entangled partner, enhancing the sensitivity of the sensing system.

3. **Quantum Coherence:** Quantum coherence refers to the ability of a quantum system to sustain a precisely specified quantum state throughout time independent of interactions. Quantum coherence aids quantum sensors to detect slight changes in environment by preserving the phase relationship between quantum states over time [4].

8.2.1.3 Classification of Quantum Sensors

1. **Quantum Magnetometers**: Quantum magnetometers use quantum properties to measure highly sensitive measurements. They are useful for monitoring geomagnetic activity, detecting underground structures, and studying space weather. Spin of subatomic particles (nuclei, unpaired valence electrons)is used by magnetometers. Particles in the earth's ambient magnetic field undergo precision due to polarization. The precession frequency obtained as result is altered to magnetic field units. Quantum results are scalar (total field intensity) as opposed to vector such as that comes fluxgate geophysical instruments or GEM's Suspended dIdD technology. Every single particle has unique spin, connected to magnetic moment which includes unpaired valence electrons and nuclei. Quantum magnetometers are classified into scalar and vector quantum magnetometers; as the conventional definition states the principle of scalar and vector, it holds the same in the case of magnetometers as well.

Scalar quantum magnetometers operate based on atomic transitions or quantum properties of materials akin to the Zeeman effect in alkali atoms. When exposed to a magnetic field, the energy levels of atoms split in a manner dependent on the field's strength. Optically pumped magnetometers use the principles of optical pumping and magnetic resonance to measure magnetic fields. Overhauser magnetometers exploit the Overhauser effect (a phenomenon in NMR where DNP is used to enhance the signal strength of nuclear spins), providing a scalar measurement of the magnetic field. This type of quantum magnetometer is applied in geophysical surveys for mineral exploration and archaeology, space missions to study planetary magnetic fields, security applications for detecting metallic objects.

Vector quantum magnetometers use multiple sensors or configurations that can resolve components of the magnetic field in different directions (three orthogonal axes). Vector components are determined by the alignment of atomic spins with the magnetic field direction. Fluxgate magnetometers incorporate quantum techniques to enhance sensitivity and accuracy. NV center magnetometers utilize NV centers in diamonds, where the quantum states of NV centers are sensitive to the magnetic field's direction, allowing for vector measurements. Applications of these magnetometers entail scientific research for studying the magnetic fields of various sources, including Earth's atmosphere, biomedical applications such as MEG for brain activity mapping, and navigation systems.

2. **Quantum Gravimeter**: Quantum gravimeters measure gravitational fields with high precision and are utilized for monitoring subsurface changes, such as groundwater levels, for detecting mineral deposits or geological formations. In a quantum gravimeter, a cloud of atoms is first trapped in a vacuum chamber and cooled down with lasers to microkelvin temperatures, prior to being propelled upward in free fall. In comparison with classical sensors, gravitational force measurements become more precise with application of quantum-based gravimeters. Use cases have potential to improve accuracy and precision in telescopes and satellites, and space probes variety of measurements across the space industry are observed. Utilization of these sensors has boosted accuracy up to 100,000,000 times more sensitive than classical sensors. Space-based quantum sensors are able to measure changes in the density and seismic status and change of the global environment, climate, and bodies in space that may be of significance to space missions through the use of Quantum Interferometry (a technique that uses quantum mechanics' principles, particularly superposition, entanglement to perform extremely sensitive measurements of physical quantities and widely applied in concepts such as metrology, sensing, and fundamental physics) [5].

3. **Quantum Clocks**: Quantum clocks measure quantum fluctuations, or narrow energy transitions, within actual strontium atoms. Thousands of ultracold strontium atoms are trapped in a lattice of intense laser light inside an ultra-high-vacuum chamber. The atoms are excited with a laser to emit near-infrared light that oscillates at 429,500 GHz – ticking at nearly 430 trillion times per second, significantly faster than the cesium clock. Over the years, advances in lasers, lattice traps, and temperature-controlled chambers have raised the stability and accuracy of strontium atomic clocks. Highly precise quantum clocks detect tiny gravitational fields caused by changes in altitude or subsurface density, useful for geodesy (a scientific discipline dealing with measurement and representation of Earth) and environmental science. A study from MIT describes that even if all noise from the outside world is eliminated, quantum mechanical effects may pose a possibility for clocks, laser beams, and oscillators to lose their stability. Quantum mechanics may limit oscillator's precision.

4. **Quantum Thermometers**: They aid in measuring temperature changes at microscopic scales or nanoscales with high precision, climate studies, monitoring industrial processes, and detecting thermal pollution. Quantum thermometers work on principles such as quantum states, energy levels, temperature-dependent properties, and the Boltzmann distribution at thermal equilibrium, where the population of particles among various quantum states follows this distribution. They are non-invasive and can operate over cryogenic temperature measurements to beyond room tempera-

tures. As far as the research pertaining to quantum thermometer, it could be used to test whether black hole emits small amounts of radiation.

5. **Quantum LIDAR**: Quantum light detection and ranging is an emerging technology that utilizes quantum mechanics principles to achieve enhanced performance in LIDAR systems. Quantum LIDAR systems utilize entangled photons to achieve higher resolution and sensitivity in mapping and surveying landscapes, vegetation, and atmospheric particles. Quantum LIDAR aims to improve the sensitivity, resolution, and range of traditional LIDAR systems by leveraging quantum properties such as entanglement, superposition, and squeezing. Quantum LIDAR systems utilize components such as photon sources, which are typically generated through non-linear optical processes like spontaneous parametric down-conversion (SPDC). Optical components such as beam splitters and phase shifters are used to manipulate quantum states of photons and generate interference patterns. High-sensitivity detectors capable of detecting single photons are essential for these systems to achieve high resolution and sensitivity. Advanced signal processing algorithms and data analysis techniques are employed to extract distance information from interference patterns generated.

8.2.2 Applications of Quantum Technology in Environmental Sensing

Quantum technology offers significant applications in environmental sensing by leveraging the unique properties of quantum mechanics to achieve enhanced sensitivity, precision, and efficiency in measuring various environmental parameters.

1. **Quantum Sensors for Climate Monitoring**:
   High-Precision Thermometry: Quantum thermometers based on properties such as entanglement and coherence offer precise temperature measurements crucial for understanding climate change and monitoring temperature variations.

   Gravimetry and Sea Level Monitoring: Quantum gravimeters detect tiny changes in gravitational changes providing insights into changes in mass distribution.

   Atmospheric Scattering: Quantum sensors measure atmospheric parameters such as humidity, pressure, composition with high precision aiding climate modelling, weather forecasting, air quality monitoring.
2. **Quantum-Enhanced Imaging and Sensing for Environmental Monitoring**:
   Quantum LIDAR systems can provide high-resolution 3D mapping of terrain, vegetation, and land cover, enabling environmental monitoring. Quantum imaging techniques involving quantum-enhanced fluorescence imaging and quantum-enhanced sensing in low light conditions, allow for improved visualization. Quantum-enhanced spectroscopic techniques enable precise identification and quanti-

fication of chemical compounds in environmental samples, facilitating pollution monitoring and contamination detection.

3. **Remote Sensing with Quantum Technologies:**
   Entangled photon sources and quantum communication systems enhance remote sensing capabilities by enabling secure data transmission, precise time synchronization, and improved signal-to-noise ratio. Integration of quantum sensors into satellite platforms allows for global-scale environmental monitoring, including measurements of ocean temperature, ice thickness, and vegetation health [6].
4. **Environmental Monitoring in Challenging Environments:**
   Quantum sensors that are robust and reliable in harsh environmental conditions make them suitable for applications in remote or hazardous locations. In addition to this, quantum technologies enable highly sensitive underwater sensing for monitoring oceanic parameters such as temperature, salinity, acidity, and detecting underwater objects and geological features.
5. **Applications in Earth Observation and Geospatial Analysis:**
   Quantum sensors, by enhancing the accuracy and resolution of geodetic measurements, aid in mapping Earth's surface, monitoring land deformation, and studying tectonic activity. Quantum technologies enhance navigation systems by providing precise positioning, which is crucial for environmental monitoring missions, as shown in Table 8.1.

Table 8.1 shows quantum computer advantages in optimization, machine learning, materials science, and physics. Quantum computers excel at difficult tasks, such as simulating molecules and factoring large numbers.

Table 8.1: Application-based quantum versus classical computers.

Segment	Applications	Likely quantum computers versus classical computers
Optimization	Find the best decision action for specific goals	New QC algorithms for many applications.
AI and ML	Solve machine learning problems	Speedup of AI learning algorithms
Encryption	Factoring very large numbers	Exponentially fewer calculation steps
Chemistry	Chemical process simulation	Not feasible on classical computers
Fundamental science	Test physics theories Analyzing high energy physics experiments	Not feasible on classical computers Tremendous speedup on QCs

8.2.3 Advantages of Quantum Technology over Classical Approaches

Classical computers have less compute power than quantum computers. The classical computers employ bits (eight units of bits is referred to one byte as basic unit of data.) whereas quantum computers use qubits. Binary writing of code by classical computers is in manner as a 1 or a 0, which is termed as serial processing, which is sequential in nature. Many computing systems use parallel processing, development of classical processing, that performs simultaneous computing tasks. Classical computers also return one result because bits of 1s and 0s are repeatable due to their binary nature. Qubits of quantum computer simultaneously represent a 1 and 0 at the same time, in relation to the number of qubits, the power of quantum computer increases exponentially. The quantum computer performs 2^N computations, where N represents number of qubits. Computations in traditional classical computer approaches is deterministic – repeating same input results in same output, while in quantum approach, computations are probabilistic – multiple possible outputs to same input. Data processing in classical computers is performed by logic, in sequential order, whereas by quantum logic at parallel instances in quantum computers. Classical approach operations are obtained through computations in Boolean algebra, linear algebra, compared to Hilbert space approach in quantum computing. Quantum computing enhances AI by increasing its speed, efficiency, and accuracy [7]. Utilizing qubits, quantum computing operates non-linearly, surmounting classical computers. This breakthrough enables quantum computing to be applied in various AI use cases. Quantum computers excel at solving optimization problems due to their ability to explore vast solution spaces efficiently. Quantum algorithms like quantum annealing and variational quantum eigensolver (VQE) offer promising approaches to address optimization tasks that are beyond the reach of traditional computers, leading to more accurate and faster AI solutions. Quantum computing holds the promise of enhancing AI capabilities in areas, such as recognizing pattern, cryptography, and natural language processing (NLP). Quantum machine learning algorithms, such as quantum support vector machines and quantum neural networks, could extract intricate patterns and correlations from massive datasets with unprecedented speed and accuracy.

In Table 8.2, the classical and quantum computing are compared, which shows that classical bits are either 0 or 1, while quantum bits can represent both at once. Quantum computers gain capability exponentially with more qubits, while classical computers gain linearly with transistors. Quantum computers are better suited for complex problems.

Table 8.2: Quantum computing versus classical computing.

Category	Quantum computing	Classical computing
Suitability	High for complex problems	Suitable for routine processing
Error rates and environment	High error rates require extreme conditions	Low error rates can operate at room temperature
Computing units	Qubits (superposition of states and entanglement)	Bits (binary states: 0 and 1)
Capacity	Exponentially increases with the number of qubits	Capacity increases linearly with the number of transistors

8.2.4 Challenges and Limitations of Implementing Quantum Technology

Although quantum technology has many advantages and applications in various sectors, there are quite a few challenges and limitations in implementing quantum technology since it is an emerging field of study, and it's been a herculean task to deal with these limitations. Researchers are orchestrating proofs to eradicate these limitations and challenges. Let us look at those challenges and limitations as well.

Quantum systems need to maintain coherence to perform computations or measurements. Decoherence, caused by interactions with the environment, can destroy quantum states. Achieving and maintaining isolation of quantum systems from environmental noise is difficult and requires sophisticated shielding and cooling techniques. Quantum bits are highly susceptible to errors from various sources – decoherence, gate imperfections. Developing effective quantum error correction is essential but challenging due to overhead of additional qubits required tom detect and correct errors without destroying quantum information. Cryogenic system's development is essential for technologies such as, superconducting qubits, sensors that require extremely low temperatures which are expected to be reliable and cost effective. Designing efficient quantum algorithms that can outperform classical counterparts for practical problems is an ongoing area of research, developing robust software tools and frameworks for programming quantum computers and managing quantum data is essential for making computing accessible to researchers and industry. Developing post quantum cryptographic methods is critical, ensuring security of quantum data – both during transmission, presents new challenges that need to be addressed. One of the limitations describes regrading output observance – data corruption may happen if output information is retrieved after a quantum computation process is completed, while the other illustrates the utilization of high-end computing applications face the limitation of high qubits and quantum states [8].

Incorporating AI technologies into quantum computing can enhance error correction, noise reduction, algorithm operation, and system reliability, making quantum computing more practical and accessible. AI algorithms, particularly machine learning, can be trained to identify and mitigate noise patterns in quantum systems. Adaptive filtering and noise prediction models can maintain coherence. Algorithms such as evolutionary and neural networks aid in discovering and optimizing quantum algorithms. NLP techniques can enable more intuitive programming interfaces, allowing users to describe quantum algorithms in natural language [9].

8.3 Components of Quantum-Enhanced Adaptive Optic System for UAVs

Quantum-enhanced AO systems consist of components such as quantum sensors and AO to be used for the purpose of **UAV** in order to achieve the outcome of quantum-supported **UAV** environmental sensing. This subchapter aims to demonstrate the construction of components, that is, quantum sensors and AO.

8.3.1 Quantum Sensors

As discussed in Section 2.1, quantum sensors for environmental monitoring leverage quantum mechanical properties such as superposition, entanglement, and coherence to achieve high sensitivity and precision in detecting and measuring various environmental parameters.

Quantum magnetometers, such as Nitrogen-Vacancy (NV) centers in diamond, are constructed by embedding NV centers within a diamond lattice – constructed in the following manner – a green laser with wavelength of about 532 nm excites NV centers. Microwave radiation manipulates spin states of NV centers. Photodetectors measure fluorescence emitted by NV centers, which changes magnetic field strength. Magnetic resonance of spin states shifts response to external magnetic fields that can be detected via changes in emitted fluorescence when excited by laser. Cold atom interferometers, a type of quantum gravimeters are designed by considering an atom source of cold atoms (rubidium or cesium). Lasers cool and trap the atoms in magneto-optical trap. A setup is designed to create an atomic interferometer involving sequence of laser pulses that split, redirect, and recombine atomic wave packets. For detection of interference pattern, a system is created by the recombination of atomic packets.

Quantum LIDARs require a source of entangled or squeezed photons, which are generated using SPDC or four-wave mixing in nonlinear crystals or waveguides or using optical parametric oscillators. A combination of lenses, mirrors, and beam splitters is used to guide the photons toward the target and capture the light reflected back. and beam splitters to direct photons toward the target and collect the reflected light. Single-photon detectors such as avalanche photodiodes or photon detectors or superconducting nanowire single photon detectors (SNSPDs) are utilized. High-precision timing electronics is used to measure the time of flight of photons. Quantum thermometers such as NV centers in diamond or quantum dots – laser excites NV centers or quantum dots, microwave radiation is used to manipulate spin states. Photodetectors measure fluorescence changes that indicate temperature variations.

Quantum LIDAR stands out directly applicable technology due to its high precision in distance measurement and environmental mapping. Quantum magnetometers can provide valuable magnetic field data, enhancing navigation and environmental sensing capabilities. Quantum thermometers are less directly related to AO, but can play supportive role in monitoring **UAV's** operating environment and health (Figure 8.3). In this approach, quantum LIDAR is considered best as per the application [10].

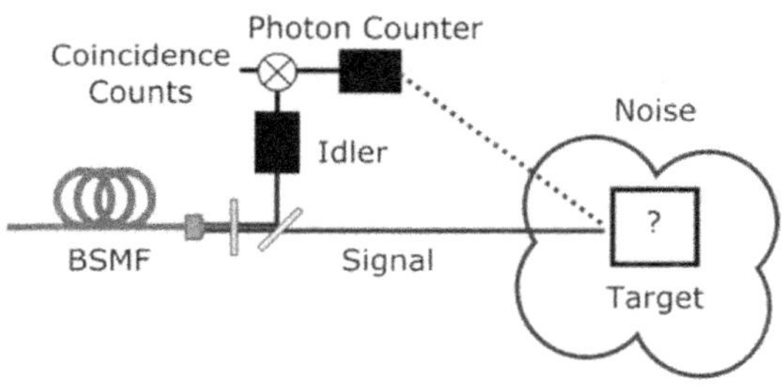

Figure 8.3: A schematic diagram illustrating a quantum optical setup for detecting a target. The setup begins with a broadband source of multi-frequency light (BSMF), followed by a beam splitter that separates the light into two paths: signal and idler. The idler path is directed toward a photon counter for measurement, while the signal path is aimed at a target located within a noisy environment represented by a cloud-like shape. The target is marked with a question mark (?). Coincidence counts are used to correlate the measurements from the photon counter and the signal path for identifying the presence of the target despite the noise (**open source licensed by CC by 4.0,** Quantum Lidar | EQOP," *EQOP*, Apr 01, 2014. https://eqop.phys.strath.ac.uk/ryd-projects/qt-lidar/).

8.3.2 Adaptive Optics

The AO system eliminates atmospheric disturbances and permits telescopes to achieve diffraction-limited imaging from the ground. This is critical in achieving a high signal-to-noise ratio, given as $S/N = (A * h/\varepsilon)^{0.5}/(FWHM)$, where A represents the area of the telescope, h is the total transmission of optics and the detector quantum efficiency, ε is the background emission, and FWHM is the full width at half maximum of the stellar image. The basic idea of AO is to first measure the amount of atmospheric disturbance, then correct for it before the light transmits through the camera. AO requires an object that is bright enough to rapidly and accurately measure the incoming wave front. A nearby bright star is required if the item of interest is not bright enough, which reduces the amount of sky coverage because there won't be any star close enough to illuminate every area of the sky. A laser guide is utilized if there isn't a nearby star. kilometer thin layer of sodium atoms at a 90-kilometre altitude is excited by a laser aimed in the same direction as the telescope in Earth's ionosphere. This provides the AO system with a point source that functions as an artificial star [11].

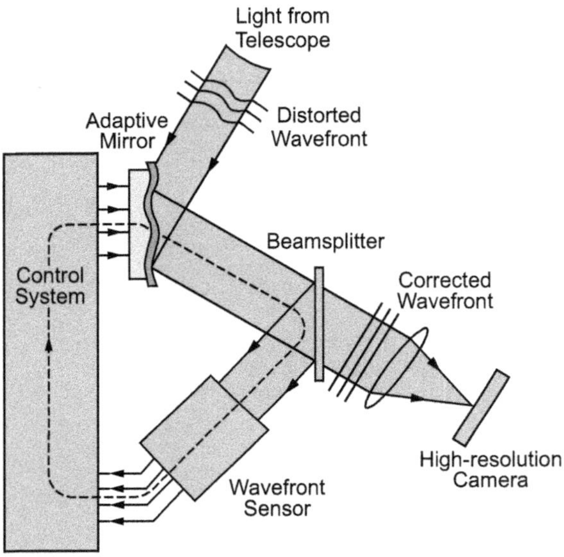

Figure 8.4: A schematic diagram illustrating the working principle of an adaptive optics system. The setup includes light from a telescope passing through atmospheric turbulence, causing distortions. A wavefront sensor detects these distortions, and a control system calculates the necessary corrections. The corrected wavefront is achieved using a deformable mirror, resulting in a higher-resolution image for improved optical performance (license number: 5963991491950: A.T. Tokunaga [3]).

Shack-Hartmann wavefront sensor is a popular type of wavefront sensor, which is designed by placing a micro-lens array in front of a CCD or CMOS detector array. Each lenslet concentrates light at center of a predetermined set of pixels in the array, which is positioned at the focal plane of the lenslet array when the wavefront is incident on the sensor. Focus areas in the lenslet's corresponding pixels are located distinctly when a warped wavefront hits the microlens array. The form of the wavefront incident on the sensor is characterized by examining the locations of each spot on the detector array (Figure 8.4). A deformable mirror is a key component in AO systems, which are used to correct wavefront distortions in real time. These mirrors have a flexible surface that can be precisely shaped by actuators to correct optical aberrations, improving the performance of imaging systems. When deformable mirrors are integrated into **UAV** systems, especially those using quantum-enhanced sensors, quantum LIDAR in this approach significantly improves imaging quality, precision navigation, and environmental monitoring capabilities. By ensuring that wavefront distortions are corrected, deformable mirrors enable UAVs to gather more accurate and high-resolution data, leveraging the full potential of quantum technologies.

AO systems provide real-time correction of wavefront distortions, which is evidently crucial for applications requiring continuous and accurate data acquisition, such as surveillance, environmental monitoring, and navigation. By correcting distortions, AO systems significantly enhance the quality of data collected by UAVs, which involves more accurate measurements and higher fidelity in sensor readings. UAVs equipped with AO can operate effectively in a wider range of environmental conditions and at higher altitudes, expanding mission capabilities. Quantum LIDAR benefits from improved optical conditions provided by AO systems, which result in better performance and more reliable data from advanced sensors [12].

8.4 Integration of Quantum Sensors into UAV Payload

Integrating quantum sensors into **UAV** payloads is an emerging field with significant potential to enhance various applications, including navigation, communication, environmental monitoring, and defense. Quantum sensors leverage principles of quantum mechanics to obtain higher levels of precision and sensitivity.

8.4.1 Design Considerations for UAV Payload Integration

Integrating quantum sensors into **UAV** payloads requires careful attention to various design considerations to ensure optimal performance and reliability. Meeting the needs of constraints is a herculean task, which requires the utilization of numerous technologies and several concepts in theory as well. Design considerations include:

Size, weight, and power constraints come into the picture essentially when there is an urge to design quantum sensors into **UAV** payloads, as such quantum sensors need to be compact and lightweight to fit within the limited payload capacity of **UAVs**. Advances in microfabrication and nanotechnology are evident considerations while designing the sensors to incorporate into the payload. Quantum sensors often require stable, significant power sources. Efficient power management systems and lightweight power supplies play a pivotal role in the design structure of sensors for **UAV** payload integration. Quantum sensors must be mounted using vibration-damping materials and isolation platforms to eradicate the effects of mechanical vibrations and shocks on **UAVs**. Integration of effective cooling systems, such as cryocoolers or thermoelectric coolers is crucial as quantum sensors need to operate at cryogenic temperatures to avoid quantum decoherence.

Shielding and grounding these sensors is essential to prevent quantum sensors from electromagnetic interference. High-performance, low-power onboard processors are essential to monitor large volumes of data that requires real-time processing. Generally, sufficient data storage capacity and reliable and secure data transmission systems are needed to handle data output and for real time data relay respectively.

Lithium-polymer (Li-Po) batteries and Lithium-ion batteries are preferred power sources for quantum sensors in **UAV**s. These batteries have essential advantages over others, such as high energy density, lightweight design, flexible form factors, and the ability to be shaped to fit into various **UAV** designs, while offering good power output for their size. Advanced Battery Management Systems are essential for monitoring and managing battery health, preventing overcharging, and ensuring safety, with features such as balancing cell charge levels, temperature monitoring, and fault detection. The utilization of DC-DC converters to step down or step up voltage as required by quantum sensors and other **UAV** systems is critical. High-efficiency converters are crucial to minimize power losses and extend battery life. Vibration damping materials, such as elastomers, foam, or gel-based materials, are utilized to isolate the sensors from mechanical vibrations. These vibration isolation platforms are specifically designed to dampen **UAV** vibrations [13]. Faraday cages, conductive coatings and grounding techniques are processed to shield quantum sensors from external electromagnetic interference (EMI). High-performance processors with low power consumption based on ARM architecture would handle real-time data processing efficiently. Integration of FPGA for specialized processing tasks aid in data processing and communication. Light weight, high-capacity SSDs or other non-volatile memory to store large volumes of data generated by quantum sensors, utilizing QKD for high-security communication links would enhance performance of quantum sensors and in successful integration with **UAV** payload (Figure 8.5).

8.4.2 Miniaturization of Quantum Sensors for UAV Applications

Miniaturizing quantum LIDAR for **UAV** applications can be achieved using micro-electro-mechanical systems (MEMS) involving several key aspects reducing size and weight of optical components ensuring efficient power management maintaining high resolution and accuracy. Utilizing MEMS technology creates tiny, fast and precise mirrors that can steer laser beam. These mirrors are much smaller and lighter than traditional mechanical scanners [14]. MEMS mirrors can replace mechanical parts providing fast, precise, and reliable beam steering. MEMS mirrors tiny, movable mirrors fabricated using semiconductor processes that can deflect laser beam in different directions. Implementing two-axis MEMS mirrors to enable full 3D scanning capabilities allows the LIDAR to scan environment in both horizontal and vertical planes. Developing micro-lasers or vertical-cavity surface-emitting lasers (VCSELs) that are small, efficient and can be integrated on chip can be achieved by MEMS technology. Miniaturizing quantum LIDAR utilizing MEMS technology to create micro-lasers or inte-

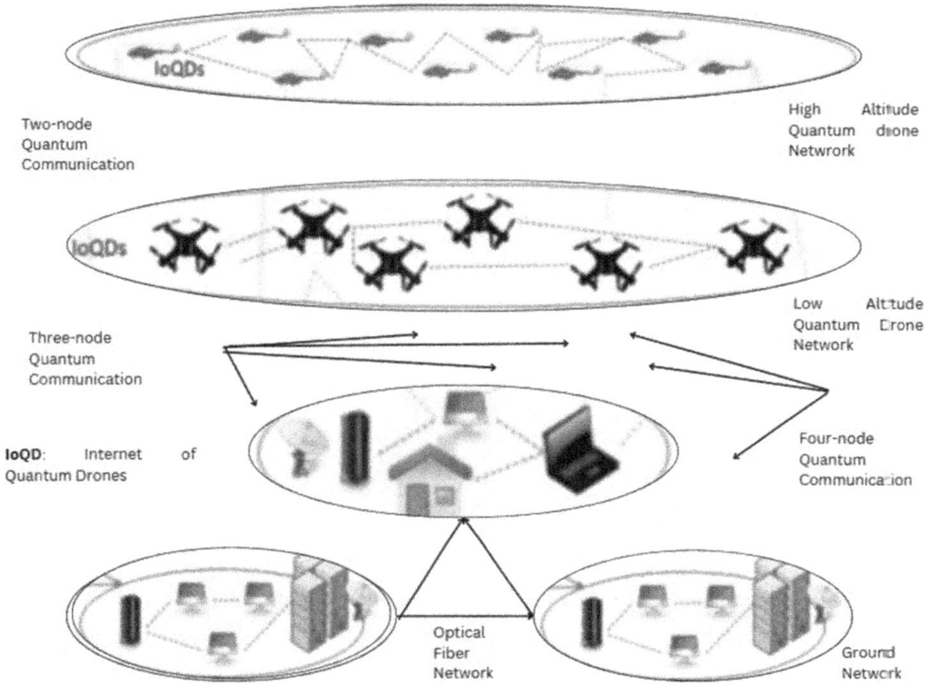

Figure 8.5: A schematic diagram of a drone network architecture. The diagram depicts UAVs equipped with quantum sensors, communication modules, and antennas. Data is exchanged with ground stations, cloud servers, and edge computing systems. The components include wireless sensor networks, vehicular communication systems, and AI/ML modules. Applications such as environmental monitoring, surveillance, and emergency response are highlighted (open-sourced license CC by 4.0: Kumar et al. [4]).

grate existing miniaturized lasers such as VCSELs on a MEMS platform is essential. Design and fabricate MEMS mirrors that can precisely control direction of laser beam. Mirrors should be able to tilt and rotate at high speeds to enable rapid scanning of environment. Implementing SPADs or SNSPDs using MEMS fabrication techniques is advised. The detectors need to be highly sensitive to single photons and have fast response times to capture return signal accurately. Designing low power electronics and efficient power management systems includes using energy-efficient drivers for MEMS mirrors and low power amplifiers for detectors. Integrating high-performance and low-power processors on same chip to handle data processing is essential which includes algorithms for distance calculation, image reconstruction and noise reduction. MEMS fabrication aids in creating extremely small components, significantly reduces the overall size of LIDAR system. Miniaturized components are lighter that makes LIDAR system more suitable for **UAV** applications since weight is critical factor. MEMS devices consume less power thus, extending operational time of **UAV**. Batch fabrication techniques used in MEMS manufacturing can lower production costs eventually making advanced LIDAR systems more accessible (Figure 8.6).

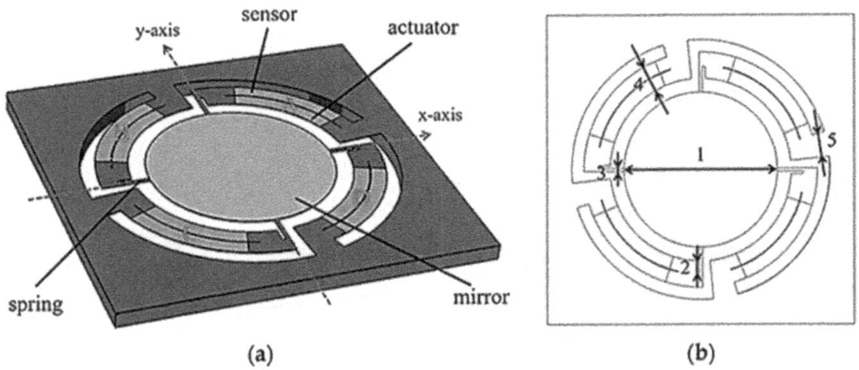

Figure 8.6: (a) The structure of the MEMS mirror *design* and (**b**) *design* parameter index (open-sourced license CC by 4.0: Liu et al. [5]).

8.4.3 Integration Challenges and Solutions

Quantum sensors often require precise temperature control to maintain quantum coherence. Fluctuations in temperature can degrade performance. **UAVs** are subject to vibrations and shocks during flight, and these mechanical disturbances interfere with the delicate operations of quantum sensors. Quantum sensors are highly sensitive to electromagnetic fields, **and UAVs** with onboard electronics and communication systems can create substantial EMI that disrupts sensor operation. Effective shielding and grounding techniques must be implemented to mitigate EMI effects. Quantum sensors generate large amounts of data that require real-time processing. Integration of high-performance, low-power data processors is essential but challenging due to limited space and power on **UAVs**. Ensuring low latency and high throughput in data processing and transmission systems is critical for real-time applications. Quantum sensors need to operate reliably under varying environmental conditions, including temperature extremes, humidity, and mechanical stress. Ensuring the long-term stability and robustness of sensors in the field is a significant challenge. Quantum sensors often require precise calibration, which can be difficult to perform in the field. Regular maintenance may be needed to ensure optimal performance. Integrating quantum sensors with **UAVs'** existing systems requires sophisticated hardware and software modifications. Ensuring compatibility and seamless operation with other **UAV** components is complex and requires thorough testing [15].

Miniaturization challenges using MEMS technology include fabrication complexity, material limitations, thermal management, signal integrity and noise, precision alignment, scalability, and reproducibility. MEMS fabrication processes are intricate and require high precision. Achieving the necessary accuracy at the microscale can be technically challenging. Integrating multiple functionalities on a single MEMS chip adds to the complexity. Materials chosen for MEMS components need to be compatible with MEMS fabrication techniques. Advanced materials such as graphene and diamond NV centers are promising but pose challenges in terms of integration and large-scale production. Miniaturized quantum sensors often generate heat, which must be managed effectively. Designing efficient thermal management systems at the micro-scale is difficult. Ensuring high signal integrity and minimizing noise in MEMS-based quantum sensors is crucial for accurate measurements. Precise alignment of optical components in MEMS-based quantum sensors is essential for their proper functioning.

Some of the solutions to the challenges in the integration and usage of MEMS technology are discussed in Section 9.4.2. Other solutions incorporating AI techniques could probably be:

AI algorithms dynamically adjust vibration damping mechanisms and can predict and counteract vibrations in real time using feedback from onboard accelerometers and gyroscopes. AI can optimize the operation of cooling systems, such as cryocoolers, which reduces power consumption while maintaining necessary low temperatures. AI-based thermal simulations during design phase to predict and mitigate hotspots can improve heat dissipation strategies. Utilizing AI-derived generative design tools to create optimized layouts for MEMS-based quantum sensors can achieve minimal size and weight while maximizing performance. AI optimizes new materials with properties suited for MEMS quantum sensors such as improved thermal conductivity or reduced electromagnetic susceptibility. Implementing AI algorithms that determine optimal placement of multiple miniaturized sensors within UAV to maximize coverage and data accuracy and those that enable distributed sensing and data fusion, allowing multiple miniaturized sensors to work together cohesively.

8.4.4 Power Management for Quantum Sensor Operation

Lithium-polymer (Li-Po) and lithium-ion (Li-ion) batteries are commonly used power sources for quantum sensor operation in UAV payloads due to their high energy density and rechargeable properties. Li-Po batteries utilize polymer electrolytes, as the name suggests, which allow for flexibility in cell design and packaging. They typically consist of multiple layers of electrodes separated by a porous polymer separator. During charging, lithium ions move to the negative electrode from the positive electrode through the electrolyte. In the discharging phase, the lithium ions trace back to the positive electrode through the electrolyte, dissipating energy that can be used to power quantum sensors such as quantum LIDAR and other devices. Li-Po batteries have higher energy density compared to Li-ion batteries, which utilize liquid electrolytes containing lithium salts. They typically consist of a graphite anode, a metal oxide cathode, and a separator soaked in electrolyte. Li-Po batteries are typically lighter than Li-ion batteries with similar energy capacities, making them ideal for applications where weight is a critical factor, as in UAV payloads. However, Li-Po batteries are prone to swelling and thermal runaway if overcharged or mishandled. If subjected to high discharge rates or frequent deep discharges, Li-Po batteries may have a shorter lifespan compared to Li-ion batteries. Li-Po batteries can be manufactured in various shapes and sizes, including flexible and custom shapes, allowing for more versatile packaging options. Enhancing the efficiency of quantum sensors requires implementing a robust battery management system (BMS) to monitor the state of charge (SoC), state of health (SoH), temperature, and current of the battery. BMS must include protection features against overcharging, over discharging and short circuiting to ensure safe operation. Li-Po battery charges with balancing capabilities to ensure each cell within battery pack is charged evenly. Avoid overcharging by setting appropriate charge limits and using smart charging techniques. Storing Li-Po batteries at cool, dry change and at approximately 50% charge if they are not used for extended period and incorporating effective thermal management solutions to dissipate heat generated during discharge and charging cycles is essential. Padding and secure monitoring solutions protect Li-Po batteries discharge cycle within 20–80% of its capacity to extend its lifespan, implementing duty cycling and power management strategies to optimize the battery's usage based on quantum sensor's power demands could enhance the performance of quantum sensors when integrated with **UAV** payload (Figure 8.7). Integrating Li-Po batteries into **UAV** prototype and conducting extensive testing to ensure reliable operation under various conditions. Test BMS and thermal management systems to validate their effectiveness. Securely mount the Li-Po batteries within **UAV** ensuring proper electrical and physical protection. Implementing monitoring, control systems to access battery's performance during operation. Utilize real time data from BMS and other sensors to monitor battery's status during flight and make adjustments based on operational data to improve battery performance and longevity [16].

Figure 8.7: The images depict a Li-Po battery and its structural arrangement design (open-sourced license CC by 4.0 [21]).

8.4.5 Communication and Data Transfer from Quantum Sensors to UAV Control Systems

Integrating **UAV**s with IoT and 5G technologies can significantly enhance their efficiency, capabilities, and overall performance. This integration enables real-time processing, low-latency communication, and better resource management. 5G networks provide ultra-low latency, allowing **UAV**s to communicate with ground control stations and other devices in real time, which is crucial for applications involving decision-making and immediate data analysis. **UAV**s can leverage edge computing to process data locally or at nearby edge servers, reducing the need to send large volumes of data back to a central server. This technique is especially useful for quantum sensors producing large datasets. 5G offers significantly higher bandwidth compared to earlier versions, enabling the transmission of high-resolution sensor data from LIDAR without delays. IoT allows for the interconnection of multiple devices, including other **UAV**s, sensors, and ground stations, facilitating coordinated missions and data sharing. IoT-enabled **UAV**s can optimize power consumption by dynamically adjusting their operations based on real-time data and environmental conditions. IoT sensors can monitor **UAV** components and predict maintenance needs, reducing downtime and extending the lifespan of **UAVs**. IoT and 5G enable more sophisticated autonomous flight capabilities by providing reliable and fast channels of communication for **UAV**s to make real-time decisions based on sensor data. These technologies allow for scalable **UAV** operations, where multiple **UAV**s can be deployed and managed efficiently for large-scale missions.

Quantum sensors, such as LIDAR, generate large volumes of data due to their high-resolution capabilities. Data from quantum sensors is highly precise and requires accurate and timely processing to be useful. Initial processing and filtering of data can be done onboard the **UAV** using edge computing. This reduces the amount of data that needs to be transmitted and allows for quick initial analysis. Applying data compression techniques significantly reduces the volume of data that needs to be sent over the network. Advanced algorithms can compress data without losing critical information. 5G can handle the high data rates required for transmitting raw or partially processed quantum sensor data. 5G's low latency and high reliability ensure that data from quantum sensors is transmitted with minimal delay, allowing for near real-time analysis. IoT platforms can manage and coordinate multiple sensors and devices, ensuring efficient data collection and transmission. IoT systems aggregate data from various sensors, preprocess it, and send the most relevant information over the 5G network. Data can be sent to nearby edge servers for more intensive processing and analysis. This reduces the load on the **UAV** and allows for more complex computations without latency issues. For long-term storage and deeper analysis, data can be uploaded to cloud platforms, which allow for extensive data analysis using AI and machine learning algorithms to extract valuable insights. The **UAV** is equipped with quantum LIDAR and configured to use 5G and IoT for communication. IoT sensors provide environmental data, which is used to optimize the **UAV**'s flight path in real time. Quantum LIDAR offers high-resolution terrain data. Initial processing of LIDAR data is essential for filtering noise and compressing data. Compressed data is sent in real time to a ground control station or edge server for further analysis. Edge servers process the LIDAR data to generate 3D maps and detect anomalies (Figure 8.8). Data from other sensors is integrated to provide a comprehensive analysis of the environment. Ground control or autonomous systems make real-time decisions based on processed data, such as adjusting the flight path or targeting specific areas for detailed scanning. Processed data is stored in the cloud for post-mission analysis and long-term use.

8.5 Benefits of Quantum-Enhanced Environmental Sensing

Quantum-enhanced environmental sensing leverages quantum mechanics principles to improve precision, sensitivity, and overall performance of sensors used to monitor various environmental parameters. Quantum sensors operate at higher speeds, which allows for real-time monitoring of dynamic environmental conditions, enabling applications such as weather forecasting and disaster response. Quantum sensors enable continuous, long-term monitoring and provide valuable time-series data to understand trends and changes in the environment over time. Quantum techniques such as squeezing and entanglement reduce noise in measurements, leading to more reliable data, which is beneficial in noisy environments where traditional sensors struggle. Quantum sensors are less susceptible to environmental interference, such as electromagnetic noise, and are useful in enhancing the collection of data from greater distances, particularly for satellite-based environmental monitoring. Quantum techniques are utilized for non-invasive monitoring of the environment, preserving the integrity of natural habitats and reducing the impact of data collection activities. Quantum-enhanced environmental sensing offers numerous benefits, including higher sensitivity and precision, greater resolution, real-time monitoring capabilities, and reduced noise and interference. Such advantages lead to better data quality and deeper insights into environmental processes. By leveraging the advanced capabilities of quantum sensors, the ability to monitor and understand the environment is significantly improved.

8.5.1 Higher Precision and Sensitivity

Entangled particles exhibit correlations that can be exploited to improve measurement precision beyond classical limits. Quantum entanglement can enhance the sensi-

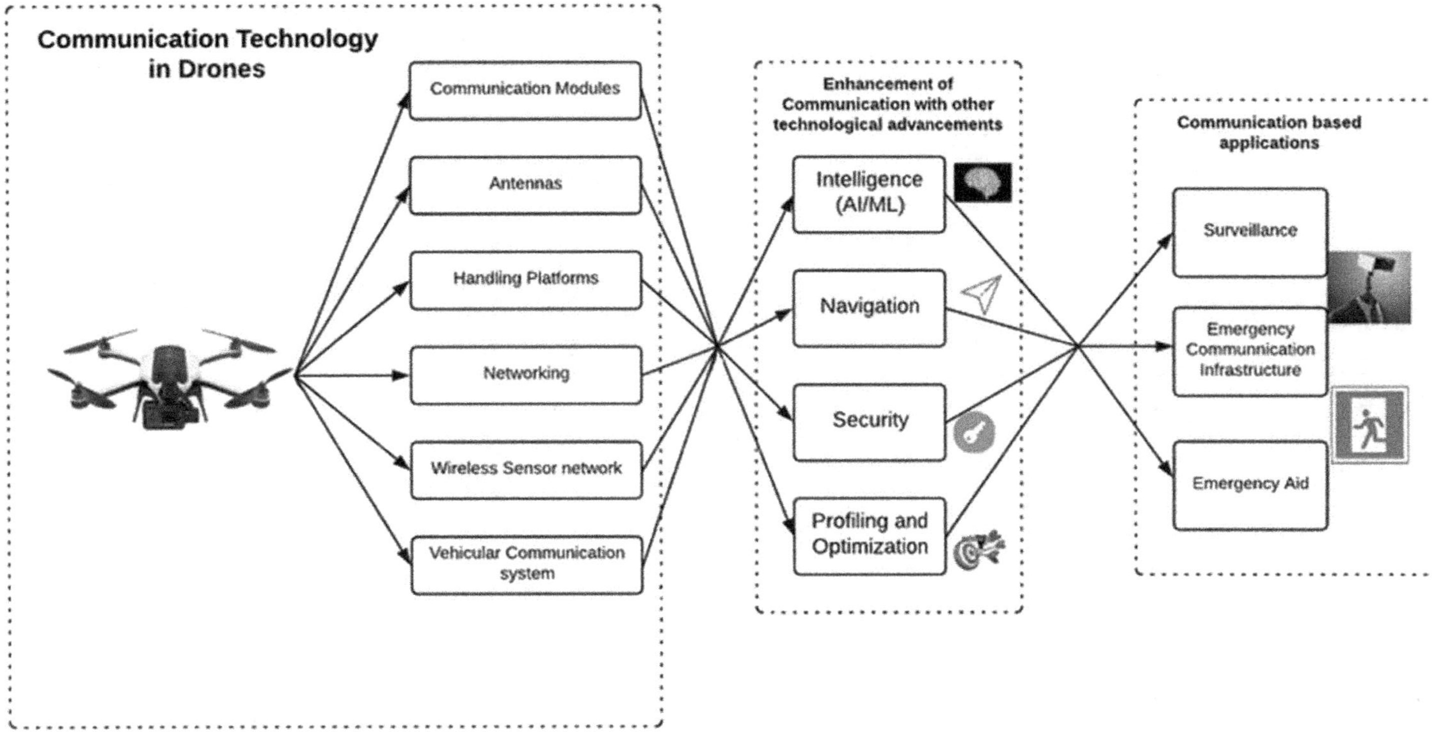

Figure 8.8: Illustration of various aspects of communication technology that enable drone operations, including communication modules, antennas, handling platforms, networking, wireless sensor networks, vehicular communication systems, and the integration of AI/ML for enhanced communication and navigation (license number: 5963700472211, A. Sharma et al. [6]).

tivity of measurements by reducing noise that affects classical sensors. Quantum metrology achieves higher precision in measurements of physical quantities such as time, magnetic fields, and gravitational fields using entangled states. Quantum sensors utilize superposition states to probe environmental parameters with greater sensitivity. Quantum squeezing reduces uncertainty in one component of a quantum system at the expense of increased uncertainty in the conjugate component; hence, the reduction in measurement noise can improve the sensitivity of sensors. Squeezed states improve the signal-to-noise ratio, enabling the detection of weaker signals that would be obscured by noise in classical sensors. Quantum sensors detect extremely small changes in potential energy landscapes.

Quantum-enhanced LIDAR systems use entangled photons to achieve higher resolution and greater range, which allows for detailed mapping of terrain, vegetation, and atmospheric profiling. Improved precision in distance measurements enables better depth perception and 3D imaging capabilities. Quantum magnetometers, based on nitrogen-vacancy centers in diamonds, detect extremely weak magnetic fields with high precision, aiding in the study of geomagnetic phenomena and the detection of underwater or underground anomalies. Quantum gravimeters use cold atoms or atom interferometry to measure gravitational fields with high precision, which entails the mapping of geological structures and the monitoring of changes in Earth's crust. Quantum sensors detect trace amounts of gases, contributing to better air quality monitoring and green house gas tracking. These sensors can be used to monitor emissions from industrial processes with high accuracy, ensuring compliance with environmental regulations.

8.5.2 Improved Data Quality Through Adaptive Optics

AO enhances the operation of optical systems by dynamically correcting distortions caused by atmospheric factors. Upon integrating with quantum-enhanced environmental sensing, AO can significantly improve data quality by ensuring quantum sensors receive clearer and more accurate signals. As discussed in Section 9.3.2, AO systems consist of three main components: a wavefront sensor that measures distortions in incoming light, a control system that calculates necessary corrections to counteract the distortions, and a deformable mirror that adjusts its shape in real time to scrutinize distortions and produce clearer images. AO corrects distortions caused by atmospheric turbulence, providing clearer and more accurate signals to quantum sensors. This is crucial for applications like quantum LiDAR, where precise distance and shape measurements are required. AO enhances the resolution of images captured by quantum sensors by correcting wavefront distortions. AO systems continuously adjust in real time to changing environmental conditions, ensuring consistent data quality in variable conditions. The rapid response of AO allows for immediate correction of distortions, which is beneficial for real-time environmental monitoring and decision-making. AO enhances the performance of quantum LiDAR systems by providing clearer images and more accurate distance measurements, which are essential for high-resolution topographic mapping and vegetation analysis. Improved signal clarity allows for better atmospheric profiling, aiding in monitoring pollution levels, weather patterns, and climate change effects more accurately. AO improves the data quality of quantum magnetometers used in geophysical surveys, leading to more precise mapping of magnetic anomalies and a better understanding of geological structures. Enhanced sensitivity and accuracy in detecting weak magnetic fields improve environmental monitoring and studies related to Earth's magnetic field. Quantum imaging systems with AO capture high-resolution images of the environment, providing detailed data for ecosystem monitoring, biodiversity studies, and habitat assessment.

8.6 Applications of Quantum-Enhanced UAV Environmental Sensing

Quantum-enhanced **UAV** environmental sensing leverages the precision, sensitivity, and unique capabilities of quantum sensors to improve the monitoring and analysis of various environmental parameters.

8.6.1 Weather Forecasting

Quantum-enhanced LiDAR offers significant improvements in atmospheric data collection, which is critical for accurate weather forecasting. Quantum LiDAR provides detailed vertical profiles of the atmosphere, including temperature, humidity, and aerosol concentration. Its high sensitivity to atmospheric particles enables better detection of cloud formation, dust, and pollutants. Improved Data Accuracy: Quantum LiDAR systems reduce noise and increase the accuracy of measurements, providing more reliable data for weather models. Enhanced precision in detecting atmospheric changes helps in forecasting severe weather events more accurately.

UAV Integration Benefits: **UAVs** equipped with quantum LiDAR can access hard-to-reach or remote areas, providing data from locations that are otherwise difficult to monitor. This is particularly useful in monitoring weather patterns over oceans, mountains, and polar regions.

Real-Time Data Collection: **UAVs** can continuously collect and transmit atmospheric data in real time, enabling up-to-date weather monitoring and immediate integration into forecasting models. Real-time data can improve short-term weather predictions and provide early warnings for severe weather conditions.

Atmospheric Profiling:
Quantum LiDAR: Equipped with quantum-enhanced LiDAR, **UAVs** can provide detailed profiles of atmospheric layers, including temperature, humidity, and aerosol concentrations. This data helps improve weather models and forecasting accuracy.

Quantum Spectrometers: These can measure the absorption of specific wavelengths of light by atmospheric gases, providing precise data on gas concentrations and helping to predict weather patterns.

Cloud and Precipitation Analysis:
Quantum Radar: Quantum-enhanced radar systems on **UAVs** can offer high-resolution images of cloud formations and precipitation. This capability allows for better tracking of storm development and rainfall intensity.

Droplet Size Measurement: **UAVs** can deploy quantum sensors to measure the size distribution of cloud droplets and ice crystals, which is crucial for understanding cloud formation and precipitation processes.

Wind Measurement:
Quantum Anemometers: These can measure wind speed and direction with high precision. **UAVs** equipped with quantum anemometers can gather data from various altitudes, providing detailed wind profiles that are essential for accurate weather forecasting [18].

8.6.2 Pollution Tracking

Quantum LiDAR can significantly enhance the tracking and monitoring of various pollutants in the environment.

Detection of Trace Pollutants: Quantum LiDAR systems can detect trace amounts of pollutants, such as NO_2, CO_2, and particulate matter, with high precision. This sensitivity allows for the identification of pollution sources and hotspots.

High Spatial and Temporal Resolution: Quantum LiDAR delivers high-resolution spatial data, enabling thorough mapping of pollution dispersion and concentration levels. Temporal resolution allows for continuous monitoring of pollution levels and trends over time. **UAVs** can be deployed quickly to monitor pollution in specific areas, such as urban environments, industrial zones, or natural disaster sites. They can fly at various altitudes to capture vertical profiles of pollutants, providing a comprehensive understanding of air quality.

Large Area Coverage: **UAVs** can cover large areas efficiently, making them ideal for monitoring regional pollution levels and tracking pollution plumes over wide areas. This capability is essential for regulatory compliance and public health monitoring.

Air Quality Monitoring:
Quantum Gas Sensors: These sensors can detect trace amounts of pollutants such as NO_2, CO_2, and particulate matter. UAVs equipped with quantum gas sensors can provide real-time, high-resolution air quality maps over urban and rural areas.

Quantum Spectroscopy: Quantum-enhanced spectrometers can analyze the composition of air pollutants with high accuracy, helping to identify pollution sources and track their dispersion.

Emission Source Identification:
Remote Sensing of Industrial Emissions: UAVs equipped with quantum sensors can fly over industrial areas to monitor emissions from factories, power plants, and other sources. This data helps in identifying non-compliant facilities and taking corrective actions.

Vehicle Emissions Monitoring: Quantum sensors can be used to track emissions from vehicles, providing data to enforce regulations and improve air quality management.

Water Pollution Detection:
Quantum Fluorescence Sensors: These can detect pollutants in water bodies by measuring fluorescence signals. UAVs can use these sensors to monitor the quality of rivers, lakes, and coastal waters.

Hydrocarbon Detection: Quantum sensors can detect oil spills and other hydrocarbon pollutants with high sensitivity, enabling timely response and mitigation efforts.

8.6.3 Natural Disaster Response

Quantum LiDAR can create high-resolution 3D maps of affected areas, providing detailed information on terrain changes, debris, and structural damage. Accurate elevation data helps in assessing flood risks and landslide potential. Quantum LiDAR can monitor atmospheric conditions, such as smoke and ash dispersion from wildfires, volcanic eruptions, and other airborne hazards. Real-time detection of atmospheric pollutants helps in assessing air quality and health risks during disasters.

Rapid Deployment: UAVs can be deployed rapidly to disaster sites, providing immediate situational awareness and aiding in the assessment of damage and needs. Quick data collection and transmission enable faster response and decision-making.

Accessibility: UAVs access areas that are difficult for ground-based teams to reach, such as collapsed buildings, flooded regions, or volcanic terrains. This capability ensures comprehensive coverage and data collection even in challenging environments.

Real-Time Data Sharing: UAVs can transmit data in real-time to emergency response teams, government agencies, and other stakeholders. Real-time information supports coordinated response efforts and resource allocation. Sensors can detect subtle changes in the Earth's gravitational field, indicating tectonic activity. **UAVs** equipped with quantum gravimeters can monitor fault lines and volcanic regions to provide early warnings of potential earthquakes or eruptions. Quantum magnetometers can detect changes in the Earth's magnetic field associated with seismic activity, aiding in earthquake prediction and monitoring volcanic activity. After heavy rains or dam failures, **UAVs** with quantum LiDAR can create high-resolution maps of flooded areas. This data is crucial for rescue operations, resource allocation, and infrastructure repair. Quantum-enhanced sensors can measure water levels practically, aiding to predict and respond to flood events efficiently. **UAVs** equipped with quantum sensors can fly into storms to collect detailed atmospheric data [19]. This information improves the accuracy of hurricane tracking and intensity predictions. Post-disaster, **UAVs** can use quantum-enhanced imaging and LiDAR to assess the extent of damage to infrastructure, helping to prioritize recovery efforts and allocate resources efficiently.

8.7 Conclusion

Quantum-supported **UAV** environmental sensing represents a significant leap forward in environmental monitoring and analysis. By leveraging the unparalleled precision and sensitivity of quantum sensors, **UAVs** can gather high-resolution data on atmospheric conditions, pollution levels, and natural disaster indicators. This enhanced capability allows for more accurate weather forecasting, real-time pollution tracking, and effective disaster response. Quantum-enhanced **UAVs** can detect minute changes in environmental parameters, providing detailed insights that were previously unattainable with classical sensors. This advanced sensing technology is crucial for addressing complex environmental challenges, informing policy decisions, improving public safety, and protecting the environment. The integration of AI with quantum-supported **UAVs** further amplifies their potential. AI algorithms can transform the large amount of information collected by quantum sensors in real time, identifying patterns and anomalies that may indicate environmental changes or threats. Machine learning models can enhance the predictive capabilities of weather forecasts and disaster response strategies by analyzing historical data and identifying trends. AI-powered image and signal processing can refine the accuracy of quantum-enhanced imaging and LiDAR data, enabling precise mapping and monitoring of environmental features [20]. Additionally, AI can optimize **UAV** flight paths and sensor operations, ensuring efficient data collection and energy use. This synergy between AI and quantum technology transforms **UAVs** into intelligent, autonomous environmental monitoring platforms that provide actionable insights and timely interventions. The combination of quantum precision and AI-driven analysis paves the way for a new era of environmental sensing, where data-driven decision-making can effectively address and mitigate the impacts of climate change, pollution, and natural disasters.

References

[1] K. A. Gilmore *et al.*, "Quantum-enhanced sensing of displacements and electric fields with two-dimensional trapped-ion crystals," Science, vol. *373*, no. 6555, pp. 673–678, Aug. 2021, doi: 10.1126/science.abi5226.

[2] S. Pirandola, B. R. Bardhan, T. Gehring, C. Weedbrook, and S. Lloyd, "Advances in photonic quantum sensing," Nature Photonics, vol. *12*, no. 12, pp. 724–733, Nov. 2018, doi: 10.1038/s41566-018-0301-6.

[3] A. T. Tokunaga, "New generation Ground-Based Optical/Infrared telescopes," in Elsevier eBooks, 2014, pp. 1089–1105, doi: 10.1016/b978-0-12-415845-0.00051-7.

[4] D. Kumar, A. D. J. Pacheco, K. Kaushik, and J. J. P. C. Rodrigues, "Futuristic view of the Internet of Quantum Drones: Review, challenges and research agenda," Veh. Commun., vol. *36*, p. 100487, May 2022, doi: 10.1016/j.vehcom.2022.100487.

[5] Y. Liu *et al.*, "AlScN Piezoelectric MEMS Mirrors with Large Field of View for LiDAR Application," Micromachines, vol. *13*, no. 9, p. 1550, Sep. 2022, doi: 10.3390/mi13091550.

[6] A. Sharma *et al.*, "Communication and networking technologies for UAVs: A survey," J. Netw. Comput. Appl., vol. *168*, p. 102739, Jul. 2020, doi: 10.1016/j.jnca.2020.102739.

[7] V. Giovannetti, S. Lloyd, and L. Maccone, "Quantum metrology," in Handbook of Quantum Information, Springer, 2010, pp. 305–352.

[8] M. K. Joshi *et al.*, "High-resolution adaptive optics imaging: Systems and applications," Optical Engineering, vol. *55*, no. 5, p. 051002, 2016, doi: 10.1117/1.OE.55.5.051002.

[9] F. Bouchard *et al.*, "Quantum metrology at the limit of classical physics," Nature Reviews Physics, vol. *2*, no. 3, pp. 150–160, 2020, doi: 10.1038/s42254-020-0151-3.

[10] R. Demkowicz-Dobrzański, M. Jarzyna, and J. Kołodyński, "Quantum limits in optical interferometry," Progress in Optics, vol. *60*, pp. 345–435, 2015, doi: 10.1016/bs.po.2015.02.003.

[11] A. A. Clerk *et al.*, "Introduction to quantum noise, measurement, and amplification," Rev. Mod. Phys., vol. *82*, no. 2, pp. 1155–1208, 2010, doi: 10.1103/RevModPhys.82.1155.

[12] R. F. García Vergara *et al.*, "Miniaturized MEMS LiDAR system for UAV-based 3D mapping," IEEE Sensors Journal, vol. *21*, no. 3, pp. 2892–2903, 2021, doi: 10.1109/JSEN.2020.3029866.

[13] A. Aspuru-Guzik and P. Walther, "Photonic quantum simulators," Nature Physics, vol. *8*, no. 4, pp. 285–291, 2012, doi: 10.1038/nphys2253.

[14] N. Takei *et al.*, "High-resolution adaptive optics imaging for ground-based telescopes," Publications of the Astronomical Society of Japan, vol. *69*, no. 4, pp. 1–12, 2017, doi: 10.1093/pasj/psx036.

[15] J. Preskill, "Quantum computing in the NISQ era and beyond," Quantum, vol. *2*, p. 79, 2018, doi: 10.22331/q-2018-08-06-79.

[16] R. Jozsa and N. Linden, "On the role of entanglement in quantum-computational speed-up," Proc. R. Soc. A, vol. *459*, no. 2036, pp. 2011–2032, 2003, doi: 10.1098/rspa.2002.1097.

[17] Y. Shikano and A. Hosoya, "Weak values with decoherence," J. Phys. A: Math. Theor., vol. *43*, no. 2, p. 025304, 2010, doi: 10.1088/1751-8113/43/2/025304.

[18] M. A. Taylor *et al.*, "Biological measurement beyond the quantum limit," Nature Photonics, vol. *7*, no. 3, pp. 229–233, 2013, doi: 10.1038/nphoton.2012.346.

[19] L. C. Bassett, F. J. Heremans, and D. D. Awschalom, "Diamond NV centers for quantum photonics," Phys. Rev. B, vol. *85*, no. 4, p. 041201, 2012, doi: 10.1103/PhysRevB.85.041201.

[20] N. Zhao, J. Honert, B. Schmid, M. Klas, and M. Schmitt, "Miniaturized portable quantum sensors for magnetic field detection," Sensors, vol. *22*, no. 5, p. 1894, 2022, doi: 10.3390/s22051894.

[21] A. O. Soge, A. A. Willoughby, O. F. Dairo, and O. O. Onatoyinbo, "Cathode Materials for Lithium-ion Batteries: A Brief review," *Journal of New Materials for Electrochemical Systems*, vol. 24, no. 1, pp. 229–246, Dec. 2021, doi: 10.14447/jnmes.v24i4.a02

V. Velmurugan, B. Elamvazhudi, K. Kamalakannan, and S. Kulothungan

9 Enhancing Mining Safety and Environmental Compliance with UAV Technology: A Focus on Tailings Dam Monitoring and Environmental Analysis

Abstract: Similar to the example of monitoring and surveillance, unmanned aerial vehicles (UAVs) are increasingly being used in the mining industry, especially for the surveillance of the deterioration of tailing dams and other structures. These UAVs, with upgraded systems such as camera systems, laser mapping, and thermal imaging sensors, are flexible and offer all-around and timely information. The above data contributes significantly to safety, operations, and the environment in general, and compliance with environmental standards in particular. Furthermore, multispectral imaging features can allow for extensive environmental analysis, as the changes reflect the impacts of mining operations on vegetation and water sources. The capability to undertake routine flights and secure precise data assists miners in risk assessment and prompt maintenance, besides considering environmental concerns, all of which contribute to safer and more environmentally friendly mining processes.

Keywords: Quantum-Systems, unmanned aerial vehicles, mining industry, environmental energy monitoring

9.1 Introduction

Mining is widely considered one of the most vital segments of the global economy owing to its capability to deliver substantial quantities of commodities that other sectors rely on, such as construction and manufacturing, among others. Nevertheless, the extraction of these valuable resources can be very challenging due to factors such as unfavorable geology, hazards to miners' lives, and environmental impacts. The methods mentioned above, albeit efficient in practice, can often be cumbersome, time-hungry, and financially demanding as they are mainly carried out manually. Such issues call for creative approaches in affairs of operation performance, safety, and environmental issues [1]. Small flying vehicles commonly referred to as drones are transforming the industries with its application ranging from agriculture, logistics, and disaster response among others. For the past few years, the mining industry has noted the possible use of unmanned aerial vehicles (UAVs) to serve the company's needs and solve specific issues. There are many advantages associated with UAV usage for example; UAVs can navigate to areas that are

https://doi.org/10.1515/9783111619323-009

hard to access, they can capture a large amount of data at one time, they can work independently hence minimizing work in risky areas [2].

One of the prominent early adopters in the field of UAVs is Quantum-Systems, which has created and launched state-of-the-art UAV technology exclusively for the mining industry. Established with the perspective of changing the existing approach to aerial data acquisition and processing, Quantum-Systems is an international team of engineers, designers, and software developers who manufacture UAVs to fit the needs of the mining industry. Software in the form of sensors, imaging, and data processing that their UAVs contain offers mining companies precise operational information as soon as, or perhaps more frequently than, it happens. Exploration is considered one of the most significant uses of Quantum-Systems' UAVs in mining [3]. Generally, exploration entails surface investigations such as aeromagnetic studies and drilling in order to establish areas with possible deposits of minerals. This process is evidently a slow one that is equally risky, especially when it has to be carried out in a relatively inaccessible region or a very unfriendly climate. This is done through obtaining high-quality images and geospatial data, which are more efficient in the preliminary assessment by geologists and mining engineers when using Quantum-Systems' UAVs. The fast speed of area coverage and high density of maps speeds up the exploration stage, which in turn helps mining firms make the correct judgments and distribute resources during the course of exploration properly.

Apart from exploration functions, UAVs developed by Quantum-Systems are used for site surveying and development. Underpinning all or most mining operations requires that mining sites are well mapped and that there is proper 3D modeling to help in laying down infrastructures and planning the layouts of mines, as well as the extraction of resources. Mounted on Quantum-Systems' UAVs, the light detection and ranging technique, or LiDAR technique, provides accurate topographical profiles and models of the terrain. This information is unique and critical for developing correct mine plans, avoiding mistakes during excavation, and decreasing expenses. Additionally, multispectral sensors, incorporating functions that help determine the chemistry of the ground and the health of plants, assist in developing strategic plans regarding the landscape. Accommodation is an important factor, especially given the nature of mining, which is a very dangerous sector that puts workers in hazardous situations. The UAVs from Quantum-Systems are useful because they provide a safe way for mines to be monitored and inspected by personnel. It is possible to make UAVs fly over rugged terrains, holes, trenches, and other areas that might be dangerous for human life, capturing video and photo streams that can be analyzed to detect threats. This capability helps minimize the chances of personnel physically accessing these areas, thereby reducing the losses that can result from accidents. Moreover, UAVs can inspect the corrosion of equipment, for example, in tailings dams and haul roads, to determine whether or not they comply with safety requirements and to avoid costly disasters. Environmental management is another important component of contemporary mining processes. Because mining entails digging, blasting, excavating, and hauling, it can destroy habitats, pollute water, and erode soil. The UAVs that Quantum-Systems offer assist with monitoring and managing the environment through data such as land use, deforestation, and water pollution. This information helps mining companies evaluate the level of influence their actions have on the environment and apply corrections to neutralize negative implications. For instance, UAVs can oversee changes in the status of reclamation projects and verify the health status of vegetation and water contamination indicators in an environment, thereby improving environmental management [4].

In addition to equipment, employees of Quantum-Systems are focused on the development of software ideas. Their UAVs are enhanced with efficient data analysis and data analytics platforms using artificial intelligence (AI) and other machine learning (ML) tools. It is a set of tools and applications that authority can use to analyze vast quantities of data without the assistance of human specialists. For example, it can identify the mineral anomalies in the aerial imagery, forecast the structural model as well as plan the ore exploration effectively. Real-time data processing improves the flexibility of the mining operation since the firms can respond as required on the field while focusing on the results and relevant data from the systems. Some case studies would highlight on the changes that Quantum-Systems' UAV technology has brought to mining. For example, one mining company in South America was able to perform the coverage of an especially remote exploration area by using Quantum-Systems' UAVs. The UAVs made photo and LiDAR surveys that were followed by image processing to produce digital terrain models [5]. This information effectively helped the company to implement better targets for minerals discovery and also to plan the drilling exercises far much better in achieving discovery of these minerals. Another instance is a mining operation in Australia which Quantum-Systems' UAVs for assessing the stability of a large tailings dam. The UAVs drove data collection on the condition of the dam; the information collected empowered the company to take the necessary actions to prevent failures. In conclusion, Quantum-Systems is a leading company specializing in UAV technology in the context of the mining industry. Proydro's sophisticated UAVs with modern sensors, picture taking techniques, and information analysis provide mining firms with an efficient instrument for prospecting, topographic assessment, security observation, and ecological control [6]. Namely, public mining firms can benefit from employing UAV technology from Quantum-Systems either way: improving their business performance and reducing negative effects on the environment while making the necessary decisions. This paper has examined how the use of Quantum-Systems is significant through mining technology which continues to upgrading its technology for use in the mining industry.

9.2 UAV Applications in Mining

The approach used in this study entails the assessment of the UAV technology utilized by Quantum-Systems and its application in the mining business. This involves an analysis of the company's structures and procedures for design and development, technologies, and actual examples of how their UAVs assist in solving practical challenges and issues. This paper also includes interviews with Quantum-Systems' clients who have applied their UAVs in mining, providing an understanding of the actual effects of those technologies in this sphere. Thus, Quantum-Systems' UAVs are involved in the exploration and surveying of prospective areas, monitoring and control during mining work, as well as environmental assessment and mitigation during and after mining. In the exploration phase, UAVs provide clear and precise images and GIS data, which aid in locating prospects for mineral resources. UAVs enhance accuracy in site surveying, mapping, and 3D modeling of land for overall planning and development. Regarding surveillance, they can monitor operational sites in real time and check for compliance with safety measures and environmental conservation.

9.2.1 Geological and Structural Analysis via Quantum-Systems

Advanced technologies are becoming essential in the mining industry in the quest for better alternatives in exploration as well as better and safer ways of conducting mining business. Of all these technologies, one which has proved to be crucial in accomplishing geophysical and structural surveys is UAVs [7]. Quantum-Systems, one of the elite UAV solution providers, has designed highly professional UAV solutions to meet the requirements of the mining industry for geological surveys, as shown in Figure 9.1. This chapter also explores how Quantum-Systems' UAVs perform geological and structural scans to help mining firms achieve accurate, effective, as well as risk-free evaluations. Geological and structural analysis is crucial in mining for several reasons: Exploration of mineral deposits is done through extensive surveys of the ground believed to contain the deposits. Geological maps assist in determining fields that contain minerals and in planning further drilling exercises. This is very important to avoid mishaps caused by the contraction of geographical features like rock formations or faults that may lead to collapses or landslides. The above methods indicate that by conducting frequent structural analyses, mine stability can be maintained. Apparently, the knowledge of the structural plan of a mining site helps to develop appropriate extraction strategies and avoid the possibility of waste while augmenting resource intake. Analysis of the structural condition of tailings dams and other structures is critical in avoiding catastrophes.

The UAVs that Quantum-Systems designed are integrated with some of the best sensors and imaging technology for geological and structural assessment. Quantum-Systems' UAVs are equipped with high-definition cameras that provide high-quality pictures. These images are necessary for producing geological maps as well as for rec-

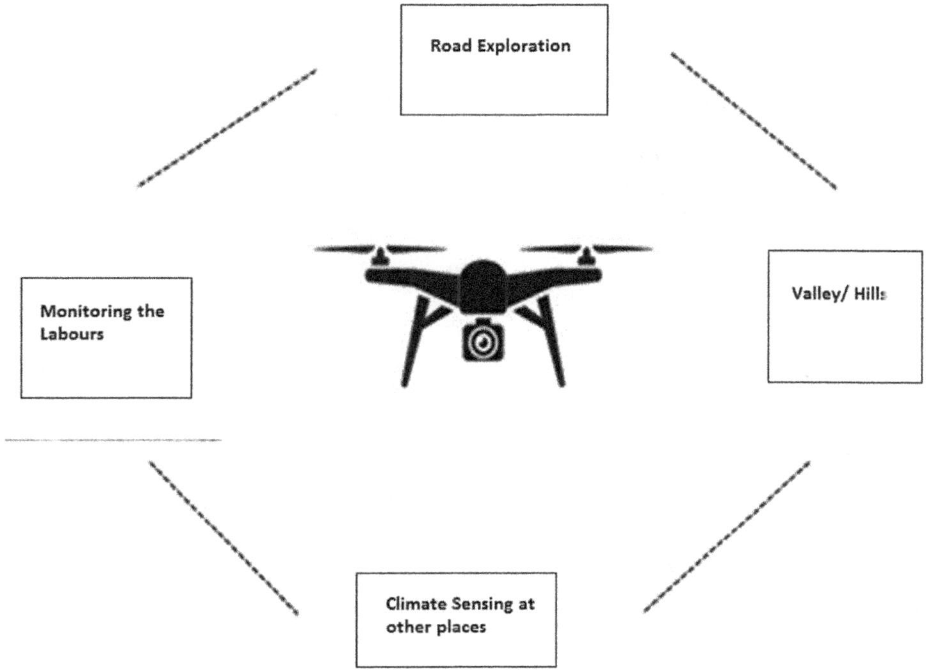

Figure 9.1: UAV geological survey.

ognizing structures on them. LiDAR is used in mapping ground features and the geology of a given area, as shown below. Topographical maps derived from LiDAR data give high density information that is necessary in identifying the geology of the mining site environment. Electronic sensors gather information at different frequencies and reveal information on the mineral content, vigor of vegetation, and the rest of the geophysical features. It is especially applicable to the differentiation of diverse kinds of minerals and the evaluation of environmental events. Firmware: Currently, AI and ML algorithms help UAVs of the company to process the gathered data. They are used to improve the efficiency of explorations in some ways, such as the recognition of geological structures, the prediction of the structural response, and the given strategies.

9.3 Applications in Geological and Structural Analysis

Quantum-Systems' UAV technology for geological and structural analysis has proven to be a game changer in the mining industry. Quantum-Systems' UAVs provide high-resolution data, advanced sensors, and AI-based analytics, which allow mining companies to get an accurate, fast, and safe overview of their assets or operations [8]. Whether it's exploration, site-surveying, safety monitoring, or environmental management you are looking at, there is nothing better than Quantum-Systems' UAV technology to help you make informed decisions or operate more efficiently. As the mining industry is looking for new technological ways to ensure cost-effective yields in the future, Quantum-Systems will continue to lead this innovation. Utilized across mining operations, Quantum-Systems' UAV technology for geological and structural analysis is deployed.

9.3.1 Exploration and Mapping

UAVs, commonly called drones, have transformed the way we map geology. UAVs can capture highly resolved aerial imagery and LiDAR data, which can be used by geologists and researchers to produce high-resolution geological maps [9]. Geological maps are relied on for a broad range of applications such as mineral exploration, environmental monitoring, land use planning, and assessing natural hazards due to their ability to provide a clear understanding of the three-dimensional distribution of subsurface rocks at the Earth's surface. The cameras on UAVs are made up of multiple sensors that capture an array of images at very high resolutions. These images map the Earth's surface so that every pixel in the image is filled with a specific color or value, creating a raster image that makes up a photograph. Photographic data is key in geological mapping as it allows for visual interpretation of geology, which can only truly be captured via optical sensors when photographing the Earth's surface. Aerial photographs taken by UAVs, as shown in Figure 9.2, can be used to generate orthomosaic images, which are high-resolution, geometrically corrected images of the Earth's surface, having consistent scale throughout. Orthomosaic images are low-altitude aerial photographs that provide the ability to see all the details on site. This is important as geologists need a large-scale overview of any geological survey area. For example, the orthographic projection of an unmanned aerial image on a horizontal plane gives the surveyor the whole picture at a glance of typically 5–30 ha (in one photograph) or 500–2,000 ha (in one flight) with an accuracy of 10–100 times better than manually or satellite-acquired photographs [10]. Geologists also study landforms: it describes how features such as mountains, valleys, and rivers appear on the landscape as well as rock types exposed on the surfaces like sand dunes. Veins of copper will be refracted differently by specific filters integrated with normal color photography that flies over areas operated by mining companies. UAVs can carry various sensors apart from normal visual spectrum cameras including multispectral and hyperspectral sensors that make use of more bands than what can be seen by human eyes, which help identify different mineral composition or plant/vegetation types' Cauchy wavelength separation spectroscopy potential for identifying or distinguishing minerals in rocks in place as they reflect sunlight differently based on their characteristics, that is, color. The response returned helps to determine if a certain mineral compound exist in those rocks detected similarly. This is done for soil samples using similar techniques because visible spectral wavelength light waves absorbed by minerals used in rocks (diagnostic absorption features) exists within these bands, that is, green chlorophyll bands [11].

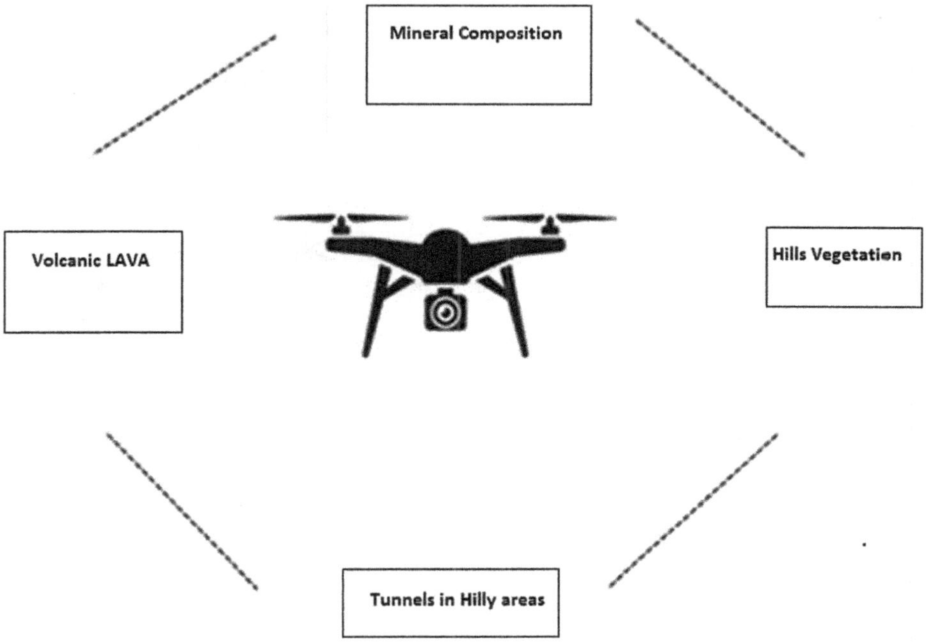

Figure 9.2: UAV in exploration.

LiDAR technology employs laser pulses to estimate distances between UAVs and the ground. This information is applied in modeling of the Earth's surface in three dimensions with high precision. For example, mapping dense vegetation terrains is easily done with the help of LiDAR since its beams can penetrate plants and provide accurate ground elevation data. The data from LiDAR undergoes a process known as digital elevation model (DEM) creation, which generates highly accurate representations of the Earth's topography [12]. Geological mapping benefits immensely from DEMs by providing precise elevation information. This helps understand geomorphology, identifying fault lines or studying slope stability among other things. LiDAR has very high accuracy and resolution enabling it to identify geological features accurately such as fractures, faults and folds hidden beneath the crust of the Earth. Additionally, it assists in identification of sedimentary structures, volcanic formations as well as other geological phenomena that have significance for geologic mapping or exploration.

9.3.2 Site Surveying and Development

The mining industry has notably benefited from advances in UAV technology, such as creating 3D models from aerial data, as shown in Figure 9.3. These models are comprehensive, accurate, and detailed, and they can be used to plan at different stages of mine development and infrastructure. The utilization of UAV-generated 3D models is advantageous in terms of enhanced safety, cost-effectiveness, and precision planning. High-resolution cameras equipped on drones, together with LiDAR sensors, capture detailed images from the air and topographic information. High-resolution pictures provide a visual description of the site, while LiDAR gives accurate elevation measurements, even through dense vegetation. Photogrammetry is among the techniques used to process captured images by UAVs into 3D models [13]. Photogrammetry software receives overlapped photos taken at various angles, which then form a three-dimensional representation of mine surfaces that have been photographed from above, thereby creating the desired image for use in decision-making by stakeholders involved in mining activities within those regions served by such modern technologies as those discussed here above (and listed below). As a result, this process creates highly precise models that mirror actual spatial dimensions and features of landforms or other surrounding areas where such surveying tools would be applied during exploration works carried out in subsequent extraction phases, following conventional methodologies employed in their kind concerning each forthcoming stage.

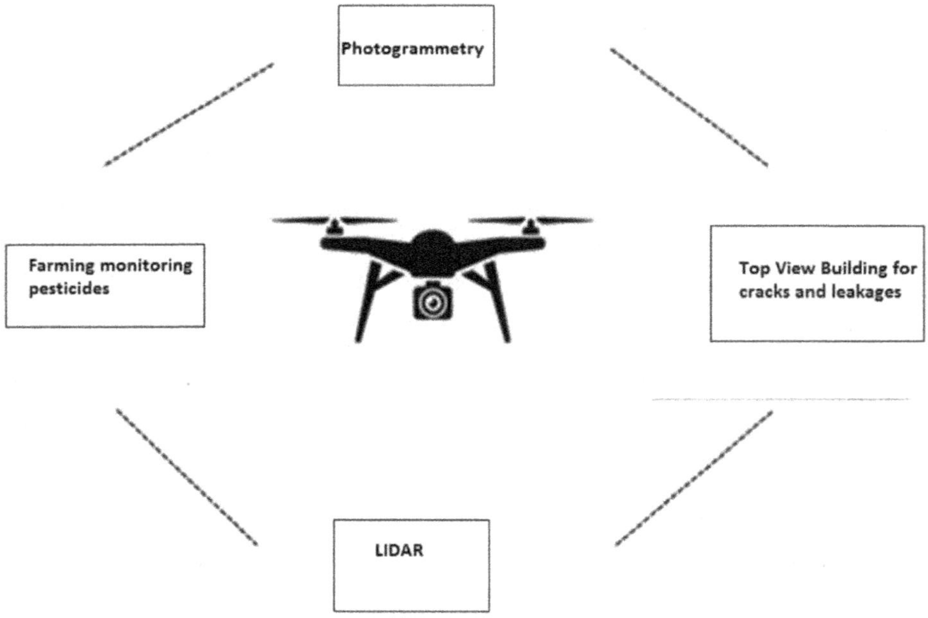

Figure 9.3: UAV site surveying.

By using high-precision UAVs to collect surveying and mapping data for 3-dimensional models, the accuracy of modeling can be improved substantially. For mining operations, the more accurately this data can be predicted or updated in real time, the more profound the effect will be on setups and changes needed for effective planning and decision-making. The collection of data is safely carried out with UAVs, avoiding the risks involved in placing human lives at the front lines near hazardous areas [14]. It improves security by reducing the risks typical of certain activities. Compared to traditional ground-based studies, UAV surveys can be carried out faster and at a lower cost. The ability to gather comprehensive data quickly reduces the expense and time needed for planning and monitoring. 3D models provide detailed insights, leading to informed decision-making. Designers and engineers can preview issues, allowing for the enhancement of designs to streamline activity output. 3D models generated by UAVs also enable accurate topographical data acquisition and feasibility studies [15].

A 3D model of the potential mining site enables geologists and engineers to study its terrain, identify potential challenges, and assess its suitability. One also uses 3D models to plan operational processes such as settings for roads, buildings, and processing plants. A 3D model of the setup, to scale, allows planners to position different elements for optimal function, reducing the costs of moving Earth and building infrastructure. Planning also ensures that the infrastructure is on solid ground, thus reducing operational losses through failures. Open-pit mining is the most common mining method, employed for both enhancing efficiency and ensuring safety, which is critical. Engineers can use 3D models to design pit slope settings and the benches. The comprehensive terrain view also enables engineers to build the necessary roads for mining equipment and solid waste dump sites [16]. Post-planning, it's paramount that one maintains a constant watch over the mining site. UAVs capture data regularly to update the 3D model, enabling the monitoring process. The continual updating ensures one can monitor changes in real-time, allowing for prompt intervention when necessary. They help in managing the site and assist in the crown design and proper running of the mine. Mining is termed a significant source of environmental degradation. The site's 3D models, taken before, during, and after mining, allow one to visualize the environmental degradation. The data obtained is used in designing the reclamation plan to restore the site to a mine-friendly state [17].

9.3.3 Safety Monitoring

Geological stability is very important for mining, both in terms of the safety of operations and the efficiency of resource extraction. Regular flights over mining sites by UAVs have become an important tool in monitoring and maintaining such stability. Advanced imaging and sensing technologies aboard UAVs provide up-to-date data about the geological condition, thus helping in change detection of rock formations, monitoring of fault lines, and hazard identification. This methodology offered by

UAVs reduces the probable risk of accidents at mining locations and ensures mine safety in general [18]. High-resolution cameras mounted on UAVs capture aerial photographs of mining sites. These images would provide a clear and full view of the terrain and would allow one to closely examine geologic features. Conducting regular flights can result in changes being detected in the images over time and thus could monitor very subtle changes in rock formations and surface condition. LiDAR technology – the very one more commonly deployed in UAVs – proves especially useful in building precise 3D models of the mining site. LiDAR systems emit laser pulses that measure distances to the ground, producing detailed topographical maps. These are used to identify shifts in the terrain, changes in rock stability, and the development of possible hazards such as landslides or subsidence [19]. Thermal cameras mounted on UAVs may perceive temperature changes on the ground. These temperature changes usually denote possible geological changes or even stress within rock formations. Monitoring these thermal anomalies would make it easier to spot an area that could be prone to collapse or any other problem of instability. Regular flights give the opportunity to track geological conditions constantly. With frequent data acquired by these devices, changes in the mining site become easy to detect. Real-time monitoring is particularly important to always have information on the state of stability of rock formations and other geological features [20]. UAV-based data is fed into sophisticated software platforms where analysis and interpretation activities go on via algorithms and ML techniques that find patterns or outliers in the data. Trends analyzed over time can indicate where in the future stability issues might become critical for geologists and engineers. The system is also capable of creating immediate alerts in case UAVs detect huge changes or potential hazards. These alerts allow for prompt response and mitigation measures, reducing the likelihood of accidents. Detailed reports that are generated from the data give actionable insights into mine management and safety teams. Regular flights by UAVs form one of the most important opportunities for the early detection of geological hazards. The capability of UAVs to identify changes in rock formations – for example, cracks or shifts, which indicate instability – enables the early identification of hazards.

The identification by UAVs in the early stages of changes allows for preventative measures against hazards to avoid them turning into real threats. Fault lines are critical areas of concern in mining operations [21]. Advanced sensor-equipped UAVs will be able to track any movement or stress within these fault lines. The continuous monitoring of fault lines through the collected data makes it easier to understand their behavior and take proper measures to avoid the catastrophe of failure. Continuous monitoring of infrastructure through the use of UAVs shall help create a safer working environment by recognizing hazards ahead of time and taking precautionary measures. Real-time detection and response to geologic changes reduce the likelihood of accidents and thereby protect personnel and equipment. A proactive approach not only ensures that accidents are kept at bay but also reduces operational downtime and the resulting costs incurred [22].

9.3.4 Environmental Management

The operations of mining companies in ensuring that tailings dams and other critical infrastructure are structurally intact are being transformed by UAVs. Using high-resolution, LiDAR, and thermal imaging technologies, these drones capture detailed, real-time data for increased safety, operational efficiency, and environmental compliance. Moreover, multispectral imaging has the capacity to provide comprehensive environmental monitoring, which helps in assessing the impact of mining activities on vegetation and water bodies. With the development of UAV technology, it is surely becoming more important in ensuring that mining operations are undertaken safely and sustainably [23]. UAVs are also equipped with high-resolution cameras, photographing tailings dams and other infrastructure in detail. Consequently, images now provide an accurate and current visual record of the structures, along with any obvious signs of wear, deformation, or damage. LiDAR, the technology used for tailings dams, is a very important tool in structural monitoring. LiDAR-equipped UAVs can create detailed 3D models by mapping and collecting data on the tailings dam topography. Such information is crucial for determining slight changes or shifts in the structure that may serve as indicators of potential failure points.

Environmental management drones, which have thermal cameras mounted onto UAVs, can capture temperature variations across the tailings dam surface, thereby showing potential seepage areas or internal erosion – crucial indications of structural instability. Multispectral sensors aboard a UAV collect data outside the visible spectrum and at different wavelengths, which is useful in charting the health of vegetation around mining sites. For example, analysis of multispectral data from UAVs allows the identification of plants under stress due to mining activities through changes in soil quality or water availability [24]. Another role that UAV-based multispectral imaging plays in monitoring water bodies near mining operations entails detecting changes in color, thereby indicating pollution or sedimentation. Through periodic monitoring of these parameters, mining firms can ensure that their operations have no negative effect on the local water resources. UAVs provide high-resolution environmental data, which assist in monitoring wildlife habitats, a critical aspect of protecting biodiversity and completing environmental policies. Among several other uses, a UAV can be employed to monitor changes in land cover and habitat conditions, thereby proving immensely useful in conservation efforts [25].

Regular UAV flights over tailings dams and other critical infrastructure greatly enhance safety. In this regard, UAVs can provide early warning signs and symptoms prior to failure, thus allowing for proactive maintenance and repairs that reduce the risks of disastrous events. UAV monitoring is cost-effective compared to traditional inspection methods [26]. In particular, UAVs are able to efficiently cover large areas and difficult-to-access locations; in addition, they reduce downtime for inspection considerations. This efficiency translates to cost savings and increased monitoring frequency [27]. Another major advantage of UAVs is the real-time data they provide.

Such immediate availability of information ensures rapid decision-making with the capability of intervening promptly in case issues are detected by the survey. Real-time monitoring is crucial in maintaining the integrity of critical infrastructure and guaranteeing continuous operational safety. UAVs are vital in helping mining enterprise entities meet the set environmental standards [28]. Basically, this is achieved through providing continuous observation of the effects on vegetation, water bodies, and habitats, thereby delivering data to prove compliance. This early monitoring thus enables avoiding or, at best, reducing negative impacts in favor of promoting sustainability in mining [29, 30].

9.4 Conclusion

Quantum-Systems is an innovation leader in UAV technology concerning the mining industry, featuring state-of-the-art sensors, imaging systems, and data processing capabilities that equip mining companies with advanced UAVs as tools for exploration, site surveying, safety monitoring, and environmental management. Fast-changing, the mining industry requires innovation and excellence; therefore, Quantum-Systems is well-placed as a key partner driving the future of mining technology. Their ingenuity in UAV technology at Quantum-Systems makes this company a game changer in mining with its efficient, accurate, and cost-effective solutions for exploration, mapping, and monitoring. This commitment to technological advancement with practical application has positioned them as leaders in this field. With the mining industry continuing to seek ways of improving efficiency and sustainability, Quantum-Systems' UAVs offer a solution that answers these demands. It is in this respect that UAV technology has been very instrumental in revolutionizing the mining industry by providing this powerful tool for monitoring adherence to the structural integrity of tailings dams and other critical infrastructure. UAVs, equipped with high-resolution LiDAR and thermal imaging technologies, provide detailed and real-time data that, in the end, candidly improve safety, operational efficiency, and environmental compliance. Early detection of the warning signs of imminent failure risks enables proactive maintenance, thus reducing the probability of catastrophic events. This multispectral imaging ability also allows for largescale environmental monitoring, thus letting mining companies understand the implications of activities on vegetation and water bodies and therefore be compliant with environmental regulations. As UAV technology continues to improve, its place in ensuring safer and more sustainable mining operations will continue to rise, and it will play a part in the more responsible and efficient procedures for mining.

References

[1] Kumar, P. Praveen, et al. “Security Analysis of UAV Communication Protocols: Solutions, Prospects, and Encounters.” The Internet of Drones. Apple Academic Press, 2022. 167–190.

[2] K. Anderson and K. J. Gaston, “Lightweight unmanned aerial vehicles will revolutionize spatial ecology,” Frontiers in Ecology and the Environment, vol. 11, no. 3, pp. 138–146, 2013.

[3] I. Colomina and P. Molina, “Unmanned aerial systems for photogrammetry and remote sensing: A review,” ISPRS Journal of Photogrammetry and Remote Sensing, vol. 92, pp. 79–97, 2014.

[4] C. Zhang and J. M. Kovacs, “The application of small unmanned aerial systems for precision agriculture: A review,” Precision Agriculture, vol. 13, pp. 693–712, 2012.

[5] D. Turner, A. Lucieer, and C. Watson, “An automated technique for generating georectified mosaics from ultra-high resolution UAV imagery,” IEEE Transactions on Geoscience and Remote Sensing, vol. 50, no. 10, pp. 4168–4174, 2012.

[6] Raj, S. Gokul, N. Srinath, and T. Ananth Kumar. “Real-Time Trespasser Detection Using GPS based UAV.” 2019 IEEE International Conference on Innovations in Communication, Computing and Instrumentation (ICCI). IEEE, 2019.

[7] F. Nex and F. Remondino, “UAV for 3D mapping applications: A review,” Applied Geomatics, vol. 6, no. 1, pp. 1–15, 2014.

[8] R. A. Zahawi et al., “Using lightweight unmanned aerial vehicles to monitor tropical forest recovery,” Biological Conservation, vol. 186, pp. 287–295, 2015.

[9] H. Eisenbeiss, UAV photogrammetry. Dissertation, 2009.

[10] S. Siebert and J. Teizer, “Mobile 3D mapping for surveying earthwork projects using an Unmanned Aerial Vehicle (UAV) system,” Automation in Construction, vol. 41, pp. 1–14, 2014.

[11] C. Gómez and H. Purdie, “UAVs and thermal imagery in mountain glacier monitoring,” Progress in Physical Geography, vol. 40, no. 1, pp. 123–145, 2016.

[12] U. Niethammer et al., “UAV-based remote sensing of the Super-Sauze landslide: Evaluation and results,” Engineering Geology, vol. 128, pp. 2–11, 2012.

[13] M. R. James and S. Robson, “Straightforward reconstruction of 3D surfaces and topography with a camera: Accuracy and geoscience application,” Journal of Geophysical Research: Earth Surface, vol. 117, no. F3, 2012.

[14] C. H. Hugenholtz et al., “Geomorphological mapping with a small unmanned aerial vehicle (UAV): Feature detection and accuracy assessment of a photogrammetrically-generated digital terrain model,” Geomorphology, vol. 194, pp. 16–24, 2013.

[15] F. Mancini et al., “Using unmanned aerial vehicles (UAV) for high-resolution reconstruction of topography: The structure from motion approach on coastal environments,” Remote Sensing, vol. 5, no. 12, pp. 6880–6898, 2013.

[16] Justin, Shekaina, et al. “An Innovative IoT-Driven Smart Streetlight Management System Design.” 2024 s International Conference Computational and Characterization Techniques in Engineering & Sciences (IC3TES). IEEE, 2024.

[17] F. Remondino et al., “UAV photogrammetry for mapping and 3D modeling: Current status and future perspectives,” International Archives of the Photogrammetry, Remote Sensing and Spatial Information Sciences, vol. 38, no. 1/C22, 2011.

[18] G. Vacca et al., “Combination of terrestrial and UAV-based photogrammetry for the 3D modeling of a historical building,” ISPRS International Journal of Geo-Information, vol. 6, no. 7, p. 188, 2017.

[19] F. Carvajal-Ramírez et al., “Evaluation of low-cost surveying methods for monitoring surface mining reclamation,” Sensors, vol. 16, no. 12, p. 2071, 2016.

[20] A. Eltner et al., “Image-based surface reconstruction in geomorphometry – Merits, limits and developments,” Earth Surface Dynamics, vol. 4, no. 2, pp. 359–389, 2016.

[21] C. Gomez and H. Purdie, "UAVs and structure-from-motion photogrammetry in glaciology: A review," Journal of Glaciology, vol. 62, no. 231, pp. 1057–1071, 2016.

[22] C. Toth and G. Jóźków, "Remote sensing platforms and sensors: A survey," ISPRS Journal of Photogrammetry and Remote Sensing, vol. 115, pp. 22–36, 2016.

[23] G. Fernández et al., "UAV photogrammetry for the characterization of high mountain environments," Geomorphology, vol. 282, pp. 104–120, 2016.

[24] B. Kalantar et al., "UAV-based landslide mapping: A case study in the Himalayas," Remote Sensing, vol. 9, no. 7, p. 646, 2017.

[25] P. Barry and R. Coakley, "Field accuracy test of RPAS photogrammetry," International Archives of the Photogrammetry, Remote Sensing and Spatial Information Sciences, vol. 40, no. 1/W2, pp. 27–31, 2013.

[26] A. Mohammadi et al., "UAV remote sensing for environmental monitoring of tailings dams," Mining Technology, vol. 124, no. 2, pp. 107–120, 2015.

[27] Kumar, T. Ananth, J. Ramapraba, P. Kanimozhi, and Sunday A. Ajagbe. "IOT based smart water quality monitoring system." In 7th IET Smart Cities Symposium (SCS 2023), vol. 2023, pp. 313–322. IET, 2023.

[28] S. Siebert et al., "UAV-based 3D mapping of surface mining and the generation of 3D printed models," International Archives of the Photogrammetry, Remote Sensing and Spatial Information Sciences, vol. 40, no. 3/W3, pp. 561–566, 2015.

[29] A. Pyakurel et al., "The role of UAVs in sustainable mining: Insights from the EU Horizon 2020 ITERAMS project," Sustainability, vol. 11, no. 22, p. 6415, 2019.

[30] H.-P. Thamm and M. Judex, "Integration of UAV-based remote sensing data into GIS for disaster management," International Archives of Photogrammetry, Remote Sensing and Spatial Information Sciences, vol. 36, no. 6, pp. 169–174, 2006.

R. Thangamani, M. Vimaladevi, and M.M. Ramyasri

10 Building Resilient Quantum UAV Networks: Cybersecurity and AI Perspectives

Abstract: The combination of quantum computing, unmanned aerial vehicles (UAVs), cybersecurity, and artificial intelligence (AI) ushers in a new era of safe and robust UAV networks. This chapter analyzes novel approaches that use quantum computing, AI, and cybersecurity principles to address the dynamic issues of UAV networks. Key target areas include quantum-resistant cryptographic algorithms, AI-driven threat detection and response systems, and UAV-specific secure communication protocols. These innovations reduce cyberrisks, protect data integrity, and increase the robustness of quantum-enabled UAV systems.

The chapter illustrates enhanced threat detection, adaptive cybersecurity, and secure data transfer through the use of cutting-edge techniques, which raise operational effectiveness and mission success rates. With features including better data protection, reduced susceptibility to attacks, and AI-enhanced decision-making, resilient quantum UAV networks find applications in military, surveillance, disaster response, agriculture, and infrastructure monitoring. This chapter concludes by discussing the revolutionary potential of AI and quantum computing in UAV networks. It emphasizes the significance of proactive cybersecurity and AI-driven resilience in shaping the development of unmanned aerial systems, fostering innovation, and ensuring the safety of critical UAV operations.

Keywords: Quantum computing, UAV networks, cybersecurity, AI-driven threat detection, resilience, secure communication

10.1 Introduction

The integration of quantum unmanned aerial vehicles (UAVs) into modern technological landscapes represents a significant advancement in cybersecurity and artificial intelligence (AI) perspectives. These resilient UAV networks combine modern quantum computing technologies with the agility and intelligence of unmanned aerial aircraft, offering dramatic improvements in security, efficiency, and operational performance. UAVs have recently become vital instruments in a variety of industries, including surveillance and reconnaissance, disaster response, and delivery services. The growing use of technology has also revealed vulnerabilities, particularly in terms of cybersecurity threats such as data breaches, hacking attempts, and information interception.

https://doi.org/10.1515/9783111619323-010

Traditional cryptographic techniques are relatively effective, but they face challenges in keeping up with the rising complexity of cyberthreats [1].

The emergence of quantum technology has resulted in a significant increase in computational power and encryption capabilities. Quantum UAV networks utilize these advancements to enhance their cybersecurity defenses, ensuring data integrity, confidentiality, and resilience against quantum attacks. Furthermore, the implementation of AI algorithms, combined with machine learning approaches, equips these networks with adaptive risk detection and response mechanisms, making them highly resilient in dynamic and hostile situations.

This sector investigates the problems of building quantum UAV networks through the lens of cybersecurity and AI. It dives into the fundamental principles of quantum processing and its applications in UAV communications and information security. It also explores the use of AI in improving cybersecurity measures within UAV networks, such as intelligent threat identification, behavioral analytics, and self-healing capabilities. This section examines the linkages between quantum processing, UAV technology, cybersecurity, and AI in order to provide insights into the difficulties, techniques, solutions, and real-world applications of establishing robust Quantum UAV networks [2]. It also seeks to highlight these networks' transformational potential for protecting important data, optimizing mission performance, and maintaining the integrity of UAV operations in complex and aggressive situations. Building durable Quantum UAV networks entails combining cutting-edge technologies such as quantum processing, AI, and cybersecurity.

Some current issues in this domain include:

- Quantum Computing Vulnerabilities: While quantum computing provides incredible processing capability, it also poses new security challenges. Current quantum techniques, such as Shor's algorithm, pose a challenge to existing encryption methods, raising questions about data security in UAV networks.
- UAVs are vulnerable to cyber-physical attacks, which can control sensors and modify flight routes. Securing UAVs against such attacks necessitates strong cybersecurity techniques.
- UAVs increasingly rely on AI for autonomous decision-making and data processing, posing new threats. However, AI systems can be vulnerable to adversarial attacks, which occur when malicious inputs corrupt AI algorithms, resulting in inaccurate choices or compromised data integrity.
- Data Privacy and Integrity: UAV networks generate significant volumes of sensitive data, including surveillance footage, position information, and communication logs. Securing data privacy and integrity across networks, especially in quantum contexts, is a critical concern.[3]
- Quantum Key Distribution (QKD) Security: Enables safe key exchange in quantum networks. However, deploying it in UAV contexts presents practical issues such as long-distance key distribution and resilience against quantum hacking.

- Standardized protocols and interoperable solutions are necessary for integrating quantum technology, AI systems, and cybersecurity. Creating industry-wide standards for safe quantum UAV networks is a critical challenge.
- Safe Human-Machine Interaction: As UAV networks become increasingly autonomous, it's crucial to ensure secure human-machine interaction.

Addressing these difficulties would necessitate interdisciplinary collaboration among quantum computer experts, AI researchers, cybersecurity specialists, and UAV engineers. Robust encryption schemes, quantum-resistant algorithms, AI-driven anomaly detection, and continuous monitoring are critical cybersecurity and AI tactics for developing robust quantum UAV networks [4, 5].

- Building durable quantum UAV networks requires continual research across multiple fronts, including cybersecurity and AI. Here are some ongoing efforts in this area:
- Researchers are developing quantum-safe encryption methods (post-quantum cryptography (PQC)) that are not affected by quantum outbreaks. These initiatives are aimed at protecting communication channels within UAV networks from future quantum computing attacks.
- QKD Protocols: They are being optimized for safe key exchange in UAV networks using quantum channels. This method generates encryption keys that are potentially impenetrable, even with quantum computers.
- AI algorithms are being used for real-time danger detection in UAV networks. ML models examine streams of data that emerge from sensors and communication channels to detect anomalies or potential cyberattacks, improving the network's resilience.
- Secure Software and Firmware Development: UAVs will include secure boot procedures, runtime integrity checks, and communication protocols. This helps prevent unauthorized access and modification of UAV systems.
- AI-based behavioral analytics help define baseline behaviors for UAVs and detect deviations that may indicate cyberattacks or system failures. This proactive technique enhances network resilience by detecting attacks early.
- Quantum-Safe Networking (QSN) Protocols: Researching and implementing quantum-resistant protocols for UAV networks to provide secure communication even with enemies.
- Collaboration among quantum computing experts, cybersecurity researchers, AI professionals, and UAV engineers is driving innovation in creating robust networks. Interdisciplinary efforts result in holistic solutions that address security concerns from different perspectives.
- Governments and regulatory agencies are developing guidelines and standards to ensure secure quantum UAV operations. This includes legislation governing data privacy, encryption standards, and cybersecurity measures unique to quantum-enabled UAV networks.

These initiatives help to construct resilient quantum UAV networks that can operate securely in demanding situations while exploiting quantum computing and AI technology.

The planned work on developing durable quantum UAV networks, with an emphasis on cybersecurity and AI considerations, comprises several key areas of research and development. Here are a few proposed initiatives [6, 7]:

- Design and implement quantum-resistant encryption solutions, including PQC, to protect against quantum-based attacks on existing encryption systems.
- Integrate QKD technologies into UAV network topologies to ensure secure quantum communication channels. Investigate the viability of applying QKD in dynamic UAV situations while maintaining key distribution robustness against quantum hacking techniques.
- Enforce safe software development techniques for UAV systems, such as coding standards, vulnerability assessments, and software integrity checks. Create procedures for safe firmware updates and fixes to reduce software-based cybersecurity concerns.
- Use AI-driven behavioral analytics to develop baseline behavior profiles for UAVs and detect deviations that could suggest cyberattacks or system anomalies.
- Develop and implement secure communication protocols for quantum-enabled UAV networks. Implement robust encryption, authentication mechanisms, and data validation techniques to protect sensitive information during transmission.
- Investigate quantum-resilient networking protocols and architectures, including QSN concepts. Create routing algorithms, network segmentation strategies, and traffic management approaches tailored for quantum UAV networks [8].
- Encourage collaboration among quantum computing professionals, cybersecurity researchers, AI specialists, UAV engineers, and regulatory stakeholders. Establish collaborative research projects, knowledge-sharing platforms, and industry alliances to stimulate innovation in the development of durable quantum UAV networks.
- Collaborate with monitoring bodies and industry organizations to develop standards, best practices, and compliance criteria for secure quantum UAV operations. Ensure compliance with data protection regulations, encryption procedures, and cybersecurity rules applicable to quantum technology.

The purpose of focusing on these proposed activities is to accelerate the development of resilient quantum UAV networks that can operate securely in dynamic and adversarial situations while leveraging the strengths of quantum computing, AI technologies, and robust cybersecurity measures.

The proposed methods aim to address numerous fundamental issues in the development of resilient quantum UAV networks from the perspectives of cybersecurity and AI. These problems include:

- Proposed solutions address quantum computing vulnerabilities by developing and deploying encryption algorithms resistant to quantum attacks, thereby maintaining data confidentiality and integrity.
- Ensuring secure communication routes in UAV networks is critical. The proposed approaches attempt to improve key distribution security by including QKD protocols, making them more resistant to quantum hacking techniques and ensuring secure quantum communications.
- Detecting Cyberthreats: UAV networks are vulnerable to unwanted access, data spills, and network disruptions [9]. The proposed AI-driven threat intelligence approaches prioritize continuous monitoring, anomaly detection, and real-time reaction to cyberattacks, which improve network resilience and security.
- Securing UAV software and firmware is crucial to prevent unwanted access, changes, or exploitation of vulnerabilities. The proposed approaches include safe software development processes, vulnerability evaluations, and methods for secure updates and patches.
- Behavioral Analysis and Predictive Security: Early detection of cyberthreats in UAV networks requires understanding normal and aberrant behavior. The suggested methods use AI-driven behavioral analytics to construct baseline behavior profiles, detect deviations, and predict potential security threats by evaluating historical data and network behaviors.
- Implementing secure communication protocols for quantum-enabled devices ensures end-to-end encryption, authentication, and data integrity, reducing the danger of eavesdropping and manipulation.
- Quantum-Resilient Networking: Creating protocols and architectures that can withstand quantum attacks, addressing the vulnerability of standard protocols in quantum environments. The suggested methods attempt to create routing algorithms, network segmentation strategies, and traffic management techniques that are optimized for quantum UAV networks.

Overall, the suggested solutions seek to address the fundamental issues of protecting quantum-enabled UAV networks from cyberattacks, including maintaining data privacy, integrity, and availability while employing AI-driven insights and effective cybersecurity measures.

The scientific contribution to the development of resilient quantum UAV networks from the perspectives of cybersecurity and AI is substantial and varied. Here are the main research contributions:

- Research is essential for developing encryption methods and protocols that can withstand quantum-based attacks, ensuring data security and confidentiality in quantum UAV networks.
- Contributed to integrating QKD protocols into UAV network topologies, developing secure communication channels, and investigating their viability in dynamic UAV situations.

- AI-Driven Threat Intelligence: Research helps build real-time threat detection approaches and adaptive security tactics for UAV networks, improving resilience against cyberattacks and intrusions.
- Implemented secure software development techniques, vulnerability assessments, and update procedures for UAV systems to reduce cyberthreats and maintain system integrity.
- Our research uses AI-driven behavioral analytics to generate baseline behavior profiles, detect deviations, and predict potential security threats in UAV networks. This enhances proactive security measures.
- Secure Communication Protocols: Tailored protocols for quantum technology provide secure data transfer, protect against unauthorized access, and maintain data privacy and integrity.
- Our research focuses on building quantum-resistant networking protocols and architectures, optimizing routing algorithms, network segmentation strategies, and traffic management techniques for quantum UAV networks.
- Research promotes collaboration among quantum computing professionals, cybersecurity researchers, AI specialists, UAV engineers, and regulatory stakeholders, resulting in industry-wide solutions and standards for secure quantum UAV operations.
- Collaborated with regulatory organizations to develop standards, best practices, and compliance criteria for secure quantum UAV operations, ensuring compliance with data protection laws and cybersecurity guidelines.

Overall, the research contributions to developing resilient quantum UAV networks from cybersecurity and AI perspectives advance the state of the art in secure quantum communication, threat detection, software security, behavioral analytics, and networking protocols, resulting in increased resilience and security in UAV operations.

Here are several use scenarios that demonstrate the applications of establishing resilient quantum UAV networks from a cybersecurity and AI standpoint [10]:

Secure Communication Channels:
Scenario: A military UAV is entrusted with gathering crucial reconnaissance information.

Use Case: QKD methods establish secure communication channels between UAVs and ground stations, ensuring that encrypted data transmissions are immune to quantum attacks.

Behavioral Analysis for Anomaly Detection:
Scenario: A UAV monitors critical infrastructure, such as power plants or pipelines.

Use Case: Implement AI-driven behavioral analytics to establish normal behavior patterns for the infrastructure and detect deviations that could indicate cyber-physical attacks or system malfunctions, enabling proactive security measures and timely responses.

Secure Software and Firmware Updates:
Scenario: A drone manufacturer releases a new firmware update for its models.

Use Case: Use cryptographic signatures, secure boot methods, and integrity checks to prevent unwanted changes or tampering throughout the software and firmware update process, thereby improving system security.

Quantum-Resilient Networking:
Scenario: A network of UAVs is deployed for collaborative missions with real-time data exchange needs.

Use Case: Create and deploy quantum-attack-resistant networking protocols that can dynamically adjust to changing network circumstances, guaranteeing safe and dependable UAV-to-UAV communication, even in the event of outages or threats based on quantum technology.

Predictive Security Powered by AI:
Scenario: In a dynamic urban setting, a UAV is performing autonomous flight operations.

Use Case: By using AI-driven predictive security models to foresee possible cyberthreats based on network patterns, environmental variables, and historical data, the UAV may proactively modify its security posture and steer clear of potential hazards [11].

Building robust quantum UAV networks by integrating cybersecurity measures, AI-driven intelligence, secure communication protocols, and regulatory compliance improves security, reliability, and UAV operations effectively across multiple domains, as these use cases demonstrate [12].

10.1.1 Building Robust UAV Networks Is Crucial in Light of Cybersecurity, AI, and Quantum Computing

It is essential to create robust UAV networks that incorporate AI, cybersecurity, and quantum computing for a number of reasons [13, 14]:

- Evolution of the Threat Landscape: As technology develops, so do enemies' capabilities. Traditional encryption techniques become outdated due to new cryptographic flaws introduced by quantum computing. To combat changing cyberthreats, resilient UAV networks use cybersecurity, AI-driven threat detection, and quantum-safe encryption.
- Data Security and Privacy: UAVs gather a vast amount of private information, including communication records, site details, and surveillance footage. Creating robust networks guarantees data security and privacy by using secure software/firmware, encrypted communication methods, and following data protection laws, which ensures that procedures and systems adhere to legal frameworks, protecting security and privacy when handling and processing sensitive or personal data.

- Mission Continuity: Mission continuity in hostile and dynamic environments is made possible by resilient UAV networks. UAVs can modify their behavior, redirect communication channels, and reduce hazards without jeopardizing mission objectives, thanks to AI-driven predictive analytics and anomaly detection, which foresee possible dangers.
- AI algorithms installed on UAVs analyze sensor data, look for irregularities, and make judgments in real time. By facilitating safe AI-driven decision-making, resilient networks help UAVs stay operationally effective, react quickly to changing circumstances, and fend off cyberattacks.
- Interconnected Operations: Smooth communication, data exchange, and coordination are made possible by robust networks in situations involving several UAVs or cooperative missions. Even in the event of quantum-based disturbances or attacks, dependable connectivity is ensured via quantum-resilient networking protocols.
- Sturdy Communication Channels: End-to-end encryption and authentication are guaranteed by quantum-safe communication channels built with QKD algorithms. This preserves the secrecy and integrity of communication within UAV networks, preventing manipulation, data interception, and eavesdropping.
- Regulatory Compliance: Resilient UAV networks adhere to industry best practices, data protection laws, and cybersecurity standards. Compliance fosters trust among stakeholders and the general public by ensuring the ethical application of AI, accountable UAV operations, and responsible data processing.
- Future-Proofing: Constructing robust networks accounts for upcoming security threats and technological developments. Adaptive networking protocols, AI-driven security mechanisms, and quantum-safe encryption enable UAV networks to remain resilient, scalable, and flexible in the face of new threats and technological advancements.

Building robust UAV networks is crucial for data protection, mission continuity, safe AI-driven operations, regulatory compliance, and future-proofing against evolving cyberthreats in the context of cybersecurity, AI, and quantum computing.

A summary of the advantages and disadvantages of quantum QAVs in the fields of cybersecurity and AI is presented in Table 10.1.

These advantages and disadvantages provide a comprehensive grasp of the benefits and challenges associated with quantum QAVs in the fields of AI and cybersecurity [15].

Table 10.1: Advantages and disadvantages of quantum UAVs in cybersecurity and AI.

Aspect	Advantages	Disadvantages
Security	Security is improved by quantum encryption, which shields communication channels from quantum-based attacks.	Quantum encryption protocol administration and implementation can be challenging and call for certain knowledge and resources.
Threat detection	Real-time cybersecurity threat identification and analysis are made possible by AI-driven threat detection algorithms, which strengthen preventive defenses.	AI-based systems have the potential to produce false positives or false negatives, which could result in missed detections or incorrect interpretations of threat signs.
Anomaly detection	Early detection of potential security breaches is made easier by an AI-driven anomaly detection system, which improves the ability to identify unusual patterns or behaviors in UAV networks.	Anomalies need to be carefully analyzed and interpreted because they may not necessarily be a sign of malevolent activity and could be the consequence of typical variations or system errors.
Incident response	AI-powered incident response tactics reduce response times and lessen the impact of security breaches by enabling quick and effective responses to cybersecurity problems.	An over-reliance on automated incident response systems could result in mistakes or unexpected outcomes, requiring human supervision and involvement.
Data privacy	Sensitive data in UAV operations is successfully protected by quantum-resistant encryption and secure data storage techniques, which also ensure compliance with privacy laws.	Continuous monitoring and modification of data protection measures are necessary to maintain compliance with evolving privacy rules and standards.
Scalability	Expanded operations and improved cybersecurity posture are supported by the scalability of quantum UAV networks to meet changing cybersecurity requirements and expanding AI capabilities.	The cost and resource allocation of scaling quantum encryption and AI systems may be affected by the need for significant infrastructure investments and upgrades.
Reliability	High uptime and dependability are provided by resilient quantum UAV networks, guaranteeing uninterrupted communication and operation even under difficult conditions or during cyberattacks.	There are possible hazards associated with system breakdowns, vulnerabilities, or compatibility difficulties when relying on quantum technology and AI systems.
Innovation	Through its innovative use of cutting-edge technologies like AI and quantum computing, quantum UAVs open the door to improved cybersecurity tactics and capabilities.	There may be compatibility, integration, and skill requirement issues with integrating new technologies and keeping up with the rapid changes in technology.

10.2 Challenges in UAV Networks

10.2.1 An Examination of the Cybersecurity Challenges That UAV Networks Face

A review of UAV networks' cybersecurity issues identifies a number of serious issues [16]:

- Data Encryption and Privacy: Private information, such as location information, conversation logs, and surveillance footage, is transmitted by UAV networks. To prevent illegal access, data breaches, and privacy violations, it becomes imperative to implement robust encryption and data protection measures.
- Secure Communication Channels: It is critical to set up secure communication channels between control centers, base stations, and UAVs. Communication protocol flaws can compromise mission integrity and operational safety by allowing data transmission to be intercepted, altered, or disrupted.
- Cyber-Physical Attacks: Cyberintrusions that alter physical components like sensors, navigation systems, or flight controls might affect UAVs. Strong cybersecurity defenses and intrusion detection systems are necessary to fend off such attempts.
- Software and Malware Vulnerabilities: Because UAVs depend on firmware and software to function, they are susceptible to viruses, malware, and cyberattacks. To reduce software-based cyberthreats, code integrity checks, timely patching, and software development process security are crucial.
- GPS Spoofing and Jamming: These types of assaults have the potential to interfere with communication signals, change flight routes, or fool UAV navigation systems. The impact of cyberthreats related to GPS can be lessened by putting anti-spoofing strategies, redundant navigation systems, and signal integrity checks into place.
- Unauthorized Access and Remote Hijacking: Malevolent actors may gain unauthorized access to or remotely hijack UAVs. Strong authentication procedures, access controls, and secure remote management protocols are necessary to prevent unwanted control or manipulation.
- Supply Chain Security: It is essential to secure the supply chains for UAV software, parts, and services since weaknesses in these supply chains can be used to inject malicious software or hardware into UAV systems, posing a significant cybersecurity risk.
- Regulatory Compliance: It is essential for UAV operators to follow cybersecurity laws, guidelines, and industry best practices. A culture of cybersecurity awareness is fostered, and compliance is ensured by adhering to specified guidelines, conducting routine audits, and providing cybersecurity training programs.

A comprehensive strategy involving encryption technologies, secure communication protocols, intrusion detection systems, software security procedures, anti-spoofing techniques, supply chain security, regulatory compliance, and ongoing cybersecurity education and training for UAV operators and stakeholders is required to address these cybersecurity issues.

10.2.2 Vulnerabilities and Threats in Quantum-Enabled UAV Environments

The nexus of cybersecurity, quantum computing, and UAV operations creates vulnerabilities and dangers in quantum-enabled UAV environments. The following are the main threats and vulnerabilities [17]:

Quantum Computing Vulnerabilities:
Quantum Threats: Within UAV networks, quantum computers have the potential to jeopardize data secrecy and integrity by posing a challenge to traditional encryption techniques.

Vulnerabilities in QKD: Long-distance key distribution, quantum hacking techniques, and flaws in the processes used to generate and distribute quantum keys are some of the difficulties that may arise when implementing QKD systems.

Secure Communication Challenges:
Quantum-Safe Encryption: Ensuring that UAV communication channels utilize encryption algorithms that are resistant to quantum attacks and provide secure end-to-end encryption.

Quantum Channel Security: Preventing quantum attackers from intercepting, altering, or listening in on quantum communication channels is referred to as quantum channel security.

Cyber-Physical Threats:
Manipulation of UAV Sensors: UAV sensor manipulation involves cyberintrusions that target onboard computers, navigation systems, or UAV sensors in order to change flight trajectories, manipulate sensor readings, or jeopardize mission objectives.

Remote Hijacking: When malevolent actors gain unauthorized access to or control of UAVs, it can result in remote hijacking, operational interruption, or misappropriation of UAV resources.

AI-Driven Threats:
Adversarial Attacks: Adversarial Attacks: Using adversarial attacks to manipulate AI systems aboard UAVs in order to trick decision-making processes, resulting in inaccurate actions or compromised data integrity.

AI Model Poisoning: The act of introducing harmful data into AI training datasets or modifying AI models to introduce biases or errors that impact AI-driven decision-making in UAV operations is known as AI model poisoning.

GPS Spoofing and Jamming:
GPS Spoofing: GPS spoofing is the practice of deceiving UAV navigation systems by manipulating GPS signals, which can result in inaccurate location information, altered flight routes, or navigational errors.

GPS Jamming: By interfering with UAV navigation, communication, or control signals, GPS jamming can result in operational disturbances or a loss of situational awareness.

Supply Chain Risks:
Malicious Components: Risks include backdoors, malware, and vulnerabilities in UAV software or components when hostile hardware or software is introduced into UAV systems through compromised supply chains.

Third-Party Risks: Reliance on outside vendors for UAV parts, software, or services, which could result in security flaws or illegal access to UAV systems.

Regulatory and Compliance Challenges:
Cybersecurity Regulations: Adherence to cybersecurity policies, procedures, and best practices that are pertinent to UAV operations, such as secure communication protocols, data protection, and encryption requirements.

Privacy Concerns: Ensuring adherence to ethical principles and data privacy laws is necessary to address privacy issues related to UAV data collection, storage, and processing.

A comprehensive cybersecurity strategy that includes quantum-safe encryption, secure communication protocols, cyber-physical security measures, AI-driven threat detection, supply chain security, regulatory compliance, and continuous cybersecurity education and training for UAV operators and stakeholders is necessary to address these vulnerabilities and threats in quantum-enabled UAV environments [18].

10.3 Quantum-Resistant Cryptography for UAV Networks

The development of quantum computing technology poses serious risks to traditional cryptography methods. Commonly used encryption techniques like RSA and ECC could be successfully broken by quantum computers, endangering the security and secrecy of data transferred between UAV networks. Quantum-resistant cryptography (QRC) is being developed to offer safe substitutes for safeguarding private data in order to reduce these hazards. This is a detailed examination of QRC for UAV networks [1, 25]:

1. Understanding Quantum Threats

- **Quantum Computing Basics**: Quantum processors use quantum binary units, or qubits, to perform calculations at speeds that traditional computers cannot match. They take advantage of concepts like entanglement and superposition.
- **Risks to Cryptography**: The security of conventional encryption techniques is compromised by quantum approaches, particularly Shor's approach, which can quickly deconstruct large numbers and evaluate discrete logarithmic functions.

2. Principles of Quantum-Resistant Cryptography

QRC focuses on developing encryption techniques that can withstand both conventional and quantum attacks. Among the fundamental ideas are:

- **Hard Problems**: QRC techniques rely on mathematical problems that are currently believed to be challenging for quantum computers to solve, including:
  - **Lattice Problems**: Problems that stem from the intricacy of lattice-based architectures include the shortest vector challenge (SVP) and the learning with error (LWE) challenge.
  - **Code-Based Problems**: One well-known example of code-based problems is the McEliece cryptosystem, which involves error-correcting codes.
  - **Multivariate Polynomial Problems**: Encryption methods that depend on solving systems of multivariate polynomial equations are known as multivariate polynomial problems.
  - **Hash-Based Problems**: The XMSS (eXtended Merkle Signature Scheme) is an example of a secure digital signature that is created using hash functions.

3. Key Features of Quantum-Resistant Algorithms

- **Post-quantum Security**: Developed to withstand quantum system attacks while maintaining effectiveness for deployment in UAV networks.
- **Scalability**: Needs to be able to adjust to varying degrees of resource limitations, which are common in UAV operations.
- **Compatibility**: Should offer improved security and blend in smoothly with current systems.

4. Examples: Quantum-Resistant Techniques

1. **Lattice-Based Cryptography**:
   - **NTRU**: An effective public-key encryption technique that is resistant to quantum attacks.
   - **Kyber**: Known for its effectiveness and security, this key encapsulation approach relies on lattice difficulties.
2. **Code-Based Cryptography**:
   - **McEliece**: A Goppa code-based encryption method renowned for its effectiveness and large key sizes, which provide defense against quantum assaults.

3. **Multivariate Polynomial Cryptography**:
   - **Rainbow**: A multivariate signature system that has demonstrated encouraging outcomes in the face of quantum attacks.
4. **Hash-Based Cryptography**:
   - **XMSS**: A stateful, hash-based signature system designed to withstand assaults based on quantum mechanics.

5. Implementation Challenges in UAV Networks
- **Limited Resources:** QRC implementation is more difficult due to UAVs' often limited processor power, memory, and battery life.
- **Latency**: For real-time operations, it is essential to ensure that QRC does not significantly increase communication latency.
- **Interoperability**: It can be challenging to switch from conventional encryption to QRC while maintaining compatibility with existing systems.

6. Best Practices for Deploying Quantum-Resistant Cryptography in UAV Networks
- **Conduct Risk Assessments**: Determine important information and the specific dangers that quantum computing poses.
- **Integrate QRC Gradually**: To provide a more seamless transition, incorporate QRC into current cryptographic systems.
- **Ongoing Training**: Inform stakeholders and UAV operators about QRC, its significance, and the proper management and operation of QRC-based systems.
- **Stay Updated**: To ensure the most modern and efficient implementations, keep up with advancements in QRC standards and research.

7. Future Directions
QRC research and development will be crucial as quantum technology advances. Strong, useful QRC solutions that can be effectively implemented in UAV networks will be developed through partnerships among government, business, and academia.

A proactive method for protecting UAV networks is provided by quantum-resistant cryptography. UAV operators may preserve the security and integrity of their data by embracing and putting into practice QRC solutions, which will enable safer and more reliable operations in an increasingly digital and networked world.

10.4 AI-Powered Threat Detection and Response

By improving threat identification and response capabilities, AI is revolutionizing cybersecurity. Organizations can enhance their ability to comprehend and respond to cyberthreats in real time by utilizing machine learning algorithms, data analytics, and automation. The methods, advantages, challenges, and strategies for implementing AI-powered threat detection and response are covered in detail below [19].

10.4.1 The Impact of AI on Identifying and Preventing Cyberthreats in UAV Networks

Their networks are becoming prime targets for cyberattacks due to the growing use of UAVs in a variety of industries, such as the military, agriculture, and delivery services. AI is essential for enhancing threat detection and prevention in order to protect these systems. An extensive summary of how AI helps protect UAV networks from cyberattacks can be found below [20]:

1. Enhanced Threat Detection

a. **Anomaly Detection**
   - **Behavioral Profiling**: AI systems examine UAVs' typical operating patterns to determine baseline behaviors. Any departure from these standards may be a sign of possible security lapses, such as illegal access.
   - **Real-Time Monitoring**: Machine learning models are able to process tremendous amounts of data from UAV sensors and communication channels in the real world, and they can promptly spot unusual activity that could indicate an assault.

b. **Predictive Analytics**
   - **Threat Prediction**: AI is able to predict potential future threats to UAV networks by analyzing historical attack data and environmental factors [21]. Preventive security measures can be implemented thanks to this proactive approach.
   - **Risk Assessment**: AI models constantly modify security measures in response to danger levels and environmental changes, continuously assessing the risks associated with UAV operations.

c. **Integration of Threat Intelligence**
   - **Data Fusion**: The ability of AI systems to compile and evaluate threat intelligence from several sources, such as industry publications, social media, and cybersecurity feeds, is known as data fusion. This combined data improves situational awareness and aids in the detection of new dangers.
   - **Contextual Threat Analysis**: AI supports UAV operators in prioritizing responses by contextualizing data to help them understand the significance of certain risks to their operations.

2. Mitigation of Cyberthreats

a. **Automated Response Mechanisms**
   - **Incident Response Automation**: Using preset protocols, AI-driven systems are able to automate incident response actions. This enables prompt threat containment and mitigation, and it shortens reaction times during attacks.
   - **Dynamic Policy Enforcement**: AI can instantly modify security rules in response to risks it detects, guaranteeing that UAVs stay within safe bounds even when cyberthreats are constantly changing.

b. **Self-Healing Systems**
   - **Autonomous Recovery**: AI can help UAV networks develop self-healing capabilities, which enable compromised systems to automatically resume regular operations following a breach. This reduces operational interruptions and downtime.
   - **Resilience Planning**: AI models can assist in creating UAV networks that are naturally resilient, meaning that attacks won't cause a major loss of functionality.

c. **Secure Communication Protocols**
   - **AI-Enhanced Encryption**: AI can assist in creating and overseeing cutting-edge encryption methods that protect data transfer between UAVs and ground control stations, making it more difficult for adversaries to intercept or alter communications.
   - **Anomaly-Based Security Protocols**: Adaptive security protocols, powered by AI, can dynamically change communication rules, making it more difficult for attackers to exploit weaknesses.

3. Improved Situational Awareness

a. **Comprehensive Monitoring**
   - **360-Degree Monitoring**: AI systems are able to continuously monitor UAV networks, identifying and evaluating threats from several perspectives (physical security, network, and data), offering a comprehensive picture of possible weaknesses [22].
   - **Real-Time Notifications:** AI systems are able to notify UAV operators in real-time when they identify risks or anomalies, allowing for a prompt response.

b. **Visual Analytics**
   - **Data Visualization**: Operators may more easily comprehend the security posture and promptly spot problems that require attention when complex security data is presented in visual representations by AI-driven analytics.
   - **Dashboards**: Detailed dashboards showing operational status, system health, and threat levels assist operators in staying situationally aware and reacting appropriately.

4. Challenges and Considerations

Even though AI has several benefits for identifying and reducing cyberthreats in UAV networks, there are a few issues that need to be resolved:
- **Data Quality and Availability**: High-quality data is essential to AI's efficacy. Inaccurate threat detection and response may result from incomplete or biased datasets.
- **Complexity and Integration**: Putting AI solutions into practice can be challenging and may necessitate significant adjustments to the current UAV system and workflow. It is crucial to ensure compatibility and integration with legacy systems.

- **Adversarial Attacks**: Cybercriminals may use complex strategies to try to trick AI systems, hence continuous learning is necessary to maintain efficacy.
- **Regulatory Adherence:** Businesses must ensure AI-powered security solutions comply with industry standards and applicable laws, particularly those pertaining to data privacy and protection.

5. Future Directions
- **Enhanced Machine Learning Algorithms**: To increase the precision of threat detection and response, future AI systems will integrate improved machine learning techniques, such as deep learning.
- **Collaborative AI Systems**: AI has the potential to provide cooperative security strategies in which several UAVs exchange threat intelligence and gain knowledge from one another's experiences, improving network security as a whole.
- **Human-AI Collaboration**: By fusing AI's analytical capabilities with human intuition, threat identification and mitigation techniques can become more successful [23]. To guarantee that operators can collaborate with AI systems in an efficient manner, training initiatives will be crucial.

The way UAV networks identify and react to cyberthreats is being completely transformed by AI. Organizations may greatly improve their security posture and attack resistance by leveraging automation, machine learning, and improved analytics. However, system integration, data quality, and the dynamic nature of cyberthreats must all be carefully taken into account for successful implementation. As AI technology develops, it will play an ever-more-important role in protecting UAV networks, allowing for safer and more secure UAV operations in a variety of industries [8].

10.4.2 ML Algorithms for Detecting Anomalies and Implementing Adaptive Cybersecurity Measures

Anomaly detection and adaptive cybersecurity solutions, which offer sophisticated techniques for spotting odd patterns and reacting quickly to new threats, depend heavily on machine learning algorithms. An extensive examination of the several machine learning methods utilized in anomaly detection and their role in adaptive cybersecurity is provided below [24]:

1. Anomaly Detection Techniques
Its main objective is to identify data points that significantly deviate from a predetermined benchmark. In cybersecurity, this method is crucial for identifying possible fraud, intrusions, and system failures.

a. **Supervised Learning Approaches**
   Using labeled datasets that contain examples of both normal and anomalous behavior, models are constructed in this learning methodology.
   - **Support Vector Machines (SVMs)**
     - **Description:** SVM differentiates between normal and anomalous data points by constructing hyperplanes within high-dimensional spaces.
     - **Application:** Network traffic is classified as normal or abnormal through the analysis of data characteristics.
   - **Random Forest**
     - **Description**: Ensemble learning achieves better forecasting accuracy by building multiple decision trees and combining their outputs.
     - **Application**: Categorizing events through learned decision criteria enables the detection of anomalies in large datasets, such as user behavior analytics.

b. **Unsupervised Learning Approaches**
   Models for unsupervised learning come into play when labeled data is absent, as they search for patterns directly within the data.
   - **K-Means Clustering**
     - **Description**: The K-means algorithm creates K clusters based on feature similarity, defining anomalies as points that remain distant from each cluster center.
     - **Application**: Network traffic anomalies are detected through clustering patterns of similar behaviors and isolating instances that stand out as outliers.
   - **Isolation Forest**
     - **Description**: The Isolation Forest algorithm detects anomalies through separation rather than by examining normal data points. Due to their rarity, anomalies need fewer splits for effective isolation.
     - **Application**: The algorithm excels at outlier detection in high-dimensional spaces, which makes it perfect for real-time network traffic anomaly detection.
   - **Principal Component Analysis (PCA)**
     - **Description**: PCA minimizes dataset dimensions while preserving important measures of variance. Data points that fail to follow primary variance directions can be identified as anomalies.
     - **Application**: Useful for identifying anomalies in user activity logs by analyzing variations in user behavior.

c. **Deep Learning Approaches**
   Deep learning methods can effectively model complex relationships in data, making them suitable for high-dimensional and unstructured data.
   - **Autoencoders**
     - **Description**: A neural network model designed to reconstruct its input. Irregularities are identified by evaluating the reconstruction error, where a significant error signifies an unusual input.
     - **Application**: Employed to detect fraud in financial transactions by learning normal transaction patterns and flagging those with high reconstruction errors.
   - **Recurrent Neural Networks (RNNs)**
     - **Description**: RNNs, particularly LSTM networks, are designed to handle sequential data, which makes them ideal for anomaly detection in temporal data.
     - **Application**: Used for monitoring system logs or network traffic over time to identify unusual patterns indicative of cyberthreats.

2. Adaptive Cybersecurity Measures

Anomaly detection's insights are used in adaptive cybersecurity measures to dynamically adjust defenses. Some methods include:

a. **Real-Time Threat Response**
   - **Automated Incident Response**: When an anomaly is detected, systems can automatically carry out predetermined actions, such as isolating affected devices or blocking a suspicious IP address.
   - **Dynamic Policy Adjustment**: Security policies can be changed in real time based on detected anomalies. For instance, access permissions may be temporarily restricted in the event unusual access patterns are discovered.

b. **Feedback Loops**
   - **Learning from Anomalies**: Anomaly detection systems can constantly adjust to new data and update their models to improve accuracy over time. This helps reduce false positives and ensures that the system can keep up with changing threat landscapes.
   - **Incident Analysis**: Post-incident analysis can provide valuable feedback for retraining models, ensuring that the system remains effective against new types of threats.

c. **Context-Aware Security**:
   - **Contextual Data Integration**: ML techniques can improve the precision of outlier detection and response strategies by incorporating situational context, such as user roles, hardware types, and previous behavior [25].
     Adaptive Authorization: Systems can modify permissions in a flexible manner in response to real-time evaluations of user behavior and detected anomalies, resulting in a more secure environment.

d. **User and Entity Behavior Analytics (UEBA)**
   - **Behavioral Modeling**: Machine learning models can create profiles of the behavior of users and businesses, spotting deviations that indicate insider risks or compromised accounts.
   - **Risk Scoring**: Security teams can effectively prioritize responses by assigning anomalies risk scores based on their likely outcomes.

3. Challenges and Considerations

Aspects of adaptive cybersecurity and anomaly detection offered by machine learning algorithms include the following [26].

- **Data Integrity and Accessibility:** The effectiveness of machine learning algorithms relies significantly on the quality and volume of the training dataset. Models that aren't accurate can result from inadequate data.
- **Complexity of Implementation**: Developing and deploying machine learning systems require expertise in both cybersecurity and data science. Organizations may face challenges in staffing and resource allocation.
- **Adversarial Attacks**: Attackers may attempt to evade detection by understanding and manipulating the models used in anomaly detection systems. As a result, it is necessary to continue conducting research and refining models [27].
- **False Positives**: Security teams can become alert-fatigued when they receive a lot of false positives. Models must be tuned to strike a balance between sensitivity and specificity.

4. Future Directions

A promising future for ML lies in adaptive cybersecurity and anomaly detection:

- **Explainable AI (XAI):** The need for transparency and interpretability in AI systems will grow in importance as they become more complex. Trust and usability will both increase when models are designed to clearly explain their choices.
- **Federated Learning:** This approach makes it possible for models to be trained across a variety of decentralized devices, preserving data privacy while still allowing for collective learning.
- **Integration with Other Technologies**: For secure logging and verification, machine learning and other technologies like blockchain improve the robustness of cybersecurity measures.

  ML algorithms are transforming the landscape of anomaly detection and adaptive cybersecurity measures by providing advanced tools that dynamically identify and respond to cyberthreats. However, in order to get the most out of these technologies, it is essential to overcome obstacles. As the domain progresses, continuous innovation and adaptation will be essential for staying ahead of emerging cyberthreats [28].

10.5 Real-Life Applications of Resilient UAV Networks

With their numerous applications and cutting-edge capabilities, resilient UAV networks are revolutionizing a variety of industries. Resilient UAV networks have a significant impact on improving efficiency, safety, and decision-making in a variety of fields, including disaster response, precision agriculture, environmental monitoring, and infrastructure inspection. These networks ensure continuous communication, data transmission, and operational resilience, enabling real-time monitoring, analysis, and response in dynamic and challenging environments. The capacity of resilient UAV networks to drive innovation and address complex challenges continues to grow as a result of advancements in technology and regulations [29]. These networks are now integral components of modern solutions in numerous real-world scenarios.

10.5.1 Case Studies and Examples of Resilient UAV Networks in Action

Resilient UAV networks are crucial in various fields, including disaster response, agriculture, military applications, and environmental monitoring. Below are several case studies and examples that illustrate how these networks have been successfully implemented and their impact [11, 30]:

1. Disaster Response: Hurricane Harvey (2017)

Overview: After Hurricane Harvey hit Texas, UAVs were used for damage assessment and search-and-rescue.

Key Features:

- **Real-Time Data Collection**: UAVs equipped with imaging and detectors provided live aerial imagery of the affected areas, making it easier for emergency responders to quickly assess the damage.
- **Network Resilience**: The UAVs used a mesh network to communicate with ground control and each other, even when the traditional communication infrastructure was down.
- **Rapid Deployment**: The ability to deploy UAVs quickly enables responders to reach inaccessible areas, significantly improving rescue efforts.

Outcome:

- Resource allocation became more effective as a result of increased situational awareness.
- Enhanced coordination among the various agencies involved in the response.

2. Agriculture: Precision Farming with UAVs

Overview: UAVs are increasingly used in precision agriculture to monitor crop health, optimize irrigation, and manage field variability.

Case Study: A farm in California utilized UAVs to monitor over 1,000 acres of vineyards.

Key Features:

- **Multi-sensor Integration**: UAVs were equipped with multispectral cameras to analyze crop health and soil moisture levels.
- **Data Sharing**: The UAVs formed a resilient network, sharing data with a centralized platform that analyzed the information and provided actionable insights to farmers.
- **Adaptive Management**: The data collected allowed for adaptive management practices, such as targeted fertilization and irrigation.

Outcome:

- Increased crop yield and reduced water and fertilizer usage.
- Enhanced resilience to pests and diseases through timely interventions.

3. Military Operations: RQ-11 Raven UAV

Overview: RQ-11 Raven UAV. The U.S. military uses the RQ-11 Raven, a small UAV, for surveillance and reconnaissance.

Key Features:

- **Networked Operations**: The Raven can function in swarms, enabling real-time coordination and communication between several units.
- **Secure Communications**: Data transmission is protected from cyberattacks thanks to encryption protocols.
- **Scalability**: The system can grow to accommodate more UAVs, improving redundancy and operational flexibility.

Outcome:

- Higher mission success rates are achieved due to improved situational awareness for ground forces.
- By delivering real-time intelligence without subjecting soldiers to enemy fire, personnel risk was decreased.

4. Environmental Monitoring: Wildlife Conservation in Africa

Overview: To monitor endangered species and prevent poaching, UAVs have been utilized in various wildlife conservation initiatives.

Case Study: UAVs were used in a Kenyan project to track elephant populations and identify poaching.

Key Features:

- **Wide Area Coverage**: UAVs enable aerial surveillance over large swaths of national parks, which are frequently inaccessible from the ground.

- **Data Fusion**: To locate poaching hotspots, the UAVs combined information from GPS tracking devices and thermal imaging cameras.
- **Collaborative Networks**: Armed with UAV data, local rangers established a robust network to efficiently plan patrols.

Outcome:
- Greater awareness and participation of local communities in conservation initiatives.
- Increased awareness and involvement of local communities in conservation efforts. Decreased poaching incidents through prompt interventions based on UAV data.

5. Urban Air Mobility: Delivery Services

Overview: Delivery Services Overview: UAV networks have been created by businesses like Zipline and Wing for urban delivery services, especially in the medical field.

Case Study: The distribution of medical supplies in Rwanda by Zipline. Important features include:

Key Features:
- **On-Demand Delivery**: UAVs provide remote hospitals with blood, vaccines, and medications, enhancing access to healthcare.
- **Network Resilience**: To guarantee uninterrupted service, a network of UAVs is controlled to adapt to air traffic and weather conditions.
- **Efficient Routing**: Timely deliveries are ensured by sophisticated algorithms that optimize flight paths and schedules.

Outcome:
- Faster healthcare delivery, especially during emergencies.
- Improved logistical effectiveness in overseeing medical product supply chains.

6. Smart City Initiatives: Traffic Monitoring

Overview: UAVs are utilized in smart city initiatives to monitor urban planning and traffic conditions.

Case Study: UAVs were used by the city of Dubai to monitor traffic in real-time.

Key Features:
- **Real-Time Data Streaming**: To analyze traffic and accidents, UAVs send real-time data to traffic management systems.
- **Interoperability**: To enhance traffic flow, UAVs can communicate with other smart city components, such as sensors and traffic lights.
- **Resilience**: By rerouting other UAVs to cover the same area, the network can adjust to failures in particular UAVs.

Outcome:
- Better emergency service response times and less traffic congestion.
- More evidence-based urban development decision-making.

These case studies demonstrate the versatility and effectiveness of robust UAV networks in a range of fields. UAVs can function well even in difficult environments by utilizing cutting-edge technologies like multi-sensor integration, secure communications, and real-time data sharing. The potential uses for robust UAV networks will grow as technology advances, opening the door for creative answers to urgent societal problems.

10.5.2 Benefits and Outcomes of Implementing Cybersecurity and AI Strategies in UAV Operations

The benefits and outcomes of implementing cybersecurity and AI strategies in UAV operations [26] are presented in Table 10.2.

Table 10.2: Benefits of implementing cybersecurity and AI strategies.

Benefits	Outcomes
Enhanced security	Reduction in cyberthreats and attacks on UAV systems.
Data integrity protection	Ensured accuracy and reliability of transmitted data.
Improved decision-making	AI-driven insights lead to faster and more informed decisions.
Automated threat detection	Early identification and mitigation of potential security breaches.
Real-time monitoring	Continuous surveillance of UAV operations ensures an immediate response to anomalies.
Operational efficiency	Streamlined processes through automation, reducing manual oversight.
Scalability	Easily adaptable to increasing UAV fleet sizes and operations.
Regulatory compliance	Adherence to industry standards and regulations regarding data privacy and security.
Cost reduction	Decreased risk of financial losses due to data breaches and cyberincidents.
Enhanced public trust	Increased confidence among stakeholders and the public regarding UAVs' safety and security.
Robust communication channels	Secure channels for data transmission, preventing unauthorized access.
Predictive maintenance	AI algorithms anticipate and address potential UAV failures before they occur.
Improved resource management	Optimization of UAV deployment based on real-time data and analysis.

Numerous advantages result from integrating cybersecurity and AI techniques into UAV operations, such as increased data security, real-time threat detection, resilience, automated security operations, cost savings, regulatory compliance, and improved reputation and trust. These outcomes enhance the overall safety, effectiveness, and dependability of UAV operations across a range of sectors and applications [13].

10.6 Future Directions and Innovations

The capabilities, security, and operational efficiency of UAV systems will be improved as quantum computing, AI, and cybersecurity technologies continue to develop. The following are some significant innovations and future directions [28]:

1. Quantum-Resistant Cryptographic Standards

- **Innovation**: To guarantee long-term data protection against quantum attacks organizations are creating standardized PQC in anticipation of the possible rise of quantum computing.
- **Future Direction**: PQC standards are being finalized by governments and organizations, such as the National Institute of Standards and Technology (NIST). These standards will eventually be embraced worldwide, protecting UAV data and communications.
- **Impact**: To reduce the risk of data breaches, even against quantum adversaries, UAV systems – particularly those employed in delicate sectors like defense, healthcare, and logistics – will rely on quantum-resistant encryption for communication, data storage, and remote control.

2. Advanced QKD

- **Innovation**: By distributing encryption keys based on the principles of quantum mechanics, QKD enables extremely secure communication.
- **Future Direction**: QKD will be able to be deployed more widely thanks to fiber optics, satellite systems, and drone-to-ground communications. UAVs will benefit from the development of more effective and scalable quantum networks, which will guarantee secure communication channels over extended distances.
- **Impact**: In fields where ultra-secure key exchanges are essential, such as intelligence, surveillance, and military operations, this development will be vital for UAVs.

3. AI-Enhanced Quantum Cybersecurity

- **Innovation**: By integrating AI algorithms into UAV networks, proactive real-time cyberthreat detection and response become possible. Anomalies or potential attacks can be promptly detected by machine learning (ML) models that have been trained on enormous volumes of UAV telemetry and network data.

- **Future Direction**: By combining AI and quantum technologies, adaptive cybersecurity systems will become possible, enabling UAVs to react to changing threats on their own, without assistance from humans.
- **Impact**: This advancement will be crucial for UAVs in domains like intelligence, surveillance, and military operations, where ultra-secure key exchanges are essential.

4. Quantum-Secure Mesh Networks for UAV Swarms
- **Innovation**: To plan and carry out missions, UAV swarms need extremely safe and dependable communication protocols. In order to protect data in swarm communications, future systems will incorporate quantum-secure mesh networking.
- **Future Direction**: Secure mesh communication between UAVs in a swarm will be made possible by quantum networking technologies, guaranteeing that data, coordination signals, and commands are impervious to cyberattacks and eavesdropping.
- **Impact**: This is especially helpful in large-scale operations where UAV swarms must remain safe while interacting with one another, such as during disaster relief, defense, or agricultural monitoring.

5. AI and Quantum Collaboration for Autonomous UAVs
- **Innovation**: By combining AI and quantum computing, major advancements in UAV autonomous decision-making may be possible, allowing for quicker and more precise judgments based on data that has undergone quantum processing.
- **Future Direction**: UAVs may be able to analyze data, make decisions, and autonomously adjust to shifting conditions or mission parameters with the aid of quantum computing, which could process enormous datasets at speeds that are impossible for traditional computers.
- **Impact**: Quantum computing will make it possible for more effective route optimization and decision-making, enabling smooth and autonomous operations in sectors like logistics, where UAVs are utilized for real-time delivery or transportation.

6. **Blockchain and Quantum Security for Supply Chain Management**
- **Innovation**: Supply chains for UAV hardware and software can be made impenetrable by combining blockchain technology with quantum-resistant cryptographic protocols.
- **Future Direction**: To prevent malicious or counterfeit hardware from entering the supply chain, blockchain will be utilized to track each UAV part, component, and software update. In a post-quantum world, the blockchain itself will remain secure due to quantum-resistant cryptography.
- **Impact**: The increased security and transparency will help sectors like defense, which need tight control over their supply chains. All UAV parts, including sensors and processors, can be tracked back to their original locations, guaranteeing complete accountability.

7. Quantum-Secure Edge Computing for UAV Data Processing

- **Innovation**: Because UAVs gather so much data, edge computing enables local processing of that data on the drone, reducing latency and improving decision-making. Even processing on the edge is protected from quantum attacks, thanks to quantum-secure edge computing.
- **Future Direction**: In order to process sensitive data on the drone itself before sending it, UAVs will eventually be outfitted with quantum-secure edge computing capabilities. This makes it possible to make decisions more quickly on the spot and reduces the risk of cyberattacks during transmission.
- **Impact**: UAVs will find this especially helpful in fields like agriculture, where they can instantly assess crop health or soil data, as well as in military applications where prompt decision-making is crucial.

8. Quantum Random Number Generators (QRNGs) for Enhanced Security

- **Innovation**: Truly random numbers, which are necessary for encryption keys and other cryptographic operations, are produced by quantum random number generators.
- **Future Direction**: To avoid predictable patterns in key generation, future UAV systems will integrate QRNG technology to produce secure and unpredictable encryption keys.
- **Impact**: QRNG will significantly increase the security of UAV encryption systems, making it much harder for cybercriminals to guess or decipher encryption keys. Secure UAV communications will benefit from this, especially during delicate military or intelligence-gathering missions.

Table 10.3: Summary of future directions and innovations.

Future innovation	Description	Impact
Quantum-resistant cryptographic standards	Developing globally adopted PQC standards for long-term data security.	Secure UAV communications against quantum threats, ensuring compliance with future encryption regulations.
Advanced quantum key distribution (QKD)	Quantum-secure key exchanges via satellite and drone communications.	Ultra-secure UAV communications, particularly in critical industries such as defense and surveillance.
AI-enhanced quantum cybersecurity	AI-driven, real-time threat detection integrated with quantum technologies.	Autonomous detection and mitigation of cyberthreats, securing UAV fleets against evolving attack vectors.
Quantum-secure mesh networks for UAV swarms	Quantum-secure communication in UAV swarms for coordinated operations.	Secure coordination of UAV swarms in large-scale operations, such as disaster response and military missions.

Table 10.3 (continued)

Future innovation	Description	Impact
AI and quantum for autonomous decision-making	Faster, more efficient decision-making through quantum processing of vast datasets.	Autonomous UAV operations with real-time decision-making capabilities, enhancing efficiency and effectiveness.
Blockchain and quantum for supply chains	Blockchain combined with quantum-resistant cryptography to secure UAV component supply chains.	Secure and transparent supply chains for UAV systems, preventing counterfeit or malicious components.
Quantum-secure edge computing	Processing sensitive data locally on UAVs with quantum-secure edge computing.	Reduced risk of data breaches and improved real-time decision-making for UAVs.
Quantum random number generators (QRNG)	QRNG technology for truly random encryption key generation in UAV systems.	Enhanced cryptographic security, preventing the prediction or cracking of encryption keys.

As shown in Table 10.3, the future of UAV systems will see greater integration of AI-driven cybersecurity measures, improved encryption protocols, and quantum-resistant technologies. AI and quantum computing will be essential in creating autonomous, safe, and robust UAV networks that can function in challenging and complex environments while reducing cyberthreats and maintaining data integrity. By adopting these innovations, organizations will be better able to maintain state-of-the-art security in their UAV operations, opening up new opportunities and applications for the industry.

10.6.1 Innovations and Ongoing Research Initiatives for Enhancing UAV Network Resilience

Numerous fascinating innovations have been brought about by the UAV networks' rapid development. However, these advancements also bring serious challenges in ensuring the robustness of UAV networks, particularly when addressing operational risks, communication interruptions, and cyberthreats. To enhance these networks' resilience, security, and effectiveness, several inventions and research projects are currently underway.

10.6.1.1 Quantum Cryptography for Enhanced Security

Using quantum principles to secure communications, especially through QKD, is the focus of the cutting-edge field of quantum cryptography. This technology uses quantum mechanics to detect any eavesdropping attempts and provides an unbreakable way to exchange encryption keys.

- **Research Example**: Space-based QKD for satellite communication has been the subject of experiments by the European Space Agency (ESA). This technology could help UAV networks that communicate with satellites by guaranteeing that private information sent between UAVs and satellites is encrypted using quantum-secure keys.
- **Impact**: UAV networks can protect data from impending quantum threats by utilizing quantum cryptography, which prevents real-time key theft and eavesdropping.

10.6.1.2 AI-Driven Threat Detection and Autonomous Response

AI, which offers sophisticated threat detection capabilities and real-time autonomous responses to cyberattacks, is essential to maintaining the resilience of UAV networks. AI systems look for anomalies that might indicate an upcoming attack by analyzing patterns in network traffic, UAV communications, and environmental data.

Innovation Example: To help UAVs identify anomalous activity or cyberintrusions, researchers are creating machine learning (ML) algorithms. For example, convolutional neural networks, a deep learning model, can be trained to identify irregularities in surveillance drone video feeds, and reinforcement learning (RL) enables UAVs to instantly adjust to novel cyberattacks.

Impact: AI can enable UAV networks to automatically detect malware, prevent unauthorized access, and fend off distributed denial-of-service attacks.

10.6.1.3 Blockchain for UAV Network Security and Trust Management

Blockchain technology is becoming a potent instrument for guaranteeing tamper-proof logging of UAV actions and communications, as well as decentralized trust. In this case, blockchain offers a distributed ledger that permanently logs UAV operations, sensor information, and mission specifications.

- **Example**: IBM has studied the use of blockchain in UAV air traffic management systems, where flight data from every UAV is recorded on a decentralized ledger to confirm mission completion, flight paths, and compliance. This guarantees that no unapproved UAV can alter the system or introduce erroneous data.
- **Impact**: By improving UAV operations' traceability and transparency, blockchain technology reduces the risk of tampered data, illegal UAVs, and control network software flaws.

10.6.1.4 Quantum-Secure Mesh Networking

For UAV fleets or swarms, mesh networks are crucial because each UAV interacts directly with its peers to exchange data and plan missions. Communication in a UAV swarm must be secure and robust, particularly when quantum attacks begin to pose a threat.

- **Research Example**: An MIT research project is investigating the creation of quantum-secure mesh networks, in which a swarm of UAVs communicates with one another via PQC protocols. These protocols guarantee that network communication remains secure, even in the event that a UAV is compromised.
- **Impact**: UAV swarms used for military surveillance or disaster relief will be protected from cyberattacks and will be able to continue operating without interruption, even in contested areas, thanks to quantum-secure mesh networking.

10.6.1.5 Resilient UAV Swarms Through Collaborative Autonomy

One major area of innovation is developing **collaborative autonomy** in UAV swarms, where each UAV is capable of independently executing tasks while also communicating with other drones in the swarm. If one UAV is compromised or disabled, others can take over its responsibilities without compromising the mission.

- **Example**: The **DARPA OFFSET (offensive swarm-enabled tactics)** initiative is a leading example, focusing on building robust UAV swarms for military applications. These swarms can autonomously navigate, perform surveillance, and share situational data in real time, with each drone able to adapt to the loss of a peer.
- **Impact**: Collaborative autonomy enhances UAV network resilience by ensuring redundancy and flexibility. UAV swarms can continue functioning effectively even if a few drones fail or are intercepted, making them highly reliable for mission-critical operations such as reconnaissance or search-and-rescue.

10.6.1.6 Adaptive Frequency-Hopping Communication

Creating collaborative autonomy in UAV swarms – where each drone can carry out tasks on its own while simultaneously interacting with other drones in the swarm – is a significant area of innovation. Others can assume the duties of a compromised or disabled UAV without jeopardizing the mission.

Research Example: DARPA OFFSET (Offensive Swarm-Enabled Tactics) program, which focuses on creating resilient UAV swarms for military uses. With each drone able to adjust to the loss of a peer, these swarms are capable of autonomous navigation, surveillance, and real-time situational data sharing

Impact: By guaranteeing redundancy and flexibility, collaborative autonomy improves the resilience of UAV networks. UAV swarms are extremely dependable because they can continue operating even if a few drones malfunction or are intercepted.

10.6.1.7 Post-quantum Cryptography (PQC) Integration

Traditional cryptographic techniques, like RSA and ECC, are becoming more susceptible to quantum-based attacks as quantum computing develops. PQC algorithms are being developed to replace weak encryption schemes and improve the resilience of UAV networks.

- **Research Example**: To find reliable PQC algorithms, the National Institute of Standards and Technology (NIST) has been holding a global competition. Even against adversaries with quantum computing capabilities, the incorporation of these algorithms into UAV systems will guarantee secure authentication and data transfer.
- **Impact**: To guarantee the long-term security of UAV communications, PQC will be crucial. These algorithms will prevent future quantum computers from decrypting private information gathered by UAVs, such as military intelligence or surveillance footage.

10.6.1.8 Edge Computing and On-Device AI

Instead of sending data to a central server for analysis, edge computing processes data locally at the source (on the UAV itself). UAVs can improve operational resilience by using edge AI to assess their environment and make decisions in real time without depending on cloud connectivity.

- **Example**: NVIDIA is creating edge computing systems for UAVs, in which AI models are installed on the drone to perform functions like path optimization, obstacle avoidance, and object detection. This makes it possible for UAVs to function in areas with inadequate or nonexistent connectivity.
- **Impact**: By ensuring that the drone can operate independently, even in the event of a communication failure with the control center, edge computing reduces latency in UAV operations and enhances resilience. This is particularly beneficial for search and rescue missions.

10.6.1.9 Multi-factor Authentication (MFA) and Zero Trust Architectures

Many organizations are implementing zero trust architectures and multi-factor authentication (MFA) to stop unwanted access to UAV networks. These methods guarantee that prior to gaining access to sensitive network resources, all users and devices – including the UAV itself – must be validated.

- **Research Example**: In its UAV systems, the U.S. Air Force has been experimenting with zero-trust architectures, in which no person or device is trusted by default. Throughout its lifecycle, a UAV authenticates and verifies each communication and action it performs.
- **Impact**: By putting MFA and zero trust principles into practice, UAV networks become much more secure, lowering the possibility of malicious actors manipulating the network, stealing data, or hijacking drones without authorization.

A concentrated effort has been made to improve the resilience of UAV networks through the advancements and current research projects mentioned above. These technologies are expanding the realm of what is feasible in terms of securing UAV operations, from AI-driven threat detection to quantum-secure communication. These developments will ensure that UAV networks are resilient, adaptable, and safe – whether in commercial, military, or humanitarian applications – even in hazardous or uncertain environments.

10.6.2 Example of UAV Network Resilience in Practice

Think about a fleet of UAVs used for border monitoring. In this scenario, the UAVs transmit live video feeds to a central command center while communicating via a quantum-secure mesh network. AI systems keep an eye on the data streams from the UAVs to look for any irregularities, like unforeseen communication patterns or system malfunctions. The system instantly reroutes communication to the other UAVs in the swarm in the event that one of them is compromised [30], allowing the surveillance mission to continue uninterrupted. Furthermore, the recorded flight data is unchangeable thanks to blockchain technology, which also provides an audit trail that ensures the mission's integrity.

This example demonstrates how advancements in mesh networking, AI, and quantum cryptography can combine to build a highly resilient UAV network that can withstand physical and cyberdisruptions.

10.7 Conclusion

Developing robust quantum UAV networks necessitates a comprehensive strategy that incorporates advances in quantum computing, AI, and cybersecurity. Because of this technological convergence, UAV networks are better equipped to protect themselves from cyberattacks, maintain secure communication, and improve operational resilience. UAV network defenses are strengthened, from a cybersecurity standpoint, by putting in place quantum-resistant encryption, QKD protocols, and AI-driven threat detection systems. These safeguards enable real-time threat response, protect against cyberattacks, and safeguard sensitive data.

AI plays a key role in improving UAV operations by enabling predictive analytics, adaptive security measures, and autonomous decision-making. The resilience and effectiveness of quantum-enabled UAV networks are increased by ML techniques, which also detect anomalies, optimize resource allocation, and enhance overall network performance. Quantum computing offers quantum-resistant AI models, quantum-enhanced sensing capabilities, and quantum-safe cryptographic solutions. These developments improve the security of UAV networks, make quantum-secure communications possible, and open up new avenues for data processing and analysis.

In conclusion, developing robust UAV networks that can handle changing threats, maintain data integrity, and support mission-critical operations in a variety of environments requires synergy between cybersecurity, AI, and quantum computing viewpoints. Adopting these viewpoints encourages creativity, improves network resilience, and creates a path for the next generation of UAVs with quantum capabilities.

References

[1] Garcia, F., et al. "Quantum computing and AI integration for UAV security." International Journal of Quantum Information, vol. 19, no. 6, 2021, pp. 1–18.

[2] John, A., et al. "Energy management and monitoring using IoT with CupCarbon platform." Green computing in smart cities: Simulation and techniques (2021): 189–206.

[3] Smith, A., et al. "Resilience in quantum computing: challenges and opportunities." ACM Computing Surveys, vol. 52, no. 3, 2019, pp. 1–34.

[4] Adams, C., et al. "Quantum-safe UAV network standards and compliance frameworks." IEEE Communications Magazine, vol. 59, no. 8, 2021, pp. 94–101.

[5] Brown, C., et al. "Quantum-safe UAV routing algorithms." Journal of Cybersecurity and AI, vol. 1, no. 1, 2021, pp. 45–56.

[6] Walker, Q., et al. "Quantum-safe UAV firmware updates." IEEE Internet of Things Journal, vol. 8, no. 5, 2021, pp. 3804–3816.

[7] Hall, Z., et al. "Quantum-secure UAV cloud integration." IEEE Cloud Computing, vol. 8, no. 4, 2021, pp. 42–55.

[8] Wilson, E., et al. "Quantum UAV networks: threat models and defense mechanisms." IEEE Transactions on Network and Service Management, vol. 17, no. 3, 2020, pp. 1501–1513.

[9] Turner, R., et al. "AI-based anomaly detection in quantum UAV networks." Journal of Artificial Intelligence, vol. 44, no. 1, 2020, pp. 78–92.

[10] Lewis, O., et al. "Trust management in quantum UAV networks." IEEE Transactions on Information Forensics and Security, vol. 16, no. 7, 2021, pp. 1759–1771.

[11] Rodriguez, H., et al. "Quantum cryptography protocols for UAV communications." Quantum Information Processing, vol. 19, no. 2, 2020, pp. 1–17.

[12] Harris, M., et al. "Quantum-safe UAV data storage solutions." IEEE Transactions on Emerging Topics in Computing, vol. 9, no. 2, 2021, pp. 462–475.

[13] Hall, Z., et al. "Quantum-secure UAV cloud integration." IEEE Cloud Computing, vol. 8, no. 4, 2021, pp. 42–55.

[14] Reed, Y., et al. "AI-driven quantum UAV network optimization." Journal of Optimization Theory and Applications, vol. 180, no. 3, 2020, pp. 789–805.

[15] Scott, A., et al. "Quantum-safe UAV network policy management." IEEE Transactions on Network Management, vol. 17, no. 6, 2021, pp. 1025–1037.

[16] Clark, P., et al. "Privacy-preserving techniques for UAV communication." ACM Transactions on Privacy and Security, vol. 24, no. 3, 2021, pp. 1–25.
[17] Martinez, B., et al. "Quantum UAV network incident response strategies." Journal of Network and Computer Applications, vol. 194, 2021, pp. 1–12.
[18] King, U., et al. "Quantum-secure UAV network protocols." Journal of Quantum Cryptography, vol. 5, no. 2, 2021, pp. 89–102.
[19] Chen, J., et al. "Quantum key distribution for secure UAV networks." Journal of Quantum Information Science, vol. 10, no. 3, 2020, pp. 1–15.
[20] Lopez, X., et al. "Quantum UAV network resilience testing." IEEE Transactions on Reliability, vol. 70, no. 2, 2021, pp. 421–435.
[21] Martinez, N., et al. "Quantum-enhanced cyber threat intelligence for UAV networks." Journal of Cybersecurity and Privacy, vol. 5, no. 1, 2020, pp. 12–25.
[22] Jones, B., et al. "AI-driven cybersecurity for UAV networks." IEEE Security & Privacy, vol. 18, no. 2, 2020, pp. 58–67.
[23] Thomas, I., et al. "Quantum machine learning for UAV network security." Journal of Quantum Machine Learning, vol. 3, no. 4, 2021, pp. 189–203.
[24] Adams, K., et al. "Quantum-secure authentication in UAV networks." IEEE Access, vol. 9, 2021, pp. 28952–28965.
[25] Hayes, W., et al. "Quantum-safe UAV network monitoring tools." IEEE Transactions on Network and Service Management, vol. 18, no. 1, 2021, pp. 32–45.
[26] Scott, S., et al. "Quantum-safe UAV traffic analysis." IEEE Transactions on Vehicular Technology, vol. 70, no. 6, 2021, pp. 5511–5524.
[27] Singh, D., et al. "Quantum-resistant cryptography for secure UAV communication." IEEE Transactions on Aerospace and Electronic Systems, vol. 56, no. 4, 2020, pp. 2681–2697.
[28] White, J., et al. "Resilient UAV networks in adversarial environments." IEEE Journal on Selected Areas in Communications, vol. 38, no. 5, 2020, pp. 1132–1145.
[29] Wright, L., et al. "AI-driven intrusion detection systems for UAVs." Journal of Artificial Intelligence Research, vol. 72, 2021, pp. 387–402.
[30] Young, T., et al. "Quantum UAV network simulation and modeling." Simulation Modelling Practice and Theory, vol. 109, 2020, pp. 1–15

Bhupinder Singh, Saurabh Chandra, and Hemant Kumar Saini

11 Unmanned Aerial Vehicle (UAV) Operations and Quantum-Safe Data Privacy: Ultramodern View to Realizing Cybersecurity with Artificial Intelligence (AI) Strategies

Abstract: The Internet of drones (IoD) is an emerging convergence framework where drones can share data through the internet, working hand-in-glove with high-speed 5G technology. Although this integration opens up a huge scope for deep collaboration across various sectors, it also brings with it a variety of cybersecurity threats and issues that need to be resolved in order to ensure safe and secure drone network operations within the IoD ecosystem. Drones will be an IoT component for the first time, and 5G networks bring a new dimension to their potential uses – from autonomously delivering packages to live-streaming video surveillance. Enter quantum computing; it augurs a bright future as it can provide cutting-edge encryption and data processing. This chapter provides exhaustive coverage of quantum-safe data privacy and unmanned aerial vehicle operations leveraging cybersecurity and artificial intelligence strategies.

Keywords: UAV, cybersecurity, IoD, data privacy, 5G connectivity, quantum computing

11.1 Introduction and Background

The benefits within smart-city scenarios make unmanned aerial vehicles (UAVs) very suitable for a wide range of initiatives. However, while those challenges may seem daunting, they also present a massive business opportunity for managed security service providers (MSSPs) and other cybersecurity providers. MSSPs can offer organizations drone security services in combination with risk assessments, tailored security policies, encryption methods and predictive threat monitoring to secure their Internet of drones (IoD) deployments. These unmanned aircraft systems provide live video feed and on-the-fly observations, providing officials with valuable information from a range of sources and locations where UAVs have unique applications in areas such as traffic control tests, environmental analysis or disaster response for public safety. With questing valleys and remote areas easily accessible to UAVs, they are highly valuable for increasing situational awareness and ensuring a fast response. But the key to digital transformation is security hard, tough security that protects drone networks and inherently vaults information they are dealing with. Data protection and

https://doi.org/10.1515/9783111619323-011

privacy are essential in the IoD. Drones collect a lot of sensitive data, so it is vital to protect this data from unauthorized access and exploitation. Whilst practical quantum computers are still a number of years away, research is currently being done to increase the . . . The use of the 5G technology has advantages as well as disadvantages. On one hand, 5G may provide higher-speed communication capabilities but on the other hand it increases the attack surface for cyberthreats. For example, man-in-the-middle (MITM) attacks, spoofing can be avoided by the use of robust network security measures like intrusion detection systems and communication channels are encrypted [1].

Though drones can be tampered with or even stolen, rendering them useless, it is also possible that unwelcome entities may tamper with the sensor data in cases of autonomous navigation. This encourages a design based on the concept of safety by design and taking measures such as including tamper-proof mechanisms and anti-jamming techniques to bolster security. The whole IoD architecture becomes fragile and is subject to be easily collapsed in a single blow, scenarios where an attacker invades components of the system to compromise its integrity as a hole have drastically grown in numbers [2]. The term "remotely controlled aerial vehicles or drones" is for commercial uses and military activities, due to its combination of sensors, actuators, and computers with wireless connectivity for data gathering and processing. Among the most common is data security and authentication, which while used in increasingly reliance but come with significant challenges cause sensitive data to be collected and transmitted over an untrusted channel such as the Internet. The current IoD communication security architecture relies heavily on factorization and the insecure nature of discrete logarithms. If you need theoretical security in the large-scale quantum computing era, it can be difficult due to Shor's algorithm for quantum computers: secure communication and no authentication. A solution to the one-way IOD communication is proposed and verified using ring learning with errors problem on lattices in this work. This approach is supposed to resist quantum attacks while keeping its computational complexity low [3].

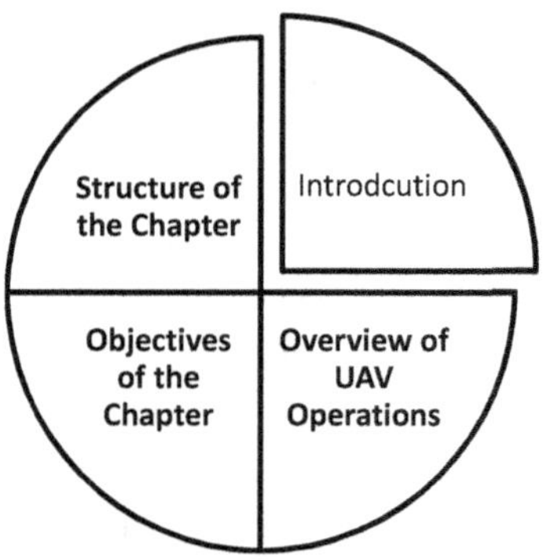

Figure 11.1: Landscapes of the introduction section and structure of the chapter.

11.1.1 Overview of UAV Operations

The UAV industry is taking off, creating a rare case of industrial and technological convergence involving all the different emerging technologies, as shown in Figure 11.1, which clarifies the landscape of the introduction section. Causes for this increase can be attributed to technological advancements and progress in areas of computerization, miniaturization, and control systems that have resulted in cheaper, dependable, safe, lightweight drones becoming a reality [4]. UAVs are superior in terms of extensively exploiting their omnidirectional advantages, such as quick mobility, vertical plant life, delay-free aerial missions, instant access to disaster-affected areas with grown vegetative hindrances, always-on operation, and controlling airborne intelligence functions using various payloads. Jetpacks certainly are a more desirable method of travel in comparison to the time and resources needed for building tunnels, digging trenches, or pouring cement, but despite all those benefits, UAVs are still operationally restricted by a range of key issues such as flying autonomy, path planning, battery life, flight duration, and small payload capacity. UAVs, as such, are of limited utility due to the difficulty in loading them with large batteries [5].

11.1.2 Objectives of the Chapter

This chapter has the following objectives:

- Potential vulnerabilities in UAV operations due to quantum computing.
- Quantum Safe cryptographic techniques should be applied to UAVs to ensure data privacy.
- Ways AI can contribute to enhancing cybersecurity for UAV operations.

Figure 11.2 expresses the objectives in a broader sense concerning this chapter.

11.1.3 Organization/Structure of the Chapter

This chapter, as shown in Figure 11.3, deeply dives into the diverse arena of quantum-safe data privacy and UAV operations, which is a futuristic view toward achieving cybersecurity and artificial intelligence (AI) strategies. Section 11.2 elaborates on the growth of UAV applications in various industries, which describes the importance of data privacy and cybersecurity in UAV operations. Section 11.3 explores quantum computing and its impact, and also its relevance and implications for cryptography. Section 11.4 outlines the potential threats quantum computing poses to UAV data security. Section 11.5 specifies the role of AI in cybersecurity, and also AI assimilation to enhance cybersecurity. Section 11.6 examines quantum computing and its impact on cryptography. Section 11.7 discusses quantum computing and cybersecurity threats.

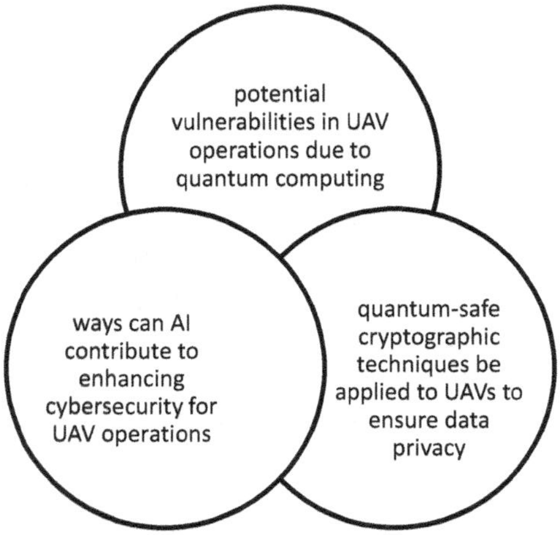

Figure 11.2: Objectives of the chapter.

Section 11.8 delves into AI and cybersecurity for UAV operations. Finally, Section 11.9 concludes the chapter with future scope.

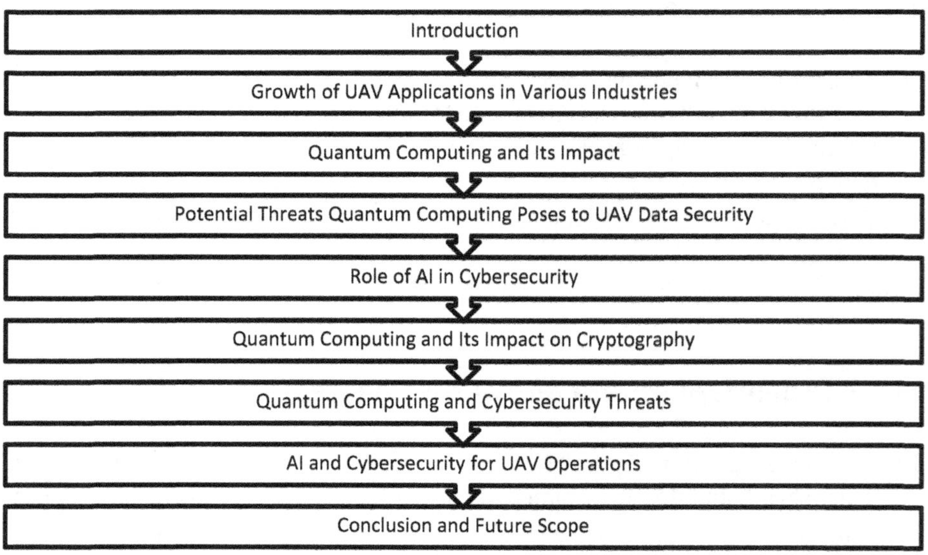

Figure 11.3: Flow of the chapter.

11.2 Growth of UAV Applications in Various Industries: Importance of Data Privacy and Cybersecurity in UAV Operations

UAVs are remotely operated aerial aircraft that can carry out a range of tasks without the need for direct human participation. Microprocessors and sensors, among other electrical equipment, can be used to remotely operate them. In dangerous situations where it would be hazardous for humans to be present, UAVs may also carry out autonomous duties. A typical UAV design incorporates communication lines for establishing a connection with ground control systems such as laptops, smartphones, or satellites [6]. UAVs can function independently, but they still need a human operator for remote control and direction. The various facets of UAV technology have been the subject of several research studies. It also explored mobile charging capabilities for drones to increase battery life. We investigated wirelessly powered receivers to reduce weight. Several wireless charging methods are available for prolonging the flight time of UAVs. The WPT-based charging system [7] focused on misalignment loss mitigation strategies.

As an increasingly digital society, digital security has grown into a necessary spectrum that binds numerous industries and technologies together. One of these is UAVs, also called drones. As a technically innovative item, they have been used for disaster management as well as in parcel delivery services. The more autonomous UAVs become, the more important it is to secure their cybersecurity. Cyberattacks pose a severe threat to UAVs, as they can impact the security, performance, and integrity of these systems. This makes it extremely important to do everything possible to implement stringent measures to protect sensitive data, secure UAVs from cyberattacks, and ensure public trust in their safe deployment. As understanding of these autonomous systems' weaknesses and the potential outcomes of intrusions continues to increase, the issue of UAV cybersecurity has garnered a lot of attention. The integration of UAVs heavily relies on the use of software, wireless connections, and data-sharing protocols, which means that they are vulnerable to cyberattacks. One of the chief concerns of UAV cybersecurity is hacking, i.e., malicious actors can attempt to gain access to UAV systems and interfere with flight controls, onboard sensors, cameras, etc. Unattended challenging access can lead to a variety of negative outcomes, up to and including uncontrolled destruction or threats to human life [8].

11.3 Quantum Computing Fit and Implications on the World of Cryptography

Quantum mechanics can be thought of as studying how very small objects – both particles and collections of particles, like electrons and photons – behave at the smallest scales. This includes concepts such as superposition, entanglement, and uncertainty, which are often non-intuitive in nature. Quantum particles such as these are employed by quantum computers to exploit such distinctive traits [9]. Quantum Information, the power of quantum information resides in qubits – a superposition of both states; simultaneously containing 1s and 0s. There are two key quantum mechanical concepts – superposition and measurement – that we need to understand for us to know anything about qubits. In the quantum physics, due to its probabilistic nature, you do not know the properties of the particle until they are measured [49]. The wavefunction $\psi(x)$, for example, describes such probability distribution of property (e.g., position of a particle). Except for one major difference: it's somewhat like a probability density function, where the square of $|\psi(x_0)|^2$ tells you how likely would be to find the particle at any particular point x_0. It is interesting that wavefunctions may be written as a sum of various smaller functions. More precisely, the rules of quantum mechanics may describe a multitude of probable states and the ways that they might supersede each other by decomposing an uncollapsed wavefunction into one rather entangled superposition of eigenfunctions. This breakdown is essential to comprehending quantum behavior and quantum computation [10].

11.4 Potential Threats Quantum Computing Poses to UAV Data Security

To ensure data security, modern encryption focuses on the computational complexity of solving particular mathematical problems. Given its exponentially larger processing power, a quantum computer could be able to calculate every solution to these issues simultaneously. This ability might crack an encryption cipher, thereby negating the effectiveness of currently used encryption methods. Companies will need to create new encryption techniques in the post-quantum computing age since the techniques used now might not be able to withstand attacks based on quantum technology. While a large portion of the current research on quantum computing is being carried out in academic environments, businesses and governmental organizations are becoming more aware of the potential concerns that this technology may present. This year's World Economic Forum saw the introduction of the National Quantum Blueprint, which aims to guide nations' investments in the development of quantum technologies while addressing related cybersecurity risks [11].

Quantum computing has become an increasingly important issue for governments and corporations alike as it transitions from an innovative tool to a possible cybersecurity danger. IBM is also researching ways to better defend cryptography against quantum attacks. IBM has created Quantum Safe, a proprietary system that works with businesses to determine their unique risks, evaluate their security maturity, and plan the steps required to set up strong security controls, such as implementing quantum-resistant cryptography standards, using its quantum computing network. IBM and Vodafone announced a collaboration in November 2022 to prepare the telecom behemoth for the upcoming wave of quantum computing challenges. They aim to protect Vodafone's assets and data in a variety of telecom scenarios using Quantum Safe. Given the enormous processing capacity of quantum algorithms, predicting the moment at which quantum computing will be useful for real-world applications is essential. Cryptographers are developing protocols that are resistant to attacks from both conventional and quantum computers at the same time [12].

Achieving a quantum advantage is the ultimate goal of quantum computing, which would enable it to perform tasks faster than classical computers, such as cracking cryptographic secrets and modeling intricate chemical operations. The time at which quantum computing becomes realistically usable is "probably still a long way off," according to experts, despite the fact that quantum computers have successfully demonstrated feats like Gaussian boson sampling, a work that would take classical computers 2.5 billion years. Some see comparisons between early machine code and the current state of quantum computing, emphasizing that software and high-level. Given the enormous processing capacity of quantum algorithms, predicting the moment at which quantum computing will be useful for real-world applications is essential. Cryptographers are developing protocols that are resistant to assaults from both conventional and quantum computers at the same time [13].

Achieving a quantum advantage is the ultimate goal of quantum computing, which would enable it to perform tasks faster than classical computers, such as cracking cryptographic secrets and modeling intricate chemical operations. The time at which quantum computing becomes realistically usable is "probably still a long way off," according to experts, despite the fact that quantum computers have successfully demonstrated feats like Gaussian boson sampling, a task that would take classical computers 2.5 billion years. Some draw comparisons between early machine code and the current state of quantum computing, emphasizing that software and high-level programming languages for quantum systems are still in the nascent stages of development [14].

11.5 Role of AI in Cybersecurity: AI Assimilation to Enhance Cybersecurity

The usage of online platforms and technical breakthroughs has led to an unprecedented rise in connectedness in the digital world. The extent of our dependence on digital systems and networks has resulted in an increase in cyberthreats. With its focus on safeguarding digital assets, data, and systems from harm, unauthorized access, and attacks, cybersecurity has emerged as a critical concern. Traditional cybersecurity techniques frequently struggle to keep up with the evolving attack methodologies as cyberthreats grow in complexity and scope. This is where AI's potentially revolutionary role in cybersecurity becomes apparent. AI is a field within computer science that aims to build computers with human-like skills. It is being utilized increasingly to support cybersecurity initiatives. AI is a valuable tool in the battle against complex cyberthreats because of its capacity for learning and adaptation [15].

11.6 Quantum Computing and Its Impact on Cryptography

Quantum computing is expected to bring profound changes to the global economy. If it becomes widely accessible, its benefits will span multiple industries, enhancing current technologies and allowing us to address complex challenges in ways that were previously impossible. Yet, these advancements come with significant risks. Quantum computing will dramatically alter the cybersecurity landscape, particularly in the realm of cryptography. Utilizing the ideas of quantum physics, quantum computing is a quickly developing technology that can solve problems too complicated for conventional computers. These problems can be resolved by quantum computers far more swiftly than by traditional computers, which would greatly advance science. They might, however, potentially be able to bypass current encryption techniques and jeopardize the security of previously protected data [16].

Commercial quantum computers that are fully functional might revolutionize a number of industries, including big data, AI, economic analysis, and other sectors needing sophisticated computations and large-scale data processing. However, there's a serious risk that the same computing capacity that spurs innovation might be exploited to compromise cybersecurity. Companies need to be proactive in mitigating the hazards associated with quantum computing's potential to undermine current cryptography frameworks. Fortunately, there is already a method to secure digital systems in a post-quantum era, thanks to technology built to fend against both conventional and quantum attacks [17].

11.7 Quantum Computing and Cybersecurity Threats

Quantum computing will be able to address issues that traditional computing is unable to, such as deciphering algorithms that safeguard our data and the Internet's infrastructure. The mathematical procedures used in current encryption techniques make them impractically difficult to break into the time constraints of modern computers. To factor ever-larger numbers, researchers have been working on creating quantum computers. For example, the outcome is simple to understand if you multiply two huge prime integers. It is far more difficult to reverse the process by dissecting a big number into its prime elements. However, these huge numbers can be factored effectively by quantum computers, which might lead to the encryption's breaking [18]. It may be approaching a "quantum leap" in the fourth industrial revolution, according to experts, given the tremendous progress made in quantum computing, which is best demonstrated by IBM's Quantum Condor processor, which has a 1,000 qubit capability. Governments and corporations are becoming more interested in quantum computers due to their exponentially greater computing capability As quantum technology moves from theoretical research to practical, commercially available solutions, this advancement opens the door to a wide range of inventive and sophisticated applications, including new drug discovery, faster and more comprehensive financial trading analytics, improved efficiency in supply chain management, and much more. Organizations must handle the cybersecurity issues raised by quantum computing in order to be prepared for all the ramifications it will have as they work to take use of these new capabilities [19].

11.7.1 Risks to UAV Operations: Potential Risks to UAV Data and Communication due to Quantum Computing

Although quantum computing presents a number of concerns, it also holds great promise for future improvements. The potential for quantum computers to breach the security protocols protecting day-to-day operations might have dire consequences. The potential for data breaches, including private health and financial information, threats to the integrity of digital assets, and a potential attack on the cryptographic underpinnings of cryptocurrencies, are all suggested by this quantum cybersecurity concern. As technology advances, UAVs have a number of advantages, but privacy, security, and safety remain major issues. Regulations and drone usage licenses are essential to addressing these problems and limiting pointless aerial photography. The majority of authorities have stringent regulations to stop unapproved aerial monitoring. Due in large part to UAVs' larger resource scope, UAV coverage is more extensive and different from that of wireless sensor networks or mobile ad hoc networks (MANETs) from the standpoint of network security [20].

Authentication, authorization, and accounting is the system that creates drone operating guidelines. Meaning, unauthorized agents cannot hijack or redirect the drone to other paths since it specifies what rights are granted to a drone owner and inclusion that we add special authentication enforcement. It also serves as anodes making drone operators traceable in case of a illegal activity or security breach. The aim of this model is to prevent privacy risks, illegal surveillance, and cyberterrorism efforts [21]. To circumvent these sinister acts a number of mechatronic engineering ways is proposed. As drones are cheap and available in market, they may be used for unlawful purposes as well [82]. UAVs can be installed with a number of different payloads, including explosives that are dangerous upon detonation. Drones can surprise as they can deliver materials surreptitiously without being detected which is what the mobility unreachable enable them.

Security and Safety Concerns: There is safety concern that drones may malfunction or crash over populated areas, leading to potential loss of life and property; not just security aspect but also in terms of protecting the liability. A number of these issues have been realized in some cases, such the UAV collision with a British Airways aircraft BA727 back in April 2016 [22].

11.8 AI and Cybersecurity for UAV Operations

Cyberattacks are constantly evolving, adapting to an ever-evolving set of security protocols and technologies. A staggering number of factors contribute to the dynamic nature of cyberthreats, from nation-state-supported attacks to the sophisticated tactics hackers are using, while the attack surface continues to expand with more connected devices powered by this era of the Internet of things (IoT). Developing effective cybersecurity strategies requires an understanding of how cyberattackers adapt their tactics, techniques, and processes. It may improve your cybersecurity abilities by taking the Understanding Cisco Cybersecurity Fundamentals (SECFND) course. Learn the foundations of applications, operating systems, and networking that are crucial for a Security Operations Center (SOC), as well as common security ideas and basic security approaches. So, prepare it to identify, look into, and handle a variety of security situations. Become an essential component of the cybersecurity initiatives of your company. If you want to further your cybersecurity profession, enroll today [23].

The cybersecurity environment is changing due to AI, which is making it more dynamic and adaptive to keep up with emerging cyberthreats [24]. The following are some benefits of applying AI to cybersecurity:

- Real-time threat detection made simpler
- Capacity to recognize patterns and signatures of known attacks
- Simplified threat detection of prevalent threats
- Prompt and anticipatory threat assessment

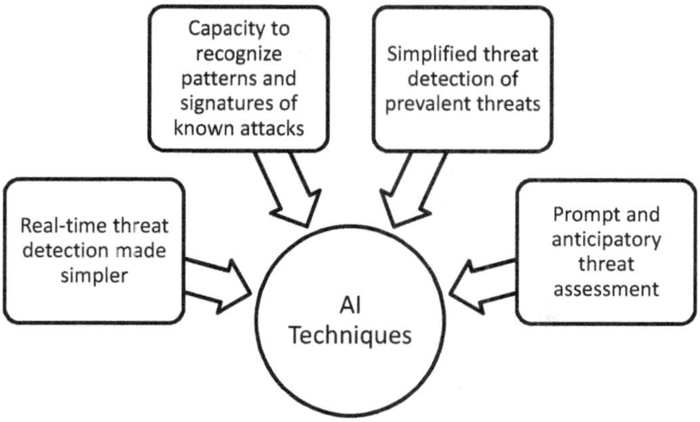

Figure 11.4: Benefits of applying AI to cybersecurity.

With reinventing cybersecurity technologies like antivirus software, these technical breakthroughs are empowering security specialists to guard against an ever-expanding range of cyberthreats as expressed in Figure 11.4. For this reason, it's becoming more and more important to invest in AI-based cybersecurity solutions. The continuous threat exposure management (CTEM) has greatly reduced the amount of time security teams need to identify and address new cyberthreats because to the integration of AI. It is like having a watchful watchdog who never stops looking out for any threats. AI isn't simply making cybersecurity risk profiles better for enterprises; it's entirely changing them [25].

Cyberattacks might target any asset, digital or physical, in an organization's vast digital ecosystem. This is when the crucial component of CTEM, intelligent asset detection, comes into play. With the implementation of AI-powered solutions, enterprises will be able to quickly discern and classify massive data in their digital ecosystem, thereby spotting every asset that could be a target of a cyber-attack. This helps the security team to invent and secure these assets, providing complete threat protection [26]. With this approach, AI and machine learning (ML) systems are used to help security teams better manage real-time information. Those machines can, of course, process data far faster than any human team and also assist in creating more effective cybersecurity strategies. Imagine a super organized librarian who knows the exact location of every book in the library, and can also foresee which books are likely to be very popular. Threat prediction is a crucial skill in cybersecurity. Predictive threat analysis is a specialty of ML, which looks at past data, spots patterns, and flags possible cyberthreats in the future. It is similar to possessing a time machine that can reveal possible threats before they materialize as that is ML's power [27].

Setting priorities is essential in a world full of potential hazards related to cyberthreats. The difficult problem of risk prioritization has been transformed by ML. By automatically classifying events according to their expected impact, ML facilitates the prioritization process and allows for a more effective use of resources by addressing the most serious concerns first. This is extended by predictive analytics. Predictive analytics can estimate possible attack vectors by analyzing historical data and current security settings. This improves the efficacy and accuracy of threat prioritization. ML is setting the standard so that businesses may focus their efforts where they really matter [28]. Cybersecurity benefits greatly from ML algorithms since they:

- identify different kinds of cyberattacks;
- identify trends that point to potential security issues; and
- enhance the categorization and identification of cyberthreats, including the detection of fraud and the classification of phishing attacks. These points are specified in Figure 11.5.

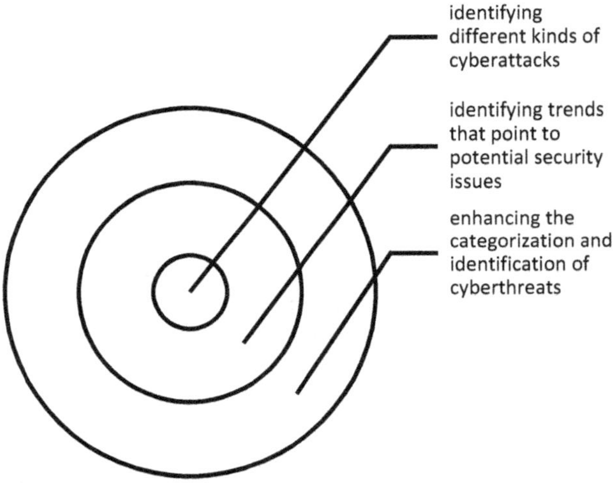

Figure 11.5: Cybersecurity benefits from machine learning algorithms.

Microsoft's attempts to mitigate the hazards associated with quantum computing are exemplified by its creation of a post-quantum transport layer security (TLS) protocol. A popular security technology that shields data being sent over the internet is called TLS. Microsoft discovered that some quantum-based attacks may be able to compromise certain asymmetric algorithms utilized in TLS exchanges [29]. The first four quantum-resistant cryptographic algorithms were revealed in July 2022 by the National Institute of Standards and Technology (NIST) of the U.S. Department of Commerce, following a six-year competition. NIST plans to publish its post-quantum cryptography standard in 2024, and these methods will be included in it [30].

11.9 Conclusion and Future Scope

Quantum computing is an emerging hardware technology that could revolutionize how the world handles large-scale and complex computations. This innovative branch of computer science is grounded in quantum theory, utilizing materials at the atomic or subatomic scale. The distinctive nature of quantum computing allows it to address problems that have traditionally been considered too complex for classical computers, thanks to its ability to process data in parallel by exploring multiple possibilities simultaneously, potentially leading to exponential growth in computational power. While governments are pouring billions into R&D initiatives to be at the forefront of quantum innovation, major firms such as Google, IBM, and Amazon are devoting significant resources to quantum technology. Quantum computing presents new and possibly serious security vulnerabilities as the technology becomes more popular and offers glimpses of revolutionary changes through ground-breaking applications. It is anticipated that quantum computing will revolutionize a number of industries, including manufacturing, banking, aerospace, and medicine. For example, the capability of quantum computers might help academics and scientists develop life-saving drugs faster, while the automotive industry could produce more robust batteries for electric cars, which would promote a more sustainable economy.

References

[1] Souilem, M., Dghais, W., &Radwan, A. (2020). Wirelessly Powered Unmanned Aerial Vehicles (UAVs) in Smart City. *Connected and Autonomous Vehicles in Smart Cities*, 437–456.

[2] Puri, V., Nayyar, A., & Raja, L. (2017). Agriculture drones: A modern breakthrough in precision agriculture. Journal of Statistics and Management Systems, 20(4), 507–518.

[3] Nayyar, A., Nguyen, B. L., & Nguyen, N. G. (2020). The internet of drone things (IoDT): future envision of smart drones. In First International Conference on Sustainable Technologies for Computational Intelligence: Proceedings of ICTSCI 2019 (pp. 563–580). Springer Singapore.

[4] Singh, B. (2024). Legal Dynamics Lensing Metaverse Crafted for Videogame Industry and E-Sports: Phenomenological Exploration Catalyst Complexity and Future. *Journal of Intellectual Property Rights Law*, *7*(1), 8–14.

[5] Singh, B. (2023). Blockchain Technology in Renovating Healthcare: Legal and Future Perspectives. In *Revolutionizing Healthcare Through Artificial Intelligence and Internet of Things Applications*. 177–186). IGI Global.

[6] Sharma, A., & Singh, B. (2022). Measuring Impact of E-commerce on Small Scale Business: A Systematic Review. *Journal of Corporate Governance and International Business Law*, *5*(1) P 15.

[7] Taddeo, M. (2019). Three ethical challenges of applications of artificial intelligence in cybersecurity. *Minds and Machines*, *29*, 187–191.

[8] Singh, B. (2023). Federated Learning for Envision Future Trajectory Smart Transport System for Climate Preservation and Smart Green Planet: Insights into Global Governance and SDG-9 (Industry, Innovation and Infrastructure). *National Journal of Environmental Law*, *6*(2), 6–17.

[9] Singh, B., Kaunert, C., &Vig, K. (2024). Reinventing Influence of Artificial Intelligence (AI) on Digital Consumer Lensing Transforming Consumer Recommendation Model: Exploring Stimulus Artificial

Intelligence on Consumer Shopping Decisions. In T. Musiolik, R. Rodriguez, & H. Kannan (Eds.), AI Impacts in Digital Consumer Behavior (pp. 141–169). IGI Global. https://doi.org/10.4018/979-8-3693-1918-5.ch006

[10] Singh, B., & Kaunert, C. (2024). Salvaging Responsible Consumption and Production of Food in the Hospitality Industry: Harnessing Machine Learning and Deep Learning for Zero Food Waste. In Sustainable Disposal Methods of Food Wastes in Hospitality Operations (pp. 176–192). IGI Global.

[11] Singh, B. (2024). Evolutionary Global Neuroscience for Cognition and Brain Health: Strengthening Innovation in Brain Science. In Biomedical Research Developments for Improved Healthcare (pp. 246–272). IGI Global.

[12] Singh, B., Kaunert, C., &Vig, K. (2024). Reinventing Influence of Artificial Intelligence (AI) on Digital Consumer Lensing Transforming Consumer Recommendation Model: Exploring Stimulus Artificial Intelligence on Consumer Shopping Decisions. In AI Impacts in Digital Consumer Behavior (pp. 141–169). IGI Global.

[13] Singh, B., & Kaunert, C. (2024). Revealing Green Finance Mobilization: Harnessing FinTech and Blockchain Innovations to Surmount Barriers and Foster New Investment Avenues. In Harnessing Blockchain-Digital Twin Fusion for Sustainable Investments (pp. 265–286). IGI Global.

[14] Thiruvikraman, P., T. Ananth Kumar, R. Rajmohan, and M. Pavithra. "A survey on haze removal techniques in satellite images." Irish Interdisciplinary Journal of Science & Research (IIJSR) 5, no. 2 (2021): 01–06.

[15] O'Sullivan, S., Nevejans, N., Allen, C., Blyth, A., Leonard, S., Pagallo, U., . . . &Ashrafian, H. (2019). Legal, regulatory, and ethical frameworks for development of standards in artificial intelligence (AI) and autonomous robotic surgery. *The international journal of medical robotics and computer assisted surgery*, *15*(1), e1968.

[16] Jackson, B. W. (2019). Cybersecurity, privacy, and artificial intelligence: an examination of legal issues surrounding the European Union general data protection regulation and autonomous network defense. *Minn. JL Sci. & Tech.*, *21*, 169.

[17] Kaushik, P. (2023). Unleashing the power of multi-agent deep learning: Cyber-attack detection in IoT. *International Journal for Global Academic & Scientific Research*, *2*(2), 15–29.

[18] Carter, D. (2020). Regulation and ethics in artificial intelligence and machine learning technologies: Where are we now? Who is responsible? Can the information professional play a role?. *Business Information Review*, *37*(2), 60–68.

[19] Wylde, V., Rawindaran, N., Lawrence, J., Balasubramanian, R., Prakash, E., Jayal, A., . . . &Platts, J. (2022). Cybersecurity, data privacy and blockchain: a review. *SN Computer Science*, *3*(2), 127.

[20] Jameel, T., Ali, R., &Toheed, I. (2020, January). Ethics of artificial intelligence: research challenges and potential solutions. In *2020 3rd International Conference on Computing, Mathematics and Engineering Technologies (iCoMET)* (pp. 1–6). IEEE.

[21] Ko, R. K. (2020). Cyber autonomy: Automating the hacker-self-healing, self-adaptive, automatic cyber defense systems and their impact to the industry, society and national security. *arXiv preprint arXiv:2012.04405*.

[22] Sugali, K. (2021). Software testing: Issues and challenges of artificial intelligence & machine learning.

[23] Singh, B., & Kaunert, C. (2024). Integration of Cutting-Edge Technologies such as Internet of Things (IoT) and 5G in Health Monitoring Systems: A Comprehensive Legal Analysis and Futuristic Outcomes. *GLS Law Journal*, *6*(1), 13–20.

[24] Alhayani, B., Mohammed, H. J., Chaloob, I. Z., & Ahmed, J. S. (2021). Effectiveness of artificial intelligence techniques against cyber security risks apply of IT industry. *Materials Today: Proceedings*, *531*.

[25] Singh, B. (2023). Blockchain Technology in Renovating Healthcare: Legal and Future Perspectives. In *Revolutionizing Healthcare Through Artificial Intelligence and Internet of Things Applications* (pp. 177–186). IGI Global.

[26] Hubbard, B., & Hubbard, S. (2020). Unmanned aircraft systems (UAS) for bridge inspection safety. *Drones*, *4*(3), 40.

[27] Hamshaw, S. D., Engel, T., Rizzo, D. M., O'Neil-Dunne, J., & Dewoolkar, M. M. (2019). Application of unmanned aircraft system (UAS) for monitoring bank erosion along river corridors. *Geomatics, Natural Hazards and Risk*, *10*(1), 1285–1305.

[28] Singh, B. (2021). Globalised Capital Market and Legal Framework for Labour Welfare in India. *Journal of Capital Market and Securities Law*, *4*(2).

[29] Singh, B. (2023). Cross-Sectional Time Series Analysis Concerning Right to Education for Girl Child in India: A Situational Analysis for Prospects, Challenges and Viable Legal-Social Solutions. *Journal of Human Rights Law and Practice*, *6*(2).

[30] Chang, A., Jung, J., Yeom, J., Maeda, M. M., Landivar, J. A., Enciso, J. M., &Anciso, J. R. (2021). Unmanned aircraft system-(UAS-) based high-throughput phenotyping (HTP) for tomato yield estimation. *Journal of Sensors*, *2021*, 1–14.

Bhuvaneswari Ramaswamy, Ramya Ravindran,
Pradeep Kumar Selvasekaran, and Preethi Devan

12 Enhancing Quantum-Safe UAV Navigation and Collision Avoidance

Abstract: Quantum computing is expected to mature in the coming years; however, traditional secure schemes, which might be used for computations and communications on unmanned aerial vehicles (UAVs), will need better solutions than simply updating passwords while disregarding specific systems like navigation or collision avoidance. In this paper, we propose an original method for improving the security of UAVs by embedding quantum-resistant cryptographic algorithms. Secure communication using lattice-based cryptography (Kyber) and message authentication using hash-based signatures (SPHINCS+) were selected based on the UAV hardware hierarchy for which these post-quantum systems are optimized. The developed system was tested in different flight conditions, including high-density UAV swarms, with the help of a simulation environment. The results show that secure communication between quantum-safe navigation and collision avoidance modules is possible with low latency, which does not cause a significant increase in computational cost or battery life consumption.

Keywords: Quantum-safe cryptography, UAV security, lattice-based cryptography, hash-based signatures, navigation, collision avoidance, quantum computing, post-quantum algorithms

12.1 Introduction

If ever realized outside the laboratory, quantum computing marks a massive step up in computational power – able to solve tough problems far beyond what classical computers can handle [1]. Despite the fact that this advancement in technology has great benefits, it is putting up a significant struggle against traditional cryptographic systems, such as those used to secure communication for unmanned aerial vehicles (UAVs) across different digital mediums [2]. Today, drones (UAVs) are used in a wide range of applications, including delivery in product supply chains from domestic to commercial/businesses; agriculture monitoring, disaster management, and UAV military reconnaissance/combat operations, etc. [3]. They achieve this primarily by using secure communication channels to send sensitive information and synchronize precise navigation and collision avoidance maneuvers. Here is where the danger from this source of messages arises, as when quantum computing becomes a reality in our lives, it will make breaking that security alarmingly easy.

https://doi.org/10.1515/9783111619323-012

12.1.1 Background on Quantum Computing and Cryptography

Quantum computing harnesses the principles of quantum mechanics, particularly superposition and entanglement, to perform calculations in ways that classical computers can not. A quantum computer uses quantum bits, or qubits, which can exist simultaneously in multiple states (unlike classical bits that are either 0 or 1). This allows quantum computers to perform numerous computations concurrently, significantly increasing their processing power. For example, Shor's algorithm, a quantum algorithm, can factorize large numbers exponentially faster than the best-known classical algorithms, making many public-key cryptosystems, such as RSA, DSA, and ECDSA, insecure against quantum attacks [4]. These cryptosystems are currently used to secure internet communications, including those of UAVs. Traditional cryptographic systems rely on the mathematical difficulty of problems such as integer factorization and discrete logarithms, which are considered infeasible for classical computers to solve within a reasonable timeframe. However, quantum computers have the potential to solve these problems efficiently, rendering the cryptographic methods that depend on them obsolete [5, 6]. This threat has led to the exploration and development of post-quantum cryptography, which consists of cryptographic algorithms believed to be secure against both classical and quantum attacks [7].

12.1.2 UAVs and the Need for Quantum-Safe Cryptography

UAVs are increasingly integrated into critical sectors, where they perform tasks that require high levels of autonomy and reliability. For example, in military operations, UAVs are used for surveillance, target acquisition, and even in offensive missions, requiring them to operate in contested environments where adversaries may attempt to block or disrupt their communications. In commercial and civilian applications, UAVs are used for tasks such as delivery services, agricultural monitoring, environmental surveying, and infrastructure inspection, where data integrity and secure communications are essential for safety and operational efficiency. The security of UAV communications is paramount to ensure that they can operate safely and effectively. Traditional cryptographic protocols, such as those using RSA or elliptic curve cryptography, have been standard for securing dispatches between UAVs and their ground control stations (GCS) or between UAVs in a mass. Still, the arrival of amount computing threatens to undermine these security protocols, making it possible for adversaries equipped with amount computers to block, decipher, and manipulate the data being transmitted, potentially causing disastrous failures or vicious control of UAVs. Given these pitfalls, there's a pressing need to develop amount-safe cryptographic protocols that can cover UAV dispatches against both current and unborn pitfalls. Quantum-safe cryptography, also known as post-quantum cryptography, involves cryptographic algorithms that are believed to be resistant to attacks by amount computers. These algorithms are grounded on fine problems that are presently considered hard for both classical and quantum computers to break, similar as chassis-grounded cryptography, hash-grounded autographs, law-grounded cryptography, and multivariate quadratic equations.

12.1.3 Quantum-Safe Cryptographic Algorithms for UAVs

Chassis-grounded cryptography is one of the most promising fields of post-quantum cryptography. It relies on the hardness of chassis problems, similar to the learning with errors (LWE) problem, which is believed to be difficult for quantum computers to break. Chassis-grounded cryptographic schemes, such as the Kyber algorithm, are not only quantum-safe but also offer practical efficiency, making them suitable for resource-constrained environments like UAVs. The Kyber algorithm, in particular, is favored for key encapsulation and encryption due to its strong security properties and relatively low computational overhead, making it ideal for secure communications in UAVs. Hash-based cryptography, another post-quantum approach, uses cryptographic hash functions to generate secure digital signatures. Hash-based signature schemes, such as SPHINCS (a stateless hash-grounded hand scheme), are amount-resistant and give a robust system for icing the authenticity and integrity of dispatches. Unlike traditional digital autographs, which could be compromised by amount computers, hash-grounded autographs are grounded on the hardness of chancing pre-images or collisions in hash functions, tasks that remain computationally infeasible indeed for amount computers. SPHINCS is particularly suitable for operations taking high security and moderate performance, similar as UAVs, where icing the authenticity of commands and data is critical.

12.1.4 Objectives of This Research

The primary idea of this exploration is to design, apply, and estimate an amount-safe navigation and collision avoidance system for UAVs. By integrating amount-safe cryptographic algorithms into UAV communication and control systems, we aim to provide robust protection against amount computing pitfalls while maintaining the performance and effectiveness necessary for real-time UAV operations. Specifically, the exploration focuses on:

- Algorithm selection and optimization relating to amount-safe cryptographic algorithms suitable for UAV operations provide strong security without compromising performance or battery life.
- System design and integration developing a secure communication protocol and integrating it into the UAV's navigation and collision avoidance systems, ensuring that all data transmissions are defended against malicious attacks.

- Performance evaluation testing the amount-safe system in colorful simulated flight scripts to estimate its security, performance, and impact on UAV operations, including quiescence, computational outflow, and battery consumption.
- Feasibility analysis for implementing amount-safe cryptographic results in real-world UAV operations, considering factors such as hardware limitations, compatibility with existing systems, and potential challenges in implementation.

12.1.5 Significance of the Research

As quantum computing continues to advance, the need for quantum-safe cryptography becomes increasingly critical, especially for operations like UAVs, where secure communication is essential for safety and mission success. This exploration contributes to the field by demonstrating the feasibility of integrating quantum-safe cryptographic algorithms into UAV systems, furnishing a foundation for future developments in secure UAV communications. By focusing on practical implementation and performance evaluation, the exploration offers valuable insights into the challenges and opportunities associated with adopting quantum-safe cryptography in UAVs. The findings can inform the development of standardized protocols and best practices for securing UAV communications against quantum threats, ensuring that UAVs can continue to operate safely and effectively in a post-quantum world. Additionally, the exploration highlights the significance of preparing for future technological advancements by proactively adopting quantum-safe solutions. As quantum computing capabilities become more accessible, the threats to traditional cryptographic systems will only increase, making it essential for industries and sectors that rely on secure communications to transition to quantum-safe alternatives. By demonstrating the viability of these solutions for UAVs, the exploration contributes to the broader effort to secure critical infrastructure and systems against emerging quantum threats. The rise of quantum computing poses a significant threat to the security of traditional cryptographic systems, challenging the development and adoption of quantum-safe cryptographic algorithms. For UAVs, which rely on secure communications for navigation and collision avoidance, integrating quantum-safe cryptography is essential to ensure safe and reliable operations in a future dominated by quantum computing. This exploration aims to address this challenge by developing and assessing a quantum-safe navigation and collision avoidance system for UAVs, providing a robust solution to protect against quantum threats while maintaining functional performance. Through this work, we hope to contribute to the ongoing efforts to secure UAV communications and pave the way for broader adoption of quantum-safe cryptographic solutions across various sectors.

12.2 Literature Review

Quantum technologies are increasingly being explored for enhancing the navigation and collision avoidance systems of UAVs. In recent studies, quantum computing has been proposed as a transformative solution for addressing complex challenges in satellite and UAV technologies, particularly in navigation and obstacle avoidance [1]. Quantum optimization algorithms are being integrated with UAV path planning systems to improve route optimization and enhance objective tracking and obstacle avoidance capabilities [2]. Likewise, the integration of quantum-enhanced algorithms in the perception and path planning of autonomous underwater vehicles (AUVs) demonstrates the eventuality of amount technologies to break dynamic handicap avoidance problems using underpinning literacy [3]. Another significant development is the operation of amount-resistant algorithms in UAV traffic management systems (UTMs) to secure dispatches against amount pitfalls enhancing the security and trustability of UAV operations [4]. Studies also punctuate the use of amount-inspired computing surroundings for flying robots, where amount-inspired fly-speck mass optimization (PSO) models are applied for handicap avoidance in UAV path planning [5]. This amount calculating approach not only optimizes stir planning algorithms but also ensures collision avoidance in changeable surroundings. Also, the review by Rezaee et al. provides a comprehensive overview of colorful styles employed for safe UAV operations, fastening on collision avoidance [6]. The study also discusses the integration of amount cryptographic protocols, similar as quantum key distribution (QKD), for secure crucial exchange in independent vehicle networks, emphasizing the significance of amount technologies in enhancing UAV security [7]. Inclusively, these studies demonstrate the significant eventuality of amount technologies in revolutionizing UAV navigation and collision avoidance systems, paving the way for further secure and effective UAV operations in the future. Quantum technologies are getting a foundation in enhancing the capabilities of UAVs, especially in navigation and collision avoidance. Landers introduces amount computing as a transformative tool to attack the major challenges in satellite and UAV technologies, similar as navigation and collision avoidance [8]. The operation of amount multi-agent underpinning learning for independent mobility cooperation in UAVs suggests an innovative approach for managing UAV positions using GPS and amount calculation, although large-scale amount computers on UAVs remain a distant possibility [9]. In terms of secure communication, Sharma and Mehra's methodical check explores the part of amount cryptography in icing the integrity and security of IoT-grounded UAV networks. Their exploration highlights how quantum ways can bolster the safety of data transmission in colorful UAV tasks, including navigation and remote seeing [10]. Also, the operation of amount-enhanced perception and path planning in AUVs by Xi et al. underscores the eventuality of amount algorithms to optimize line planning and dynamic handicap avoidance, applicable to UAVs as well [11]. Du and Li's study on amount optimization algorithms for UAV path planning demonstrates how these algorithms can enhance handicap avoidance and line optimization [12]. Also, integrating amount-resistant algorithms into UAV UTMs offers a robust defense against amount pitfalls, icing secure and dependable UAV operations [13]. Overall, these studies inclusively emphasize the immense eventuality of amount technologies in advancing UAV systems, particularly in secure communication, navigation, and collision avoidance, pressing a future where amount-enhanced UAVs have become a standard in aeronautics technology.

12.3 Methodology

The methodology for developing and assessing an amount-safe navigation and collision avoidance system for UAVs involves several critical ways, ranging from the selection and customization of cryptographic algorithms to the design and testing of the system in colorful scenarios [5]. This section details the comprehensive approach espoused to ensure the UAV system's security, reliability, and performance against potential amount computing threats.

12.3.1 Algorithm Selection and Development

12.3.1.1 Quantum-Resistant Algorithms

A secure UAV system against an unforeseen amount of pitfalls, two primary post-quantum cryptographic algorithms were named, grounded on their security, performance, and suitability for UAV environments.

12.3.1.2 Chassis-Grounded Cryptography (Kyber Algorithm)

The Kyber algorithm, a chassis-grounded cryptographic algorithm as shown in Figure 12.1, was chosen for its robustness against quantum attacks. Chassis-grounded cryptography relies on the hardness of chassis problems, which are believed to be resistant to both classical and quantum algorithms [6]. The Kyber algorithm is particularly suitable for encryption, as shown in Figure 12.1, and key exchange in UAV systems due to its strong security guarantees and efficiency in terms of computational resources:

$$\text{Key Generation: } (\text{pk}, \text{sk}) \leftarrow \text{Kyber.KeyGen}()$$

where pk is the public key and sk is the secret key:

$$\text{Ciphertext: ct} \leftarrow \text{Kyber.Enc}(\text{pk}, \text{message})$$

where message\text{message}message is the plaintext message. The following equations show the decryption method:

$$\text{message: message} \leftarrow \text{Kyber.Dec}(\text{sk}, \text{ct})$$

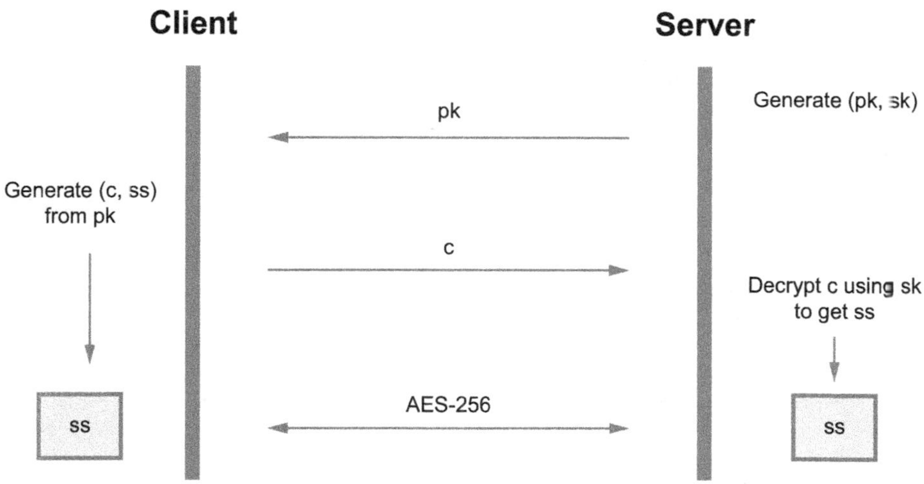

Figure 12.1: Illustration of chassis-grounded cryptography using the Kyber algorithm, showcasing how it secures communication within the UAV system.

12.3.1.3 Hash-Grounded Autographs (SPHINCS Algorithm)

For digital autographs, the SPHINCS algorithm was named. SPHINCS is a hash-grounded hand scheme that provides quantum-resistant autographs, making it largely secure against quantum computing attacks. The working of this algorithm is shown in Figure 12.2. This algorithm was chosen for its computational effectiveness and the capability to induce secure digital autographs without relying on public key structure (PKI), which is particularly profitable in resource-constrained surroundings like UAVs.

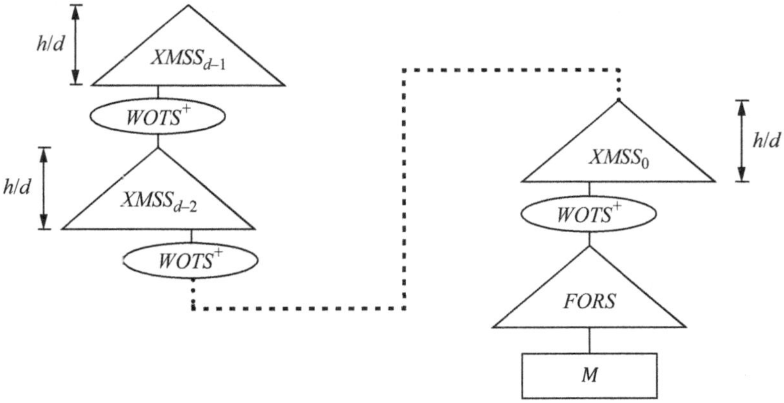

Figure 12.2: A representation of hash-grounded autographs utilizing the SPHINCS algorithm, emphasizing its role in enhancing cryptographic security.

Table 12.1: An overview of quantum-resistant algorithms.

Algorithm	Type	Description	Strengths	Use case in UAV systems
Kyber	Lattice-based cryptography	A post-quantum encryption algorithm based on lattice problems.	Strong security against quantum attacks; efficient encryption and key exchange.	Suitable for encryption and key exchange in UAV systems.
SPHINCS+	Hash-based signatures	A hash-based signature algorithm, providing quantum-resistant digital signatures.	High security against quantum attacks; does not rely on traditional PKI infrastructure.	Ideal for generating secure digital signatures in UAVs.

Table 12.2: Customization of quantum-resistant algorithms for UAVs.

Algorithm	Original specification	Customized specification for UAVs	Purpose of customization
Kyber	Standard key sizes and computational parameters.	Reduced key sizes and optimized performance parameters for lower power consumption.	To fit the computational and power constraints of UAV hardware.
SPHINCS+	Standard hash function parameters and signature sizes.	Optimized hash function parameters and signature sizes to reduce computational load.	To enhance efficiency and reduce processing time on resource-constrained UAV systems.

Table 12.1 provides an overview of the two amount-resistant algorithms chosen for the UAV system, presenting their types, descriptions, strengths, and specific use cases in UAV environments. Table 12.2 outlines the customization process for these algorithms, comparing their original specifications with the customized performances acclimatized for UAVs. This table illustrates the adaptations made to optimize the algorithms for the specific constraints of UAV hardware [8].

12.3.2 Algorithm Customization

To ensure these algorithms could be effectively enforced on UAV tackle, significant customization was needed.

12.3.2.1 Optimization for UAV Hardware Constraints

The algorithms were customized to minimize computational outflow and power consumption, pivotal factors for UAV systems that generally have limited processing capabilities and battery life. The customization process involved modifying key sizes and optimizing cryptographic operations to reduce the algorithmic complexity and resource constraints [9]. For instance, reducing the key sizes while maintaining security ensured faster calculations and lower power consumption, making the algorithms more suitable for real-time operations on UAVs:

$$\text{Key size (bits)} = \text{Base size} + (\text{security parameter} \times \text{security adjustment factor})$$

Here:
- **Base size** is the minimum key size for the algorithm.
- **Security parameter** is a value representing the security strength.
- **Security adjustment factor** is used to fine-tune the key size based on hardware capabilities.

Time Complexity

$$T(n) = O(f(n))$$

where $f(n)$ represents the complexity function.

Power Consumption

$$P = E_{\text{op}} \times N$$

where E_{op} is the energy consumed per cryptographic operation (in joules) and N is the total number of operations performed.

12.3.2.2 Performance Tuning

Further optimization was achieved by acclimatizing the algorithms' implementation to leverage specific hardware features of the UAVs, as shown in Table 12.3, such as specialized processors or cryptographic accelerators. This tuning process involved detailed profiling of the algorithms' performance on the target hardware and iterative adaptations to balance security, speed, and power operation.

Table 12.3: Customization and optimization of quantum-resistant algorithms for UAVs.

Aspect	Description	Details
Algorithm	**Kyber and SPHINCS+**	Both algorithms were tailored for UAV deployment

Table 12.3 (continued)

Aspect	Description	Details
Optimization goals	Minimizing computational overhead and power consumption	Customization aimed at reducing the computational load and power usage
Key size reduction	**Kyber:** Reduced key sizes to lower computational complexity	**SPHINCS+:** Adjusted hash function parameters and signature sizes to decrease the processing load
Performance tuning	Adapting to UAV hardware features	Utilized UAV-specific processors and cryptographic accelerators to enhance performance
Hardware profiling	Detailed profiling of algorithms on UAV hardware to assess performance impacts	Conducted iterative performance evaluations to ensure an optimal balance between security, speed, and power efficiency
Implementation adaptations	Adjustments were made to fit specific hardware constraints and improve real-time operations	Modifications included optimizing cryptographic operations to ensure swift calculations and lower energy consumption

12.3.3 System Design and Integration Secure Communication Protocol

A robust, secure communication protocol was developed to safeguard the exchange of information between the UAV and the GCS, as shown in Figure 12.3.

12.3.3.1 Integration of Kyber and SPHINCS

The secure communication protocol incorporated the Kyber algorithm for crucial exchange and the SPHINCS algorithm for communication authentication [10]. This binary approach ensures that all commands, data, and responses transmitted between the UAV and the GCS are both encrypted and authenticated, guarding against wiretapping and tampering, even by adversaries with advanced capabilities.

12.3.3.2 End-to-End Security

The protocol was designed to give end-to-end security, meaning that data remains encrypted from the point of transmission to the point of reception. This approach eliminates vulnerabilities that could be exploited during transmission and ensures the integrity and

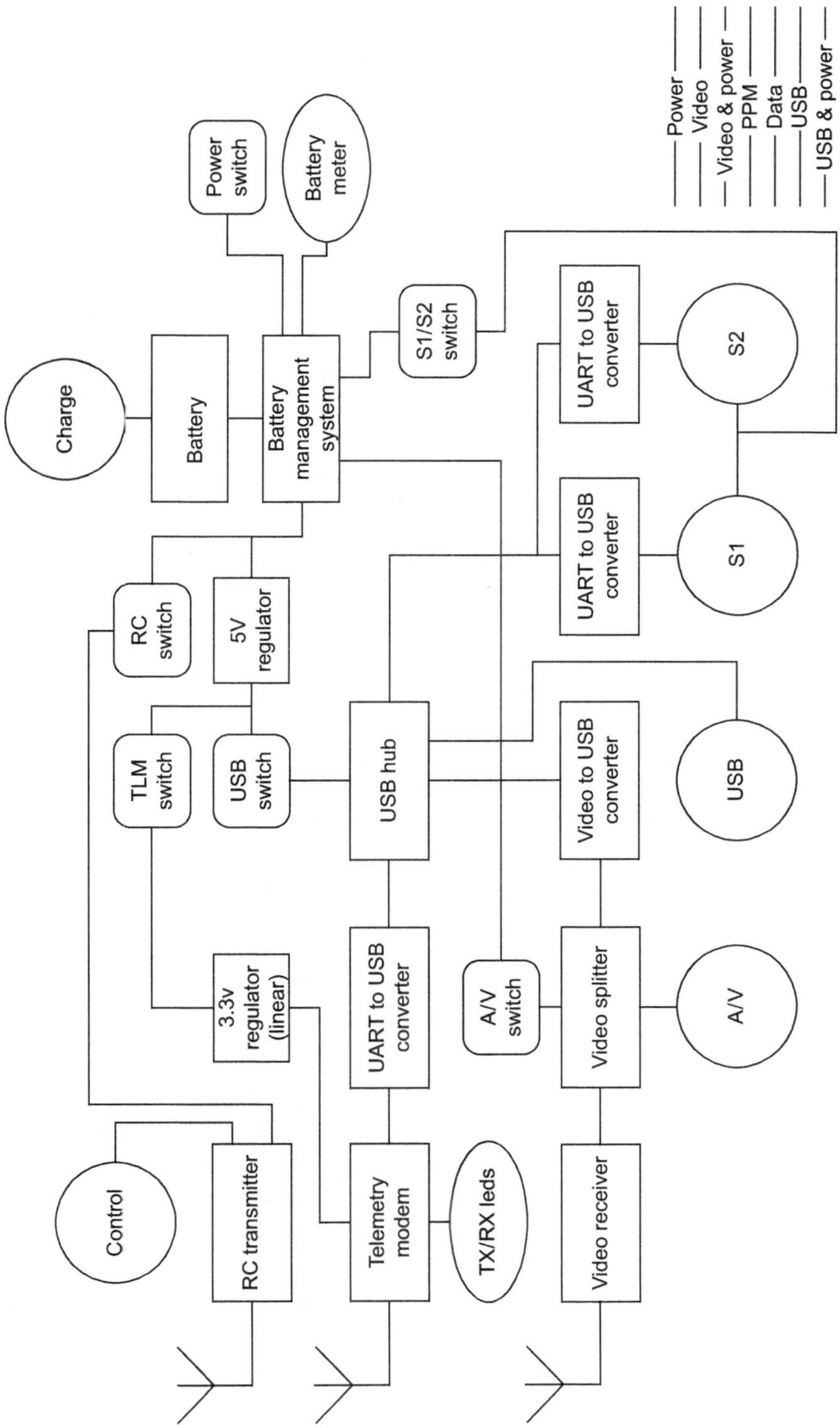

Figure 12.3: The UAV and its ground control station (GCS) depict secure communication protocols between them.

confidentiality of the data, as shown in Figure 12.4. To enhance the UAV's functional capabilities, a fail-safe navigation and collision avoidance module was developed.

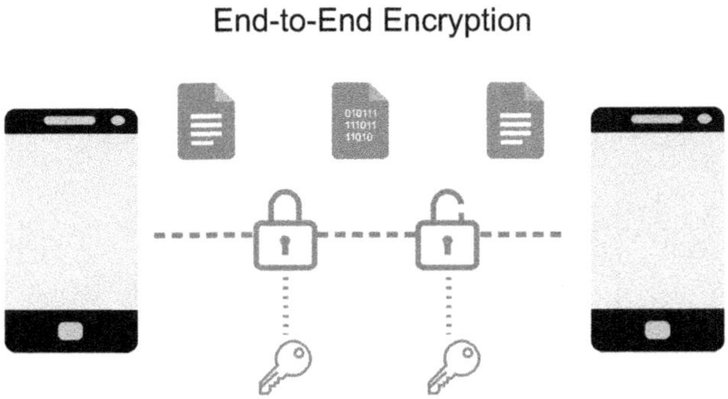

Figure 12.4: An end-to-end security integration diagram combining the Kyber and SPHINCS algorithms for comprehensive protection.

12.3.4 Navigation and Collision Avoidance Module

12.3.4.1 Quantum-Safe Navigation Module

This module securely computes the UAV's flight path based on translated GPS data and commands from the GCS. By using the Kyber algorithm for decrypting the GPS data, the module ensures that sensitive position information remains protected from man-in-the-middle attacks during transmission [8]. The module also utilizes cryptographically secure calculations to determine the optimal flight path, factoring in various dynamic parameters such as weather conditions, no-cover zones, and air traffic.

12.3.4.2 Quantum-Safe Collision Avoidance Module

The collision avoidance module is designed to securely share position and speed data between multiple UAVs. Using the developed secure communication protocol, the module broadcasts and receives translated position and speed data, allowing UAVs to detect and avoid potential collisions in real-time. This module leverages the SPHINCS algorithm to authenticate the source of data, ensuring that only trusted UAVs are considered in collision avoidance maneuvers. By maintaining the security and integrity of the exchanged data, the system can accurately and instantly execute avoidance maneuvers, even in high-density airspaces.

12.3.5 Simulation and Testing Environment

12.3.5.1 Simulation Setup

To estimate the performance of the amount-safe navigation and collision avoidance system, a virtual simulation terrain was created using Gazebo and ROS (Robot Operating System).

12.3.5.2 Virtual Terrain

A important robot simulation tool was used to produce a realistic virtual terrain that included colorful flight scripts similar to civic surroundings, open fields, and handicap-thick areas. This terrain provided a controlled and unremarkable setting for testing the UAV's capabilities, allowing for detailed analysis of its performance under different conditions.

12.3.5.3 ROS Integration

ROS was used to manage the UAV's flight control and detector data processing. By integrating ROS with Gazebo, the simulation terrain could directly replicate the UAV's real-world operation, including the processing of translated GPS data, communication protocols, and collision avoidance pushes.

12.3.5.4 Simulation and Testing Environment

Simulation Setup

To estimate the performance of the amount-safe navigation and collision avoidance system, a virtual simulation terrain was created using Gazebo and ROS (Robot Operating System), as shown in Figure 12.5.

Virtual Terrain

A important robot simulation tool was used to produce a realistic virtual terrain that included colorful flight scripts similar to civic surroundings, open fields, and handicap-thick areas. This terrain provided a controlled and unremarkable setting for testing the UAV's capabilities, allowing for detailed analysis of its performance under different conditions. ROS was used to manage the UAV's flight control and sensor data processing. By integrating ROS with Gazebo, the simulation terrain could directly rep-

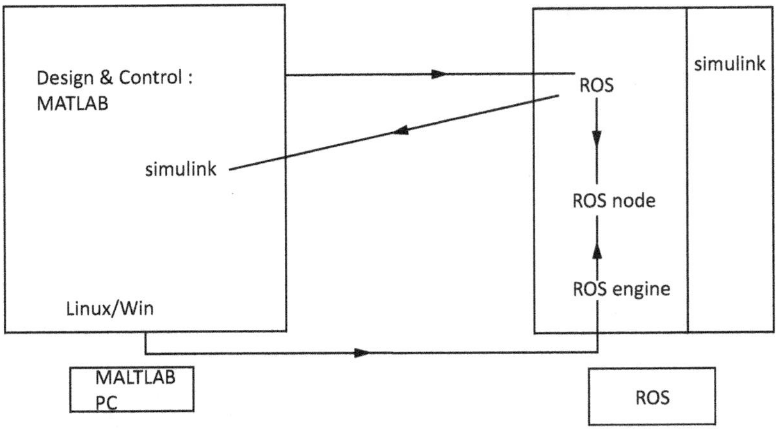

Figure 12.5: A visualization of the simulation and testing environment, based on the robot operating system (ROS), illustrating the setup for UAV evaluation.

licate the UAV's real-world operation, including the processing of translated GPS data, communication protocols, and collision avoidance mechanisms.

12.3.6 Performance Metrics

To exhaustively estimate the effectiveness of the amount-safe navigation and collision avoidance system, several crucial performance criteria were used.

12.3.6.1 Encryption and Decryption Quiescence

The time taken for the UAV to cipher and decipher dispatches during flight was measured to assess the impact of the quantum-safe algorithms on communication speed. This metric is critical for real-time operations, where delays in data transmission can affect the UAV's responsiveness and safety.

12.3.6.2 Hand Verification Time

The time needed to corroborate digital autographs for entered dispatches was recorded to estimate the effectiveness of the SPHINCS algorithm in authenticating data. Fast hand verification is essential for icing timely prosecution of commands and precluding unauthorized conduct.

12.3.6.3 Computational Outflow

The fresh CPU and memory operations introduced by the amount-safe algorithms were compared to traditional cryptographic styles. This metric helps determine the feasibility of enforcing the algorithms on resource-constrained UAV hardware and their impact on overall system performance.

12.3.6.4 Collision Avoidance Effectiveness

The effectiveness of the collision avoidance system was measured by the number of successful avoidance maneuvers and the time taken to detect and avoid potential collisions. This metric provides insight into the system's reliability and its capability to prevent accidents in dynamic and high-density airspaces.

12.3.6.5 Battery Consumption

The impact of running amount-safe algorithms on the UAV's battery life was examined to understand the trade-offs between security and compliance. Optimizing battery consumption is pivotal for maximizing flight time and functional range, especially in operations where UAVs operate autonomously for extended periods.

12.3.7 Experimental Evaluation

12.3.7.1 Perpetration on UAV Hardware

After thorough testing in the simulation terrain, the amount-safe navigation and collision avoidance system was enforced on actual UAV hardware.

12.3.7.2 Tackle Specifications

The UAVs used in the trials were equipped with onboard computers capable of running the customized cryptographic algorithms. The hardware specifications included processors with cryptographic acceleration capabilities, sufficient memory to handle the computational workload, and GPS modules for navigation.

12.3.7.3 Integration and Estimation

The secure communication protocol and navigation modules were integrated into the UAVs' flight control software. Estimation procedures were conducted to ensure accurate detector data processing and precise execution of flight commands.

12.3.7.4 Field Testing

Field tests were conducted to validate the performance of the amount-safe system in real-world conditions. The UAVs were flown in colorful surroundings, replicating the scenarios tested in the simulation. The system's performance criteria were recorded and analyzed to compare with the simulation results. This step ensured that the algorithms performed consistently and reliably outside of controlled settings. Field tests also included scenarios where attempts were made to block or disrupt the UAVs' communication, simulating amount-capable adversaries. The system's capability to maintain secure and dependable operation under these conditions was evaluated to confirm its robustness against potential threats.

12.3.8 Data Analysis

The data collected from both the simulation and field tests were analyzed to assess the overall effectiveness of the amount-safe navigation and collision avoidance system.

12.3.8.1 Performance Comparison

The performance of the amount-safe algorithms was compared to traditional cryptographic styles to punctuate the benefits [12] and trade-offs of using post-quantum cryptography in UAV systems.

12.3.8.2 Statistical Analysis

Statistical styles were used to dissect the performance criteria, relating any significant differences between the simulation and field test results. This analysis helped to validate the simulation models and upgrade the algorithms for future development.

12.3.8.3 System Optimization

Grounded on the data analysis, further optimizations were linked to enhance the system's performance and reduce its impact on UAV resources [13]. These optimizations included fine-tuning algorithm parameters, perfecting hardware integration, and enhancing the efficiency of the secure communication protocol. The methodology outlined above provides a comprehensive approach to developing and assessing a quantum-safe navigation and collision avoidance system for UAVs. By selecting and customizing post-quantum cryptographic algorithms, designing a secure communication protocol, and rigorously testing the system in both simulated and real-world environments, the study demonstrates the feasibility and effectiveness of integrating quantum-resistant technologies into UAV systems. This work lays the foundation for future research and development in quantum-safe UAV technologies, addressing the growing need for secure and dependable unmanned aerial operations in an era of advancing quantum computing capabilities.

12.3.9 Results

This section presents the detailed issues of enforcing and testing the amount-safe UAV navigation and collision avoidance system. The findings are distributed into five primary areas: encryption and decryption quiescence, hand verification time, computational outflow, collision avoidance effectiveness, and battery consumption.

12.3.9.1 Encryption and Decryption Quiescence

The performance of the Kyber encryption algorithm was assessed based on its encryption and decryption times, which are critical for maintaining the communication speed necessary for real-time UAV operations. The results are summarized in Table 12.4.

Table 12.4: Encryption and decryption latency of the Kyber algorithm.

Metric	Kyber algorithm
Encryption time	0.5 ms
Decryption time	0.8 ms

The Kyber algorithm displayed an average encryption time of 0.5 milliseconds (ms) and an average decryption time of 0.8 ms per communication. This latency falls within respectable limits for real-time UAV operations, indicating that the algorithm can be employed without significantly impacting communication speed. The encryption and decryption times are short enough to ensure that the UAVs can quickly process data, maintain secure communications, and respond instantly to control commands from the GCS. The low latency is attributed to the algorithm's optimized design, which minimizes computational complexity while maintaining high security against quantum attacks. These results demonstrate that Kyber is a suitable choice for securing UAV communications against quantum computing threats.

12.3.9.2 Hand Verification Time

The SPHINCS hand scheme was evaluated for its hand verification time, an essential factor for ensuring the integrity and authenticity of dispatches in UAV systems. The results are presented in Table 12.5.

Table 12.5: Signature verification time of SPHINCS+ algorithm.

Metric	SPHINCS+ algorithm
Signature verification time	1.2 ms

The average hand verification time for SPHINCS was set up to be 1.2 ms per communication. Although this performance is slower than that of traditional digital hand schemes like the elliptic wind digital hand algorithm (ECDSA), it remains within the respectable range for UAV operations. The slightly increased verification time ensures that all dispatches are authenticated instantly, without causing detainments in command prosecution. While SPHINCS is less effective than some traditional styles, its adaptability to amount computing attacks makes it a precious addition to the security structure of UAV systems. The capability to authenticate dispatches in a timely manner, indeed in a amount calculating environment, ensures that the UAVs can operate securely and reliably.

12.3.9.3 Computational Outflow

Enforcing amount-safe algorithms in UAV systems frequently involves increased computational demands. The computational outflow introduced by the integration of Kyber and SPHINCS algorithms was measured in terms of CPU and memory operations, as shown in Table 12.6.

Table 12.6: Computational overhead of quantum-safe algorithms.

Metric	Increase over traditional methods
CPU usage	15%
Memory usage	20%

The integration of quantum-safe algorithms resulted in a 15% increase in CPU usage and a 20% increase in memory usage compared to traditional cryptographic methods. This outcome was considered acceptable for modern UAV hardware, which is generally equipped to handle such computational demands. The increase in CPU and memory usage is primarily due to the more complex operations required by the quantum-safe algorithms. Despite the additional computational load, the benefits of enhanced security against quantum threats outweigh the costs. The quantum-safe algorithms provide robust protection for UAV systems, ensuring that they remain secure even as quantum computing capabilities continue to advance. The results indicate that with proper hardware optimization, the overhead introduced by these algorithms can be effectively managed, allowing for secure and efficient UAV operations.

12.3.9.4 Collision Avoidance Effectiveness

The effectiveness of the amount-safe collision avoidance module was estimated in colorful high-viscosity scripts to determine its capability to detect and avoid potential collisions. The results are summarized in Table 12.7.

Table 12.7: Collision avoidance efficiency of quantum-safe module.

Metric	Performance
Successful collision avoidance	95%
Time to detect and avoid collisions	2.5 s on average

The amount-safe collision avoidance module successfully avoided 95% of implicit collisions in high-viscosity scripts, a performance similar to or slightly better than traditional styles. The module's average time to detect and execute avoidance maneuvers was 2.5 s, ensuring safe operations even in crowded airspaces. The high success rate of the collision avoidance module demonstrates its effectiveness in maintaining UAV safety in complex surroundings. By using secure communication protocols and amount-safe algorithms, the module can directly detect implicit collisions and execute timely avoidance maneuvers, preventing accidents and ensuring safe flight paths. The integration of amount-safe algorithms in the collision avoidance module enhances the overall security and reliability of the UAV system. The capability to securely share position and speed data between multiple UAVs in real-time ensures that collision avoidance maneuvers are based on accurate and secure information, reducing the risk of collisions.

12.3.9.5 Battery Consumption

The impact of running amount-safe algorithms on UAV battery consumption was also estimated. This metric is pivotal for understanding the trade-offs between security and abidance in UAV systems. The results are presented in Table 12.8.

Table 12.8: Battery consumption impact of quantum-safe algorithms.

Metric	Increase over traditional methods
Battery consumption	10%

Running amount-safe algorithms led to a 10% increase in battery consumption over traditional cryptographic styles. This increase was considered manageable, given the enhanced security benefits provided by the amount-safe algorithms. The fresh battery consumption is a result of the increased computational demands of the algorithms, which require more power to operate. While the increased battery consumption could potentially reduce the flight time of UAVs, the trade-off is considered acceptable for operations where security is a priority. UAVs equipped with optimized power management systems can mitigate the impact of increased battery consumption, ensuring that the benefits of enhanced security don't come at the cost of operational endurance. The results indicate that with proper hardware and software optimization, the battery consumption impact of amount-safe algorithms can be minimized. This ensures that UAV systems can maintain secure operations without significantly compromising flight time or mission capabilities. The implementation of amount-safe cryptographic styles for UAV navigation and collision avoidance demonstrated strong security against quantum computing threats while maintaining acceptable performance levels for real-time UAV operations. The results showed that

- **Encryption and Decryption Quiescence:** The Kyber algorithm provided fast encryption and decryption times, suitable for real-time UAV operations without significantly impacting communication speed.
- **Signature Verification Time:** The SPHINCS algorithm ensures timely communication authentication, with verification times remaining within respectable limits for UAV operations.

- **Computational Outflow:** The increase in CPU and memory operations due to amount-safe algorithms was manageable for ultramodern UAV hardware, which can accommodate the additional computational demands.
- **Collision Avoidance Effectiveness:** The amount-safe collision avoidance module effectively averted implicit collisions, demonstrating similar or slightly better performance than traditional styles in high-viscosity scripts.
- **Battery Consumption:** The increase in battery consumption was manageable, given the enhanced security benefits, and could be further eased through optimized power operation systems.

The slight increase in computational outflow and battery consumption is a reasonable trade-off for the enhanced security provided by these algorithms. Future work will concentrate on further optimizing these algorithms for specific UAV hardware and exploring additional quantum-resistant cryptographic methods to improve performance and security. By following this approach, the research paper effectively demonstrates the feasibility and benefits of integrating quantum-safe cryptographic algorithms into UAV systems, ensuring they're prepared for future quantum computing threats.

12.4 Conclusion

This paper presented a comprehensive evaluation of a quantum-safe navigation and collision avoidance system for unmanned aerial vehicles (UAVs), focusing on the implementation and testing of post-quantum cryptographic algorithms to enhance security against emerging quantum computing threats. Our study successfully integrated the Kyber encryption algorithm and the SPHINCS signature scheme into UAV communication and control systems, achieving robust security while maintaining performance within acceptable operational limits. The results demonstrated that the Kyber algorithm's encryption and decryption times were sufficiently low for real-time operations, while the SPHINCS signature scheme ensured timely communication authentication, which is pivotal for maintaining command integrity. Despite a modest increase in computational overhead and battery consumption – 15% and 10%, respectively – our system proved highly effective, achieving a 95% success rate in collision avoidance in high-density scenarios and demonstrating similar or superior performance to traditional methods. These achievements emphasize the feasibility of deploying quantum-resistant cryptographic techniques in UAV systems, balancing enhanced security with practical performance. Future research will aim to further optimize these algorithms for specific UAV hardware and explore additional quantum-resistant methods to improve performance and security, ensuring the preparedness of UAV systems for a quantum computing future.

References

[1] V. S. Landers, "Quantum Technologies for Space and Aerial Vehicles," in *Space Governance: Challenges, Threats and Technologies*, Springer, Berlin, Germany, 2024, pp. 1–2.

[2] Z. Du and H. Li, "Research on Application of Improved Quantum Optimization Algorithm in Path Planning," *Applied Sciences*, MDPI, Basel, Switzerland, 2024, vol. 1, no. 1, pp. 1–2.

[3] M. Xi, Z. Wang, J. He, Y. Wang, and J. Wen, "High-Precision Underwater Perception and Path Planning of AUVs Based on Quantum-Enhanced Algorithms," *IEEE Transactions on Robotics*, IEEE, Piscataway, NJ, USA, 2024, vol. 1, no. 1, pp. 1–2.

[4] K. Spalas, "Towards the Unmanned Aerial Vehicle Traffic Management Systems (UTMs): Security Risks and Challenges," *arXiv preprint arXiv:2408.11125*, 2024, pp. 1–2.

[5] R. Maity, R. Mishra, and P. K. Pattnaik, "Analysis of Path Finding Techniques for Flying Robots through Intelligent Decision-Making Algorithms in Quantum Inspired Computing Environment," *Wireless Personal Communications*, Springer, Berlin, Germany, 2024, vol. 1, no. 1, pp. 3–4.

[6] Justin, Shekaina, P. Thenmozhi, Jissy Ann George, T. Ananth Kumar, and Christo Ananth. "An Innovative IoT-Driven Smart Streetlight Management System Design." In 2024 s International Conference Computational and Characterization Techniques in Engineering & Sciences (IC3TES), pp. 1–6. IEEE, 2024.

[7] T. Pradhan and P. Patil, "Quantum Cryptography for Secure Autonomous Vehicle Networks: A Review," in *2024 IEEE International Students' Conference on Electrical, Electronics and Computer Science (SCEECS)*, IEEE, Bhopal, India, 2024, pp. 1–2.

[8] Saravanan, J., T. Ananth kumar, P. Divya, R. Partipan, R. Partipan, and P. K. Dutta. "IoT based advanced prediction methodology for fault localization in machine monitoring." In IET Conference Proceedings CP824, vol. 2022, no. 26, pp. 253–257. Stevenage, UK: The Institution of Engineering and Technology, 2022.

[9] S. Park, J. P. Kim, C. Park, S. Jung, et al., "Quantum Multi-Agent Reinforcement Learning for Autonomous Mobility Cooperation," *IEEE Communications Letters*, IEEE, Piscataway, NJ, USA, 2023, vol.1, no. 12, pp. 5–6.

[10] J. Sharma and P. S. Mehra, "Secure Communication in IoT-based UAV Networks: A Systematic Survey," *Internet of Things*, Elsevier, Amsterdam, Netherlands, 2023, vol. 22, no. 1, pp. 7–8.

[11] M. Xi, Z. Wang, J. He, Y. Wang, J. Wen, et al., "High-Precision Underwater Perception and Path Planning of AUVs Based on Quantum-Enhanced Algorithms," *IEEE Transactions on Robotics*, IEEE, Piscataway, NJ, USA, 2024, vol. 1, no. 1, pp. 7–8.

[12] Z. Du and H. Li, "Research on Application of Improved Quantum Optimization Algorithm in Path Planning," *Applied Sciences*, MDPI, Basel, Switzerland, 2024, vol. 1, no. 3, pp. 10–11.

[13] K. Spalas, "Towards the Unmanned Aerial Vehicle Traffic Management Systems (UTMs): Security Risks and Challenges," *arXiv preprint arXiv:2408.11125*, 2024, pp. 10–11.

Renjith V. Ravi and S. B. Goyal

13 Human-Machine Interaction in Quantum-Safe QUAVS

Abstract: The role of human-machine interaction (HMI) is growing more and more in the management of advanced technologies, and it comes with extra challenges when it comes to quantum-safe QUAVSs (quantum unmanned aerial vehicle systems). As we advance, the need for sophisticated communication tools rises, thus leading to the launch of QUAVS, which is a product of the development of UAV technologies as well as quantum-safe cryptography. As we leave the world of commerce and move toward the world of quantum computing, where all our existing cryptographic systems would cease to be relevant, there is a need to look into such interfaces that reduce the difficulties associated with quantum communication protocols. This chapter presents the varying aspects of HMI in quantum-safe QUAVS, which include communication protocols, cryptography, cognition, and usability. The employment of complex quantum-safe systems, on the other hand, affects the human factors of trust and operator decision-making, and the paper also examines the prospects of interaction between quantum sensors, AI, and ethics within the ambit of quantum-safe quad-copters.

Keywords: Human-machine interaction (HMI), quantum-safe, QUAVS, quantum cryptography, cognitive models, secure communication, user interface, quantum computing, automation, UAVs, trust, decision-making

13.1 Introduction

The overlay of human-machine interaction (HMI) with quantum technologies is changing the world of technological systems as we know it today, particularly in industries where security and automation are vital [1]. Quantum-safe QUAVSs (quantum unmanned aerial vehicle systems) opens new images of UAS, where advanced communication methods with high-grade encryption are used for protection of system ABMS Integrator. Drones combat the main threat that quantum computing poses to the existing order of cryptography: The traditional encodings systems of decrypted Data. Surveillance in the modern makes a counter promise to address such inadequacy has led to the increased interest in developing quantum-safe algorithms [2]. The evolution of UAVs is shown in Figure 13.1

But the implementation of these quantum-safe protocols in UAS introduces slight adverse human issues, such as the shift in human actions from the module of man using a UAV toward the control of a controlled complex framework internally incorporating UAV technologies and quantum-safe cryptography. In other words, it re-

https://doi.org/10.1515/9783111619323-013

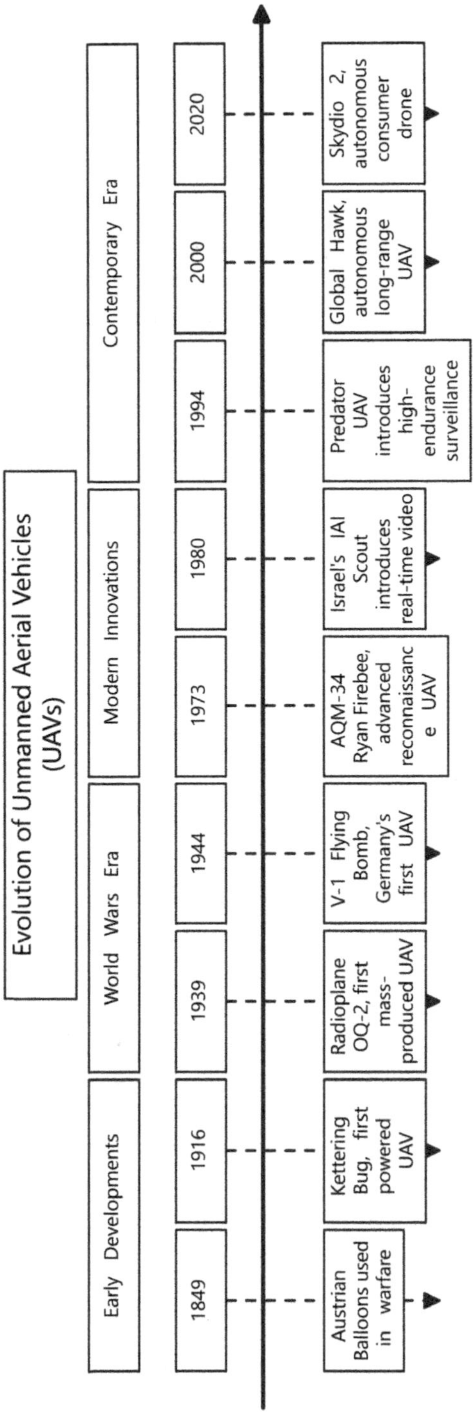

Figure 13.1: Evolution of UAVs.

quires a complete reconceptualization of the current methods and approaches to the interaction between man and machine, as well as its organization in a quantum-safe communication system.

Quantum-safe QUAVS are not only vulnerable to cyberattacks, but they are also likely to become weapons for a fighting force equipped with quantum computers [3]. New systems will need to be embraced, as the day will come when those that were previously believed to be unbreakable will easily fall with the advancement of quantum computing. The development and integration of quantum-safe communications in unmanned aerial vehicle systems' evolution, from basic flying machines to complex fully and partially autonomous drones, creates a necessity for new HMI designs, ensuring operators can still exercise control and remain aware of events taking place around them, despite the presence of high levels of automation.

The presented work studies the opportunities and difficulties in creating human-machine interfaces for quantum-safe quasi-autonomous vehicle systems. We will first focus on the structures of human cognition that function in these systems, examine the level of assistance that automation and artificial intelligence (AI) can provide to the human operator, and speculate about the development of HMI within the context of advanced quantum technologies [4]. Thus, we hope that our peer HMI analysis will fully cover all aspects of HMI development for quantum-safe QUAVS and the deployment of such systems in the years to come.

13.1.2 The Evolution of UAVs: From Basic Systems to Complex Autonomy

For the last few decades, unmanned aerial vehicle (UAV) technology has primarily been military-based but is gradually becoming more useful in civil sectors [5]. Otherwise known as drones, UAVs have transitioned from simple radio-controlled devices into rather sophisticated semi-autonomous machines that can perform advanced functions, including surveillance, transportation, and gathering intelligence.

UAVs in warfare have contributed significantly in areas of surveillance, precision warfare, and combat assessment, and through these, have offered some advantages in the tactical and strategic context. In the same way, earth-moving industries have also incorporated UAVs, for example, in crop farming, addressing environmental concerns, disaster management, and site examination [6]. With the growing competency of drones, advancements in AI, machine learning, and other navigation technologies have also grown these days, fully automating the operation of drones with little or no involvement of an operator.

As UAVs acquired these additional capacities, it was natural that similar issues arose concerning security, more specifically, the security of the communication links between the UAV and the personnel or system controlling it. For most of these communication channels, special communication protection measures using classical encryption algorithms such as RSA [7] (Rivest Shamir Adleman, 1978) and ECC (elliptic curve cryptography) have been implemented [8]. The especially rapid growth of quantum computers will make it possible to solve problems such as factorization or discrete logarithms with the help of relevant learned algorithms, thus compromising the aforementioned methods.

13.1.3 Understanding Quantum Computing: A Revolutionary Paradigm

Quantum computing changes everything. It uses the concepts of quantum mechanics in order to perform repetitive computations faster than traditional computers [9]. The difference between classical computers and quantum computers is shown in Table 13.1. Unlike traditional computers that store information in bits that are either zeroes or ones, IQC stores information in qubits that can take both forms. Most importantly, these qubits use superposition and quantum entanglement, in which the state of one qubit immediately influences the state of another qubit, to enhance the effectiveness of quantum computers.

Table 13.1: Comparison between classical computing and quantum computing.

Feature	Classical computing	Quantum computing
Basic unit of information	Bit (0 or 1)	Qubit (0, 1, or both simultaneously, due to superposition)
Processing power	Grows linearly with the number of bits	Grows exponentially with the number of qubits
Computational model	Deterministic, sequential processing	Probabilistic, parallel processing
Algorithms	Classical algorithms (e.g., sorting and searching)	Quantum algorithms (e.g., Shor's algorithm, Grover's algorithm)
Strengths	Well-established, widely used, and excellent for everyday tasks	Potential to solve complex problems currently intractable for classical computers (e.g., drug discovery and materials science)
Limitations	Limited ability to handle exponentially complex problems	Currently in the early stages of development, prone to errors, and requires specialized hardware
Applications	General-purpose computing, data processing, and simulations	Optimization, cryptography, and quantum simulations
Future	Continues to improve but faces physical limitations	Holds promise for groundbreaking advancements in various fields

For certain problems, like integer factorization, which is essential for the security of traditional encryption systems based on RSA, quantum computers can provide answers significantly faster than conventional ones. For example, with Shor's algorithm, a polynomial-time approach to large-number factorization is possible using quantum computers, and hence it enables them to break RSA encryption in a fraction of the time a classical computer would take. This is a serious concern, bearing in mind the construction of quantum computers, since it threatens many of the current means of encryption that safeguard internet operations, mobile phones, Gmail, and even UAV systems.

13.1.4 The Need for Quantum-Safe Cryptography in UAV Systems

With the progress of quantum industries, it is clear that the threat of quantum attacks on today's cryptography must not be ignored. This is particularly true for cryptography, which is defined as quantum-safe or post-quantum cryptography (PQC) which attempts to counteract attacks using both classical and quantum computers [10]. They are said to be post-quantum because they make use of underlying mathematical problems, such as lattice-based, hash-based, or code-based, that can be considered safe from quantum onslaught.

When considering UAV systems, quantum-safe cryptography is essential in this scenario to protect communication between the UAV and its operator. As explained above, in the absence of secure inter-communication interfaces, an adversary with experience or access to the possibility of quantum computing can intercept and hijack UAV communications, proceed to control the UAV, or even steal sensitive information.

13.1.5 Quantum-Safe QUAVS: Merging UAV Technology with Quantum Cryptography

Quantum-safe QUAVS are systems incorporating quantum-safe technologies for anti-quantum warfare purposes [11]. These systems are structured in such a way as to guarantee that the communication between the UAV and its control system remains unbroken, even in communication scenarios that incorporate quantum commutative abilities.

Due to this advancement of quantum-safe cryptography into UAV systems, new trends in human-machine relations arise. The operators are required not only to pilot the UAV in flight tasks but also to maintain the quantum-safe communication channels [12]. For this, knowledge of UAVs and quantum cryptography is not enough, and critical thinking is also required in developing robust plans of action for security circumvention.

One of the key challenges in quantum-safe QUAVS is coping with the limitations of the human brain as system complexity increases [13]. Vanilla HMI systems, which were designed for classical systems, can no longer be relied upon as they do not possess the capacity to handle the complexity brought about by quantum-safe communication. This calls for new HMI frameworks that will enable operators to manage the UAV and its quantum-safe, comprehensive, and reliable communication system.

13.2 Quantum Computing and Its Impact on UAV Systems

13.2.1 An Overview of Quantum Computing and Its Threat to Cryptography

Quantum computing works on a completely different principle compared to classical computing because it uses qubits rather than bits and makes use of quantum phenomena such as superposition and entanglement. The immense processing capability of quantum computers is expected to bring a significant impact to diverse industries like cryptography, materials science, AI, and so on [14]. Nevertheless, this great strength also represents a significant risk to existing encryption mechanisms, which depend on certain hard mathematical problems that require enormous computational power, such as integer factorization or discrete logarithm problems.

The two most popular algorithms for cryptography today are the RSA algorithm and the ECC algorithm. The RSA security relies on how difficult it is to factor large numbers, while the ECC security is based on how difficult it is to solve the discrete logarithm problem. Such problems can be marked as computationally hard for classical computers, as they run into exponential time for key sizes that are large enough. With a quantum computer employing Shor's algorithm, these problems can be solved in polynomial time, hence rendering both RSA and ECC practically useless in terms of security [15].

13.2.2 The Emergence of Post-quantum Cryptography

In response to the quantum threat, a number of scientists have created PQC algorithms, which they hope will withstand a quantum attack. These algorithms are based on the conjecture that certain mathematical problems will remain hard even for quantum computers. Some of the most advanced PQC algorithms are as follows:

- **Lattice-Based Cryptography**: Concerning lattice-based cryptography, it is based on the hardness of a few lattice-related problems such as the shortest vector problem and learning with errors (LWE) [16]. Such problems are considered to be secure against the quantum threat and thus lattice-based encryption systems are therefore ranked among the most efficient in post quantum encryption.

- **Hash-Based Cryptography**: In hash-based cryptographic schemes, the security is based on hash functions, for example, in the Merkle signature scheme [17]. Because of this, hash-based schemes are expected to be safe against quantum computers since they do not make any assumptions about the hardness of factoring or discrete logarithms.
- **Code-Based Cryptography**: Code-based cryptography, such as the McEliece cryptosystem, does not require high-level mathematics and is focused on the decoding of certain classes of linear codes [18]. This theory has been explored for decades and it is known that this type of cryptography is still secure against quantum computers; however, it imposes the usage of large key sizes.
- **Multivariate Cryptography**: Multivariate public key cryptosystems are based on multivariate polynomials with respect to the number of variables, where systems of equations are solved over finite fields [19]. This is, nevertheless, not the case with the multivariate schemes, whereby, in fact, these schemes display encouraging ability to withstand quantum attacks, although they have not been fully exploited.
- **Isogeny-Based Cryptography**: Cryptosystems that employ isogenies are built on the assumption that it is hard to compute isogenies between two elliptic curves [20]. Though it is still nascent, isogeny-based cryptography provides small key sizes and is likely to be quantum safe as well.

For UAV systems, post-quantum encryption algorithms must be incorporated in order to retain the safety of the communication channels in the foreseeable future. Nevertheless, PQC has brought additional difficulty that may obstruct the HMI as the human operators need to handle more severe human factors such as managing large amount of encryption keys, ensuring the integrity of the communications and the timing to avoid responses to attacks that can be inflicted with a quantum system.

13.2.3 Quantum Threats to UAV Security

The advent of quantum computing brings about numerous visible dangers to UAV systems, especially concerning the security of communications. The standard methods of encryption utilized in UAV communications, encryption standards such as RSA and ECC, can easily be circumvented by quantum attacks. The implication here is that those who make use of quantum computers will be able to listen in on the communication, decipher it, and alter it.

One of the strongest risks is the ability of the attacker, who possesses quantum computing capabilities, to perform man-in-the-middle (MITM) attacks [21]. In such an attack, the attacker eavesdrops on a UAV and its operator communication, and acts as a quantum computer in the middle de-coding the messages then forwarding the re-coded ones to the UAV operator. This attack method could give the attacker control of the UAV or retrieve confidential information without the concern that the operator could ever know that the communication is at threat.

Another potential threat is the risk of key compromise. In quantum-safe UAV systems, there is a need for the operating personnel to handle numerous encryption keys and certificates, which are essential for protecting the communication links. Key compromise may occur either due to negligence or a security breach, and if this happens, the whole system is at risk. This creates enormous problems for HMI design because operators must be trusted with cryptographic key management without risking security loss.

13.2.4 Quantum-Safe Communication Protocols in UAV Systems

The use of quantum-safe cryptographic protocols in UAV systems requires an additional in-depth assessment of the efficiency versus the security of the systems deployed in UAVs [22]. Although quantum-safe cryptography provides a higher level of security against quantum computing threats, it is accompanied by high processing power and communication bandwidth overhead, which could adversely affect the operation of the UAV systems.

Lattice-Based Cryptography in UAVs: Lattice-based schemes for cryptography, including lattice problems like LWEs, have gained popularity as one of the best tools in quantum-safe UAV communications. A lattice-based scheme provides a good level of security and trust, where it is anticipated to offer none to very little protection against quantum attacks [23]. Thus, content such as key generation and computation in lattice-based schemes is increasingly higher compared to conventional methods of cryptography. This poses a problem for resource-constrained UAV systems with respect to resources like bandwidth and power consumption.

Hash-Based Cryptography: Merkle signature schemes for example are classified as hash-based strategies which are practiced as quantum-safe approaches to protecting UAV communications [24]. Hash-based approach schemes are very cost effective and do provide adequate levels of security. Nonetheless, the measure of the signature size from hash-based approach schemes tends to be very high and increase the communication cost in instances where bandwidth is constrained.

Key Management in the Quantum-Safe UAV Systems: The management of encryption keys and certificates is one of the most difficult tasks in the entire realization process in UAV systems that use quantum-safe cryptographic protocols [25]. All forms of secret and public key systems based on quantum-safekeeping practices use larger and more complex components and structures regarding key management than other traditional forms. This poses potential difficulties in HMI, as the operator must handle all these keys without exposing security risks or complicating the system to an uncomfortable level.

In quantum-safe QUAVS, key management systems should be developed to eliminate human errors as much as possible by allowing operators to create, share, or cancel encryption keys without exposing themselves. This can include applying and employing KMS that is equipped with IT capabilities and can perform tasks such as routine key creation and key distribution, all under the oversight of the operator.

13.3 Human-Machine Interaction (HMI) and Quantum-Safe Systems

13.3.1 Traditional HMI Models in UAV Systems

The majority of UAV systems are operated with collaborative human-machine interfaces, allowing the typical operator to proficiently control the UAV while performing real-time operational execution [26]. The HMI framework is mostly composed of a collection of control surfaces that include operator devices (joysticks, touch screens, and keyboards) with their operator-computer interface visual, auditory, and touch feedback devices. These interfaces are designed to assist operators in ensuring that the UAV is effectively flown along a certain trajectory and for specific performance and mission objectives, while also being able to adapt to any external influences or system failures.

The main task in considering the traditional HMI models has been the ease of performance of the operator's functions, as well as decreasing the information load so that the person is capable of acting in a reasonably timely manner. This becomes more pertinent in sensitive situations that require quick and accurate decisions, for example, during military operations or disaster management, where making a wrong decision could lead to failure.

13.3.2 Unique Challenges in Quantum-Safe QUAVS HMI

The implementation of quantum-resistant technological components within UAV systems makes it possible to propose new advances in the design of the HMI. Such systems place new demands on operators since it is no longer only about the UAV flight, but also the maintenance of high security for communication channels and encryption keys, as quantum systems are much more complicated than any standard systems. This additional requirement often leads to an increased level of information overload on operators, which makes it quite difficult for them to manage and supervise the operations of the system.

Complexity of Quantum-Safe Protocols: One of the requirements that is often overlooked in the management of a quantum-safe system is the complexity of the quantum-safe protocol itself [27]. Cryptographic algorithms of the quantum-safe type cannot be managed so easily, as they come with larger key sizes, more advanced methods of distributing keys, and the constant establishment of communication. They face the risk of focusing more on the system's complexity rather than on how to manage these cryptographic aspects effectively.

Key Management and Distribution: For the operators of a quantum-safe system, the integration of key management is one of the most difficult activities, if not the hardest [28]. In modern quantum-safe systems, communication channels are secured by means of underlying quantum-safe cryptographic algorithms that impose regular rekeying and distribution of keys. Encryption keys have to be properly created to avoid human error, distributed to other users who shall not misuse them, and removed when appropriate. This calls for simple, basic key management tools that ease the operators' load but do not limit their ability to respond to any given situation, which is a security concern.

Quantum-Safe Communication Monitoring: Effective management of quantum-safe communication means constant supervision of such communication at all times [29]. Whereas operators have communication restrictions that are normal due to the complexity of quantum-safe protocols, potential security breaches like MITM attacks or compromising a key must not be homogenized with communication delays. Development of cognitive models, on the other hand, enables real-time monitoring of feedback from the system, helping operators understand how to respond to potential threats.

13.3.3 Cognitive Models for Quantum-Safe HMI

There exist cognitive models for the exercise of designing every psycho-physical human-machine interface, as they elucidate the thought processes that operators employ, availing themselves to see, comprehend, and react to the system feedback. Considering quantum-safe QUAVS, there is a need to also be concerned about cognitive models, as they must consider how the management of communication endures polarized control of quantum-safe communication within the operational limits of the UAV.

13.3.3.1 Situational Awareness in Quantum-Safe QUAVS

Situational awareness is pertinent to the operators controlling UAV systems. It refers to the operator's ability to notice certain elements within the environment of the UAV, appreciate their meaning, and anticipate their development. In the case of quantum-safe QUAVS, the factors that make situational awareness difficult extend further, as the internal control must also ensure the monitoring of quantum-safe communication channels, encryption key security, and information related to potential threats.

Cognitive models for quantum-safe QUAVS must assist operators in managing and directing their attention to multiple and competing objectives and tasks, such as governance over UAV movement, realization of the mission, and communication security. Applying effective situational awareness enables operators to react appropriately in the shortest periods when the system's semiautonomous status is compromised or dysfunctional. For example, any communication blackout, a breakdown of the encryption upon detecting a covert twist by a quantum-powered enemy, and so on.

The Situational Awareness Pyramid Concept also Applied in Quantum-Safe QUAVS: A three-tier chrono-structure of scene description comprehension may be used.

Perception of System Status: The operators are always required to keep track of the UAV terrain, battery status, the course, as well as the general status of the mission being performed, including the operational quantum-safe links.

Comprehension of System Implications: Operators do not simply rely on the system performance status metrics; they have to analyze what such system performance status metrics mean. For instance, when there is a communication failure, it is not prudent to sit back and overlook it as just a service-down issue.

Projection of Future States: The operators are required to also be able to estimate the responses of the system in consideration of the feedback they receive at the present time. For example, a busy operator who sees that the ninth article is being omitted might venture that there is a quantum-based insurgent assault in progress and do something about it.

13.3.3.2 Decision-Making in High-Complexity Systems

The complexity of quantum-safe QUAVS imposes considerable constraints on the decision process and the actors that work with them. Operators may even have to decide in a matter of seconds whether or not to abandon the particular mission, or devote themselves to another course or retaliatory actions against the assumed intrusion of the communication channel [30]. The integrating brains of the cognitive models should encompass the different phases of decision-making as regards:

- **Recognition**: The operators have to identify the "What is wrong?" scenarios; that is the focus of the problem. Users must be provided with alerts that are easily understood and act on deviant communication behavior or abnormal behavior of the encryption keys.
- **The Development of Options**: The operator is responsible for creating alternative solutions once a problem becomes evident. For instance, if some communication is delayed, the operator will have the option to defend the UAV, change encryption keys, or redirect communication.
- **Risk Assessment**: Associated with each and every judicious action within the quantum-safe QUAVS are some risks. The cognitive models should be able to assist the operators in such matters very swiftly. For example, encryption keys are replaced while on the move, posing further risks of miscommunication, and in instances such as cybersecurity threats, taking too long to confirm the attack gives the foes enough time to take advantage of the unlocked door.
- **Implementation and Control**: Finally, once a decision is made, the last action for the byte operator is to undertake the selected range of actions and ensure that the system is functioning in the manner intended.

The foregoing ladders of the decision-making process are made more complex by the existence of quantum-safe communication protocols. There are hurdles in the understanding of quantum cryptography, and thus the operators cannot appreciate the technicalities of it, which implies that the cognitive models should be able to assist lay persons well as experts in decision-making.

13.3.3.3 Managing Cognitive Load

The impact of this vulnerability is even worse since quantum-safe cryptographic systems are very complex, creating a high cognitive load. Cognitive load is defined as the total amount of mental effort that is used to process information and make decisions. In quantum-safe QUAVS, however, cognitive load is increased since there is a need to keep track of secure communication, encryption keys, flight paths, and mission objectives at the same time.

There are three types of cognitive load that HMI designers have to take into account.

- **Intrinsic Load**: Intrinsic load is the level of difficulty that exists in the task per se. Thus, the decoding and management of quantum-safe protocols in UAV remote facility operations, which mainly include dealing with huge encryption keys and key rotation, are very extrinsic. HMI frameworks should eliminate this reliance and conventional focus on manual processes (like encryption key distribution), which should be addressed by replacing the non-novel tasks.
- **Extraneous Load**: Added to the other two are the loads that involve time and effort wasted through inefficiencies, for example, in system design. Excessive clut-

ter within UIs, inconsistencies in feedback, and unnecessary notifications all serve to enhance extraneous mental load. Reducing the extraneous load requires careful attention to UIs and how information is delivered, ensuring there are no unnecessary distractions.
- **Germane Load**: This is the amount of energy spent on creating and managing mental models, as well as the required knowledge and experience for task performance. Training programs for the quantum-safe QUAVS operators should include comprehension-enhanced training on how effective quantum-safe communication is, as well as the possible threats posed by quantum adversaries.

When developing cognitive systems for quantum-safe QUAVS, it is necessary to define the scope of the activities that will be delegated to the machine and those that will be performed by operators.

13.4 User Interface (UI) Design for Quantum-Safe QUAVS

User interface (UI) can be described as one of the most integral areas of human-computer interaction, especially in complicated systems such as quantum-safe QUAVS. The UI is the main tool by which operators control the UAV, its communication systems, monitor its condition, and make other operational decisions. In the case of quantum-safe communication, the UI will have to be very utilitarian, as it is necessary to provide the required information quickly without unnecessary complexity.

13.4.1 The Challenge of Designing for Quantum-Safe Systems

Designing UIs in quantum-safe cryptography would be another daunting task because of its inherent complexity [31]. The operators will have to look after quantum channels, encrypt and manage encryption keys, and detect security breaches within a span of time. This implies that there is a need for a UI that encapsulates more complex information and more information than necessary in a less complex manner.

When formulating UIs for quantum-safe QUAVs, several design principles are taken into account, as shown in Figure 13.2:
- **Simplicity and Clarity**: Irrespective of the degree of sophistication of the system, the UI should always convey the information as simply and clearly as possible. Operators should be able to instantly understand the state of the UAV, its communication devices, and the protection of its keys. The need to provide stirring information to the operator may lead to overloading, which, in turn, will affect the level of decision-making ability.

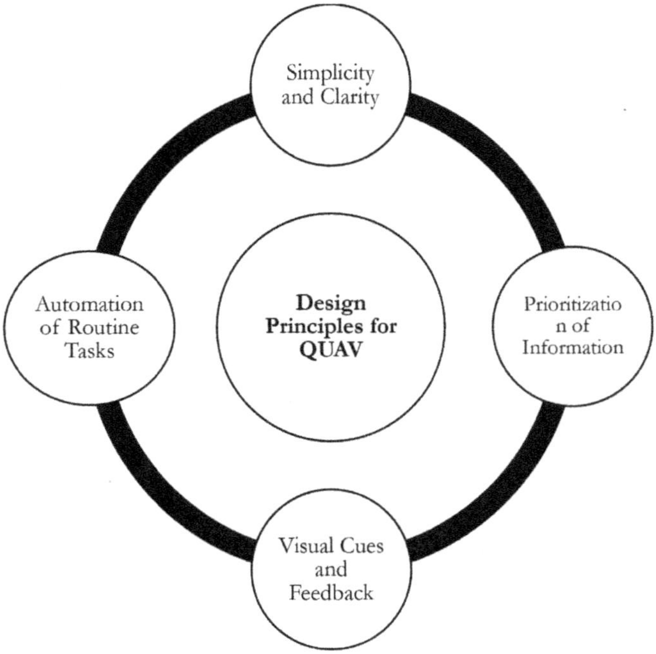

Figure 13.2: Design principles for QUAV.

- **Prioritization of Information**: However, it is also true that not all information is equally important throughout the process, and those that are more critical should be highlighted. When designing UIs, the information hierarchy should follow the most salient features of the current and potential possibilities of its use at the moment. For instance, in normal operation, the flight path and mission objective will take precedence in the UI. But when there is any form of communication anomaly, emphasis must be placed on the status of communication channels and encryption keys.
- **Visual Cues and Feedback**: Visual feedback on the opposite side, such as a color status indication or a percentage of a progress line, may be useful for operators in understanding the current conditions of the system. For example, a green indicator may inform the user of the secure state of the communication channel, and a red one may indicate the detection of a security abnormality. While this does not completely eliminate the need for data penetration, it does reduce the amount of effort required to do so, since the system can simply be observed and the answer deduced, rather than searching for answers within complex data.
- **Automation of Routine Tasks**: In quantum-safe systems, there are many elements of the system that are labor-intensive and repetitive, one being the management of encryption keys. A sufficient amount of time should be allocated for

the UI to eliminate such tasks, as they should not be performed by the operators. For the case in example, encryption of data may be automatically done by the system, with the keys generated and disseminated, and only in extreme cases would the operator be required to do it physically.

13.4.2 Real-Time Feedback Mechanisms

In order to ensure that the operators are aware of the situation and are able to manipulate the system in an appropriate way, it is important to provide feedback in real time. Feedback mechanisms in quantum-safe QUAVS should track, in real time, the status of the UAV, its communication links, and its cryptographic quality.

Taking these measures into account, the observer is reasonably well-equipped to work with the system interface, which provides the following feedback mechanisms, as shown in Figure 13.3:

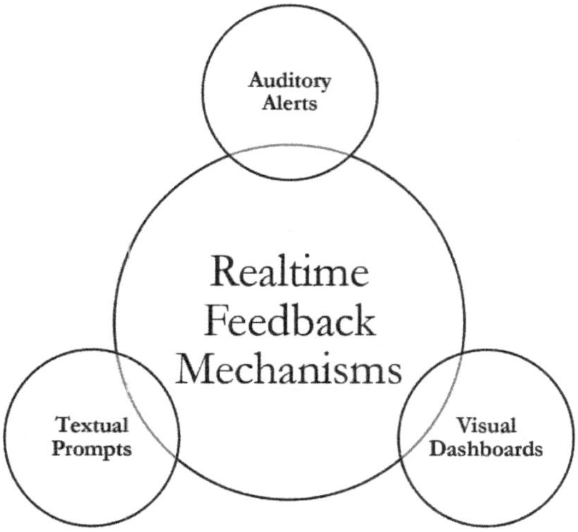

Figure 13.3: Real-time feedback mechanisms in UAVs.

- **Auditory Alerts**: Such guidance is important during the operator's performance, as there may be several time-sensitive problems on the horizon, such as communication timeouts, keys going out of scope, and system security being compromised. Nevertheless, such auditory alerts shall be used selectively in order to strike an optimal balance and avoid either excessive burden on the operator or excessive alertness fatigue.
- **Visual Dashboards**: Moreover, a visually oriented front-end can enable system users to gain a glimpse of the system's status by displaying real-time information on the position of the "UAV," mission status, communication with the system and

its integrity, and the status of current encryption keys. Additionally, visual elements like charts, graphs, or gauges can be employed to represent complex information in simple, visualized forms.

- **Textual Prompts**: However, from time to time, the need arises when such further information is required. Hence, such prompts would provide more complicated tasks or related configurations necessary for the operators. For example, if the system predicts a possible security breach based on user activities, the UI may freeze and display instructions and a graphical representation of threats and possible countermeasures for operators.

13.4.3 Designing for Cognitive Load Reduction

The UI design should always focus on limiting the effort placed on the user's mental processing capabilities and organizing the content in a logical manner. The following design principles will help:

- **Reduce Excess Information**: A UI should not present excessive and duplicated information that is irrelevant to the operator. The operator should only be presented with the most necessary information at that immediate time, while the rest will be made accessible when required.
- **Color and Layout**: The use of colors and layout can also be effective in drawing the operator's attention to certain elements within the UI. Some critical alerts, like warnings, could be shown in red, while other less critical information could be displayed using neutral colors like gray or blue.
- **Progressive Disclosure**: Instead of overwhelming operators' minds with all the possible forms of details right at the start, the UI can apply the progressive technique of disclosure by providing information only when called for. A case in point is the operator opening the status panel and looking at the system's summary, only to be presented with more details of the synopsis when they click on particular items of interest.

13.4.4 User Customization and Flexibility

Because of the intricacy of quantum-safe QUAVS, various operators may wish to have the information organized in different ways. Enabling operators to change the UI to suit their own needs can help satisfy users and improve the performance of the system.

For instance, some operators may prefer an interface that is data-centric and provides more information on communication latency, while other operators may prefer a versatile interface that only gives high-level mission objectives. By giving the operators the freedom to change the arrangement of the items, the colors, and the details presented in the interface, many sorts of users with different expertise can be served within the system.

13.5 Human-Machine Interaction Frameworks: A Path Forward

Inquiries into the prospects of HMIs in quantum-safe QUAVS will entail seeking new systems that can facilitate such communication in a safe manner without compromising the safety, efficiency, and security of UAV operations. Envisioned frameworks need to implement progress in cognitive science, automation, AI, and user interaction.

13.5.1 Modular HMI Frameworks

The use of modular HMI architectures is one of the strategies designed to tackle the problem of quantum-safe QUAVS cybersecurity complexity [32]. The modular approach segments the system into components, each of which performs independently with regard to monitoring and control. This enables operators to operate specific parts of the system, lessening cognitive load and enhancing situational relearning.

For example, the structure of modules targeted for quantum-safe QUAVS may include:

- Operations in Flight: Overseeing the UAV's trajectory, altitude, velocity, flying mode, and battery levels.
- Mission Management: Supervision of mission changes in progress and overseeing the handling of the mission.
- Communication: Assurance of communication with and encryption of ready-to-use quantum-resistant channels.
- Communication: Readiness and event reports related to security breaches, such as MITM attacks or compromised encryption keys.

Through modularization of the system, operators only expend their attentional resources on the most important parameters concerning the system at any time, but still have the option of seeking detailed information when necessary.

13.5.2 Adaptive HMI Frameworks

Another perspective under the 'more practical' category in the management of the complexity of quantum-safe QUAVS is to augment the hardware using adaptive HMI frameworks [33]. Dynamic HMIs alter the quantity and type of information available to the operator according to the situation at hand. They help prevent information overload for operators during standard operations with the help of information deference but still allow for operational details during emergencies.

Through an example, imagine that while performing regular surveillance UAV missions, instead of displaying the entire flight path and objectives of the UAV's mission, an adaptive framework will display the positional and mission updates of the UAV's flight in a more compact, enduring, and concerned position. But as soon as it detects a breach warning, it may switch to a more situational presentation, informing about the state of the communication links and their encryption keys rather than the flight of the UAV, its mission, and the procedures being flown on it.

13.5.3 AI-Enhanced Decision Support Systems

In quantum-safe QUAVS, AI can facilitate HMI through augmented decision-making and the performance of menial tasks [34]. The operators can benefit from AI technologies that analyze a huge amount of information, identify certain risks, and advise on possible actions to take in response.

For instance, the system could be programmed to alert an operator to any 'irregularities' likely to be exhibited by a system, even in the absence of a cyber-security incident. After this, the AI could propose a chain of actions that includes changing the encryption keys, flying the UAV to a place where the communication link is free of ESC, or changing the direction of the communication line.

As operators are required to work with AI to achieve the operational objectives, they can relieve themselves of the embarrassment or challenges of continuous monitoring and decision-making and devote their efforts to higher-level strategic and operational thinking.

13.5.4 AI Collaboration in Quantum-Safe QUAVS

AI technology transforms how humans and machines, including complex things like quantum-safe QUAVS, function together. While AI technologies help human operators by taking over repetitive actions, working through large amounts of data processed on the go, and providing decision support within the details of the ongoing situation, AI is perceived as a tool for extending the operator's scope of activities without over stressing the operator's mind.

13.5.4.1 AI for Real-Time Threat Detection

This task of real-time threat detection is among the most important applications of AI in quantum-safe QUAVS. Safe cryptographic systems add extra layers of communication complexity to the interaction between the UAVs and their operators. Yet, AI has the ability to help combat these harmful conditions by constantly monitoring the channels and detecting any anomalies or irregularities that signal a cyberbreach or a quantum attack in action.

In advance, AI will be able to detect a sudden change in latency, too much packet loss, or a change in the key encryption patterns more often than normal, among many others [35]. Over time, AI would memorize these patterns and understand the mildest signs, which may be attempts to harm the communications, whether at the stage of interception or manipulation. When a possible threat is identified, the AI then raises an alarm for the operator and provides a threat severity scale on how best to carry out the response.

It is even more acceptable in quantum-safe systems since quantum threats have been or are likely to be carried out in ways that will be hard for operators to notice and eliminate. If such attacks are possible, quantum-empowered adversaries might resort to tactics that are impossible to achieve in the presence of egalitarian structures, such as an MITM attack on the key exchange, in which the adversary pretends to be one of the communicators. However, these anomalies can be detected by AI systems much more quickly and accurately than human operators could.

13.5.4.2 AI-Assisted Decision-Making

In quantum-safe QUAVS, decision-making is largely about weighing the security of the communications against the mission [36]. AI can work with human operators by supporting decision-making during crises when time and mental resources are limited.

For instance, if AI identifies an anomaly in the communication channel, based on the operator's request, it can show some measures on how to remove the identified threat. These could include additional re-encryption of the communication with other cryptographic keys, switching to other, more secure channels, or temporarily ending the process until the risk is taken care of.

So this helps the operators make these decisions efficiently and improve the overall productivity of the operators by removing the need for them to think about every little detail. Also, based on the methods each operator's option falls into, the operators can accurately strategize and decide which measure to focus their attention on and which to disregard, considering the risk level. This mode of interaction with AI extends the possibilities for the operator and enables them to adequately respond to the rapid changes in the situation.

13.5.4.3 AI for the Automation of Routine Tasks

Specialized tasks that need to be performed in quantum-safe QUAVS, such as the generation of encryption keys, performing key rotation, or the distribution of certificates to users, are often menial tasks, yet they remain critical for the security of the system [37]. However, such tasks are necessary, as they exist in the span of intense scenarios where human decision-making reaches the level of making decisions regarding the mission, which are probably time-intensive and draining.

Many of these situations, however, concern tasks that require human action and, therefore, can be regarded as good practice in augmenting safety – any risks to security or confidentiality are transferred to AI. This improves operator efficiency, since human factors are, to some extent, the source of weakness within the system. These improvements result in perceiving other aspects of operation that require a higher level of decision-making: reconnaissance and patrolling areas by a UAV, and eliminating threats that have potentially materialized.

Still further, there are additional benefits to the automation of systems that are information-centric: they ensure that the prescribed cryptographic measures are observed without fail and without anyone having to remind the users. This is especially true in quantum-resistant systems, where even the slightest negligence in the handling of keys or rotating certificates may invite attacks from highly sophisticated quantum-based systems.

13.6 Real-World Case Studies

It is beneficial to analyze actual case studies in order to comprehend how HMI paradigms can be implemented in quantum-safe QUAVS. These examples demonstrate not just problems but also advantages that will arise when introducing quantum-safe solutions and AI into UAVs, thus providing a perspective of what might come in the design of HMI.

13.6.1 Military UAV Systems and Quantum-Safe Cryptography

Due to the distance to the conflict, the commander of the military UAV expects warm and secure shielded communication technology is crucial to accomplish a good outcome in the assigned mission as well as safeguard the people engaged in the risk. In response to such an increasing threat, many of the military structures are already looking for ways to incorporate quantum-safe ciphers into their UAV systems [38]. This entails developing a secure communication architecture, which can be used by military drones flying in a contested environment where the opponent can have quantum computing capabilities.

In this case study, a modular human-machine interface (HMI) framework was developed for the operators to support them in managing the additional difficulties brought about by quantum-safe communication. Properly implemented, the system relieved operators of the need to oversee communication during operations, as it was able to automatically furnish them with communication status regarding communication channels, including encryption keys, threats to security, and the chances of the system's utilization by another party. Within the developed interfaces, it was also possible to incorporate AI support interfaces that provided time-sensitive responses to threat situations.

The operators' cognitive load, especially in cases that demand split-second decisions, was one of the difficulties highlighted in this case study. In order to counter this, the HMI framework included AHM features which auto queued the level of inform mat ion displayed to this operator based on the time's security threat level. For example, during non-critical phases of operation, the information system provided low information content so that the normal users do not lose carer targets. However, once a malicious activity was spotted, the system automatically assisted the operator by displaying more information about the channels of communication along with the encryption keys required to take action.

This case study also showed the advantages of using modular HMI frameworks combined with AI support tools that help determine decisions in the complex environment of quantum-safe cryptography. The system decreased cognitive load and increased operator feedback in real time, which improved the operator's ability to manage secure communication within an operationally challenging environment.

13.6.2 Civilian UAV Systems in Disaster Response

As for the use of UAVs for emergency management, their role is also expanding in civil use activities. In such instances, uninterrupted and secure contact between the operator and the UAV is very important for ensuring the effective supply of relief materials and the assessment of the caused destruction [39]. While quantum-resistant cryptography is being adopted by many others, it is also necessary to include these cryptographic measures in civilian UAV systems to prepare for cybersecurity threats.

In the case study involving disaster response UAVs, the operators lost a lot of time because there were very few means of securely integrating communications into the process of searching for and rescuing victims. The issue of quantum-safe cryptography was incorporated into the system, requiring operators to manage the issuance and maintenance of encryption keys while remaining vigilant against communication channel tampering.

With the purpose of solving such problems, the changing human-computer interface system was elaborated, which applied AI for routine operations such as key handling and communication supervision to be performed automatically. The system also generated alarms whenever there was a communication blackout or abnormal delay in communication, enabling the operator to evaluate the current scenario and take required actions promptly.

The consequences of this case study underlined the need for the reduction of cognitive overload in critical situations, such as disaster response. Harnessing AI and delegating petty activities to machines offloads decision-making functions to operators in the HMI framework, whereby the focus remains on the critical operational decisions, enhancing the effectiveness and safety of the UAV system.

13.7 Ethical Considerations in Quantum-Safe HMI

With the continuing advancement of AI and quantum-safe cryptography in UAV systems, there will be an increasing need to consider the ethics in the design of interfaces between man and machines [40]. These issues concern who should be trusted and held responsible, as well as the consequences of removing human judgment from the decision-making process.

13.7.1 Trust in AI-Driven Systems

Among AI systems' critics, one of the most prominent issues is that users may not always feel comfortable trusting or even following the AI's advice. The optimal human + AI interaction will be where humans and AI trust each other based on the application and context [41]. For example, when AI is soliciting action from its human operators to manage security or use any insights to support decisions. Being an advanced Intelligent Agent, if the user does not have faith in the AI's decision, there are risks in omitting the implementation of recommendations, which could be beneficial, or even in decision-making, especially in toward context of threat recognition.

It is important to earn trust if such systems are to be widely used in practice. It entails being comprehensible regarding what has been done and the steps undertaken to reach such conclusions. Acceptable justifications in the form of AI reasoning will even lead to better human-operator interaction with the system.

The record of previous performance is another issue that bolsters trust and especially affects such systems. Even in situations where AI makes recommendations and decisions, trust will be earned from operational personnel only if the given recommendations are correct. Therefore, if any recommendations or decisions are intended to be made concerning the AI, it is important that regular updates on its efficiency are conducted.

13.7.2 Accountability in Automated Systems

An important issue that has arisen as a result of the rising automation of some routine tasks in quantum-safe QUAVS is accountability. There is always an aspect of responsibility in the operation of machines. What if a particular task that is automated using the AI system, such as a key rotation or communication rerouting, were to fail and lead to security breaches? Who is to blame? Is it the human operator who permitted it, even after being shown that the AI would outperform him, the internal staff who designed the AI system, or the AI itself?

Decision-making is also complicated when machines are applied in the decision-making processes; accountability hence becomes even more complicated when such machines replace and make decisions [42]. For instance, when the UAV is faced with a threat and the AI system decides to change the communication channel due to the threat, it might be possible for the operator to ask whether that focus on communication compromise is necessary. Then, who is to be blamed for that failure?

In line with these problems, HMI structures have to be formulated so as to guarantee that critical decision-making is retained by the human operators. While it is positive that AI will be an excellent source for assisting with the decision-making processes, the operators should determine, as a last resort, if a particular action is to be taken. This way, the management of complicated systems will allow some degree of responsibility to be assigned to the machines without the direct loss of human responsibility.

13.7.3 Ethical Use of Quantum-Safe UAV Systems

Activities, including warfare, public transport, and industries, combined with the employment of UAV systems, tend to raise further cultivable issues related to misuse [43]. The security of the system rider–UAV communications is guaranteed through the use of quantum-safe schemes, but such technologies are equally susceptible to use in the concealment of vices.

In this case, a rivaling adversary who possesses a quantum-safe UAV system would also be able to perform any other quintessence of covert operations without translating communications and risking interception. The responses to these challenges pose a big dilemma for the majority of policymakers and regulatory authorities, as they have to promote efforts geared toward the development of secure systems but are also wary of their abuse.

So, since HMIs have always been designed with the most important major – the end-user – in mind, such designers must therefore hold beliefs in fair bureaucratic procedures pertaining to social justice. For instance, if an AI task management system is made to overly prioritize certain objectives of concern, such as mission success orientation at the expense of minimizing human casualties, this could pose adverse challenges in instances of warfare scenarios or emergencies. There is a compelling need to ensure that ethical approaches and considerations are incorporated into all AI decision-making processes in order to avert abuses and ensure the proper application of quantum-safe UAV technologies.

13.8 Future Trends in Quantum-Safe HMI and QUAVS

The future of HMI in quantum-safe QUAVS will be defined by several emerging trends in the evolution of quantum computing. Such trends will include automation, AI cooperation, improved capabilities of quantum sensors, and details regarding PQC [44]. Each of those will contribute to the development of the secure UAV systems.

13.8.1 Increased Automation in Quantum-Safe Systems

Managing the intrinsic complexities expected in future quantum-safe systems can be made easier with the necessity of increasing levels of automation. More advanced UAV systems will demand more of the operators in regard to the management of routine activities such as key distribution and analysis, as well as optimal routes adopted for flight and surveillance, and this will, in turn, lead to more outcomes handled by the cognitive system. This will relieve mental stress and allow the operators to concentrate on strategic tasks.

On the other hand, the same factor of increased automation comes with an occupational hazard that has to do with the extent to which participants continue to be involved in decision-making, especially in cases where a person's input is most required. In these future HMI designs, some consideration has to be made to provide sufficient automation without eradicating human control of highly useful systems, so that operators can use them in case intervention has to happen while the efficiency of the system continues to be utilized.

13.8.2 Integration of Quantum Sensors in QUAVS

The advent of technology, such as quantum sensors, would enhance the existing UAV systems very significantly. Such sensors take advantage of quantum effects, such as superposition or entanglement, to achieve sensitivity and precision that have never been accomplished before. Quantum sensors can be onboard UAV systems for several purposes, including navigation, monitoring the environment, and detecting threats.

There will be HMI design challenges since these sensors will be integrated into quantum-safe QUAVS, where the operators will be required to deal with the additional complexity associated with those sensors. Information regarding these quantum sensors will also have to be properly managed, leaving some degree of control over other systems onboard the UAV.

13.8.3 AI-Enhanced Human-Machine Collaboration

With the advancement of technology, the quantum-safe QUAVS of the future will entail better collaboration between humans and AI systems. Within these scenarios, where all relations would be quantum-safe and the deployment of museum communication control is provided, AI will dominate the management of complex situations through a decision-making support system and the automation of most ordinary workspaces. In this situation, ensuring that human decision-makers can control important decisions made within that chain will be properly accounted for and trusted.

One of the paramount fields of concern pertaining to human and AI interaction that has gained momentum in the recent past is the explainable AI (XAI) systems. This is to provide people in charge with the necessary information coefficients. It is also useful in trust building on AI systems when making decisions in particular situations regarding the proposed recommendations by them.

13.8.4 Ethical and Regulatory Challenges

As quantum-safe QUAVS become common, concerns associated with the use and regulation of these systems will become more significant. FREEGUIDE, on the other hand, such systems must be kept within ethical and regulatory boundaries for both military and civilian use.

Control over quantum encryption, drone security tactics, as well as AI, should be addressed by governmental bodies and other regulatory organizations for the legal and legitimate application of these technologies. Such trends in martial technologies will not develop in isolation; rather, HMI specialists, engineers, and policymakers, all in their various ways, ensure the next generation of quantum-safe QUAVS not only guarantees safety but also respects ethics.

13.8.5 The Role of Quantum Networking in Quantum-Safe QUAVS

The application of quantum networking will also be crucial in the field of QUAVs. In particular, quantum networking is useful for the secure communication of multiple UAVs operating in a swarm or a coordinated mission [45]. Sustained functionality of quantum networking enables the transmission of cryptographic keys over vast distances using quantum entanglement, which is believed to be eavesdrop-secure in nature.

Use of classical techniques to encrypt communication within UAV systems is the most secure form of communication; however, this will become obsolete with the advent of quantum computing. In quantum networks, however, this is achievable without the risk of interception. This is where quantum key distribution (QKD) comes into play.

13.8.5.1 Quantum Key Distribution in UAV Networks

QKD is one of the advanced techniques for providing security in QUAVS communications, where quantum-resistant techniques are used. QKD enables users to share cryptographic keys securely through quantum mechanics [46]. BB84 protocol, which employs the photon's polarization technique to distribute secure keys, is the most common QKD method known to people.

QKD can be employed to protect communication between the UAVs in the swarm and also between the UAVs and the ground control station in the case of UAV networks. The added advantage of using QKD on UAV platforms would be the accessibility of instant messaging systems that are safe from interference, even in scenarios where the enemy has access to quantum computers.

Nevertheless, there are some technical issues that arise from the implementation of QKD in UAV networks. First, it consists of encoding information in quantum bits, known as qubits, which need to be sent over a communication channel. This transmission usually requires a line of sight. Drones operating at high altitudes or in standalone areas may experience problems with maintaining a stable quantum communication link. The encoding of qubits, on the other hand, is very much prone toward noise and interference, which can lead to attenuation of the quantum communication channel.

To resolve these issues, scientists opt for the use of quantum repeaters, which are devices used to boost and distribute quantum signals over long distances. These quantum repeaters might be embedded in UAVs to accomplish quantum communication at further ranges, thus facilitating inter-UAV intercommunication in extensive distributed networks.

13.8.5.2 Quantum Entanglement in Distributed UAV Operations

Another interesting feature, which has the possibility of radically changing the system of secure communications, is quantum entanglement in UAV systems. With entangled particles, there exists a link in the sense that knowing the state of one particle allows one to predict the state of the other particle, irrespective of the distance [47]. This property can be exploited for the provision of super-secure interactions between UAVs in a network because any communication within the UAVs could be monitored, and if any effort to tamper with or intercept such communication is applied, the entanglement would obviously be lost.

In distributed operations involving UAVs, quantum entanglement could be handy in synchronizing the actions of several UAVs without the use of conventional radio communication, which is prone to jamming and confinement attacks. With this communication method, UAVs would always be able to reach targets that previously required significant time and effort because information could be transferred securely through the air, even over long distances.

Nevertheless, just like in the case of QKD, there are several technical issues to be resolved with regard to the practical realization of entanglement-assisted communications within the networks of UAVs. Quantum entanglement is very difficult to sustain, as every researcher in the field knows the particles are very sensitive, and this makes it hard to maintain the entangled particles over long distances without great control of the system. Solutions to such challenges are currently being developed, such as expanding existing quantum networks and employing quantum teleportation techniques that will address how to communicate entangled particles over longer distances.

13.8.5.3 Quantum Networking and HMI Design

Integrating quantum networking into the quantum-safe QUAVS will involve extensive modification of HMI frameworks. New tools will be created for the appreciation of sophisticated quantum communication channel networks, which will include the current state of quantum key distribution surveillance and entanglement control [48]. This should, however, ensure that the information presented to the operator concerning the quantum network is not rendered overwhelming to the operator.

HMI frameworks for quantum networking are expected to incorporate the display of the quantum communication links between the UAVs, their status (secure, compromised, interrupted), and the UAVs themselves. Operators should be able to assess the operational status of the quantum network at a glance and intervene in case some communication terminals are excised.

Decision support, managed with the help of AI, will also be one of the factors that gravitate toward the control of the quantum network since it will help the operators troubleshoot the quantum communication links. To illustrate, in case an actual quantum link has been established, transported, and feedback is rendered but in some cases interrupted due to signal degradation, the AI would recommend certain steps to be taken, such as turning routers in optimal directions or installing quantum repeaters.

13.8.6 Policy and Regulatory Considerations for Quantum-Safe QUAVS

As the adoption of quantum-safe QUAVS increases, new concerns will emerge, and policymakers and regulatory bodies will have their work cut out for them in managing the risks associated with their use. The integration of quantum-safe encryption, AI-based decision support, and advanced human-machine interfaces is complex in nature and needs effective regulation to avoid abuse of the system and ensure it is used responsibly.

13.8.6.1 Regulatory Challenges in Quantum Cryptography

Quantum cryptography brings about various regulatory issues, especially with regards to the implementation of quantum cryptography in a responsible manner. There is a need to determine the specifics relating to the safe usage of quantum cryptography within UAV systems, and the relevant authorities responsible for policymaking should come up with regulations concerning key lifecycle management, communication security, and encryption key turnover [49].

A major problem with regard to the regulation of quantum-related cryptography is that it involves many parties. The threat that arises from the development and use of quantum computers and quantum cryptography is not confined to any country. In order to combat these threats, it is likely that international agencies, such as the International Telecommunication Union (ITU) or the International Civil Aviation Organization (ICAO), will, in the future, provide guidelines for quantum-safe cryptography for UAVs and other related systems.

Furthermore, such regulation would see policymakers setting aside policy papers that argue quantum cryptography might be used for foul play. For example, quantum-safe encryption could be employed by crime syndicates or unfriendly states to conduct clandestine operations or launch cyber-attacks without the risk of being intercepted. The development of legal regimes requires full-fledged consideration of threats as well as the development and promotion of such technologies that are quantum-safe.

13.8.6.2 Ethical Use of AI in Quantum-Safe QUAVS

Application of AI techniques in the quantum-safe QUAVS generates risks and dilemmas regarding ethical issues such as accountability, transparency, decision-making, etc. With the increased deployment of AI systems in charge of managing secured communications, flight activities, and threat assessment, the need to incorporate an ethical approach toward the design and use of these systems escalates [51].

Attribution of responsibility for the actions taken in AI-driven systems is one of the primary ethical concerns. No matter which party takes the action – for instance, an AI makes a decision to divert and transmit intercepted communication over a different network, or an agent decides on combat evasion – there remains an important party who bears the ultimate accountability for that decision [50]. The human operator should take charge of all important choices, while AI systems should assist the decision-making process rather than the other way around.

The ethical issue of transparency is also paramount. Any AI strategy needs to be able to provide detailed answers regarding any assumptions that led to a particular outcome so that the operator can understand how such a decision was made. This is not the case in low-threat environments where the AI system is deployed in a trust-but-verify mode.

At last, when making policies, the authorities should also think about the feasibility of AI-based UAV systems being more than what people expected and how they can be used unethically. For instance, in a military context, strike-on-call operations may utilize UAVs that can function autonomously, which makes one question the morality of applying AI to perform killing. Legal frameworks have to encapsulate measures aimed at guaranteeing that AI-based systems operate in a safe and ethical manner with the requisite checks and balances.

13.8.6.3 Legal Implications of Quantum-Safe Communication

The development of quantum-safe communication systems also creates legal issues, especially around data protection and privacy concerns. As UAV systems incorporate quantum-resistant encryption methods, the communication between UAVs and their respective ground control stations is likely to go beyond the scope of spying by governments and law enforcement agencies [52]. This, however, provides better security but raises concerns as well about whether quantum-safe communication may one day fall into the hands of criminals.

However, criminal structures or adversary states may wish to employ quantum-resistant UAV systems for secret missions in the cyberdomain, unashamedly within their infrastructure, feeling free from the interception of their communications. This poses great risks to enforcement authorities, who are likely to find it hard to monitor and tackle such vulnerabilities.

Policymakers, in such a case, will become obligated to ensure the development of legal orders insofar as security communication is concerned, while law enforcement agencies will be able to receive very important information when there's a need for it. In such situations, it may be beneficial to consider the possibilities of developing special laws which would permit access to quantum-safe channels in the course of investigating certain crimes or threats to national security.

Besides, international agreements will be required to set out the standards for the use of quantum-safe communication in civilian and military spheres. Those agreements ought to delineate the grounds for the employment of quantum-safe communication technologies, and the obligations of states, institutions, and individuals in regard to best practices and legal enforcement of these technologies.

13.9 Technical Challenges in Implementing Quantum-Safe QUAVS

The opportunities that quantum-safe QUAVS provides are enormous; however, there are also significant technical hurdles that must be cleared before any success can be achieved with these systems [53]. These challenges are diverse, owing to the fact that they extend to areas such as cryptography, communication infra structures, hardware factors, as well as system integration.

13.9.1 Cryptographic Overhead and Resource Constraints

The use of quantum-safe crop rotation in the design of UAV systems poses yet another critical drawback, which is the high level of its encryption aimed at post-quantum algorithms [54]. Post-quantum lattice-based or hash-based cryptography algorithms can require larger key sizes and more resources than their EC or RSA counterparts.

This additional cryptography creates a problem for UAV systems, which are usually limited in processing and communication bandwidth as well as battery life. Onboard UAVs may have to perform many and various types of real-time cryptographic operations, which are considered to be performance-intensive, without compromising other features such as motion performance or mission objectives.

For this reason, many researchers are also considering the possibility of trimming the weight of PQC algorithms to make them suitable for low power consumption. Taking this approach might entail optimization of the algorithms into more effective ones that, in most cases, may minimize power consumption, or it could call for the integration of external devices that will perform the cryptographic functions, ensuring the weight on the main processor is low.

13.9.2 Communication Infrastructure for Quantum Networking

With the integration of quantum networking technologies such as QKD and quantum entanglement into UAV systems, there will be a need for a new communication infrastructure to support these technologies. First, in the case of quantum communication, one needs to transmit qubits, which are very critical as they do not tolerate noise or interference. Secondly, it is very complex to create and maintain a steady quantum communication link between a UAV and its operator or between multiple UAVs in a network.

For instance, quantum communication generally relies on a signal in a direct line – equally termed as line-of-sight techniques – and this may prove to be a challenge, especially in regions that have impediments or where the weather is not favorable [55]. Similarly, the conveyance of qubits over great expanses usually presents considerable challenges since there are demands for quantum repeaters that transmit graciously strengthened quantum signals.

There is great buzz among researchers about providing the required enhancements in UAV systems, such as adding quantum repeaters and extending the use of satellite-based quantum communication networks. However, these technologies are still in their adolescent phases, and substantial progress is needed for their acceptance in UAV systems within practical frameworks.

13.9.3 Hardware Limitations in Quantum-Safe UAVs

Lack of adequate hardware is another downside related to the fusion of quantum-safe cryptography as well as quantum networking within UAV systems. Generally, most UAV structure designs tend to be light in mass and energy consumption with regard to onboard computing resources [56]. However, further developments involving quantum-safe encrypted algorithms, quantum communication hardware, and specifically designed quantum sensors would significantly increase the mass and power levels of the UAV.

In order to address these hardware achievements, the researchers have started working on developing quantum communication devices that are smaller and consume less power. For instance, small QKD modules could be used in UAVs without any additional weight or size increments to the UAVs themselves. Moreover, the integration of advanced quantum sensors and quantum computing components would create devices that function optimally in UAV systems.

13.9.4 System Integration and Interoperability

This successful deliverable will ensure that the projected quantum-safe QUAVS will be able to deliver services because it will be integrated smoothly across several technologies, such as quantum-safe cryptography, quantum networks, AI-supported decision-making, and HMI, amongst others [57]. One of the most daunting technical tasks in this regard is to make sure these technologies are integrated in a manner such that they perform in conjunction, even when most of these operate on different layers of the system [58, 59].

As an example, the policy on the incorporation of quantum-safe cryptography [60] into the communication layer should plan the coordination between this system function and the AI-driven decision support systems embedded at the operational layer, ensuring that the operators can effectively manage time-critical responses to perceived security risks. Similar considerations apply to the HMI framework, which must offer a continuous view of the system to the operators.

13.10 Conclusion

The development of HMI in quantum-safe QUAVS offers both possibilities and problems, given the growing danger of quantum computing in the context of UAV systems innovation. Robust and efficient operation of such systems will require the integration of quantum-safe encryption, the quantum internet, AI, and advanced UIs. The application of cognitive models and maladaptive interface designs may effectively manage the intricacy of quantum communications, thereby minimizing the cognitive load and improving the situational performance of operators. Modular and AI-supported HMI designs will empower the operator to promptly and accurately make choices in response to system and security risks and abnormalities. Over the coming years, the rapid progress in quantum computing and networking will serve as the blueprint for the next generation of quantum-safe QUAVS. Nevertheless, the emerging phenomena in the field of quantum imaging and quantum entanglement expose new possibilities and challenges in the field of HMI. These include a shift in focus and ongoing discussions in the development of HMI. Furthermore, the ethical, legal, and regulatory dimensions of quantum-safe QUAVS are equally significant. It is the responsibility of engineers, ethicists, and industry champions to ensure that these technologies are utilized responsibly and with appropriate control.

In conclusion, HMI in quantum-safe QUAVS represents a multidisciplinary challenge that spans fields as diverse as cryptography, cognitive science, AI, and quantum mechanics. By addressing these challenges, we can ensure that quantum-safe QUAVS are both secure and operable, paving the way for a new era of UAV systems that are resilient against the emerging threats posed by quantum computing.

References

[1] N. D. Kumar and S. K. Kota, "Effectiveness of Human Machine Interaction (HMI) for Development of Communication and Interaction between a Human and a Machine to Control Machines," *Technoarete Transactions on Industrial Robotics and Automation Systems*, vol. 2, no. 1, Mar. 2022, doi: 10.36647/ttiras/02.01.a005.

[2] O. Grote, A. Ahrens, and C. Benavente-Peces, "Paradigm of Post-quantum Cryptography and Crypto-agility: Strategy Approach of Quantum-safe Techniques," in *Proceedings of the 9th International Conference on Pervasive and Embedded Computing and Communication Systems*, SCITEPRESS – Science and Technology Publications, 2019, pp. 91–98. Accessed: Sep. 18, 2024. [Online]. Available: http://dx.doi.org/10.5220/0008162800910098

[3] Y. Baseri, V. Chouhan, and A. Hafid, "Navigating quantum security risks in networked environments: A comprehensive study of quantum-safe network protocols," *Computers & Security*, vol. 142, p. 103883, Jul. 2024, doi: 10.1016/j.cose.2024.103883.

[4] D. S. Tan and A. Nijholt, *Brain-Computer Interfaces: Applying our Minds to Human-Computer Interaction*. Springer Science & Business Media, 2010.

[5] "Unmanned Aerial Vehicles for Military Applications," in *Theory, Design, and Applications of Unmanned Aerial Vehicles*, Taylor & Francis Group, 6000 Broken Sound Parkway NW, Suite 300, Boca Raton, FL 33487-2742: CRC Press, 2016, pp. 47–101. Accessed: Sep. 18, 2024. [Online]. Available: http://dx.doi.org/10.1201/9781315371191-3

[6] D. Glade, *Unmanned Aerial Vehicles: Implications for Military Operations*. 2000.

[7] R. A. Mollin, *RSA and Public-Key Cryptography*. CRC Press, 2002.

[8] A. Enge, *Elliptic Curves and Their Applications to Cryptography: An Introduction*. Springer Science & Business Media, 2012.

[9] S. Akama, *Elements of Quantum Computing: History, Theories and Engineering Applications*. Springer, 2014.

[10] D. J. Bernstein, J. Buchmann, and E. Dahmen, *Post-Quantum Cryptography*. Springer Science & Business Media, 2009.

[11] R. Kuang and M. Barbeau, "Quantum permutation pad for universal quantum-safe cryptography," *Quantum Information Processing*, vol. 21, no. 6, Jun. 2022, doi: 10.1007/s11128-022-03557-y.

[12] V. Vallivaara *et al.*, "Quantum-Safe Communication Research in Finland," in *NATO Science for Peace and Security Series – D: Information and Communication Security*, IOS Press, 2024. Accessed: Sep. 18, 2024. [Online]. Available: http://dx.doi.org/10.3233/nicsp240009

[13] M. Ohya, "COMPLEXITY IN QUANTUM SYSTEM AND ITS APPLICATION TO BRAIN FUNCTION," in *Quantum Information II*, WORLD SCIENTIFIC, Jul. 2000, pp. 149–160. Accessed: Sep. 18, 2024. [Online]. Available: http://dx.doi.org/10.1142/9789812792761_0011

[14] D. Vlahek, "The problem of quantum computers in cryptography and post-quantum cryptography," in *StuCoSReC. Proceedings of the 2018 5th Student Computer Science Research Conference.*, University of Primorska Press, Oct. 2010, pp. 61–64. Accessed: Sep. 18, 2024. [Online]. Available: http://dx.doi.org/10.26493/978-961-7055-26-9.61-64

[15] *A Quantum Computer for Shor's Algorithm*. 2004.

[16] J. Zhao, "From learning with errors (LWE) problem to CLWE problem," *Theoretical and Natural Science*, vol. 26, no. 1, pp. 286–298, Dec. 2023, doi: 10.54254/2753-8818/26/20241119.

[17] I. von Maurich, *Efficient Implementation of Code- and Hash-based Cryptography*. 2017.

[18] M. Baldi, *QC-LDPC Code-Based Cryptography*. Springer Science & Business, 2014.

[19] J. Ding and B.-Y. Yang, "Multivariate Public Key Cryptography," in *Post-Quantum Cryptography*, Berlin, Heidelberg: Springer Berlin Heidelberg, pp. 193–241. Accessed: Sep. 18, 2024. [Online]. Available: http://dx.doi.org/10.1007/978-3-540-88702-7_6

[20] B. Drzazga and Ł. Krzywiecki, "Review of Chosen Isogeny-Based Cryptographic Schemes," *Cryptography*, vol. 6, no. 2, p. 27, May 2022, doi: 10.3390/cryptography6020027.

[21] A. Schrottenloher and M. Stevens, "Simplified Modeling of MITM Attacks for Block Ciphers: New (Quantum) Attacks," *IACR Transactions on Symmetric Cryptology*, pp. 146–183, Sep. 2023, doi: 10.46586/tosc.v2023.i3.146-183.

[22] H.-K. Lo and T.-M. Ko, "Some attacks on quantum-based cryptographic protocols," *Quantum Information and Computation*, vol. 5, no. 1, pp. 40–47, Jan. 2005, doi: 10.26421/qic5.1-4.

[23] P. Bagchi *et al.*, "Public Blockchain-Envisioned Security Scheme Using Post Quantum Lattice-Based Aggregate Signature for Internet of Drones Applications," *IEEE Transactions on Vehicular Technology*, vol. 72, no. 8, pp. 10393–10408, Aug. 2023, doi: 10.1109/tvt.2023.3260579.

[24] P. Lafrance, *Digital Signature Schemes Based on Hash Functions*. 2017.

[25] E. Y. Bushuev, "Comparative Analysis of Quantum Key Distribution Schemes in Quantum Communication Channel and Quantum Key Distribution Process Simulation," in *2023 Systems of Signal Synchronization, Generating and Processing in Telecommunications (SYNCHROINFO*, IEEE, Jun. 2023, pp. 1–5. Accessed: Sep. 18, 2024. [Online]. Available: http://dx.doi.org/10.1109/synchroinfo57872.2023.10178490

[26] F. Zoric, G. Vasiljevic, M. Orsag, and Z. Kovacic, "Towards Intuitive HMI for UAV Control," in *2022 International Conference on Smart Systems and Technologies (SST)*, IEEE, Oct. 2022, pp. 325–332. Accessed: Sep. 18, 2024. [Online]. Available: http://dx.doi.org/10.1109/sst55530.2022.9954775

[27] Y. Baseri, V. Chouhan, and A. Hafid, "Navigating Quantum Security Risks in Networked Environments: A Comprehensive Study of Quantum-Safe Network Protocols," Elsevier BV, 2024. Accessed: Sep. 22, 2024. [Online]. Available: http://dx.doi.org/10.2139/ssrn.4687851

[28] I. Meijers, "Challenges and Solutions for Secure Key Management and Monitoring: Review of the Cerberis3 Quantum Key Distribution System," *Quantum Reports*, vol. 6, no. 3, pp. 426–435, Aug. 2024, doi: 10.3390/quantum6030027.

[29] O. Grote, A. Ahrens, and C. Benavente-Peces, "Paradigm of Post-quantum Cryptography and Crypto-agility: Strategy Approach of Quantum-safe Techniques," in *Proceedings of the 9th International Conference on Pervasive and Embedded Computing and Communication Systems*, SCITEPRESS – Science and Technology Publications, 2019, pp. 91–98. Accessed: Sep. 22, 2024. [Online]. Available: http://dx.doi.org/10.5220/0008162800910098

[30] S. Fukuda, "Pattern Driven Decision Making Approach to Engineering Systems Design: Concept," in *Volume 11: Systems, Design, and Complexity*, American Society of Mechanical Engineers, Nov. 2014. Accessed: Sep. 22, 2024. [Online]. Available: http://dx.doi.org/10.1115/imece2014-37989

[31] A. Christo, K. Ananth T., and I. K. Osamah, *Real-World Challenges in Quantum Electronics and Machine Computing*. IGI Global, 2024.

[32] J. Schönig and A. Terfloth, "A modular approach for the -development of executable HMI specifications," *ATZelektronik worldwide*, vol. 6, no. 5, pp. 26–29, Oct. 2011, doi: 10.1365/s38314-011-0049-y.

[33] J. Ying, Y. Zhou, G. Jiang, S. Han, and B. Jing, "Disturbance Observer Based Adaptive Fault-Tolerant Attitude Control of a QUAV Using FNTSM," in *2023 China Automation Congress (CAC)*, IEEE, Nov. 2023, pp. 1646–1651. Accessed: Sep. 22, 2024. [Online]. Available: http://dx.doi.org/10.1109/cac59555.2023.10451573

[34] K. Coussement, M. Z. Abedin, M. Kraus, S. Maldonado, and K. Topuz, "Explainable AI for enhanced decision-making," *Decision Support Systems*, vol. 184, p. 114276, Sep. 2024, doi: 10.1016/j.dss.2024.114276.

[35] Bright Ojo and Chukwudi Tabitha Aghaunor, "AI-driven cybersecurity solutions for real-time threat detection in critical infrastructure," *International Journal of Science and Research Archive*, vol. 12, no. 2, pp. 1716–1726, Aug. 2024, doi: 10.30574/ijsra.2024.12.2.1401.

[36] M. Wu, Y. Niu, M. Gu, and J. Cheng, *Proceedings of 2021 International Conference on Autonomous Unmanned Systems (ICAUS 2021)*. Springer Nature, 2022.

[37] H. Chen, X. Wang, and Y. Li, "A Survey of Autonomous Control for UAV," in *2009 International Conference on Artificial Intelligence and Computational Intelligence*, IEEE, 2009. Accessed: Sep. 22, 2024. [Online]. Available: http://dx.doi.org/10.1109/aici.2009.147

[38] Colbeck, M, "Quantum Cryptography and Military Communications," Institute of Marine Engineering, Science and Technology (IMarEST), Nov. 2023. Accessed: Sep. 22, 2024. [Online]. Available: http://dx.doi.org/10.24868/11070

[39] J. Luman and B. Luman, "Exploring the Utilization of Small Unmanned Aerial Vehicles (UAV) Known as Drones in Early Phase Disaster Response," *Prehospital and Disaster Medicine*, vol. 34, no. s1, pp. s132–s132, May 2019, doi: 10.1017/s1049023x19002875.

[40] A. B. Tiwari, D. A. Rai, A. M. Singh, and K. J. Jaiswal, "Ethical and Security Considerations in the Era of Quantum Computing: A Framework for Responsible Implementation Strategies," *International Journal of Research Publication and Reviews*, vol. 5, no. 5, pp. 7266–7271, May 2024, doi: 10.55248/gengpi.5.0524.1295.

[41] D. Pham *et al.*, "A Case Study of Human-AI Interactions Using Transparent AI-Driven Autonomous Systems for Improved Human-AI Trust Factors," in *2022 IEEE 3rd International Conference on Human-Machine Systems (ICHMS)*, IEEE, Nov. 2022, pp. 1–6. Accessed: Sep. 22, 2024. [Online]. Available: http://dx.doi.org/10.1109/ichms56717.2022.9980662

[42] E. Marsden, *Control and accountability in highly automated systems*. FonCSI, 2011.

[43] M. Orsag, C. Korpela, P. Oh, and S. Bogdan, *Aerial Manipulation*. Springer, 2017.

[44] "Quantum Computation – Future of Microelectronics?," *Future Trends in Microelectronics*, pp. 19–27, Jan. 2007, doi: 10.1002/9780470168264.ch2.

[45] I. B. Djordjevic, "Basics of quantum information, quantum communication, quantum sensing, and quantum networking," in *Quantum Communication, Quantum Networks, and Quantum Sensing*, Elsevier, 2022, pp. 1–30. Accessed: Sep. 22, 2024. [Online]. Available: http://dx.doi.org/10.1016/b978-0-12-822942-2.00002-9

[46] M. Mehic, S. Rass, P. Fazio, and M. Voznak, "Modern Trends in Quantum Key Distribution Networks," in *Quantum Key Distribution Networks*, Cham: Springer International Publishing, 2022, pp. 209–223. Accessed: Sep. 22, 2024. [Online]. Available: http://dx.doi.org/10.1007/978-3-031-06608-5_8

[47] F. J. Duarte, "Quantum entanglement and quantum computing," in *Fundamentals of Quantum Entanglement*, IOP Publishing, 2019. Accessed: Sep. 22, 2024. [Online]. Available: http://dx.doi.org/10.1088/2053-2563/ab2b33ch24

[48] R. V. Meter, *Quantum Networking*. John Wiley & Sons, 2014.

[49] M. Kop *et al.*, *Towards Responsible Quantum Technology*. 2023.

[50] A. Joshi, "Ethical AI and Data Integrity: Ensuring Responsible AI Innovation Through Quantum – Resistant Algorithms and Federated Learning Paradigms," *International Journal of Science and Research (IJSR)*, vol. 12, no. 1, pp. 1297–1304, Jan. 2023, doi: 10.21275/sr24615150919.

[51] A. Rejeb, K. Rejeb, S. Simske, and H. Treiblmaier, "Humanitarian Drones: A Review and Research Agenda," *Internet of Things*, vol. 16, p. 100434, Dec. 2021, doi: 10.1016/j.iot.2021.100434.

[52] A. Minbaleev, S. Zenin, and K. Evsikov, "Prospects for Legal Regulation of Quantum Communication," *BRICS Law Journal*, vol. 11, no. 2, pp. 11–54, Jul. 2024, doi: 10.21684/2412-2343-2024-11-2-11-54.

[53] A. Kumar, D. Augusto de Jesus Pacheco, K. Kaushik, and J. J. P. C. Rodrigues, "Futuristic view of the Internet of Quantum Drones: Review, challenges and research agenda," *Vehicular Communications*, vol. 36, p. 100487, Aug. 2022, doi: 10.1016/j.vehcom.2022.100487.

[54] R. Aissaoui, J.-C. Deneuville, C. Guerber, and A. Pirovano, "A survey on cryptographic methods to secure communications for UAV traffic management," *Vehicular Communications*, vol. 44, p. 100661, Dec. 2023, doi: 10.1016/j.vehcom.2023.100661.

[55] V. K. Ralegankar *et al.*, "Quantum Cryptography-as-a-Service for Secure UAV Communication: Applications, Challenges, and Case Study," *IEEE Access*, vol. 10, pp. 1475–1492, 2022, doi: 10.1109/access.2021.3138753.
[56] Padmapriya, N., T. Ananth Kumar, R. Aswini, R. Rajmohan, P. Kanimozhi, and M. Pavithra. "IoT Based Energy Optimization in Smart Farming Using AI." Hybrid Intelligent Approaches for Smart Energy: Practical Applications (2022): 205–223.
[57] B. Rana and Y. Singh, "Internet of Things and UAV: An Interoperability Perspective," *Unmanned Aerial Vehicles for Internet of Things (IoT)*, pp. 105–127, Jul. 2021, doi: 10.1002/9781119769170.ch6.
[58] M. Asim and A. A. Abd El-Latif, "Intelligent computational methods for multi-unmanned aerial vehicle-enabled autonomous mobile edge computing systems," *ISA Transactions*, vol. 132, pp. 5–15, Jan. 2023, doi: 10.1016/j.isatra.2021.11.021.
[59] M. Asim, C. Junhong, A. Muthanna, L. Wenyin, S. Khan, and A. A. A. El-Latif, "A novel simulated annealing trajectory optimization algorithm in an autonomous UAVs-empowered MFC system for medical internet of things devices," *Wireless Networks*, vol. 29, no. 7, pp. 3163–3176, May 2023, doi: 10.1007/s11276-023-03370-0.
[60] R. Aissaoui, J.-C. Deneuville, C. Guerber, and A. Pirovano, "A survey on cryptographic methods to secure communications for UAV traffic management," *Vehicular Communications*, vol. 44, p. 100661, Dec. 2023, doi: 10.1016/j.vehcom.2023.100661.

Nithish P. U., Jayapandian Natarajan, Allen Sabu, and Korhan Cengiz

14 Quantum Computing Algorithms for Disaster Recovery Using Genetic and Fuzzy Adaptive Method

Abstract: Since quantum computing is based on ideas of superposition and entanglement, it will open new horizons in the field of disaster recovery and relief. Based on principles like superposition and entanglement, quantum computing holds immense and revolutionary potential in disaster relief operations. This is because, in the massive volumes of data produced during any kind of disaster, combined with the high level of complexity, response times and accuracy are often out of reach using traditional methods of computing. Quantum computing plays a revolutionary role in providing optimizations to the several stages of disaster management: prediction, readiness, response, and recovery. Quantum computers can be used to generate complex simulations using quantum parallelism and predict extreme weather conditions or several other disasters that can prove to be destructive to life and property. This can also be used to implement efficient evacuation in a timely manner. Quantum algorithms can also make resource allocation more efficient and ensure that essential supplies reach affected regions faster. Real-time processing and analysis of huge datasets enable good coordination among response teams and support dynamic decision-making. This article gives a very broad overview of the applications of quantum computing in disaster management protocols, providing insights into how quantum-enhanced algorithms can be used for generating simulations and for real-time analysis. These results are focused on developing efficient disaster management techniques that can better serve communities in the face of the increasing occurrence of natural and man-made disasters.

Keywords: Disaster management, earthquake, quantum computing, classical algorithms, genetic algorithm

14.1 Introduction

A disaster is a sudden catastrophic phenomenon which can cause significant damage to life and property. Since the dawn of recorded time, Earth has been plagued with several disasters. Some were natural, like the Gujarat Earthquake (2001), the Indian Ocean earthquake and tsunami (2004), Kerala floods (2021), Cyclone Fani (2019), while some were man-made, like the Bhopal Gas Tragedy (1984) and Teesta Dam Collapse (2023), to name a few. Over the years, we have adopted several techniques, advanced technology, and infrastructural modifications to predict, prevent and combat these disasters [1]. Disaster recovery includes the principles, policies, and processes put in

https://doi.org/10.1515/9783111619323-014

place to combat such disasters and restore normal conditions [2]. The disaster recovery extends way beyond the immediate response to disasters and includes every step that needs to be taken for the swift return to normalcy. Over the years several engineers and scientists have come up with several algorithms that have proven to be immensely useful in this process [3]. Moreover, over the years, this community of engineers and scientists has constantly strived to make this process more and more efficient. Initially, rule-based systems were used, and these systems relied on various pre-established rules to analyze data and perform actions [4]. Then, we developed Linear Programming and Network Optimization, which consisted of algorithms used to enhance resource allocation during emergency scenarios [5]. At last we started executing artificial intelligence and machine learning principles to predict, evaluate, and assist with well organized decision making throughout the process of disaster recovery [6]. Several models like support vector models started being used to analyze historic data and predict the probability of these disasters. Following the disaster, principles of data analytics and deep learning, especially convolutional neural networks (CNNs) [7], started being used largely to develop satellite images and drone footages and derive data from them for systematic analysis, which would additionally help in better decision making. Nowadays, one such computational principle, which has a lot of potential and is being researched to a great extent, is quantum computing. Quantum computing bridges the connection between mathematics, computer science and physics [8]. It offers a groundbreaking approach by using principles of quantum mechanics such as superposition and entanglement to process information in radically new ways. Using quantum mechanics to improve the efficiency of computation is the key area of interest with regards to quantum computing [9]. On the basis of Landauer's principle, all information is ultimately physical, in classical computing all data stored is in bit streams of 1s and 0s it should be recorded somewhere, either it be on paper or silicon. Since all atoms are composed of nuclei and electrons, these methods are used to store information in semiconductor material in the form of qubits [10]. The qubits by itself stores only one bit of information exactly as much as switchable circuits but using a phenomenon of quantum physics called entanglement the special properties are stored in it [11]. Instead of using high and low voltages to 1s and 0s of binary data it can represent a 0, a 1 or any proportion of 0 and 1 in superposition of both states with a certain probability of being a 0 and a certain probability of being a 1. Where classical computers use familiar silicon-based chips, qubits can be made from trapped ions, photons, artificial or real atoms, or quasiparticles. Depending on the architecture and qubits systems, some implementations need their qubits to be kept at temperatures close to absolute zero. Multiple Qubits can exhibit quantum entanglement [12]. Entangled qubits always correlate with each other to form a single system. Even when they are infinitely far apart, measuring the state of the other, without needing to measure it directly [13]. Entanglement is required for any quantum computation and it cannot be efficiently performed on a classical computer. Applications include factoring large numbers and solving search problems.

A computer whose memory is exponentially larger than its apparent physical size, a computer that can manipulate an exponential set of inputs simultaneously – a whole new concept in parallelism; a computer that computes in the twilight zone of Hilbert space is a quantum computer. Relatively few and simple concepts from quantum mechanics are needed to make quantum computers a possibility. The subtlety has been in learning to manipulate these concepts. Whether such a computer is inevitable or will be too difficult to build remains uncertain. This evolving form of computing offers significant potential in addressing various challenges faced in disaster management, where traditional computers often prove inadequate. By manipulating quantum parallelism, quantum computers can create complex simulations and improve decision-making processes that are crucial during disasters. Even though the potential of quantum computing changes swiftly across numerous dimensions, its application in the area of disaster management remains largely unexplored. Such awareness about how methods of quantum computing may be used to improve disaster response and recovery efforts has led this research to attempt to fill a gap [14]. It is hoped that, by probing into the feasibility and benefits of applying quantum computing in this context, it could help develop much more resilient, robust, and effective disaster management plans [15]. Quantum computing will, therefore, elevate disaster management to a whole new level with regard to computational power [16]. Among the areas most relevant and open to significant impact by quantum computing, such as high-fidelity simulations, are those that can contribute toward better preparedness in terms of disaster response [17]. Currently, computer systems are already powerful, but they face significant challenges in handling complex details at large scales to simulate actual disastrous situations [18]. For instance, computing the effects of a super hurricane or an earthquake involves highly diverse factors, including weather patterns [19]. Quantum computing not only improves the accuracy of simulations but also ensures that resource allocation and logistics in response to any catastrophe are optimized [20]. Classical optimization algorithms, such as linear programming and network optimization [21], are quite limited by the complexity and dynamism inherent in the nature of disasters [22]. Quantum algorithms, on the other hand, can handle millions more variables and constraints simultaneously. For example, the quantum approximate optimization algorithm (QAOA) applied to complex optimization problems in responding to disasters [23], which road route will be most effective for emergency vehicles and through which distribution channels the relief supplies will be received the quickest. In effect, quantum computing may look for the best strategies that support optimal resource allocation, reduce response times, and thus are significantly more impactful in efforts at relief. This is very valuable in situations where time matters, and any delay might result in huge losses.

This also means that quantum computing can leverage its processing velocity and precision of data in post-disaster analysis. Analysis of various kinds of data, from satellite images to drone footage to social media feeds, does make up most of the consequences of disasters. The processing speeds involved in classical data processing techniques are inadequate or unimaginative and may not even be able to capture all the variations in the data. Quantum computation operations enable the intense and high-speed computation of many complicated calculations that may potentially change the process of analyzing data. This, therefore, means that algorithms for machine learning will intensely analyze high-resolution pictures as well as large amounts of data for a better assessment of damage and needs. This, therefore, enhances more informative decisions and speeds up recovery further [24]. Another potential advantage that quantum computing can provide is in predictive models, which can be built for future disasters as historical data can eventually develop into recognizable patterns and trends that otherwise would have gone unnoticed with traditional methods. With the entry of quantum computing into disaster management, it may altogether transform the way in which challenging issues characterizing disaster responses and recoveries are approached. Increasing simulations, optimizing resource allocation, and enhancing data analysis will considerably improve capabilities in preparation for and recovery from disasters. Research and development in quantum computing will continue to ensure that advanced capabilities become more of a reality [25]. When quantum technology is properly developed, its application in disaster management will likely be highly important in terms of finding new avenues for building resilience and reducing the impacts of future disasters.

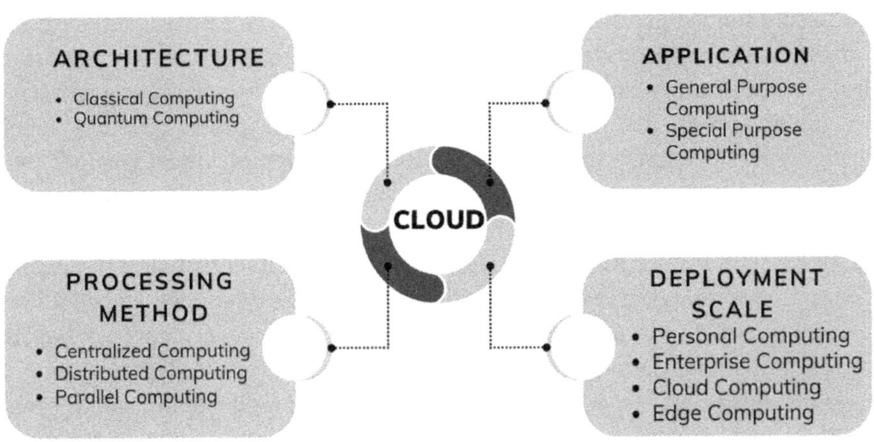

Figure 14.1: Types of computing.

Figure 14.1 classifies several forms of computing on the basis of four basic factors: architecture, application, method of processing, and scale of deployment. Each one of the following factors provides a unique viewpoint regarding classification and comprehension for computing systems [26]. Based on the architecture, computing is divided into two categories: quantum computing and classical computing. Classical computing is based on conventional binary logic where bits are used to stand for 0 s and

1 s. This kind of computing is the most widely used computing in servers and desktop computers and a combination of gadgets. Qubits can exist in more than one state at the same time in superposition due to the laws of entanglement and superposition; these are the building blocks of quantum computing. Quantum computing is going to be significantly advantageous in increasing processing power and speed for a several problems [27]. By application, computing can be categorized into general-purpose computing and special-purpose computing. General-purpose computing systems have multifunctional capabilities to perform an array of tasks. Examples are personal computers, which, running a variety of software for different applications, can do many things. Special purpose computing is customized in nature for specific tasks and optimized for particular functions. Examples include the embedded systems used in automotive control systems and the graphics processing units to render images. Processing method refers to how data is being handled and how the computations are being performed inside a system. Centralized computing takes place in one location. It usually occurs in the mainframe computer, where all data processing is done in a single system. On the contrary, distributed computing spreads data and processing tasks across many systems, referred to as nodes, which are mostly connected through a network. This method enhances processing power and reliability through multiple computing resources. Parallel computing refers to computing in which several processors perform computations at the same time. It can increase the speed of processing data considerably. It is quite common in high-performance computing environments. To this regard, scale of deployment would refer to how far the setting or computing resources range from small devices to large systems. By personal computing, what comes to mind is devices used by one person, like desktops, laptops, and mobile phones. Enterprise computing generally refers to large systems that organizations use in running businesses primally servers and data centers. Cloud computing provides computing resources on demand over the internet to make computing power scalable and flexible. Edge computing places computing resources close to where data will be coming from to make real-time processing possible and decrease latency, important in many applications requiring instant analysis of data.

14.2 Literature Review

Disaster management is the totality of actions that involve reduction, preparedness, response, and recovery efforts used to deal with natural as well as man-made disasters [28]. Effective disaster management starts with an accurate analysis of the risks and possible measures put in place for managing these disasters. It comprises the identification of problems and their respective vulnerabilities [29]. Technological innovations play a crucial role during the prediction, prevention, and response stages of disaster management. Efficient algorithms for data collection and analysis can help predict numerous disasters, thus enabling the implementation of efficient measures to deal with the disaster [30]. Infrastructural modifications can help in preventing these disasters or at least ensure minimal impact upon their occurrence. Advanced systems and tools also play a part in the response phase of the disaster. For instance, GIS permits spatial analysis and mapping of areas prone to disaster; proactive planning and emergency response are promoted. International dimensions acquire significant importance in the context of disaster management as practices vary significantly between countries and regions [31]. There are various comparative analyses and studies that underscore the approaches to disaster management across countries. These studies draw attention to the cultural, political, and institutional factors that influence procedures and implementations. Further, global cooperation and knowledge sharing are essential for an effective disaster management approach [32]. UN agencies, international organizations, and global partnerships are crucial in exchanging information, capacity building, and mobilizing resources across countries. From the computing standpoint, disaster management demands super-fast real-time decision-making under immense pressure [33], considering several variables and efficiently analyzing the huge amounts of incoming raw data. Developing efficient disaster management protocols requires vulnerability assessments and simulated training programs [34]. It also requires the establishment of clear and secure communication channels with high-level encryption. Large volumes of such data cannot be analyzed by traditional computing systems in order to obtain useful information on time to find optimized solutions to manage the disaster at hand [35]. They cannot support simulations that replicate disaster scenarios as realistically as they occur, nor can they establish communication channels that are fully immune to disruptions and interceptions. This brings us to the requirement of a stronger computing model in this field. This is where quantum computing comes into the picture. The potential that quantum computing holds in various fields is tremendous. For example, in material science, quantum computing models can be used in the production of much more accurate simulations. This would subsequently greatly quicken the discovery of novel materials with required properties. These can efficiently simulate the complicated interactions of materials and do not require experiments to be conducted in real time, therefore saving lots of resources, especially for experiments that would need rather specific conditions that are hard to replicate in the real world. Coupled with machine learning, quantum computing principles can analyze even the largest datasets much faster and more accurately as they can process information in superposition. Additionally, certain calculations that would be impossible with classical computers can be performed with a quantum computer. Quantum computing has also been found to have potential applications in financial modeling, where it can be used to significantly improve risk analysis, which will, in the long run, lead to portfolio optimization. It can also be put to use for establishing much more secure communication when used in cryptography. In disaster management, quantum computing has the potential to transform the various phases of disaster management. From early predictions before disaster strikes to response after disaster strikes, quantum computing could provide a big leap in the optimization of the overall process of disaster management. Optimization of resource utilization can also be done in disasters and rescue operations by using quantum computing algorithms. It may streamline critical tasks and even suggest human actions, like the order of the timely dispatch of emergency vehicles; thus, it can significantly improve the efficiency of the process and do so much faster. As discussed earlier, quantum computers may be applied to develop much more efficient and faster simulations. Such simulations could be used to plan evacuation procedures and assess what might happen in various scenarios, thus reducing the general response time. Such simulations can also predict and mitigate disasters much more accurately and promptly than would have been possible without them. Used in conjunction with machine learning algorithms, they can predict extreme weather conditions with very high accuracy. Quantum computing principles can be applied to establish highly secure communication channels free from disruptions and interceptions, thus creating avenues for efficient information sharing between the responding teams and the public during disasters. In the recovery phase after the disaster, quantum computers can process large amounts of data much faster than conventional computers and determine the best reconstruction strategies. The analysis of such enormous data in a timely manner, along with factors like damage to the infrastructure and population displacement, could help identify areas of maximum need for reconstruction and plan accordingly, rather than going through the reconstruction phase in an unplanned manner. Thus, it becomes absolutely important that quantum computing power be rightly applied to disaster management as a transformative tool. Real-time data synthesis and analysis play a great role in disaster management, and quantum computing should enable record stream transformations – from satellite imagery to weather forecasts to social media updates and emergency analysis. This capability can dynamically change disaster management plans with the arrival of newly available records, ensuring choices reflect the latest detailed statistics. It allows for much greater efficiency and the potential to alter disaster management plans through real-time processing and consolidation of this information. Series of records makes less time for selection and planning. Besides, quantum computing should accelerate the pace of predictive methodology development by simulations of the high accuracy and scale driven by quantum algorithms. These kinds of models have to be more inclusive of variables with sophisticated interactions than conventional models and, therefore, provide better predictions, such as an earthquake or flood, where complex quantum simulations are required to model the detailed geology and special hydrological process elements. The nature of this standardization phase will spur preparedness, thus resulting in effective development of risk mitigation strategies that will save lives and minimize long-term financial loss.

14.3 Problem Statement

Quantum computers can be used not only in business situations but also as a disaster remedy. In the event of a disaster such as an earthquake or tsunami, it is important to evacuate to safety as soon as possible. However, the concern is that certain areas may become crowded with evacuees, creating a dangerous situation. The incorporation of quantum computers is expected to establish ideal evacuation paths for residents. The combination of multiple simulations along with situational awareness intelligence helps emergency management professionals make more comprehensive response plans. Applied quantum computing can be improved by quantum-enhanced AI. It could be used for mission-scale simulations that provide real-time or near real-time analysis for ideal preparedness. For example, a company in Japan developed a quantum-hybrid application to facilitate planning related to tsunami response. Disasters pose a significant threat to communities worldwide, causing loss of life, property damage, and disruption to the economy. While traditional disaster management approaches have improved, there is still a critical need for more successful prediction, efficient resource allocation, and optimized response strategies. This research looks into the potential of quantum computing to reduce these challenges and revolutionize disaster management. The main problem lies in making use of the unique capabilities of quantum computers, like superposition and entanglement, to overcome disadvantages in traditional computing and significantly improve disaster preparedness, response, and recovery efforts.

Specific research questions that would be answered include:

- How might quantum computing be applied to improve the accuracy and timeliness of disaster predictions, specifically in weather forecasting and earthquake simulation?
- Can quantum computing algorithms improve the process of resource distribution and logistics during a disaster so that the important supply lines reach the affected areas efficiently?

By exploring these questions, this chapter aims to bridge the gap between theoretical potential and practical application. The ultimate goal is to establish a framework for utilizing quantum computing in disaster management, ultimately leading to a more resilient future for communities facing natural hazards. Q-CTR claims to use quantum techniques such as quantum gravimetry, quantum magnetometry, and quantum-enhanced logistics optimization to deliver previously impossible insights and solve previously intractable problems. Last year the firm announced a partnership with Australia's Department of Defence to develop quantum sensors that will deliver quantum-assured navigation capability for military platforms.

In Figure 14.2, quantum noise detector array will form the first stage, which is bound to capture the data. The term "quantum noise" informs the reader that the data may have originated from a quantum system introducing intrinsic noise.

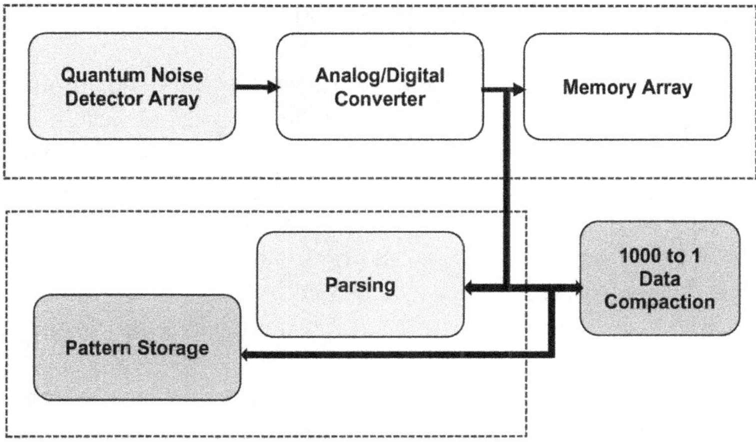

Figure 14.2: Quantum computing working architecture.

Analog/Digital Converter: The captured data, presumably in analog form, will be converted into digital format for further processing, hence fitting into most computational systems of today.

Memory Array: The digitized data is then stored in a memory array. It can serve as a buffer to soak up data rates or as a staging area for further processing.

Parsing: At this stage, the stored data is probably analyzed, with possible extraction of meaningful patterns or features. The type of parsing depends on the nature of the data and the objectives of the system.

Data Compaction: 1,000 to 1: This clearly is a major step of data reduction. The system will compress this data and reduce it to nearly one-thousandth of its size. This becomes very necessary in terms of storage, and sending or receiving this data efficiently.

Pattern Storage: The final stage is storing the compressed and probably processed data for future use or analysis. This could be long-term storage or an archive.

14.4 Proposed Methodology

With disasters becoming more frequent in the current situation, both natural and man-made disasters, management is one of the key aspects to maintain and coordinate efficiently. To solve these obstacles, solid decision-making procedures backed by thorough data analysis are required. The huge volumes and complexity of data building up during a crisis is not compliable for traditional computing, whose frequency limits speed and accuracy of responses. The baseline foundation of computational problem-solving and algorithms has become nondispensable instruments for increasing recovery rate and response to disasters. This is an essential tool for disaster management at every level, from anticipating disasters to effectively utilizing limited resources. The algorithms will reduce loss of life and property by improving through pattern understanding of real-time and historical data. But the need for highly efficient computational methods is developing as disasters are getting bigger and more complex. In this sense, quantum computing offers a promising horizon with capabilities that make the possibility of ideal catastrophe management. Quantum computers use quantum bits or qubits, in contrast with classical computers, which processes data using bits (0 or 1). Because these qubits may acquire the value of numerous states at once, it can analyze huge amounts of data and carry out intricate processing of data, in less time.

The area of disaster management is still in its infancy when it comes to implementing it with quantum computing, and ongoing research is mainly concentrated on creating and improving quantum algorithms that are suited to these particular problems. To fully utilize this forward-thinking technology, maximum cooperation between researchers, emergency response organizations and specialists in quantum computing is necessary.

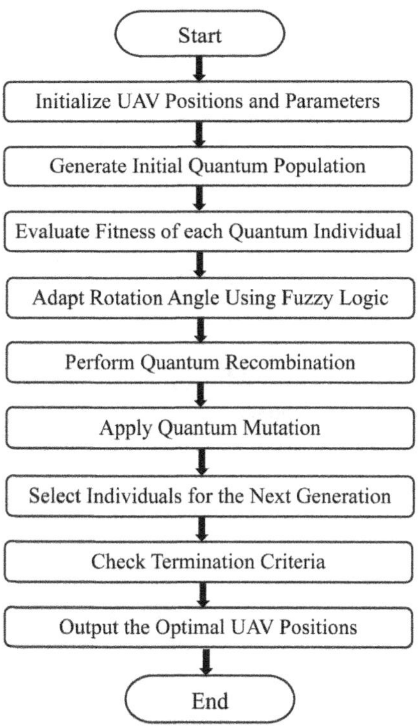

Figure 14.3: Proposed flow architecture diagram.

The approach that we will discuss here is the hybrid quantum genetic algorithm with fuzzy adaptive rotation angle (HQGAFARA). It is designed to optimize the placement of unmanned aerial vehicles (UAVs) in disaster-stricken areas. It uses the Deutsch-Jozsa quantum circuit to blend classical and quantum computing methods to produce quantum populations that function as haploid recombination and mutation operators. The Deutsch-Jozsa quantum circuit is used in the context of HQGAFARA to efficiently generate all possible solutions at high speeds, and thus assisting in the investigation of several choices for drone placement in disaster zones as in Figure 14.3. In genetics, haploid regeneration denotes the vivid presence of a single pair of chromosomes, while recombination will be the process of combining genetic elements to form different combinations. In order to generate new and potentially better placement options, HQGAFARA combines various potential solutions using a version of quantum mechanics, some genetic algorithms use techniques called mutation operators, these operators create random modifications to solutions generated, which will facilitate the discovery of new possibilities. These mutation operators add variations to drone placement solutions within HQGAFARA, so it can continue exploring more possibilities and avoid being trapped on a less-than-ideal solution. This approach makes use of quantum entanglement to offer features that are both exploratory and extensive, to improve the algorithm's capacity to generate innovative approaches. A phenomenon known as quantum entanglement occurs when two particles get entangled to a point where their sates can easily influence one another due to the distance between both. Quantum entanglement is employed in HQGAFARA to map several regions of solution space, facilitating this investigation of several solutions would increase efficiency.

In order to adjust the quantum gate rotation angle, a fuzzy heuristic is included, providing flexible and efficient disaster management with complex settings. Heuristics are a type of rules of thumb for addressing problems, and fuzzy logic works with approximations of reasoning as opposed to rigid ones. HQGAFARA's fuzzy heuristic allows the algorithm to dynamically modify and improve by adjusting the rotation angle of quantum gates according to the use cases. Similar to logic gates in classical computing, quantum gates are the fundamental components of quantum circuits. These rotation angles determine how the gates are changed to alter quantum bits, or qubits.

14.5 Results and Discussion

The experimental setup for evaluating the HQGAFARA algorithm comprises tools such as Qiskit for the simulation of the quantum circuit, Python for overall algorithm management, and MATLAB for additional computational and visualization tasks. The experiment can be carried out on a Windows 11 operating system with an NVIDIA GeForce RTX 3080 GPU running CUDA Version 11.6 to exploit the computational capacity. The dataset used in this experiment includes many different simulated disaster cases, which we use to represent various complications, and further, the testing of this algorithm will be done in a very comprehensive way.

Table 14.1: Experimental setup parameters.

Parameters	Values
Dataset size	Varies depending on the simulated disaster scenario
Software	Qiskit, Python, MATLAB
Programming language	Python 3.10.11, Qiskit 0.36.1
Packages	NumPy, Scipy, Qiskit, PyQtGraph, Matplotlib
Operating system	Windows 11
GPU chip	NVIDIA GeForce RTX 3080, CUDA Version 11.6

Table 14.1 deliberately outlines the experimental setup. Software packages NumPy, SciPy, Qiskit, PyQtGraph, and Matplotlib are the tools needed for scientific computation, quantum algorithm implementation, plotting UAV placement, and the presentation of results, respectively. The programming environment is developed using Python version 3.10.11 to ensure proper compatibility with the required libraries and frameworks. Therefore, HQGAFARA is subjected to numerous tests against a classic genetic algorithm, where it is primarily evaluated based on factors such as accuracy, precision, and efficiency of the obtained UAV coverage in a network of obstacles.

Table 14.2: Proposed level comparison.

Model	Accuracy	Precision	Recall	F_1 score
HQGAFARA	0.90	0.88	0.94	0.91
CGA	0.82	0.79	0.87	0.83

The comparison between the HQGAFARA procedure and the classical genetic procedure (CGA) reveals significant effectiveness and improvement with HQGAFARA across all evaluated metrics, as shown in Table 14.2. HQGAFARA achieves higher truth preciseness and F_1 score compared to CGA. Specifically, HQGAFARA's precision of 0.90 surpasses CGA's 0.82, indicating that HQGAFARA is more effective in correctly identifying both positive and negative instances. These results lead to a higher F_1 score of 0.91, reflecting a well-balanced effectiveness between precision and recall. Figure 14.4 illustrates the accuracy and precision level comparison for CGA and HQGAFARA.

Figure 14.5 deliberately illustrates the recall and F_1 score level comparison for the existing and proposed models. In practical terms, the improved effectiveness metrics of HQGAFARA suggest it is a more reliable and effective tool for scenarios requiring high precision and recall, such as disaster management. The hyperbolic truth and stable F_1 account mean that HQGAFARA not only improves general anticipation characteristics but also ensures that less important instances are not lost, minimizing incorrect bold

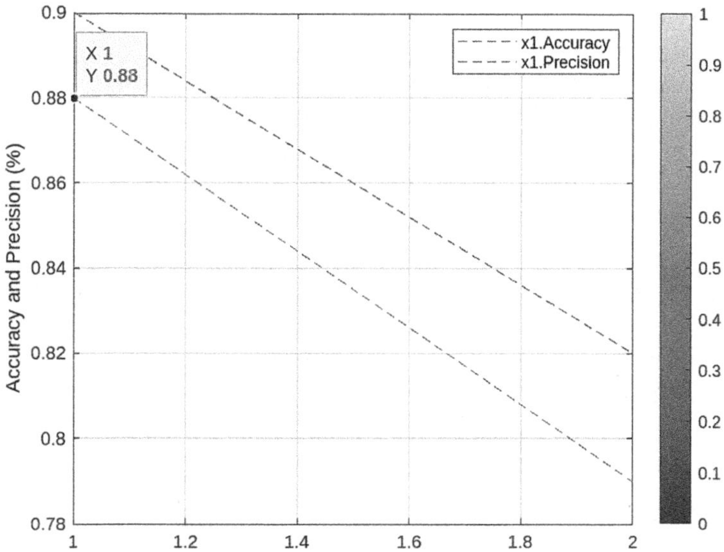

Figure 14.4: Accuracy and precision level comparison.

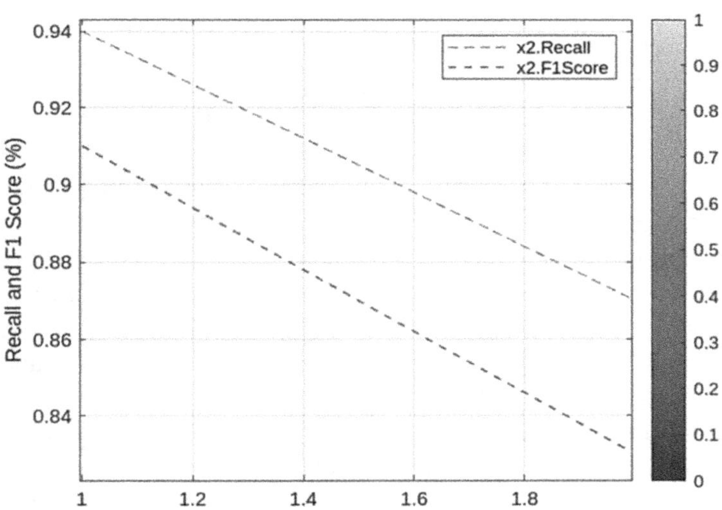

Figure 14.5: Recall and F_1 score level comparison.

identifications. This can lead to more accurate and timely decision-making in challenging applications, making HQGAFARA a superior choice over CGA for complicated optimization problems where both precision and comprehensive coverage are essential.

14.6 Conclusion

While it opens up possibilities where traditional computation falls short, quantum computing could fundamentally change the way disaster management is conducted. The results of this study highlight how quantum algorithms can improve real-time decision-making, optimize resources, and predict disasters. A combination of quantum computing and classical systems in disaster management will ensure a sharp improvement in the accuracy and timeliness of general predictions, while in simulations that involve weather and seismic activities in particular, quantum computing guarantees very fast processing of enormous volumes of data and complex calculations hitherto not possible with any classical system. Quantum computers inherit two special properties: quantum parallelism and entanglement.

14.7 Future Direction

It allows, for example, good management in quantum-enhanced AI and optimization algorithms under conditions such as the distribution of essential supplies at the right time during emergency situations to help reduce response time and mitigate the effects of disasters on affected populations. Quantum-classical hybrid systems, the integration of quantum computing into disaster management systems, offer an opportunity for a significant rise in the accuracy and timely delivery of disaster predictions, especially in weather forecasting and simulations of seismic activities. With properties like quantum parallelism and entanglement, quantum computers can handle massive amounts of data at very high speeds and carry out complex calculations that were once impractical using classical systems. It reduces the period of aid delivery and lessens the burden on affected communities by allowing efficient management through quantum-enabled AI and optimization algorithms in cases of distributing essential supplies at the right time to emergency sites. Methods from hybrid quantum-classical approaches, such as HQGA-FARA, can explain the pragmatic use of quantum computing in crisis and realistic scenarios through the positioning of UAVs among other logistical issues. Application of such breakthroughs in quantum computing paradigms to disaster management is still, however, in its infancy. More research in the field of quantum computing is needed, with coordination among experts in the said field and other professionals in crisis management, and even policymakers for translating these visions into reality. Scalable, robust quantum algorithms can be developed and integrated within frameworks that already exist for disaster management. These will constitute important milestones for achieving the goal of resilient communities. This is the case with quantum computing in disaster management: the difference it makes between life and death, economic loss, and enhanced societies' resiliency. The more the advance of quantum technology is pushed, the more viable this powerful technology will be applied to disaster management, enabling a prospect of communities coping with and recovering from disasters.

References

[1] A. Khan, S. Gupta, and S. K. Gupta, "Multi-hazard disaster studies: Monitoring, detection, recovery, and management, based on emerging technologies and optimal techniques," International Journal of Disaster Risk Reduction, vol. 47, p. 101642, 2020.
[2] D. A. Wilhite, M. V. Sivakumar, and R. Pulwarty, "Managing drought risk in a changing climate: The role of national drought policy," Weather and Climate Extremes, vol. 3, pp. 4–13, 2014.
[3] D. A. McEntire, Disaster response and recovery: strategies and tactics for resilience. Hoboken, NJ, USA: John Wiley & Sons, 2021.
[4] Q. Liu, V. Hagenmeyer, and H. B. Keller, "A review of rule learning-based intrusion detection systems and their prospects in smart grids," IEEE Access, vol. 9, pp. 57542–57564, 2021.
[5] S. Gupta, S. Modgil, A. Kumar, U. Sivarajah, and Z. Irani, "Artificial intelligence and cloud-based collaborative platforms for managing disaster, extreme weather, and emergency operations," International Journal of Production Economics, vol. 254, p. 108642, 2022.
[6] D. T. Bui, T. A. Tuan, H. Klempe, B. Pradhan, and I. Revhaug, "Spatial prediction models for shallow landslide hazards: A comparative assessment of the efficacy of support vector machines, artificial neural networks, kernel logistic regression, and logistic model tree," Landslides, vol. 13, pp. 361–378, 2016.
[7] K. Muhammad, J. Ahmad, and S. W. Baik, "Early fire detection using convolutional neural networks during surveillance for effective disaster management," Neurocomputing, vol. 288, pp. 30–42, 2018.
[8] M. Coccia, S. Roshani, and M. Mosleh, "Evolution of quantum computing: Theoretical and innovation management implications for emerging quantum industry," IEEE Transactions on Engineering Management, vol. 71, pp. 2270–2280, 2022.
[9] H. A. Bhat, F. A. Khanday, B. K. Kaushik, F. Bashir, and K. A. Shah, "Quantum computing: Fundamentals, implementations and applications," IEEE Open Journal of Nanotechnology, vol. 3, pp. 61–77, 2022.
[10] A. Chatterjee et al., "Semiconductor qubits in practice," Nature Reviews Physics, vol. 3, no. 3, pp. 157–177, 2021.
[11] M. R. Wasielewski et al., "Exploiting chemistry and molecular systems for quantum information science," Nature Reviews Chemistry, vol. 4, no. 9, pp. 490–504, 2020.
[12] R. K. Vaddy, B. Dhamodharan, and A. Jain, "Quantum computing applications in real-time route optimization for supply chains," in Quantum Computing and Supply Chain Management: A New Era of Optimization, IGI Global, 2024, pp. 113–124.
[13] M. Möller and C. Vuik, "On the impact of quantum computing technology on future developments in high-performance scientific computing," Ethics and Information Technology, vol. 19, pp. 253–269, 2017.
[14] A. Majot and R. Yampolskiy, "Global catastrophic risk and security implications of quantum computers," Futures, vol. 72, pp. 17–26, 2015.
[15] Arunmozhiselvi, S., T. Ananth Kumar, P. Manju Bala, S. Usharani, and G. Glorindal. "A systematic approach to agricultural drones using a machine learning model." In Machine Learning Approaches and Applications in Applied Intelligence for Healthcare Data Analytics, pp. 41–60. CRC Press, 2022.
[16] O. O. Olatunji, P. A. Adedeji, and N. Madushele, "Quantum computing in renewable energy exploration: Status, opportunities, and challenges," in Design, Analysis, and Applications of Renewable Energy Systems, 2021, pp. 549–572.

[17] B. Merz et al., "Impact forecasting to support emergency management of natural hazards," Reviews of Geophysics, vol. 58, no. 4, p. e2020RG000704, 2020.
[18] N. Agrawal and N. Agrawal, "Defining natural hazards–large scale hazards," in Natural Disasters and Risk Management in Canada: An Introduction, 2018, pp. 1–40.
[19] P. Sajwan and N. Jayapandian, "Challenges and opportunities: Quantum computing in machine learning," in 2019 Third International Conference on I-SMAC (IoT in Social, Mobile, Analytics and Cloud) (I-SMAC), IEEE, 2019, pp. 598–602.
[20] V. Topno, T. Kundu, and M. K. Dehury, "Role of quantum computing in government and the defence sector," in Digital Technologies in Modeling and Management: Insights in Education and Industry, IGI Global, 2024, pp. 296–312.
[21] F. Bova, A. Goldfarb, & R.G. Melko, "Commercial applications of quantum computing," EPJ quantum technology, vol. 8, no. 1, pp. 1–15, 2021.
[22] L. Fan and Z. Han, "Hybrid quantum-classical computing for future network optimization," IEEE Network, vol. 36, no. 5, pp. 72–76, 2022.
[23] R. Fakhimi and H. Validi, "Quantum approximate optimization algorithm (QAOA)," in Encyclopedia of Optimization, Cham: Springer International Publishing, 2023, pp. 1–7.
[24] J. Preskill, "Quantum computing 40 years later," in Feynman Lectures on Computation, CRC Press, 2023, pp. 193–244.
[25] A. Saji, A. R. Thomas, A. M. Benny, and N. Jayapandian, "Rescue operation with RF pose enabled drones in earthquake zones," in 2023 International Conference on Sustainable Computing and Smart Systems (ICSCSS), IEEE, 2023, pp. 1333–1338.
[26] V. B. Kasyap and N. Jayapandian, "The world of communication & computing platform in research perspective: Opportunities and challenges," in 2021 3rd International Conference on Signal Processing and Communication (ICPSC), IEEE, 2021, pp. 289–293.
[27] S. S. Gill et al., "Quantum computing: A taxonomy, systematic review and future directions," Software: Practice and Experience, vol. 52, no. 1, pp. 66–114, 2022.
[28] P. K. Das, "A study in disaster management," Journal of Emerging Trends in Economics and Management Sciences, vol. 14, no. 2, pp. 76–83, 2023.
[29] A. Suppasri et al., "Cascading disasters triggered by tsunami hazards: A perspective for critical infrastructure resilience and disaster risk reduction," International Journal of Disaster Risk Reduction, vol. 66, p. 102597, 2021.
[30] M. Erdelj, E. Natalizio, K. R. Chowdhury, and I. F. Akyildiz, "Help from the sky: Leveraging UAVs for disaster management," IEEE Pervasive Computing, vol. 16, no. 1, pp. 24–32, 2017.
[31] K. Uddin and M. A. Matin, "Potential flood hazard zonation and flood shelter suitability mapping for disaster risk mitigation in Bangladesh using geospatial technology," Progress in Disaster Science, vol. 11, p. 100185, 2021.
[32] L. Elmhadhbi et al., "An ontological approach to enhancing information sharing in disaster response," Information, vol. 12, no. 10, p. 432, 2021.
[33] N. Chen et al., "Application of computational intelligence technologies in emergency management: A literature review," Artificial Intelligence Review, vol. 52, pp. 2131–2168, 2019.
[34] B. D. Williams and G. R. Webb, "Vulnerability and disaster: Practitioner strategies for addressing the needs of vulnerable populations," Journal of Homeland Security and Emergency Management, vol. 17, no. 2, p. 20180063, 2020.
[35] C. P. Chen and C. Y. Zhang, "Data-intensive applications, challenges, techniques and technologies: A survey on Big Data," Information Sciences, vol. 275, pp. 314–347, 2014.

Rishav Srivastav, Khyati Sinha, Aryaman Gautam, Niha Kamal Basha, and Christo Ananth

15 Quantum Cryptography for UAVs: Enhancing Security and Collision Avoidance in Next-Generation Aerial Systems

Abstract: Unmanned aerial vehicles (UAVs) have accordingly been successfully integrated into different industries, including military, healthcare, and education, because of their diverse applications and sophisticated capabilities. Nevertheless, UAVs are increasingly confronted with significant security and operational issues related to communications and collision avoidance. This chapter will discuss in detail the convergence of quantum cryptography with UAVs for enhanced security against these problems. Starting with a critical examination of the literature on the subject at hand, the paper traces the historical evolution of cryptography and discusses the present trends in UAV technology. The existing security standards are discussed, including their shortcomings, paving the way for the need for stricter ones. The principles of quantum cryptography, including quantum key distribution (QKD), are briefly summarized below in relation to classical cryptography. In the case of UAV communications, challenges include uniqueness threats, the real-time context, and the recently emerging problem of collision, all of which should be addressed. The proposed layered architecture here will include quantum cryptography to guarantee data security while enhancing communication efficiency. This will provide a framework for future development by explaining in detail the design, implementation, and security mechanisms of the proposed solution. Its performance will be evaluated in various case studies of military, healthcare, and education environments. The performance evaluation parameters have been addressed in detail to emphasize the strengths of the system and areas for improvement. The research raises the newly arising collision problem with UAVs and offers quantum solutions to improve the collision avoidance mechanism. This chapter will describe in depth how quantum cryptography merges with UAVs for increased protection against these issues. Beginning with a critical study of the available literature on the topic at hand, the paper follows the historical development of cryptography and discusses the current trends in UAVs. The prevailing security standards are discussed, along with their limitations, leading to the necessity for stricter ones. The quantum cryptographic principles, such as QKD, are briefly summarized below as applied to classical cryptography. Uniqueness threats, the real-time context, and the recently emerging problem of collision are challenges that should be addressed in the case of UAV communications.

https://doi.org/10.1515/9783111619323-015

Keywords: Collision avoidance mechanism, quantum cryptography, unmanned aerial vehicles (UAVs), quantum key distribution (QKD), UAV technology, layered architecture, UAV communication

15.1 Introduction

Unmanned aerial vehicles (UAVs) are widely used in various sectors, ranging from healthcare and logistics to agriculture and the armed forces, as a result of advancements in today's technology. UAVs, or drones, are utilized for various purposes such as monitoring crop activity, delivering commodities, spying, and providing disaster relief [1]. UAVs enable us to perform perilous operations across distant areas without a significant loss of human lives. They are an asset of great worth to society in the modern era. The challenge posed by mounting dependence on UAVs is communication security. The data that UAVs transmit and receive is generally sensitive, and interception or modification could lead to disastrous, apocalyptic consequences. The increasing amount of cybercriminal activity, particularly with the emergence of quantum computers that can render most existing encryption algorithms obsolete, is a concern running rampant. The central problem is how susceptible UAV communications are to a wide array of cybercrimes, including snooping, signal jamming, and web spoofing. These are some examples of cybercrimes that compromise the overall integrity and availability of transmitted data. Some of the currently available controls do not offer the desired guarantees and are at risk of being ineffective against skilled adversaries. These finer-grained systems communicate with contemporary quantum technologies:
1. The susceptibility of UAV communication systems to hacking and eavesdropping.
2. The limitations of current cryptographic methods in securing UAV communications in the era.
3. The demand for enhanced security features with solid protection against conventional and next-gen attacks remains high.

The principal goals of this study are: (1) UAV communications security assessment – Investigate the cryptographic techniques used in UAV communications and their weakness and limitations. (2) Next-generation cryptographic solutions – Explore next-generation technologies (e.g., quantum cryptography) for improving UAV communication security. (3) Design a security framework developing an end-to-end security solution integrating state-of-the-art cryptographic techniques with conventional methods. (4) Solutions testing – Examining the proposed security system through simulation and real-case application to confirm that it suits the demands of real-time communication of UAVs.

15.2 Existing Work

15.2.1 Historical Evolution of Cryptography

Cryptography is a set of policies and practices that have shifted over centuries to adapt to communication strategies. It encompasses a range from ciphers such as substitution ciphers and the Caesar cipher employed by the Romans. During the Roman domination, much confidential information was transferred, and it had to be secured using contraction methods, thus substitution ciphers was the foundation of maintaining secret information confidential [2]. The Arab world cryptologist Al-Kindi developed the evolution of frequency analysis as a way to crack codes through letter frequencies during the ninth century. This, like all things, has advantages and with it, another advancement in cryptographic methods. Germany's utilization of the Enigma machine in World War II was one of the largest investigations performed in a world of cryptography. The quantity of work the Allied forces had to crack the Enigma code was crazy, and the devastation that emerged from Bletchley Park was significant to Britain's domination in the war. The creation of computer devices in the mid-1920s introduced new and innovative methods of encryption algorithms making it far more robust. The development of public key algorithms entered the spotlight due to Shannon's theory, the center of the new generation of cryptography.

15.2.2 Advancements in UAV Technology

UAVs have come a long way, from remotely controlled basic aircraft to advanced systems that can carry out autonomous missions. UAVs were initially used mainly for military reconnaissance and target practice. Early UAVs were quite basic and required direct human control. But with technological developments in sensors, computing, and AI, autonomous flight has become a reality. Current UAVs can execute sophisticated missions in the form of surveillance, mapping, and logistics transport with minimal or no input from humans. Communications in real time have progressed by leaps and bounds, with UAVs sending high-definition video and sensor feeds to ground stations continuously. New technologies like 4G/5G networks and satellite communications provide the necessary bandwidth and low latency that these missions need. Currently, UAVs are used in a variety of applications such as agriculture for monitoring crops and the application of pesticides, logistics for the delivery of goods, and emergency response for search and rescue operations. The various applications call for strong and secure communication systems to ensure mission success.

15.2.3 Existing Security Solutions and Limitations

The security of UAV communication systems is of utmost importance due to their critical uses. Solutions available include encryption, secure communication protocols, and new technologies such as quantum cryptography. Conventional encryption techniques like AES and RSA are commonly used to encrypt UAV communications. These techniques use mathematical algorithms to secure data from unauthorized parties. But they are susceptible to being attacked with quantum computers that can decrypt these encryptions by efficiently solving difficult mathematical problems [4]. Quantum key distribution (QKD) provides an added level of security by applying the principles of quantum mechanics. QKD is such that any eavesdropping effort will be detectable; hence, communication interception will become very hard to do without detection. Though promising, QKD exists in an experimental stage and comes with challenges that include implementation as well as scalability. Blockchain offers the potential of increasing the security of UAV missions by ensuring the existence of a tamperproof record of transmitted data. Blockchain can ensure unaltered communications logs and hinder data tampering. The interfacing of blockchain with UAVs is complex, and a lot must still be done work on that to be able to scale or perform at par. Intrusion detection systems (IDS) can also be used to scan UAV communication networks for malicious activities, which may be early signs of cyberattacks. By analyzing patterns of data, machine learning algorithms can identify possible threat sources. The IDS is well at finding established attack vectors, although new threats sometimes prove difficult to identify, and it has to be updated frequently with new detection algorithms.

15.3 Quantum Cryptography Fundamentals

Quantum cryptography relies on the principles of quantum mechanics in order to encrypt communication channels such that the integrity and confidentiality of information can be safeguarded against attacks of a classical or quantum nature. In this section, the various mechanisms are discussed in comparison to classical methods and their advantages.

15.3.1 Principles of Quantum Mechanics

In their turn, quantum cryptography rests mostly on aspects surrounding quantum mechanics. For them, wave-particle duality is a concept where particles such as electrons and photons possess wave-like properties and particle-like properties too. This duality is vital for the operation of quantum cryptographic protocols [5]. (1) Superposition: A quantum system exists in many states. For instance, a photon can be in a superposition of horizontal and vertical polarization states. (2) Entanglement: Entangled particles share a state such that the state of one (no matter the distance) instantly influences is crucial for certain QKD protocols. This underpins security of quantum cryptography by ensuring that any eavesdropping attempt will introduce detectable anomalies.

15.3.2 Quantum Key Distribution (QKD)

QKD is a technique for securely distributing cryptographic keys using quantum mechanics. The most notable QKD protocol is BB84, proposed by Bennett and Brassard in 1984. Here's how QKD works, and the diagrammatic representation is shown in Figure 15.1.

Quantum Communication:
(1) Preparation and Transmission: Alice prepares quantum states (typically photons) to encode random bits. These states are transmitted through a quantum channel to Bob.
(2) Discrete-Variable QKD: Uses discrete quantum states such as the polarization states of photons (e.g., $|0\rangle$, $|1\rangle$).
(3) Continuous-Variable QKD: The continuous quantum variables used are quadrature amplitudes of light modes.

Classical Post-Processing: Further Bob's reception of quantum states, classical post-processing is performed by both parties for the following:
(1) Parameter Estimation: To check the integrity of the quantum channel, Alice and Bob estimate parameters, that is, noise levels and transmission efficiency.
(2) Error Correction (EC): Reconciles any inconsistency that has arisen in raw key strings due to noise or possible eavesdropping. (3) Privacy Amplification (PA): Alice and Bob subsequently truncate their key by hashing in order to minimize the actual information that could have been eavesdropped, thereby ensuring the security of the remaining key.

Reconciliation Techniques: These decomposition reconciliation techniques are adopted based on the protocol:
(1) Direct Reconciliation (DR): Bob estimates Alice's data based on his measurements.
(2) Reverse Reconciliation (RR): Alice estimates Bob's data; usually with this technique applied for continuous-variable QKD.

Sifting: Alice and Bob sift the data to retain only that information which was measured in compatible bases; anything else is discarded. This step is vital to ensure that both parties actually share the same key.

Entanglement-Based Protocols: Some QKD protocols depend on the presence of shared entangled quantum states by Alice and Bob. One such instance would be the E91 protocol, which uses the violation of Bell's inequalities as a security certificate. The mathematical details can be found in the following equations :

$$\text{Error rate calculation: } Q = N_e / N \tag{15.1}$$

Here, Q is the quantum bit error rate, N_e is the number of errors, and N is the total number of bits:

$$\text{Privacy amplification:} \quad K\_\text{final} = K_{raw} * (1 - H(e)) \tag{15.2}$$

Here, K_final is the final key length, K_{raw} is the raw key length, and $H(e)$ is the binary entropy function of the error rate.

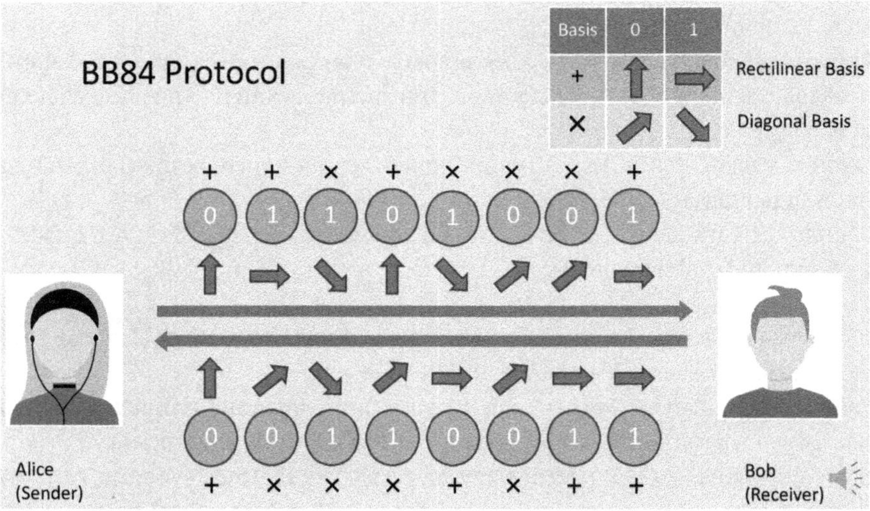

Figure 15.1: BB84 QKD protocol diagram.

Table 15.1: Comparison of classical and quantum cryptography.

Feature	Classical cryptography	Quantum cryptography
Security basis	Computational difficulty	Quantum mechanics
Vulnerability to quantum attacks	High	Low
Eavesdropping detection	Indirect and delayed	Direct and immediate
Key distribution method	Public key algorithms	Quantum key distribution (QKD)

15.3.3 Advantages over Classical Cryptography

Quantum cryptography has many advantages over classical cryptographic systems, and they include the following: (1) Unconditional security – Classical cryptography relies on the infeasibility of solving certain mathematical problems, which is subject to change under algorithmic improvements or developments in quantum computing, whereas quantum cryptography is furnished with unconditional security. (2) Eavesdropping detection – Eavesdropping attempts on quantum keys create disturbances observable by any user due to the no-cloning theorem and the superposition principle. This property allows honest users to detect and thwart any attempts of eavesdropping. (3) Atomic resistance – PHP on atomic resistance – Quantum cryptography is forcibly resistant and will resist attacks even from quantum computers that are able to break classical cryptosystems. In summary, it promises to give security against future attacks for a long time. Such attributes are highlighted for ease of reference in Table 15.1. Generally, quantum cryptography, primarily QKD, serves as a backbone of security and a formidable resistance to newer challenges. The system is constantly overcoming practical issues that hinder mass implementation.

15.4 UAV Communication Challenges

This transformational technology in the field of UAVs has found application in a multitude of highly diversified sectors such as agriculture, defense, and MedTech. However, UAVs almost completely depend on robust communication systems for fulfilling real-time demands and countering security threats to be efficiently operated. The real-time communication necessity that UAVs require, as well as the security threats they face, will be discussed herein. The requirements have been explained in Table 15.2. Successful real-time connection is very vital for UAV operation, and this comprises: (1) Low latency: The communication scheme should have little to no delay in cases where an instant response is required, for example, military missions or emergency services. As an example, there are promises to achieve 1-millisecond latencies for such types of critical operations in 5G networks. (2) High bandwidth: With the majority of data transmission performed by unmanned autonomous vehicles, this includes the likes of high-definition video and sensor data. Such high data volumes need to have reliable communication channels to support them. The crop-scanning drones in agriculture and air-quality monitoring drones, for instance, yield considerable data that should be transferred in real time over communication channels. (3) Reliability: Reliable and consistent communication links are crucial, especially for cases where operations are very critical, such as in healthcare or military reconnaissance. Reliable communication ensures that all UAVs maintain continuous links with either ground control stations or with other UAVs, avoiding a mission loss of control or data that could have otherwise endangered the operation. (4) Scalability: The communication infrastructure should be scalable and capable of operating with many drones at the same time without degrading performance, as UAV operations will have an increased scope. This is especially important in urban air mobility initiatives or extensive agricultural monitoring, where multiple UAVs are likely to operate closely together [6]. (5) Security: Because of the sensitive nature of much of the data, and because many applications of UAVs are so critical, it is necessary that they have secure channels of communication.

Table 15.2: Real-time communication requirements for UAVs.

Requirement	Description	Example applications
Low latency	Minimal delay for immediate feedback	Military operations and emergency response
High bandwidth	Large data transmission capability	Agricultural imaging and environmental monitoring
Reliability	Stable communication links	Healthcare delivery and military surveillance
Scalability	Accommodation of multiple UAVs	Urban air mobility and large-scale agriculture
Security	Protection against cyberthreats	All applications

Currently, UAV communication systems face a multitude of security threats, including: (1) Eavesdropping, which is the unauthorized access to transmitted data between UAVs and control stations and may lead to unauthorized exposure of sensitive information, such as recording live video streams from a surveillance UAV, revealing significant details regarding a secure facility or operation. (2) Jamming and spoofing: Jamming refers to interference in the communication channels of UAVs, while spoofing involves sending signals to mislead the UAV. Both can cause a failure or crash of the UAV. One of the most common forms of this attack is GPS spoofing, in which fake GPS signals deceive the UAV into believing it is in a different location. (3) Man-in-the-middle (MITM) attacks between UAV and control ground station: The attacker intercepts and possibly modifies the communication between the UAV and ground control station and may, therefore, gain control of the UAV. This can cause hijacking of the UAV, transmission of false commands, or manipulation of the data being communicated. (4) Data tampering: It could also mean unauthorized tampering with data communicated by UAVs leading to false or misleading information in the hands of the operators. Tampering surveillance data, for instance, might be disastrous as it could lead to wrong assessment of a situation and thus inappropriate responses.

Here are the following details which are secured by these measures: (1) Encryption: UAVs are protected from weak threats by employing strong encryption protocols. For example, employing sensitive UAVs that capture imagery to their control stations [7]. Encryption allows protection by preventing unauthorized reading or modifying of the

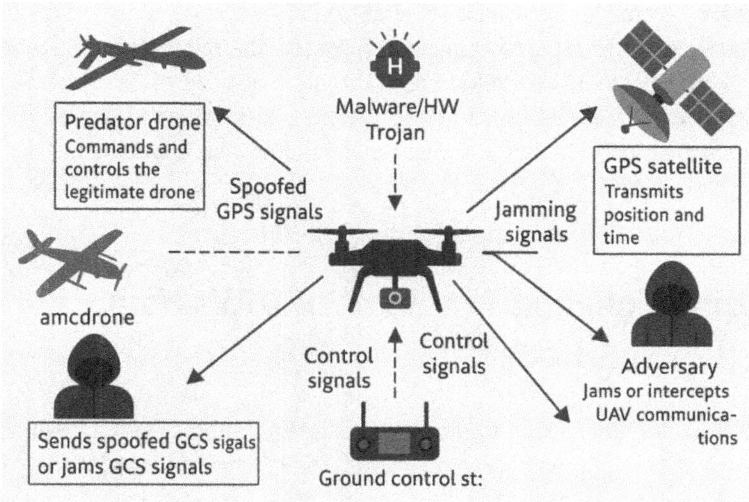

Figure 15.2: Security threats to UAV communication.

data that is captured, transmitted, or intercepted without proper decryption keys. (2) QKD: Quantum encryption offers powerful security features provided by quantum mechanics encryption keys and intercept detection. If communication occurs, it would be noticed, allowing the communicating parties to take appropriate action. (3) Blockchain technology: Integrating blockchain technology also has the potential to create secure UAV transmission data that is accessible and tamper-free for every data packet transacted or relayed. By using blockchain, data is stored verifiably while preventing unauthorized alterations of UAV flight paths and operations. Logging is done using the blockchain. (4) IDS: IDS could aid in the detection and response to unauthorized access or other suspicious activity in the UAV communication network. Monitoring UAV communication for patterns of jamming or spoofing attacks will enable an IDS to send alerts to operators. The illustration in Figure 15.2 depicts the remaining protection.

Table 15.3: Enhanced security measures for UAV communication.

Security measure	Description	Example applications
Encryption	Strong protocols to protect transmitted data	All applications
Quantum cryptography	Utilizes quantum mechanics for enhanced security	Military, critical infrastructure
Blockchain	Immutable ledger for data transmissions	Logging UAV flight paths, securing data handling
Intrusion detection systems	Detects and responds to unauthorized access or anomalies	All applications

UAV technologies are becoming increasingly integral to various sectors, offering significant benefits in terms of efficiency and capability. However, the reliance on real-time communication exposes UAVs to several security threats that need to be addressed through advanced security measures such as encryption, quantum cryptography, and blockchain technology. Ensuring secure and reliable communication is essential to harness the full potential of UAVs, as mentioned in Table 15.3.

15.5 Emerging Collision Problems in UAVs Using Quantum Solutions

The use of UAVs is steadily increasing in the commercial and military sectors, which has resulted in a corresponding increase in air traffic. The concern over potential collisions with other UAVs, crewed aircraft, or even ground obstacles is rising due to the increase in UAVs. Collision avoidance systems (CAS) are crucial to maintaining safe operation of UAVs in the airspace. The main problems of a UAV's collision avoidance and detection are: There are many challenges in sensing and recognizing obstacles. Cameras have lower performance in low visibility conditions, certain weather conditions, and even lens obstructions in form of dust or dirt. Primarily for visual navigation and obstacle detection. LiDAR (light detection and ranging) has a relatively high-power consumption, allowing heavy rain, snow, or fog to diminish its performance. This provides high-resolution 3D mapping, helpful for precise obstacle detection and navigation. Radar has obstacles of its own in the form of detecting low reflective and small objects; LiDAR requires substantial processing power. This is effective in ranging larger obstacles at farther distances and under various weather conditions. Ultrasonic sensors have low sensitivity to environmental noise on top of having a limited range; extreme temperatures can also obstruct performance. Commonly speaking, they are used for short-range indoor navigation. Infrared sensors have two dimensional sensors they respond to.

15.5.1 Collision Avoidance Techniques

Reactive methods of collision avoidance techniques are obstacle detection and avoidance. It detects an obstacle in real-time using onboard sensors and performs immediate evasive maneuvers. Path planning and dynamic rerouting are used to avoid obstacles. Predictive method → Predicting future positions of obstacles and forecasting possible collisions: (1) Path prediction: The safe paths are calculated by algorithms based on the predictive movements of nearby obstacles. (2) Inter-UAV communication: UAVs share position, velocity, and intended flight paths to collectively avoid collisions. UAVNet, IEEE-WCNC, IEEE-GLOBECOM, UAAI, UF-ACM, SCOPE, and IUT-PAN are used in multi-UAV environments like drone swarms or urban air mobility scenarios. (3) Integration with ground control: Integration with ground controllers, such as air traffic control (ATC), improves situational awareness and compatibility with crewed aircraft. (4) Machine learning and AI: Sophisticated algorithms are used. These algorithms are taught to learn from earlier flights and improve over time. SLAM (simultaneous localization and mapping): In the case of SLAM algorithms, they create a map of the environment and, simultaneously, track the UAV location in that map. This is helpful for navigation in GPS-denied environments. Deep reinforcement learning: Firstly, UAVs use reinforcement learning technique to learn optimal avoidance strategies via exploration-based learning (trial and error). It allows to deal with complex and changing environment. Figure 15.3 shows the diagrammatic representation of collision avoidance system.

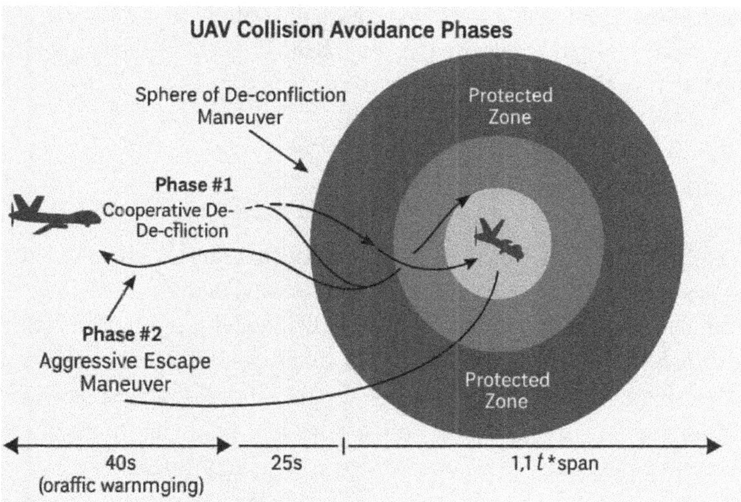

Figure 15.3: UAV collision avoidance system.

15.5.2 Communication and Coordination Challenges

Inter-UAV Communication, one type of communication, currently has no standard protocol [3]. Latency impacts real-time decision-making, while limited bandwidth can hinder the quality of data exchange.

Data Exchange: Sharing position and velocity data is essential for collision avoidance. This requires effective data compression and transmission methods to prevent saturating the communication channels. Incorporating ATC integration as one of the necessary aspects of aviation regulation requires that all UAVs be affiliated with ATC primarily when interacting with controlled airspace. Comma separation is used between reason and the result when multiple such results are mentioned.

Situational Awareness: Up-to-date information from ATC on positions and movements of nearby aircraft is vital. This is where ATC comes in to resolve conflicts between UAVs and manned aircraft. Coordination in UAV Swarms (1) Formation flying: Continuous communication is needed for adjustment of position. Coordination mechanisms avoid collision during complex maneuvers. (2) Dynamic allocation of tasks among UAVs based on the capability and status of each. Cooperative behavior requires awareness of mission objectives and support roles by all team members. Environmental factors cover signal interference, multipath interference, and weather conditions. Buildings, trees, and other obstacles are degrading or blocking the communication signals. This situation shall cause the connectivity failure. Reflections off surfaces shall give multiple paths to the signal receiver, reducing clarity in communications. Adverse weather conditions are scattering and absorbing signals, degrading the obstacle detection and navigation capacity. High winds are going to push UAVs out of their intended trajectories for flight; hence, complicated communications link.

15.5.3 Impact on UAV Operations

Collisions can severely hamper the functioning of such UAVs in terms of safety, economics, regulatory compliance, and public perception. Such incidents may include either fatalities or injuries to people caused by UAV collisions with manned aircraft. Most UAV collisions are truly disastrous. Property damage resulting from collisions inflicts financial losses and primarily culminates in the threat of legal liabilities. Loss or destruction of UAVs damages operational capabilities and financial investments. Repair and replacement of damaged UAVs and infrastructure incur huge costs.

Collision causes delays in missions, negatively affecting very important key critical tasks such as delivery service, surveillance, and agricultural monitoring. Such shutdowns heavily impact revenue flows and levels of customer satisfaction. In terms of regulatory repercussions, noncompliance with aviation law may lead to punishments, fines, and increased scrutiny by regulators. Lengthy and costlier licensing and certification processes are additional burdens for the operators since they increase compliance time and cost. Trust and people's feelings matter too; accidents occur, damaging morale in UAV technology, which may lead to community relations issues and significant public mistrust toward UAV missions. These accidents lead to advancements in technology and subsequently result in further safety measures and highly crash-resilient UAV designs. Data and analytics-wise: careful deliberation of incidents establishes forecast modeling and preemptive measures. Lawsuits, compensation, and increasing insurance rates are other forms of judicial consequences of collisions. New policy guidelines and rules create more complications and burdens on operators in terms of compliance.

15.5.4 Quantum Solutions to Collision Problems

Impactfully and elegantly, quantum computing is carving a path to harness the opportunities – they expectantly promise to be a solution to UAV collision conundrums.

Source Potential Solution: Quantum computing prevents collisions between UAVs. Given that large amounts of simultaneous data processing will be done due to enhanced processing power, thereby allowing collision-prediction and avoidance computations to be run faster and with more accuracy, the area of quantum algorithms is of significant concern. Two advanced algorithms are quantum path planning and quantum machine learning. These quantum algorithms can aid the machine-learning model on the perspective of bettering detection and classification of obstacles [8]. Collision-free trajectories may be optimized by quantum algorithms in a manner that would allow multiple path options to be explored simultaneously. Quantum-level communication and effective and safe data exchange protocols secure data transfer and minimize the risk of interference and breach. QKD implements unbreakable encryption for UAVs and control centers communication. Higher sensitivity and accuracy offered by quantum sensing facilitate obstacle detection and navigation through intricate settings. Quantum radar systems have good detection capability of low-reflective objects and work equally well during inclement weather. Possible benefits and challenges improved collision avoidance capabilities with faster, more accurate calculations; enhanced data integrity and communication security; more robust sensing systems that are more sensitive and accurate; quantum systems are expensive and complex; integration with already existing infrastructure and UAV technologies; making of uniform quantum protocols for the operation of UAVs. New collision concerns in UAV operations challenge public acceptability, safety, and viability. Hence, quantum computing and communication also offer potential avenues to enhance UAV operations and collision avoidance capabilities. Therefore, further research and development are needed to effectively merge quantum technologies and tackle all implementation-related issues.

15.6 Proposed Architecture

Integrating quantum cryptography with a layered security model for UAVs strengthens and secures the operational environment. Since UAVs are used for applications ranging from commercial delivery to military operations, the architecture addresses the need for an increasing number of end-to-end security solutions. This chapter proposes the architecture and discusses aspects of the design and implementation, quantum cryptography integration, and the layered security approach.

15.6.1 Layered Security Approach

Defense-in-depth or layered security provides a comprehensive solution to protecting UAV systems against every kind of threat. It ensures that other layers safeguard the system even when one of the layers is compromised. Physical security serves as the first line of defense and protects UAVs and infrastructure from unauthorized physical access.

ID Cards and Secure Badges: Issue ID cards or secure badges for staff entry. Deploying observation cameras will aid in monitoring access points and identifying unauthorized attempts. Providing clear evidence of tampering, tamper-evident seals will help identify unauthorized physical entry to UAVs or their components.

Access Panel Seals: Place seals on UAVs' critical access panels.

Inspection Procedures: Regularly inspect for signs of tampering with seals on UAVs' critical access panels and replace seals when necessary. Network security protects data communications and exchanges between UAVs and ground stations. All traffic coming in and out of the network is monitored and controlled by firewalls with pre-established security rules. Security for external access at the network border is enforced by installing perimeter firewalls.

Internal Firewalls: To isolate the network and secure key systems from an attacker's lateral movement, use internal firewalls.

Signature-Based IDS: Identify and alert on patterns using a signature database of known attacks [9].

Anomaly-Based IDS: Monitor network traffic for abnormal patterns that may signify unknown attacks. Non-availability of encryption on communication between the UAVs and the control stations makes sensitive data susceptible to unauthorized access and eavesdropping. Backed up by the principle of quantum mechanics is QKD, an unconditionally secure way of exchange of keys for encryption.

Transmitter/Receiver of Quantum Information: QKD systems shall be installed in the ground stations and drones for secure key exchange.

QKD Protocol: Apply protocols like BB84 or E91 in key distribution. Encryption guarantees that the data transferred between control stations and drones is secure and confidential.

Quantum-Safe Encryption Algorithms: Algorithms resistant to quantum attacks, that is, lattice-based ones, should be used.

End-to-End Encryption: The other way to safeguard data while in transit is to encrypt it from the source itself to the destination. Application security provides security and freedom from vulnerabilities for the software and applications utilized in UAV systems. The more frequent the code review, the higher the chance of finding and fixing vulnerabilities. Manual code evaluation is the act of having security professionals review the code being written for possible issues.

Automated Tools: Use automated tools for spot checks of code against known vulnerabilities and weaknesses. To safeguard against known vulnerabilities, software must regularly be updated with the latest patches.

Regular Updates: Create a routine for regular software updates to seal new vulnerabilities.

Automated Patch Management Systems: Implement automated systems for effective patching and patch management. To keep UAV operations secure, operational security encompasses procedures and guidelines. Security staff will need to be trained on security in the optimum procedures and practices to create a safe UAV environment. Frequent security awareness training sessions will acquaint staff with emerging threats and recommended procedures.

Role-Specific Training: For every employee to understand their individual specific security role, conduct role-specific training.

Recurrent Exercises: Conduct drills to ensure personnel are aware of response procedures and can respond in the correct manner [10]. A multilayered security plan offers a solid foundation for securing UAV systems from all forms of attacks. By addressing physical security, network security, communication security, application security, and operational security, organizations are able to develop an all-inclusive defense strategy.

15.6.2 Integration of Quantum Cryptography with UAVs

A totally secure way of exchanging encryption keys is provided by quantum cryptography in the form of QKD. UAV communication security is improved considerably with this integration, as data will be kept confidential and unbreakable. The building blocks to incorporate quantum cryptography into UAV systems are discussed elaborately in this section. Secure key exchange among communicating parties is made possible through QKD systems, which depend on the concept of quantum mechanics. Unlike normal cryptography processes, QKD delivers a safer system since its protocol and components assure that eavesdropping attempts will be discernible. To install QKD, there must be quantum transmitters and receivers: Equipment mounted on UAVs that produce and transmit quantum bits, or qubits, to ground stations are called quantum transmitters. Receivers of quantum information are equipment that detect and recognize the qubits sent by the UAVs at ground control stations. The quantum receiver and transmitter together establish a secure communication channel. The qubits will be altered in case of an interception [11]. The protocol for secure key exchange is specified by QKD protocols. Owing to the no-cloning theorem of quantum mechanics, it guarantees that any attempt at eavesdropping will be discovered. Protocol E91: It is a quantum entanglement-based protocol introduced by Artur Ekert in 1991. Secure key storage and key distribution of keys produced by QKD are based on a key management system (KMS). During the creation to destruction lifecycle, the KMS ensures keys are handled in the correct manner. Secure key storage: To avoid unauthorized access, quantum keys produced by QKD systems are stored in extremely secure environments. Hardware security modules (HSMs) or another safe storage space with the goal of safeguarding sensitive cryptographic information may be used here. Authorized key distribution: KMS governs the quantum keys' issuing to authorized organizations. This requires securely distributing the keys and giving them to the right people solely. Keys upon delivery may be safeguarded via techniques like transport layer security (TLS) and safe key enveloping. Post-quantum cryptography or quantum-resistant cryptography uses algorithms which are quantum adversary-resistant. In spite of high-power quantum attacks, these algorithms will still prove to be secure. Cryptography after quantum mechanics is:

(1) Lattice-based cryptography: It is among the most promising post-quantum cryptography solutions. Lattice-based cryptosystems, such as learning with errors (LWE) and ring-learning with errors (ring-LWE), are supposed to be quantum attack-resistant.
(2) Code-based cryptography: Code-based cryptoschemes, for example, based on decoding problem for random linear codes, are another candidates. Niederreiter and McEliece cryptosystems are examples of code-based post-quantum algorithms. A crucial component of modern UAV security infrastructure, QKD networks, trusted key management, and quantum-resistant encryption algorithms enable UAVs to achieve a higher degree of security and ensure integrity.

15.6.3 Design and Implementation

There are several steps involved in designing and implementing a secure UAV system with integrated quantum cryptography, and each is essential to guaranteeing the stability and dependability of the finished product. Conceptual design, detailed design, prototype development, integration, deployment, testing, validation, and operational deployment are all included in this process. Here, we provide a thorough guide by delving into detail about each step. By establishing the security requirements and creating a high-level architecture, the conceptual design phase lays the foundation for the entire project. The requirements for security are: Determine the risks and weaknesses: Conduct a comprehensive risk assessment to identify potential dangers and vulnerabilities unique to UAV systems [12]. Precisely state your security objectives, and ensure the availability, confidentiality, and integrity of your information.

Standards and Compliance: Take relevant legal and industry standards into account, for example, ISO/IEC 27001 for information security management systems.

Layered Security Concept: Integrate layered security into high-level architecture development, addressing operational, network, communication, application, and physical security.

Points of Integration: Identify the key points of integration among the different elements of quantum cryptography and the overall architecture [13]. The exact components, specifications, and procedures needed to implement the conceptual design are defined in the detailed design phase:

Hardware Components: Outline the exact hardware components required, such as secure storage facilities, key management systems, and quantum transmitters and receivers.

Software Components: Discuss various software elements, including intrusion detection systems, QKD protocols, and encryption algorithms. Communication Protocols: Specify protocols that UAVs and control stations may use to secure data sharing.

Security Protocols: Define protocols for system updates, incident response, and key management. The functional model of the UAV system is built during the prototype development stage for testing and refining the design:

Hardware Assembly: Assemble all the components required for a working UAV prototype.

Software Integration: Ensure interoperability and compatibility of the software elements through integration.

Functionality Testing: Perform initial tests to validate the fundamental functioning of the prototype.

Security Testing: Conduct security testing to identify potential flaws and vulnerabilities. Integrating and deploying the solution involves preparing the system for real-world implementation:

System Integration: Combine all software and hardware components into one cohesive system to ensure seamless functionality and communication.

Integrating Quantum Cryptography with Layered Security: Implement the layered security design. The control system shall be staged as a staging environment to simulate conditions and apply extensive testing to achieve realistic results.

Final Deployment: Complete and deploy the system in a live environment according to standard deployment procedures after testing [26]. Testing and validation ensure that the system meets all expectations and criteria for security requirement satisfaction:

Functional Testing: Once functional testing has been done on the system, the system must fulfill very necessary functionalities. Conduct full security testing, including vulnerability scans and penetration testing, to identify and remediate security flaws. Conditions approve the system to fulfill stated security objectives and requirements. Compliance verification ensures that a system meets all necessary industry standards and legal criteria. Operational deployment involves actively putting the system to use, with maintenance and support continuously in place:

User Training: Train system users to ensure that operations staff and security staff understand their responsibilities and roles.

Incident Response Education: Train staff on how to handle security incidents. Regular Maintenance: Establish procedures for periodic hardware checks and software updates for the entire system [26].

Support Infrastructure: Set up a help desk and technical support team to resolve issues that arise during system operation. The design and development of a secure UAV system utilizing quantum cryptography is highly challenging because meticulous planning, execution, and validation are required. Organizations can build safe and reliable UAV solutions by following the process model for best security standards, systematically and sequentially addressing conceptual design, detailed design, prototype development, integration, deployment, testing, validation, and operational deployment.

15.7 Collision Avoidance in UAVs

Obstacle detection and avoidance are critical operations of any UAV system. They enable UAVs to fly in an entirely autonomous way in diverse situations by enabling them to identify likely obstacles along their flight trajectory and avoid collisions by flying around them. Some of the techniques and technologies are described in this section for the viable detection and avoidance of obstacles. UAVs detect obstacles using different types of sensors with different strengths and weaknesses. The use of different sensors could afford a more complete overview of the world by fusing all of them together.

LIDAR: The use of LIDAR creates an understandable three-dimensional model of the environment by measuring the distance to an object using laser light. It has the advantage of being able to create very high-accuracy, high-precision 3D maps, and its applications are well suited to the detection of obstacles in cluttered environments.

Limitations: High power usage, costly, and may be weather dependent. The LIDAR system is shown in Figure 15.4.

Radar Capability: In order to ascertain an object's range, velocity, and bearing, radar technology employs radio waves. The advantages are that radar can detect both metallic and non-metallic targets and is efficient in a wide range of weather conditions.

Usage: Suitable for obstacle detection from a distance.

Disadvantages: Heavier, bulkier, and lower resolution than LIDAR. Ultrasonic sensors: These sensors use sound waves to measure the range to objects near them. Advantages are minimal cost, mobility, and ease of integration.

Uses: Often used for obstacle detection in close range.

Drawbacks: Range and resolution limitations can be influenced by environmental noise.

Operation of Optical Cameras: Visual information is captured by cameras, which is processed to recognize obstructions.

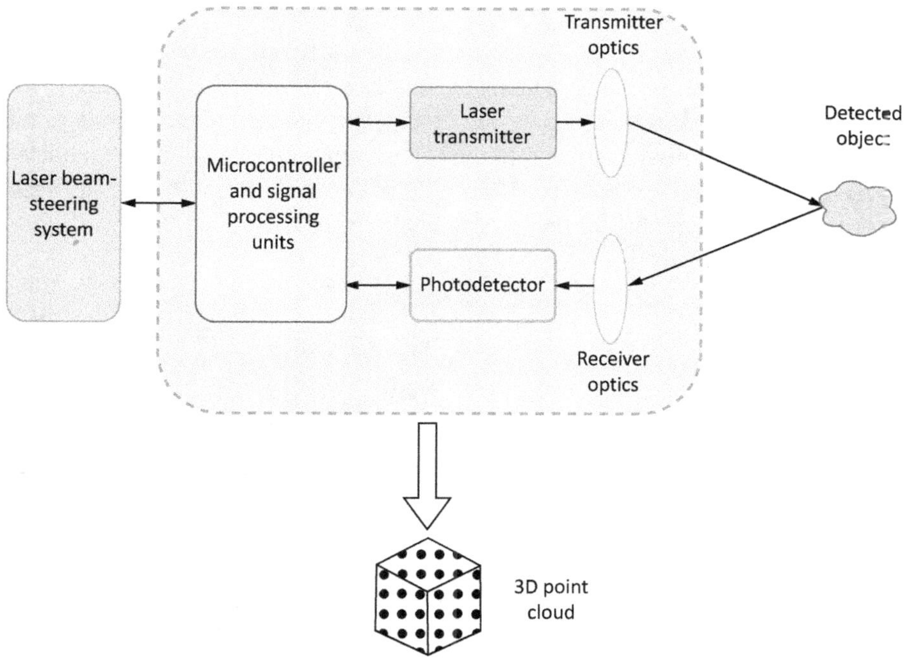

Figure 15.4: LIDAR system for obstacle detection.

Advantages: Color and texture information can be accessed at high resolution.

Uses: Useful in recognizing obstacles and visual identification.

Restrictions: Dependent on illumination and demands vast amounts of processing power to analyze in real time.

Sensor fusion: Merging data from a variety of sensors, sensor fusion increases obstacle detection system reliability and accuracy. Sensor fusion algorithms analyze and combine data from various sensors to provide a more accurate and detailed representation of the environment.

Kalman Filter: A mathematical algorithm that blends forecasts from multiple sensors to estimate the state of the system.

Particle Filter: A sequential Monte Carlo technique employed for state estimation of a system from noisy observations.

Bayesian Networks: Probabilistic models that integrate sensor data based on conditional dependencies to enhance detection accuracy. The Kalman filter is a recursive algorithm employed to estimate the state of a dynamic system from noisy data. Here's an explanation of the terms used in the Kalman filter equations as shown below:

$$\widehat{\boldsymbol{X}}_{\cdot k|k-1} = \boldsymbol{F}\widehat{\boldsymbol{X}}_{\cdot k|k-1} + \boldsymbol{B}\boldsymbol{u}_k \quad (15.3)$$

$$P_{k|k-1} = FP_{k|k-1}\,F^T + Q \tag{15.4}$$

where $X.k|k\text{–}1$: Predicted state estimate at time k, given the state at $k - 1$; F is the state transition matrix, which models how the system evolves from one state to the next; B is the control matrix, applied to the control input (uk); $P_{k|k\text{–}1}$ is the predicted covariance of the state estimate, representing the uncertainty; and Q is the process noise covariance, accounting for the uncertainty in the model:

$$K_k = P_{k|k-1}\,H^T\left(HP_{k|k-1}\,H^T + R\right)^{-1} \tag{15.5}$$

$$\widehat{X}_{\cdot k|k} = \widehat{X}_{\cdot k|k-1} + K_k\left(z_k - H\widehat{X}_{\cdot k|k-1}\right) \tag{15.6}$$

$$P_{k|k} = (I - K_kH) * \,P_{k|k-1} \tag{15.7}$$

where K_k is the Kalman gain, determining the weighting between prediction and measurement update; H is the observation matrix, linking the observed measurements to the system's state; z_k is the observation at time k; R is the measurement noise covariance, representing the uncertainty in measurements; I is the identity matrix, used to update the covariance; $X._{k|k}$ is the updated state estimate after measurement; and $P_{k|k}$ is the updated covariance of the state estimate.

Sensor data processing in real time is of paramount importance in efficient obstacle detection and avoidance. This entails applying sophisticated algorithms for interpreting sensor information, anticipating collision possibilities, and calculating the best avoidance maneuvers.

Image Processing Algorithms:

Edge detection, feature extraction, and object recognition techniques are employed to detect obstacles from image data captured from cameras [14].

Machine Learning Models: Neural networks and other machine learning models can be trained to identify and classify obstacles from sensor data.

Path Planning Algorithms: These algorithms compute the best path to bypass detected obstacles while keeping on the desired trajectory. The comparison of sensor technology has been shown in Table 15.4.

Table 15.4: Comparison of sensor technologies.

Sensor type	Range	Resolution	Cost	Power consumption
LIDAR	Long	High	High	High
Radar	Long	Medium	Medium	Medium
Ultrasonic	Short	Low	Low	Low
Optical camera	Medium to Long	High	Medium	High

There are many challenges that need to be addressed when incorporating effective obstacle detection and avoidance systems into UAVs:

Processing power: It requires a tremendous amount of computing power to analyze enormous amounts of sensor data in real time. The design of any safety measures must consider latency, meaning that the time between sensing the obstacle and executing avoidance actions must be very low. Environmental conditions include any variations in the performance of sensors or detection due to changes in terrain, illumination, or weather conditions. Integration: to ensure that the multiple sensors and processing systems function together under the weight and space limitations placed on UAVs.

15.7.1 Collision Avoidance Algorithms

UAVs have the capability to autonomously navigate through their environment and avoid potential collisions with objects or other UAVs due to collision avoidance algorithms. This section analyzes the various collision avoidance algorithms, what theory is behind each of them, and what application they are utilized for. Reactive algorithms react immediately to the detected obstacles and rely on the sensor data in real time. The environment of the UAV is simulated by the potential field method as a potential field in which goals have the attractive forces and obstacles have the repulsive forces. Positive aspects include implementation simplicity and computational efficiency. Use: Ideal for dynamic environments with mobile obstacles. Negative side: May trap the UAV in local minima [15]. The vector field histogram employs a polar histogram representing obstacle density around the UAV and finds a path that uses the minimum number obstacles.

Benefits: Minimizes the chances of collisions and is indeed very effective in dense environments. Use: Best when moving through very dense areas.

Disadvantages: It may incur substantial computation for high-resolution histograms.

Bug Algorithms: Bug1 and Bug2 navigate the environment with very simple obstacles and reach their goals, easy implementation without promises for a path to exist for its reaching goal in practice.

Applications: Suitable to the simple environment, well-defined barriers.

Limitations: Not suitable in a complex environment crowded with barriers. Collision avoider, intentional algorithms implemented by planning and making decisions.

It is an algorithm for graph search that avoids obstacles to find the shortest route, starting from point A to reach destination point B.

Advantages: discovering the shortest path; it is optimal and exhaustive.

Uses: Well-suited for static obstacles in known environments.

Disadvantages: Computationally intensive in large search spaces. For example, rapidly exploring random trees link new nodes to the nearest current node using random space exploration to build up a tree of possible paths. Advantages include dealing with sophisticated environments and high performance in high-dimensional spaces. Applications: Efficient planning traces out online in changing conditions.

Disadvantages: Smoothing is required for smooth paths because paths may be suboptimal. An extension of A*, D* performs well when replanning routes because of changes in the environment.

Advantages: Adaptive and efficient in changing conditions.

Utilizations: Appropriate for variable barrier or variable situation environments. Disadvantages: Harder to implement than A*. In some UAVs, various robots assist one another in avoiding collisions and reaching common goals by employing cooperative algorithms. For exchanging information about locations, velocities, and planned courses, multi-agent systems exchange information and collaborate.

Advantages: Enhanced situational awareness and reduced collision likelihood.

Applications: Coordinated missions using swarm robotics.

Limitations: Requires strong protocols for communication and coordination. Consensus for collision avoidance and formation, the algorithms come to an agreement on speeds and positions.

Advantages: Naturally well adapted to close training, it ensures motion coordination.

Applications: Cooperative mapping and formation flight.

Limitations: Highly prone to failures and delays in communication. Predictive models are used in prediction to simulate the future locations of other UAVs and plan their flight. Advantages are reducing the risk of collision and facilitating simple route planning.

Applications: Multiple UAVs in dynamic environments.

Disadvantages: Needs to have good predictive models and is computationally intensive.

In the hybrid algorithms collision avoidance is accomplished by the combined effect of cooperative methods, deliberative methods, and reactive methods. Hybrid algorithms are A* and D* borrow some of their features from each other for their optimal path planning and dynamic replanning respectively.

Advantages: Flexible and efficient within dynamic environments. Applications: Dynamic cases in complex environments.

Disadvantages: More complex than single algorithms.

Hierarchical Algorithms: These hierarchical algorithms have a hierarchical structure in planning paths within those hierarchical levels of abstraction and segregating the environment into regions. All of this reduces computational requirements for large domains. Applications include long-distance navigation and wide-area navigation. Disadvantages include a requirement for environment segmentation and state-of-the-art mapping. Model predictive control predicts and plans optimal actions in a finite time horizon by optimization methods.

Advantages: Efficiently deals with dynamic environments and constraints. Applications are for online navigation in complex environments.

Disadvantages: Requires exactness in its model representation and is computationally intensive. On such a basis, UAV collision avoidance research and development are pretty much on demand; hence, UAVs are expected to safely and efficiently maneuver in various diverse environments. By using advanced sensor technologies and algorithms, UAVs will navigate in densely populated areas, undertaking their missions while minimizing chances of collision through cooperation.

15.8 Case Studies: Applications of UAVs

15.8.1 Military Applications

Today, drones (UAVs) are a crucial military instrument for warfare. The whole dimension of modern military operations has opened up, enabling entry into enemy territory, performing reconnaissance work, and carrying out missions while keeping human lives out of danger. This case study researches the various military applications and effects of UAVs in depth. Intelligence-surveillance-reconnaissance (ISR), or advanced UAS, provides this capability for the military. This real-time surveillance system provides information using advanced sensors and cameras, which allow for generating high-definition images and streaming live video.

Advantages: It can monitor extensive areas for long periods without being encumbered by the presence of humans in the air. Areas of application include monitoring insurgent activity, surveillance in battle zones, and border monitoring.

Case Study: In Iraq and Afghanistan, U.S. forces carry out ISR missions on the battlefield using the MQ-9 Reaper UAV. Signal intelligence (SIGINT): intercepting and processing an opponent's radar date signal and electronic signals.

Benefits: Gathers intelligence without direct confrontation. Application includes locating radar installations for jamming signals and locating enemy communication networks. For example, the RQ-4 Global Hawk UAV is employed for SIGINT mission work at high altitude and long ranges. Target acquisition and designation identifies and designates targets for precise engagement by ground systems or manned aircraft. Advantages include reduced collateral damage and improved strike precision. Applications consist of continuous tracking of targets, damage assessment post-strike, and reconnaissance pre-strike.

Case Study: Laser designation for guided munitions supported by the MQ-1 predator drone. Strike and close combat increasingly used in a direct combat role; armed UAVs conduct combat missions as well as precision strikes. A precision strike consists of armed UAVs using bombs and missiles to hit enemy targets.

Advantages: These strikes reduce collateral damage, allow operation in high-threat environments, and reduce risk to human pilots. Applications include close air support (CAS), counterterrorism, and attacks on high-value targets.

Case Example: terrorist leaders targeted with precision strikes employing MQ-9 Reaper, armed with laser-guided bombs and Hellfire missiles. CAS provides the sort of support that attacks enemy forces and provides ground troops with crucial situational awareness in real time.

Advantage: Increased soldier safety and thus the increased efficiency of ground operations. Examples of its applications include troop support, convoy protection, and urban operations. For instance, the MQ-1C Gray Eagle UAV provides continuous ground troop support in contested places.

Suppression of Enemy Air Defenses (SEAD): Locate and destroy enemy air defense systems to provide safe passage for manned and manned-associated assets.

Benefits: It ensures operational success for the engrossing air operations and minimizes risk for manned aircraft. In applications, it covers destruction of anti-aircraft artillery, radar installations, and surface-to-air missile (SAM) sites.

Case Example: The MQ-9 Reaper has electronic warfare capabilities for SEAD missions [16].

Logistics and Supply: Enhance supply chain and logistics delivery actions within an installed military environment.

Role for Cargo Delivery: UAVs transport supplies, munitions, medical hardware, and various other necessities to remote or hostile locations. Benefits include reduced risk for human convoys, assurance of timely delivery, and operation under special conditions. Applications such as aiding humanitarian efforts; resupplying forward operating bases; or transporting priority supplies during crises are also encompassed.

Example Case: This is replacing ground convoys for move cargo in Afghanistan. Battlefield medical evacuations are performed to take injured soldiers to hospitals. Facilitation of timely evacuations, less risk to the health care workers, and the better scope for the injured soldiers' survival. Applications include disaster relief, war, and humanitarian missions.

Case History: A hybrid UAV called the Black Knight transformer is in development with potential missions for medical evacuation. Surveillance coupled with route planning for logistics convoy will improve the discovery of safe routing and hazard controls.

Benefits: This reduces threats of ambushes and IDDs and thus improves logistical operations' security and effectiveness. Applications are to include road condition identification, the convoy movement monitoring, and the supply route planning. One of these is the RQ-7 Shadow UAV, used for route reconnaissance missions and giving useful information for logistics planning. Training pilots and operators are called in a realistic scenario training program of drone pilots and operators.

Benefits: Reduction in costs of training; operational proficiency improvement; realistic practice. Uses pilot training programs, emergency response simulations, and rehearsal of various missions. For example, MQ-9 Reaper training programs instruct operators for real-world missions through mission simulation. Combat simulation assesses the tactics and reactions of soldiers in the replication of real environments Efficiency regarding enhanced production of tactical competency, augmented combat readiness, and safe training under hazardous conditions.

Use: Urban warfare training, joint military exercises, and live-fire training.

Case Study: The Joint Unmanned Aerial Systems Center of Excellence developed activity around using UAVs in training operations aimed at improving fighting effectiveness and simulating enemy numbers. The making of useful information offers tactical analysis and planning possible improvements to post-mission analysis and tactical planning. Mission success improves, and deeper insight and decision-making are fostered through analysis, planning, and post-action evaluation through intelligence gathering. An instance here is the RQ-11 Raven UAV for tactical reconnaissance and real-time data provision for analysis and use in mission planning. With respect to fighting and intelligence gathering, logistics as well as training; UAVs have been an integral pillar of modern military operations. Independent action and real-time intelligence reception for precision strikes have changed military operations, improving troop effectiveness and safety. The use of UAVs is expected to increase larger scale in various aspects for military endeavors as technology continues to innovate.

15.8.2 Healthcare Applications

Novel applications of UAVs are increasingly being discovered within the health sector – some of which offer solutions to medical logistics, emergency services, and even patient care. There are various innovations flagged by drone technology that considerably increase the productivity, reach, and effectiveness of health professionals. This case study narrates some of the health applications of UAVs and the impact and advantages of such applications. Vaccine and drug delivery to remote or isolated places, as well as other essential medical supplies, are guaranteed delivery.

Benefits: one-day delivery, access to health care, and maintenance of the cold chain for temperature-sensitive drugs. Uses include vaccination campaigns, delivery of lifesaving drugs, and emergency medical equipment. For example, Zipline brings essential medicines, blood, and vaccines to distant health clinics in Rwanda and Ghana, with delivery time reduced dramatically and enhanced patient outcomes. Transportation of blood and organs is crucial for overcoming logistics hurdles and maximizing the delivery of healthcare by moving blood products and organs for transplant. Transport efficiency allows organs and blood units to be transported rapidly and effectively from donor centers to hospitals. Advantages include reduced transit time, enhanced delivery reliability, and assured on-time delivery of critical resources. Uses include regional blood bank assistance, organ transplant supply, and emergency transfusion of blood. Example of a case: Matternet's partnership with Swiss medical centers shows the ways in which UAVs can be used to transport blood and organs to reduce time and enhance patient outcomes. Disaster relief and response constitute a crucial role in disaster response activities, delivering prompt help and relief to disaster-hit zones. Quick triage and assessment swiftly evaluate disaster-affected areas, gathering data and aerial imagery to help emergency responders with resource allocation and triage. Benefits include faster reaction times, increased situational awareness, and improved decision-making skills. Applications include large-scale emergency situations, triage in conflict zones, and natural disaster response.

Case Example: UAVs were employed to map affected areas after the 2015 Nepal earthquake, gathering vital data for rescue and relief operations [39]. Provide emergency aid, including food, water, and medical kits, to remote or inaccessible locations through supply drops and medical evacuation.

Advantages: Enables medical evacuation missions, ensures rapid resource deployment, and surmounts logistical challenges. The uses include critical care evacuation, remote area resupply, and humanitarian aid. As an example, Project Premonition, a collaboration between Microsoft and universities, enhanced disaster response through the deployment of UAVs to deliver vital aid to remote communities in Africa [17].

To enable telemedicine consultations and remote monitoring, telemedicine infrastructure develops communication networks and interfaces with remote healthcare facilities.

Advantages: Promotes delivery of healthcare to marginalized communities, improves access to healthcare services, and enhances patient outcomes. Applications employed include management of chronic diseases, rollout of telehealth infrastructure, and remote consultation.

Case Study: UAVs establish telemedicine networks to deliver services to isolated communities in Papua New Guinea.

Sample Collection and Laboratory Test: Swabbing and blood are collected as biological samples for rapid diagnosis and treatment, which are sent to the laboratory for testing.

Advantages: It is conducive to disease surveillance activities, enhances diagnostic capabilities, and reduces transport time for samples. Its applications are in public health research, chronic disease management, and infectious disease surveillance.

Example of a Case: Drones were used to transport baby blood samples for early diagnosis of HIV in Malawi, cutting turnaround times by half and improving patient outcomes. By facilitating more effective medical logistics, greater emergency response capacity, and greater access to health care services in rural communities, drones are transforming the delivery of healthcare in the industry. The healthcare sector is changing as they can provide medical supplies in a timely fashion, offer situational awareness in real time, and assist telemedicine programs. It introduces new modes of improving patient health and care.

15.8.3 Educational Applications

UAVs, or drones, are being used for new purposes in education, changing the manner of learning and teaching in many different fields. Used to improve STEM education or offer new dimensions in environmental studies and the arts, UAVs provide new opportunities for experiential learning and discovery. The case study in this paper is a discussion on some of the educational uses of UAVs, their impact, and benefits. Improving STEM education has emerged as a beneficial aid in science, technology, engineering, and mathematics (STEM) education, giving students practical experience and relevance.

Students may design, build, and code drones using engineering and robotics, where they enhance their computer science, robotics, and engineering knowledge.

Advantages: Provides practical uses for academic knowledge and develops creativity, problem-solving, and teamwork. Areas of application involve programming courses, engineering design challenges, and robotics competitions. For example, students are able to undertake UAV design and programming projects at the University of Southern California's Drone Lab as preparation for engineering and robotics careers [18]. By conducting experiments and projects, physics and aerodynamics allow students to explore the principles of flight, aerodynamics, and physics. The advantages include enhanced understanding of challenging concepts, active learning, and stimulation of scientific inquiry. Applications range from flight dynamics projects to aerodynamics studies and physics experiments. For example, Stanford University's Aeronautical Engineering program incorporates UAVs into its curriculum so that students can study aerodynamics and flight mechanics. Students are able to explore the natural world due to the new perspectives and data collection tools that environmental and earth sciences education provides. Environmental monitoring and conservation collects data on ecosystems, wildlife habitats, and environmental status for use in ecological studies and conservation efforts. Benefits include enhanced fieldwork capacity, provision of real-time information, and development of environmental awareness. Uses involve research on climate change, habitat mapping, and wildlife monitoring. For example, the University of Exeter is using UAVs in environmental sciences to monitor coastal erosion and determine the impact climate change has on ecosystems. It is possible to study topography, land-use patterns, and landscapes through the use of high-resolution imaging and geospatial data in geography and geospatial analysis.

Advantages include enhancing geographic awareness, developing spatial analysis skills, and providing experiential learning with geospatial technology. Applications are in geographic information systems (GIS) projects, land use planning, and cartography.

Case Study: In order to provide students with real-world experience using geospatial technologies, the University of California, Berkeley offers remote sensing and GIS courses utilizing UAVs for data acquisition and analysis.

Creative Arts and Media are revolutionizing creative arts and media education, providing new tools and approaches to artistic storytelling and expression. Photography and Filmmaking create distinctive aerial photos and video shots, allowing students to develop new creative applications in photography and filmmaking.

Benefits: Offers distinct views, increases storytelling potential, and builds creativity and innovation.

Uses: Aerial photography assignments, filmmaking class instruction, and multimedia narration.

Case Example: Savannah College of Art and Design has a course on aerial cinematography, educating students on how to employ UAVs to achieve breathtaking aerial footage and enhance creative storytelling techniques. Architecture and design help capture architectural features, examine urban areas, and visualize design projects.

Benefits: Increases design visualization, aids architectural analysis, and offers new means of creative exploration.

Applications: Architectural surveys, urban planning studies, and landscape architecture.

Case Example: Harvard Graduate School of Design integrates UAVs into their architecture and urban planning studies so that students can push the boundaries of design visualization and analysis. UAVs are revolutionizing learning by opening up new opportunities for experiential learning, exploration, and creativity. Their uses in STEM education, environmental science, and art provide new opportunities for increasing student engagement, creativity, and practical skill development. As technology evolves daily, the application of UAVs in teaching and learning is expected to expand, providing new opportunities for innovative pedagogy and learning experiences.

15.8.4 Delivery Drones in Urban Areas

The quick rise of UAVs or delivery drones is set to revolutionize logistics in cities. Drones are popular because they constitute a quick, effective, and green last-mile delivery method that eliminates pollution, hectic traffic jams, and time constraints in goods delivery. This study discusses the viability of drone delivery within urban settings, bringing to light its benefits, challenges, and the promise it holds. Delivery drones are increasingly accepted for urban freight due to their benefits in comparison to conventional delivery. The speed and efficiency of delivery drones allow them to bypass all possible obstructions with the fastest speed directly in the line of delivery. They offer numerous benefits: increased customer satisfaction, efficiency, and lower delivery times. Uses include emergency supply drops, same-day delivery services, and last-mile delivery.

Case Study: To significantly reduce delivery times, Amazon Prime Air has been experimenting with delivery drones that are capable of delivering packages within 30 min after an order has been made [19].

Environmental Impact: In contrast to traditional delivery vehicles, drones do not produce emissions and consume less energy. Benefits include reducing air pollution, minimizing carbon emissions, and encouraging sustainable urban development. Examples of applications include carbon reduction schemes, green logistics initiatives, and eco-friendly delivery services. For example, let us talk about Zipline, a drone delivery company that has been reducing emissions and enhancing environmental sustainability in Rwanda through the use of UAVs to deliver medical supplies. The reach and accessibility of drones enable them to deliver items to areas inaccessible to ground transportation, such as remote or hard-to-reach locations.

Benefits: Increased delivery reach, enhanced service availability, and catering to poor communities.

Uses: Rural deliveries, disaster response operations, and long-distance area logistics. Example in use: Wing, a company from Alphabet, has been conducting drone deliveries in the rural terrains of Australia and supplying critical items to areas that are almost out of reach of conventional delivery systems. Although promising, delivery drones face a number of issues and factors that will have to be weighed before achieving public acceptance on a wide scale.

Regulatory and Safety Issues: Handling complex regulatory arrangements, maintaining airspace safety, and addressing privacy-related issues. Solutions include collaboration with authorities to formulate regulations, safety measures, and effective privacy policies.

Technical Limitations: Restricted payload capacity, battery-life limitations, and weather sensitivity. Solutions include high-tech batteries, low-consumption drones, and weather-resistant designs. Example: Matternet, another drone logistics firm, has been striving to create UAVs using high-tech battery technology and weather-resistant capabilities to bypass such technological limitations.

Public Acceptance and Trust: Addressing community concerns about noise, privacy, and safety. Solutions include community engagement, public awareness campaigns, and showcasing safety and reliability. Case Example: Project Wing implemented a public outreach program in Australia, interacting with local communities to hear concerns and establish trust in drone delivery services. The future of urban delivery drones is bright, with numerous opportunities for expansion and innovation.

Integration with Smart Cities: Delivery drones can integrate with smart city infrastructure, enabling real-time logistics and improving urban planning.

Case Example: Singapore's Smart Nation strategy will integrate drone delivery into its smart city environment, using technology to enhance urban logistics and quality of life.

Growth of Services: Expanding drone delivery services to carry a broader variety of goods, serving various industries and uses.

Case Example: UPS is studying the application of delivery drones for medical sample transportation, broadening the horizon of drone applications beyond package delivery.

Case Example: IBM is working on the application of AI-driven drones for autonomous delivery and navigation, seeking to enhance efficiency and reliability. Delivery drones are full of potential in revolutionizing city logistics with a quick, efficient, and green means of last-mile delivery. Although there are challenges, continuous innovation in technology, regulatory structures, and societal acceptance are opening doors for widespread use of delivery drones in cities. With cities growing and developing further, delivery drones are set to be an integral part of the future of urban logistics and sustainable development.

15.9 Performance Evaluation

Evaluation Metrics: Various fundamental metrics are considered vital when evaluating the performance of a security solution for a UAV communication system. Latency: This metric indicates the time required to transmit and receive data. Real-time communication is impacted by the presence of high latency in UAV operations. This is especially applicable to military surveillance and disaster recovery, wherein every second matters and may lead to disastrous effects if there's a significant delay. Studies show that for real-time applications in UAVs, latency needs to be reduced to below 10 ms, which is becoming achievable with the introduction of 5G and edge computing technologies [20].

Throughput: This defines the quantity of data length that is successfully transmitted in a particular communication channel over a specific period. Throughput evaluation reveals the potential for high data transmission without causing congestion and also ensures the prevention of data loss. For instance, in agricultural monitoring, thousands of UAVs need to carry several gigabits per second.

Error Rate: Error rate measures the occurrence of errors during data transmission. A smaller error rate means that the communication is more reliable and is crucial in determining the significance attached to data, especially when the application is critical from a security point of view, such as in military operations. Quantum bit error rate (QBER) is a specific metric in quantum cryptography used to assess the efficiency of key distribution.

Energy Efficiency: Energy efficiency becomes very crucial as UAVs only have a limited energy supply. It helps in evaluating how much energy a system uses during data transmission and security processing. Energy-efficient communication protocols increase the operational time of UAVs, which becomes especially important for missions with long durations.

Scalability: Scalability refers to how well a given communication and security solution can accommodate larger numbers of UAVs operating under that communication system without degradation in performance. Communication performance generally scales with the capability of the deployed UAV system, especially in urban settings, since increasing numbers will inevitably translate into increased traffic, which should still guarantee security and efficiency. Security analysis involves performing robustness tests on the proposed cryptographic solutions by simulating scenarios of different cyberattacks the system has been subjected to:

Resistance to Eavesdropping: Eavesdropping is one of the most significant threats in UAV communication. Because any interception in quantum communication changes the quantum state, it provides a direct measure of how secure the system is when evaluated through the QBER. A low QBER, combined with privacy amplification techniques, will guarantee that the final key remains safe even if some parts of the communication are compromised.

Protection Against Spoofing and MITM Attacks: Significant advantages of quantum cryptography are evident in the case of spoofing and man-in-the-middle (MITM) attacks. Quantum entanglement and the no-cloning theorem ensure that any attempt at communication interception or impersonation will, in principle, always be detected. The security analysis involves simulating these attacks and testing whether the system can retain secure communications during such scenarios. Security analysis must also measure the resistance of the proposed solutions against quantum attacks. The combination of lattice-based cryptography with QKD provides a layered defense against both classical and quantum adversaries. Efficiency and scalability are critically important for the practical deployment of secure UAV communication systems.

Efficiency: This is characterized by the amount of computational and communication overhead introduced by the security protocols. Although quantum cryptographic solutions provide strict security measures, they often have a much higher demand for computational resources. Hence, they need to be evaluated to determine how best the efficiency of protocols can be paired with trade-offs between security and resource consumption. Improvements in error correction, key management, and privacy amplification should focus on reducing overhead without compromising security.

Scalability: As its name suggests, the number of UAVs in operation will increase, necessitating scalability in communication and security infrastructure. The scalability of the proposed solutions has been tested with simulations of large-scale UAV networks, and their performance has been measured in terms of latency, throughput, and security. The system's ability to withstand a dense operational environment – from urban air mobility to swarm-based operations – will frequently be the focus of scalability assessments.

15.10 Challenges in UAV Collision Avoidance

15.10.1 Detection of Small and Moving Obstacles

So far, small and dynamic obstructions have posed considerable challenges in collision avoidance with UAVs. Such UAVs need to be equipped with sophisticated sensors and algorithms that can detect objects of different sizes and speeds in real time, such as monocular camera-based systems and deep learning architectures like Mask R-CNN.

These methods are known to significantly improve UAV detection capabilities, enabling small or fast-moving intruders to be easily avoided.

Key techniques are as follows:

Monocular Camera-Based Systems: They use a camera to perceive obstacles and estimate their distance. They are small and inexpensive enough to fit on small UAVs.

Deep Learning Models: Mask R-CNN is a model that detects and segments obstacles in real time, essentially providing information related to the position and movement of the obstacles. Minimization of path deviations and costs is important for UAV operations.

Cost Function Computation (eq. (15.8)): A given path planning algorithm should achieve a balance between safety and efficiency, reducing deviations that cause a route to involve additional travel time and energy expenditure. These include velocity obstacle algorithms and dynamic path planning methods for optimizing flight routes while ensuring safe separation from obstacles.

Key techniques are as follows:

Velocity Obstacle Algorithms: These algorithms predict potential collisions based on the velocities of both the UAV and obstacles, allowing adjustments to the flight path dynamically.

Dynamic Path Planning Methods: Continuous path updates for UAVs, based on real-time sensor data, guarantee the best possible route while avoiding obstacles:

$$\text{Cost function: } J = \sum_{i=1}^{n} (c_i * d_i + e_i * t_i) \tag{15.8}$$

where c_i is the cost of path deviation, d_i is the distance traveled, e_i is the energy consumption, and t_i is the travel time.

Limits of Computation and Processing in Real Time: Collision mitigation systems are highly computationally intensive, which means they need to respond very quickly and process sensor data in real-time. The computations required are provided in Table 15.5. Strong onboard processing capacity or the use of edge computing is essential for UAVs to perform these functions effectively. The requirements for high data throughput and low-latency communication especially exacerbate the problem in complex environments with many moving obstacles. Solutions at the level of onboard processors include fast processors that handle data collection and processing in real-time as well as utilize complex algorithms. Edge computing is the practice of shifting computation to adjacent edge servers as a means of boosting processing speed and reducing latency.

Adaptability and Environmental Conditions: Environmental influences like:

- climatic conditions
- lighting
- terrain variation will all considerably affect the operation of collision-avoidance systems.

Table 15.5: Computational requirements.

Task	Processing power	Latency
Object detection	High	Low
Path planning	Medium	Medium
Communication management	Low	High

UAVs should have the flexibility that allows them to adapt to various conditions so that they can stabilize. A great deal of experimentation is on-going over sensor fusion techniques, which incorporate data from several sensors such as LiDAR, cameras, and radar, aimed at improving the robustness and adaptability of UAV CAS. A few among them include: Sensor Fusion: Acquisition and fusion of data from different sensors to provide a comprehensive idea of the world [21] (Figure 15.5). Adaptive Algorithms: Algorithms that automatically vary during their usage by modifying their parameters based on real environmental input.

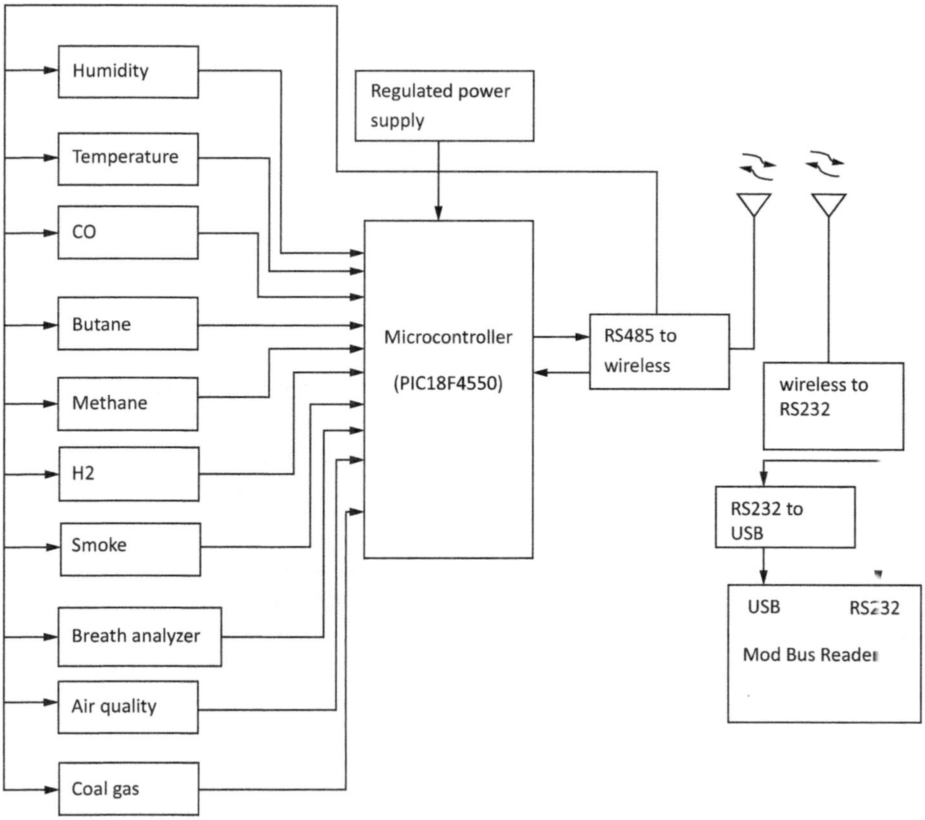

Figure 15.5: Sensor fusion.

Safety and Regulatory Compliance: Safety and regulatory compliance are considered two important variables for any operation involving UAVs, as depicted in Table 15.6. CAS shall follow all aviation standards and regulations as designated by the Federal Aviation Administration (FAA) [22] and other regulatory authorities. This involves installing reliable detect-and-avoid (DAA) technologies for BVLOS operations, the integration of respect for air traffic management systems, and the maintenance of safe separations from other aircraft.

FAA Regulations: Guidelines for safe UAV operations in national airspace serve as fundamental regulations. DAA technologies are ones that are responsible for autonomous operation of UAVs to detect and avoid any other aircraft.

Table 15.6: Regulatory requirements.

Requirement	Description	Regulatory body
Safe distance maintenance	UAVs must maintain a safe distance from other aircraft.	FAA
BVLOS operations	UAVs must be capable of safe operation beyond visual line of sight.	FAA, EASA
Air traffic compliance	UAVs must comply with existing air traffic management systems.	FAA, ICAO

15.11 Discussion

Classical cryptographic solutions are widely used and have a long history. They are able to work with the current communication system. Fewer resources and computations are needed compared to quantum approaches. However, they are susceptible to developments in computing power, particularly quantum computing, and vulnerable to specific attack types, including side-channel and brute-force attacks. The purpose of post-quantum cryptography solutions [23] is to protect against quantum attacks. These solutions make use of mathematical problems that are challenging for computers, both classical and quantum. They frequently entail greater computational overhead and larger key sizes and are still in the development and testing phases. While they guarantee a safe key exchange, they require specialized infrastructure and hardware, which raises the cost. Additionally, they are sensitive to environmental influences, necessitating cautious setup and upkeep. Table 15.7 presents a comparison of various cryptography types.

Table 15.7: Comparison of different types of cryptography.

Feature	Classical cryptography	Post-quantum cryptography	Quantum cryptography
Security	Vulnerable to future quantum attacks	Resistant to quantum attacks	Theoretically unbreakable
Computational overhead	Low to moderate	High	High
Infrastructure requirement	Existing infrastructure	Existing infrastructure	Specialized hardware
Maturity	Well-established	Experimental/ Developmental	Emerging

Strengths and Weaknesses of the Proposed System

Enhanced Security: By integrating quantum cryptography, the system provides theoretically unbreakable security, making it highly resistant to various types of attacks.

Real-Time Communication: The proposed system is designed to meet the real-time communication requirements of UAVs, ensuring timely and secure data exchange [24].

Layered Security Approach: The system employs a multilayered security architecture, combining classical and quantum methods. You should, therefore, develop methods in the design and implementation of a system to improve the general level of security.

Quantum Computing Power Required: All quantum mathematical algorithms in cryptography, particularly QKD, place enormous demands on computer resources, thereby straining the onboard processing capabilities of UAVs.

Environmental Sensitivity: Quantum cryptographic systems may be susceptible to outside interference from changes in temperature, electromagnetic interference, and atmospheric conditions.

Cost and Complexity: Setting up and running a quantum cryptographic system is a very costly affair and requires considerable expertise and tools.

Theoretical Improvements Being Investigated: Error Corrections: Better and more efficient error-correction techniques would reduce computation overheads, thus improving QKD protocol performance.

Advanced Privacy Amplification Techniques: Resource frustration can be circumvented by the optimized use of privacy amplification techniques, resulting in enhanced security.

AI for Anomaly Detection: The proposal establishes the use of AI and machine learning for detecting anomalies and potential security issues instantly, thus improving the robustness of the systems.

Realistic Testbeds for Testing: Building real testbeds for quantum cryptographic systems for UAVs may help in understanding practical issues and system reliability.

Regulatory Compliance Standards: Standardization of protocols for the installation of quantum cryptographic systems on-board UAVs would ensure regulatory compliance.

Interoperability: Interoperable quantum cryptographic systems integrated into the existing communication infrastructure will ensure greater acceptance.

15.12 Future Directions and Emerging Trends

With more focus on increasing performance levels and the accuracy of the algorithm using QKD protocols, which are secure and tend to be computationally intensive, efforts have been made to optimize such QKD protocols, thereby achieving efficiency while conserving resources and maintaining security. New and better error correction methods and privacy amplification mechanisms have been developed to increase performance. Research by Yakushiji et.al. on continuous-variable quantum cryptography with coherent states presents major breakthroughs in this field [25]. Key techniques are: Error Correction: Improved error correction techniques minimize the computational burden without compromising data integrity, as demonstrated in Figure 15.6. Privacy Amplification: Improved techniques guarantee that even if there is leakage of some information, the ultimate key remains secure. The security level computation is shown below:

$$\text{Security level} = f(\text{QKD protocol, communication network}) \tag{15.9}$$

Integrating Advanced AI Techniques: AI is increasingly contributing to the development of quantum cryptographic systems. Machine learning algorithms [26] can aid in predictive analytics, anomaly detection, and adaptive security responses. AI can improve QKD protocols by anticipating and preventing potential security vulnerabilities. Work by Papanastasiou and Pirandola [22] shows how AI methods can be applied to continuous-variable quantum cryptography to improve security. Key Techniques are: Anomaly Detection: Machine learning models identify odd patterns in data, pointing toward possible security breaches. AI anticipates potential future threats and adjusts cryptographic protocols accordingly. A comparison of the security algorithms is illustrated in Table 15.8.

Developing Real-World Testbeds: Creating real-world testbeds is crucial for quantum cryptography in UAVs. These testbeds allow researchers to tackle practical problems such as signal interference, environmental effects, and real-time data processing; for instance, "variational quantum algorithm applied to collision avoidance of unmanned aerial vehicles." Signal Interference Mitigation: Real-world testbeds help identify and solve signal interference issues. Environmental Adaptability: Testbeds help researchers measure the effects of environmental factors on quantum cryptography systems.

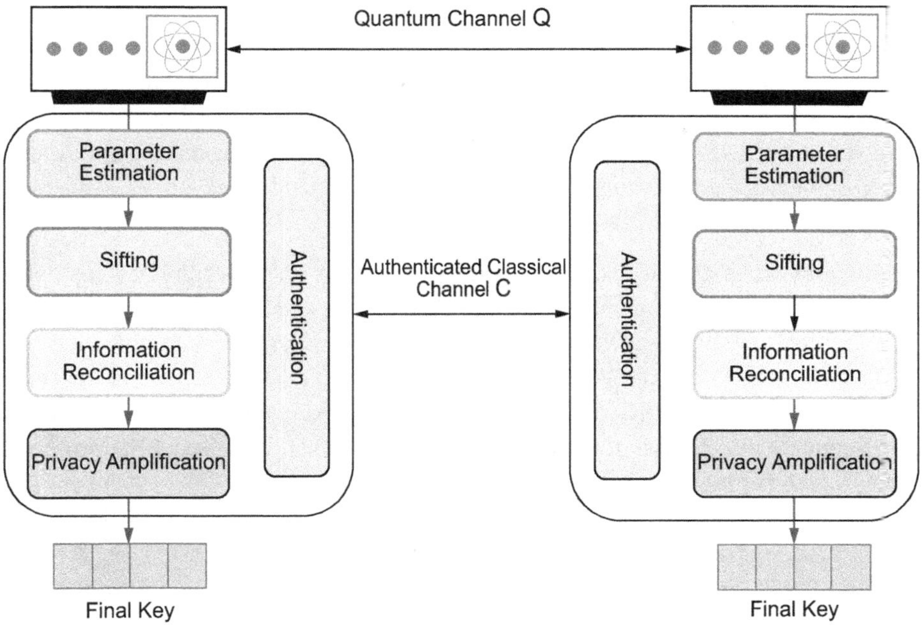

Figure 15.6: Error correction mechanisms.

Security Issues Related to UAV Systems: Essentially, because of their wireless/mobile nature, UAV networks are greatly prone to being affected by security threats. Quantum-resistant algorithms provide strong security against such attacks. Work has been concentrated mainly on enhancing QKD protocols and developing post-quantum cryptographic methods. A thorough investigation into collective symmetric attacks and the efficacy of quantum cryptography to foil them has been carried out in works such as Pirandola (2008) [27]. Major Techniques: Algorithms that resist quantum attacks are termed post-quantum cryptography. Mitigation of Symmetric Collective Attacks: Ways to nullify attacks that target multiple communications.

Table 15.8: Comparison of security algorithms.

Algorithm	Strengths	Weaknesses
QKD	High security, proven robustness	High computational resource demand
Post-quantum cryptography	Resistant to quantum attacks	Complexity in implementation
Symmetric collective defense	Mitigates multi-node attacks	Increased system complexity

Collaboration Between Academia, Industry, and Regulators: Close cooperation among academia, industry, and regulators is vital to push forward the development of quantum cryptography for UAV applications. Cooperation can facilitate the development and standardization of quantum cryptographic protocols to meet the needs of practicality and compliance with the law. The effort on the "State-of-the-Art Survey of Quantum Cryptography" project is an excellent example of effective cooperation between research organizations and industry players, and it has produced substantial progress in the field. Major Advantages are: Standardization: Collaborative efforts result in standardized procedures which are globally accepted. Regulatory Compliance: Ensures that new technologies are legally compliant.

Advancements in Quantum Cryptography: Quantum cryptography developments [28] need to maintain pace with evolving UAV security measures. Hundreds of secure protocols are on the anvil, and the integration of quantum communication networks. Significant work in Gaussian-modulated coherent states QKD led to the realization of a secure and efficient method for communication among UAVs. The salient developments are: Gaussian-Modulated Coherent States: Enhanced QKD protocols providing higher security.

Quantum Communication Networks: Integration of quantum networks for secure communication across multiple UAVs. Security levels are computed using eq. (15.10):

- The equation states that the security level will be a function of the QKD protocol and communication network.
- QKD Protocol – QKD Protocol: It refers to the method by which quantum keys are securely exchanged. The security offered by different QKD protocols like BB84 or E91 depends on their specific principles of quantum mechanics.
- Communication Network: The degree of strength and assurance in data transmission. It encompasses factors such as encryption features, communication latencies, and the possibility of attacks against the network itself that could affect security.

Potential Applications Beyond the Traditional Domains: Quantum cryptographic principles have numerous applications beyond UAVs and military uses. For instance, quantum cryptographic methods can protect sensitive patient data by ensuring privacy in healthcare and preventing breaches. Andersen et al.'s study indicates that quantum cryptographic principles are potentially applicable in many fields, creating room for broad acceptance and novel applications [29]. The main applications are healthcare, securing patient data and medical records; and finance, protecting transactions and communications related to finance.

15.13 Conclusion

This chapter comprehensively explains how quantum cryptography is combined with UAVs to enhance security and solve new operational challenges. The research's main findings are summarized as follows:

Quantum-Actions-All Capable of Security Levels: Quantum cryptography has an almost perfect security level in UAV communications. QKD adapts to all problems relating to integration and possible solutions. There are numerous computational and environmental issues related to the general use of quantum cryptography systems in UAVs. The means of increasing the performance and adaptability of the system include AI-based adaptive protocols, privacy amplification, and other improved error-correcting solutions.

Design a Testbed for Real-World Testbeds and Practical Implementation: In order to mitigate real-world problems such as signal interference and real-time processing, it is essential to apply real-world testbeds for quantum cryptography systems in UAV networks.

Testing Crypto Techniques Across the Board: Comparative analysis shows that while traditional cryptography is less costly with years of experience, its vulnerability to attacks from quantum systems is evident. On the contrary, post-quantum cryptography is computationally intensive but highly resistant to quantum attacks. Quantum cryptography actually provides ultimate security; however, it is vulnerable to environmental factors as well as being dependent on specialized hardware.

Performance Analysis and Case Studies: In the case studies conducted in military, healthcare, and academic environments, it is clear that the performance of the proposed system enhances security and communication efficiency. While lowering computational demands and improving environmental adaptability are areas for further development, performance evaluation metrics show that the system's strong security features are its main strength. Efficiency is defined as the ratio of the number of resources (like time, computing power, or energy) used to maintain security to the level of security that is offered. The objective is to reduce resource usage and increase security.

Future Research Implications for Algorithm Efficiency Enhancements: These should aim to lower resource usage without sacrificing security. Quantum cryptography systems can be made much more flexible and secure with advanced AI integration. Robustness can be improved with AI-driven anomaly detection, predictive analytics, and adaptive security measures. The creation of extensive testbeds will validate working quantum UAV systems. By guaranteeing compatibility between various quantum cryptography systems and current communication infrastructure, testbeds will be developed and used in the future to address real-world issues and improve system performance.

References

[1] Pirandola, S., Braunstein, S. L., Lloyd, S., "Advances in quantum cryptography", Advances in Optics and Photonics, vol. 4, pp.1012–1037, 2020.

[2] Khan, L. U., Saad, W., Bennis, M., Debbah, M., "Edge computing for UAVs: Resource allocation and computation offloading", IEEE Internet of Things Journal, vol.7, pp.3724–3735, 2020.

[3] Shi, W, Cao, J, "Security and privacy in edge computing: A survey", IEEE Communications Surveys & Tutorials, vol.2, pp. 1234–1256,2024.
[4] Deng, Z., & Liu, Z, "Quantum machine learning for UAV path planning and obstacle avoidance", Quantum Information Processing, vol.18(12), pp.1–15,2019.
[5] Lloyd, S., Mohseni, M.,Rebentrost, P, "Quantum algorithms for supervised and unsupervised machine learning", arXiv, pp.1307.0411, 2013.
[6] Paul, C., Giordano, F,"UAV networks and communications", Cambridge University Press, 2016.
[7] Bennett, C. H., & Brassard, "Quantum cryptography: Public key distribution and coin tossing", Proceedings of IEEE International Conference on Computers, Systems and Signal Processing, pp.175–179,1984.
[8] Bernstein, D. J., Lange, T, "Post-Quantum cryptography", Springer International Publishing,2017.
[9] Ekert, A. K., "Quantum cryptography is based on Bell's theorem". Physical Review Letters, vol.67(6), pp.661,1991.
[10] Kolodny, L, "The importance of layered security in UAV systems", Journal of Aerospace Security, vol.12(4), pp.199–212, 2019.
[11] Peikert, C, "A decade of lattice cryptography", Foundations and Trends in Theoretical Computer Science, vol.10(4), pp.283–424, 2016.
[12] Pirandola, S, "Quantum communication with coherent states and linear optics. Nature Photonics", vol.9(12), pp.724–733, 2015.
[13] Kanimozhi, P., Kannan, M., Ananth, C., Kumar, T. A., & Parkavi, S. (2023, September). Design of IoT Based Agricultural Farm Protector Drone Using Raspberry Pi. In 2023 6th International Conference on Contemporary Computing and Informatics (IC3I) (Vol. 6, pp. 1175–1179). IEEE.
[14] Kumar, P. P., Kumar, T. A., Muthu, P., Rajendirane, R., & Samuel, R. D. J. (2022). Security Analysis of UAV Communication Protocols: Solutions, Prospects, and Encounters. In The Internet of Drones (pp. 167–190). Apple Academic Press.
[15] Adams, B., Friedland, C. J,"A survey of unmanned aerial vehicle (UAV) usage for imagery collection in disaster research and management", Proceedings of the 9th International Workshop on Remote Sensing for Disaster Response, 2011.
[16] Amukele, T. K., "Can unmanned aerial systems (drones) be used for the routine transport of chemistry, hematology, and coagulation laboratory specimens", Journal of Clinical Microbiology, vol.53(10), pp.3222–3226,2015.
[17] Anderson, K., & Gaston, K. J, "Lightweight unmanned aerial vehicles will revolutionize spatial ecology", Frontiers in Ecology and the Environment, vol.11(3), pp.138–146,2013.
[18] Clarke, R, "Understanding the drone epidemic: Technology, law, and ethics", Computer Law & Security Review, vol.32(3), pp.229–246, 2016.
[19] Dorling, K., et al., "Vehicle routing problems for drone delivery", IEEE Transactions on Systems, Man, and Cybernetics: Systems, vol.47(1), pp.70–85, 2017.
[20] Grosshans, F., Grangier, P, "Continuous variable quantum cryptography using coherent states", Springer,2002.
[21] Otto, A., & Agatz, N, "The impact of delivery drones on urban logistics: Cost, time, and environmental benefits", Transportation Research Part C: Emerging Technologies, vol.94, pp.132–153, 2018.
[22] Papanastasiou, P., & Pirandola, S, "Machine learning for continuous-variable quantum cryptograph", Springer, 2020.
[23] Pirandola, S, "Symmetric collective attacks in quantum cryptography", Springer, 2008.
[24] Remondino, F., & Nex, F, "UAV photogrammetry for mapping and 3D modeling – Current status and future perspectives", International Archives of the Photogrammetry, Remote Sensing and Spatial Information Sciences, vol.40(1), pp.25, 2013.

[25] Yakushiji, F., Yakushiji, K., Murata, M., Hiroi, N., Takeda, K., & Fujita, H, "An evidential study of the unmanned aerial vehicle conveyance of transfusion material", Journal of Clinical Medicine, vol.9(4), pp.1234–1246, 2020.

[26] Opiłka, F., Niemiec, M., Gagliardi, M., & Kourtis, M. A., "Performance analysis of post-quantum cryptography algorithms for digital signature", Applied Sciences, vol.14(12), pp.4994, 2024.

[27] Telli, Khaled, et al, "A comprehensive review of recent research trends on unmanned aerial vehicles (uavs)." Systems, pp.8–11,vol.4, 2023.

[28] Kumar, Manish, "Post-quantum cryptography Algorithm's standardization and performance analysis", Array pp.15,vol.10,2022.

[29] Liang, X., Wang, H., Cao, M., Guo, T, "On-line path planning for UAV in dynamic environment", In Proceedings of the 2011 2nd International Congress on Computer Applications and Computational Science, pp. 9–19, 2011.

Arun Anoop M. and Karthikeyan P.

16 Quantum Machine Learning for Sensor Fusion

Abstract: Combining machine learning with quantum computing provides a promising approach for image processing, medical imaging, and machine learning to ensure accuracy and speed. The utilization of quantum machine learning allows the same machine learning classifiers, which provide improved speed and performance accuracy rates for any kind of images or different modalities. The term "sensor fusion" refers to the integration of different imaging modalities or cross-domain features like spatial, frequency, metadata, and physical attributes to enhance detection capabilities. As we know, in classical computers, bits are the data storage units where values are either 0 or 1. In the case of quantum computers, bits are replaced by the term qubits or quantum bits. Qubits are managed by quantum computers, which are capable of solving more complex problems. In quantum machine learning (QML), modules like Qiskit and AerSimulators (which are quantum simulators) play a role in managing training and prediction before accuracy score generation. Primarily, QML works based on fidelity (similarity between two quantum states, meaning similarity between classical and quantum predictions), samplers, zzfeaturemap() (which converts classical features into a quantum-capable format), and adhoc kernels, along with classifiers before the training or fit function. Here, the sensor fusion we have considered involves categorical and numerical values from the Titanic survival dataset. In this chapter, we mainly concentrate on the phases required for sensor-used feature classification, and finally, primer evaluation will also be conducted with the help of the Titanic survival dataset to highlight the importance of the quantum machine learning branch in handling sensor fusion.

Keywords: Titanic survival dataset, quantum machine learning classifiers, sensor fusion selection

16.1 Introduction

Quantum computing originates from computer science and utilizes physics concepts from quantum theory to solve problems that cannot be solved by classical computers. Based on data storage, there are some differences between classical and quantum computers. Bits, which are either 0 or 1, are used in classical computers [1]. There is a term named qubits or quantum bits, which is used to denote in quantum computers. Quantum devices can solve complex problems and that has the capability to pass information to each other. In machine learning there are some additional concepts like

https://doi.org/10.1515/9783111619323-016

quantum computing algorithms is called quantum learning to improve speed and accuracy of learned data. Some examples of quantum machine learning are quantum-based neural network (QNN), quantum-based support vector machine (QSVM), and quantum-based reinforcement learning. In quantum classification, there is a term called fidelity [2], which represents a class label that matches the expected state. Quantum neural networks are quantum concepts with classical neural networks that uses gates called quantum gates and concept called circuits to learn and process the information. QSVM uses exactly the same like support vector machine classical version, where sigmoidal term is in kernels, and in QSVM uses quantum kernels [3] to find separating hyperplanes (same concept in classical SVM) in a high-dimensional quantum feature space [4].

Sensor fusion enhances accuracy and robustness, and it involves integrating different imaging modalities. It is the utilization of different types of images like RGB, grayscale, and infrared in order to gather comprehensive data. Another example includes integrating features from various domains such as spatial, frequency, physical, or metadata to enhance detection capabilities. Terms are classical (**C**) in the case of classical computing and quantum (**Q**) in quantum computing. Classical data using classical computers is denoted by CC, classical data using quantum machine learning algorithms is denoted by CQ, quantum data using classical machine learning algorithms is denoted by QC [5], and quantum data using quantum machine learning algorithms is denoted by using QQ.

The following are the programming languages or frameworks that are commonly used in the field of quantum computing:

- Intel quantum simulator (Intel-QS), also known as the quantum high performance software testing environment called Q-HiPSTER [6], which is a simulator of quantum circuits optimized and is with multicore and multinodes architectures.
- Authored by softwareQ Inc. under the MIT License, Staq is a modern C++ 17 library for the synthesis, transformation, optimization, and compilation of quantum circuits.
- Simon Benjamin's quantum technology theory group (qtechtheory) was developed by QuEST [7] and its e-research center (OERC) at the University of Oxford.
- Scaffold programming language is called ScaffCC, and it is written using open-source infrastructure. It is mainly used for writing and analyzing code for quantum computing applications.
- C++ quantum bit and gate simulator called Qrack, with the ability to support arbitrary numbers of entangled qubits.
- Universal quantum circuit simulator called The QX Simulator, developed at QuTech by Nader Khammassi, is a simulator that allows quantum algorithm designers to simulate quantum computers with the execution of their quantum circuits.
- Simulations of up to 40 qubits on a 90-core Intel Xeon workstation with the help of QSim integration with Cirq [8].

- Modern C++ general-purpose quantum computing library called Quantum++, which consists of template header files.
- Quantum imperative programming language called LanQ.
- C library for the simulation of quantum mechanics named lib-quantum, which is specially focused on quantum computing.
- Cloud-based platform named IBM quantum experience, which allows its customers to experiment with real quantum hardware, is the product of IBM.
- Open-source software development kit called Qiskit [9], which is for quantum computing, is developed and maintained by IBM.
- Open-source software framework for quantum computing, named ProjectQ, supports high-level programming languages, mainly Python.
- Another open-source software library for quantum computing is Cirq, which is developed by Google.
- Open-source software library for quantum computing is PyQuil, developed by Rigetti Computing.
- Open-source framework for quantum algorithm design is called Yao.
- High-level quantum programming language is Silq, developed by ETH Zurich.
- High-level programming language called Qrisp, mainly used for creating and compiling quantum algorithms.
- Open-source toolkit called Mitiq, which is for implementing error mitigation techniques on quantum computers.
- Quantum machine learning tool, Paddle Quantum, developed based on Baidu's Flying Paddle.
- Extensible quantum information and learning architecture, named Tequila.
- Linux machine image, Bayesforge that curates the very best open-source software for quantum computing.
- Python-based software framework for universal quantum computing, named Blueqat [10], developed by Blueqat Inc., is an open-source quantum computing framework for design and simulation based on quantum algorithms.
- Web-based graphical user interface, the Quantum Programming Studio, is designed to allow users to construct quantum algorithms.
- Open-source high-level API called Qibo, provided by Qilimanjaro, written in Python and capable of running quantum algorithms.
- The Berkeley Quantum Synthesis Toolkit is a super-optimizing quantum compiler.
- Variational Quantum Circuit Simulator, Qulacs for quantum computation research.
- The Quantum Programming Studio is a web-based quantum programming IDE and simulator, supported by the Unitary Fund and Rigetti Computing.
- Extensible compilation framework, XACC for hybrid quantum-classical computing architectures.
- Quantum Circuit Analyzer Tool name is QuCAT, QuTIP is an open-source quantum computing software, and TensorFlow quantum is a quantum machine learning library.

16.2 Important Terms

Quantum computing consists of some important terms that are explained below as follows:

16.2.1 Vanilla and Natural Gradient Descent

Gradient descent is an optimization algorithm. Gradient descent is also called "the deepest downward slope algorithm [15]," which is used to minimize the cost function in machine learning and deep learning. There is a concept called quantum circuit, which consists of quantum states and gates. In quantum states, there is a concept named Hilbert space [8], and it is denoted as psi between a vertical line and the > symbol. In quantum gates, there is a concept that consists of unitary operators that transform q-states. Vanilla gradient descent uses the standard first-order derivative gradient of the loss function for updating model parameters. Natural gradient descent is built based on the Fisher information matrix, and it has a concept named Riemannian manifold.

16.2.2 Pauli Expectation

Example of Pauli vector visualization using qisit [16], as shown in Figure 16.1:

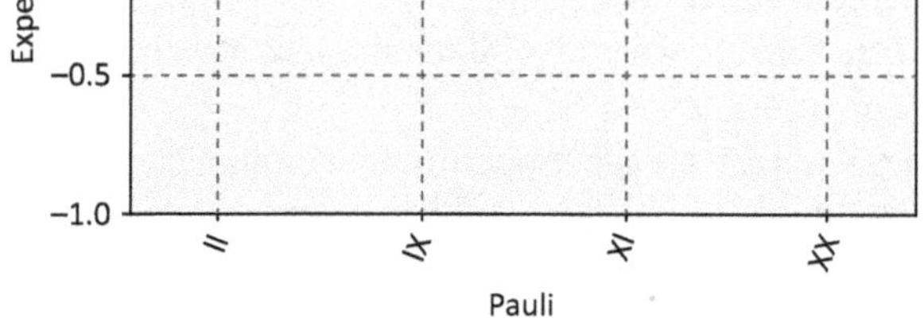

Figure 16.1: Pauli vector visualization.

plot_state_paulivec() function parameters are

Table 16.1: Pauli vector parameters.

Parameter	Brief description
State	It is a state vector that represents the state of all qubits in the quantum circuit.
Title	It is a string that is used for the plot title.
Color	It is a string or a list of strings that contains the color of the bars.
Figsize	It is a tuple that contains the size of the figure in inches.

Above, the focus is mainly on the arguments of pulivec(). Four attributes are clearly discussed in Table 16.1. Paulivec() description:

- Qiskit's library is used to create and simulate a quantum circuit with some parameters defined as in Table 16.2.
- Initializes a quantum circuit with two qubits. [11]
- The Hadamard (H) gate is applied to both qubits.
- Backend for the simulation is set to the statevector simulator using BasicAer library.
- Executing the quantum circuit using the first and third steps.
- Result function is used in the fourth step for result output from the job.
- Retrieves the state vector of the quantum circuit's final state.
- Visualizes the state vector using a Pauli vector plot.

Table 16.2: Qiskit parameters with description.

Qiskit parameters	Description
QuantumInstance	It is a utility class that allows joint configuration, and it provides functions at a higher level of abstraction, which is part of a quantum computing library, such as Qiskit. Setting up a QuantumInstance for executing quantum algorithms in a quantum computing framework, such as Qiskit, where a quantum simulator is the backend.
CircuitSampler	In quantum machine learning (QML), a quantum sampler is an element that generates samples from a quantum circuit's parameters.
Quantum Fisher Information	The **quantum Fisher information** (QFI) is a fundamental quantity of interest [17].

Table 16.2 (continued)

Qiskit parameters	Description
Simultaneous perturbation stochastic approximation (SPSA)	It is an optimization algorithm that was invented by James C. Spall. This method is usually used for handling multiple unknown parameters. It falls under the category of stochastic approximation algorithms, and its main application is modeling, including atmospheric modeling. SPSA consists of the term simulated annealing. Another stochastic approximation algorithm is FDSA, which is finite difference stochastic approximation algorithm. In SPSA, main mathematical concepts include Newton-Raphson algorithm, Hessan matrix, and stochastic gradients.
SPSA algorithm in Qiskit **[13]**	This concept is mainly used for simulation optimization. One simulator named quantum assembly language (QASM) simulator is recommended for SPSA algorithm. In that maximum iteration must be declared while keeping SPSA as a function. There are two attributes available: learning rate and perturbation. If learning rate is set, then perturbation is must, and if perturbation is set, learning rate is also a must.
Quantum natural SPSA (QN-SPSA) optimizer **[14]**	QN-SPSA is the Quantum Natural Simultaneous Perturbation Stochastic Approximation algorithm. It belongs to the family of gradient descent methods. It uses a circuit called ANSATZ, and QNSPSA applies the fidelity concept by utilizing the circuit already mentioned.
Qubits **[18]**	In Section 16.1, CC uses "Bit" and QC uses "Qubit." Qubits [18] are called quantum bits, and they calculate and represent 0 and 1 at the same time. QC stands for quantum computing. It is the fundamental unit of information in quantum computing. The bit representation of a quantum high-performance computer is a qubit. Classic computing, which calculates with transistors, can represent either 0 or 1.
Quantum feature map and kernels **[19]**	Normally, it can be used based on nonlinear problems. It utilizes feature mappings to transform data into higher-dimensional spaces for better classification. The kernel function is a tool that is used for measuring distances in feature space without feature vector calculation.
QSVM **and** ***QKSVM***	QSVM is an enhanced version of CSVM by utilizing quantum computer to manage complex datasets. CSVM is used to separate data points using hyperplanes. QKSVM is QSVM with kernel functions.
PegasosQSVM	It is used to train for larger training sets.
Quantum convolutional neural network (QCNN) [10]	QCNN combines principles of quantum mechanics with the convolutional neural networks concept.

Table 16.2 (continued)

Qiskit parameters	Description
Quantum autoencoder [12,13]	It is a QNN model, and the term used is quantum states.
IBM QISKIT machine learning framework	– SHOR's: is a factoring algorithm used for QC. – GROVER's: is a searching algorithm mainly using to solve unstructured search problems in QC. – DEUTSCH-JOZSA algorithm: is used to solve multiple and deterministic computational problems. – DEUTSCH's: check if the function is constant or balanced. – BERNSTEIN-VAZIRANI: is a restricted version of the Deutsch-Jozsa algorithm, and it provides information about "blackbox function." – SIMON algorithm: is used to break classical cryptosystems, and it is called a quantum algorithm.

16.2.3 Quantum Machine Learning Classifiers

Quantum machine learning classifiers has primarily focused on *quantum support vector machine (QSVM), quantum neural networks,* and *quantum k-nearest neighbor,* which have been used for the performance evaluation of application-based tasks. For the application, the Titanic Survival Prediction dataset has been used. Main functions like ZZFeatureMap() has used to convert classical features into quantum capability format. Quantum states are generated through a concept called sampler. After that step, fidelity is calculated and it is used to measure the similarity between quantum state and classical predictions. Each classifier may be used for fine-tuning; during that time, kernel variants can be used. So here kernel methods have applied and trained using the transformed features for fitting and prediction process. Algorithm1 includes an algorithm for the proposed step. The methods have utilized some predefined methods and functions.

Algorithm 1: Quantum machine learning classifier for Titanic survival prediction

1. **Input:** Input Image
2. **Output:** Classification Accuracy
3. **Start**
4. **Input Data:** Titanic Dataset
5. **Feature Extraction:** Classical features
6. **ZZFeatureMap():** Convert classical features into quantum-capacity format.
7. **Sampler:** Generate quantum state.
8. **Fidelity calculation:** Measure similarity between quantum states and classical predictions.

9. **Adhoc kernels:** Apply kernel methods.
10. **Classifiers:** Train classifiers using transformed features.
11. **Fit function:** Train the model.
12. **Output:** Predictions on Titanic Survival
13. **Stop**

16.3 Materials and Methods

16.3.1 Data Collection

The dataset is the Titanic survival dataset, as shown in Table 16.3, which consists of attributes mainly survived(0,1), pclass(1,3), name(passenger names), and sex(male,female) requiring a mapping process to numerical values since machine learning requires numerical values for evaluation: age (passenger's age), sibsp(0,1), parch(0,1), ticket(options), fare(money), cabin(rooms representations), and embarked(options). The total number of data entries in this dataset is 892 with 11 attributes.

Table 16.3: Titanic survival dataset attributes and description.

Variable	Definition	Key
Survival	Survival	0 = NO, 1 = YES
Pclass	Ticket class	1 = 1st, 2 = 2nd, 3 = 3rd
Sex	Sex	
Age	Age in years	
Sibsp	# or number of siblings or spouses aboard the Titanic	
Parch	# or number of parents or children aboard the Titanic	
Ticket	Ticket number	
Fare	Passenger fare	
Cabin	Cabin number	
Embarked (C,Q,S)	Port of embarkation	Queenstown

Pclass: A proxy for socioeconomic status, which consists of 1st = Upper, 2nd = Middle, and 3rd = Lower.

Age: Age is fractional if it is less than 1. If the age is estimated, it is in the form of xx.5.

Sibsp: The dataset defines family relations in this way: Sibling and spouse. Sibling mainly refers to brother, sister, stepbrother, and step sister. Spouse mainly refers to husband or wife.

Parch: The dataset defines family relations in this way: Parent, child, and some children traveled only with a nanny; therefore, parch = 0 for them. Parent refers mainly to mother and father. Child refers mainly to daughter, son, stepdaughter, and stepson.

16.3.2 ZZFeatureMap() and Mathematical Explanation of ZZFeatureMap()

The ZZfeatureMap() function is a representation available in QML [1,9], and it is used to map classical kinds of data into a quantum space with higher dimensions. Sensor fusion is the integration of multiple sensor data in order to improve the quality of the information. A feature map is used to embed classical types of data into a quantum state, which is the representation of classical sensor data. Quantum circuits can be used to manipulate quantum states [3].

Classical Data Representation: Assume we have classical data $X = (x_{1,} x_2, \ldots\ldots x_n)$ gathered from different sensors. The feature map $\varnothing$: $R^n \rightarrow C^2$ transforms this vector into a state suitable for quantum systems.

Quantum States: A quantum state can be represented as a linear combination of basis states. For a single qubit, a state can be expressed as: $|\psi\rangle\ \alpha|0 + \beta|1\rangle$

$$\text{where}|\propto|^2 + |\beta|^2 = 1$$

Embedding Function:
The feature map *ZZfeatureMap*(**x**) could have the following form:

ZZfeatureMap$(x) = \cos(\theta)|0\rangle + \sin(\theta)|1\rangle$ Here, θ is a function of the features in **x**, $\theta = f(x_{1,}\ x_2, \ldots\ldots, x_n)$ where f is an embedding function, which can be derived from the original data **x**.

Multi-qubit Cases: When dealing with multiple sensors, multiple qubits are involved. For instance, with two qubits:

$$|\varphi\rangle_{12} = U(\mathbf{x})(|0\rangle \otimes |0\rangle)$$

where U(**x**) is a unitary operation derived from *ZZfeatureMap*(**x**).

Quantum Gates: Different quantum gates can be applied based on the embedded data, allowing quantum operations.

16.4 Mathematical Explanation of Sampling

A sampler in QML refers to a quantum process that can generate distributions over quantum states. In sensor fusion applications, samplers have a major role in combining and evaluating data from multiple sensors effectively.

16.4.1 Quantum States and Measurement

A quantum state can be represented as $|\psi\rangle = \sum_i c_i|i_i$
where c_i is the probability amplitude and $|i\rangle$ is the basis state.

16.4.2 Probabilities

The probability of measuring a state $|i\rangle$ from$|\psi\rangle$ is given by $P(i) = |c_i|^2$.

16.4.3 Quantum Sampling Process

When sampling, a quantum measurement is performed. If a quantum state $|\psi\rangle$ is measured, it is based on the probability distribution $P(i)$.

16.4.4 Quantum Circuits

In practical QML applications, a quantum circuit is designed to represent the data from sensors. A parameterized circuit $U(\boldsymbol{\theta})$ is used as $|\psi\,(\boldsymbol{\theta})\rangle = U(\boldsymbol{\theta})\ |0\rangle^{\otimes n}$.

16.4.5 Sampling Strategy

The output from the circuit can then be sampled to determine the measurement results, which provide sensor data.

16.5 Fidelity in Quantum Machine Learning and Mathematical Explanation of Fidelity

In sensor fusion, fidelity can evaluate how well a QML model represents the true distribution of the fused sensor data.
Pure States: For two pure states $|\psi\rangle$ and $|\phi\rangle$, the fidelity F is defined as

$$F(|\psi\rangle, |\phi\rangle) = |<\psi|\phi\rangle^2$$

This ranges from 0 to 1, where $F = 0$ indicates that they are orthogonal, and $F = 1$ indicates that the states are identical.

16.5.1 Mixed States

For mixed states represented by density matrices ρ and σ, the fidelity is given $F(\rho, \sigma) = \left(Tr\sqrt{\sqrt{\rho\sigma}\sqrt{\rho}} \right)^2$.

16.5.2 Fidelity in Sensor Fusion

In a sensor fusion context, if we have a target distribution ρ derived from classical sensor readings, and a quantum state σ representing the QML outcome, we can evaluate: $F(\rho, \sigma) = \left(\mathrm{Tr}\sqrt{\sqrt{\rho\sigma}\sqrt{\rho}} \right)^2$.

16.6 Ad Hoc Kernels and QML Classifiers

Quantum support vector machine (QSVM), quantum neural networks, and quantum *k*-nearest neighbor.

For demonstrating the sensor fusion concept, the titanicsurvival.csv dataset file has been used to showcase the performance of classical and quantum support vector machine systems. Initially, you can observe the dataset loading, the first five data values, and the performance and time taken for classical and quantum computing performance evaluations. The confusion matrix has been plotted based on Table 16.4

df = pd.read_csv('CSV file location')

df.head(5) lists the first five entries from the dataset.

Table 16.4: Titanic survival dataset sample values.

Pclass	Sex	Age	Fare	Survived
3	Male	22.0	7.2500	0
1	Female	38.0	71.2833	1
3	Female	26.0	7.9250	1
1	Female	35.0	53.1000	1
3	Male	35.0	8.0500	0

Titanic survival dataset is used to demonstrate multi-modal values fusion. For that, the Pandas library is utilized to read the csv dataset, and the head() function is used to display the first five values. The head function, when called, displays five attribute values from the same dataset.

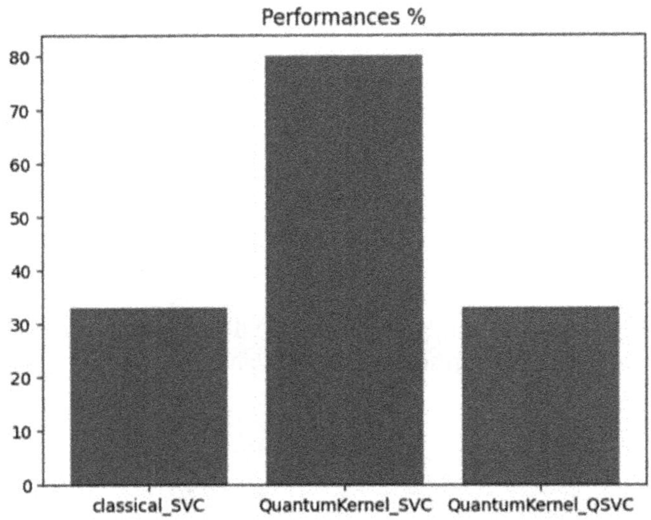

Figure 16.2: Performance of C_SVC, QK_SVC, and QK_QSVC.

The performance of classical and quantum computing is plotted above, which considers the performance difference between classical and quantum computers in terms of evaluation by considering different support vector machines: classical SVC versus quantum kernel [7] SVC and its variant QSVC [6]. In the above plot in Figure 16.2, it is very clear that quantum kernel SVC's performance accuracy is higher than all other

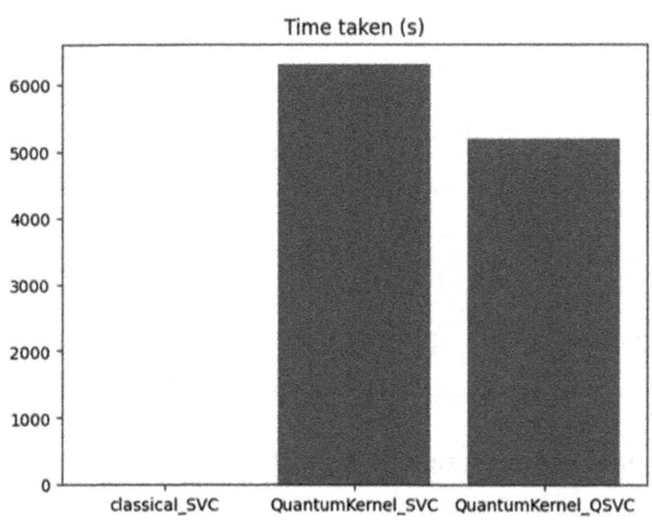

Figure 16.3: Time taken measurement of C_SVC, QK_SVC, and QK_QSVC.

methods. Quantum kernel SVC achieved a performance accuracy of 70–80%, while other methods achieved below 50%.

Time data of classical and quantum computing have been plotted in Figure 16.3. Mainly, support vector machine is considered for the work. In this, different kernels and different variants of SVM, like quantum kernel SVC and quantum kernel QSVC, have been added. The above plots compare the results of classical and quantum computers based on machine learning classifiers. QK_SVC took 6,000, while the other method, QK_QSVC, took 5,000.

Confusion matrix for sensor fusion is plotted based on Titanic survival dataset attributes, as shown in Figure 16.4, and values of corresponding attributes. Mainly categorized as died and survived in the above plot, those values were collected from the Titanic survival dataset. Also, it is the fusion of different modalities of data and the concept of sensor fusion.

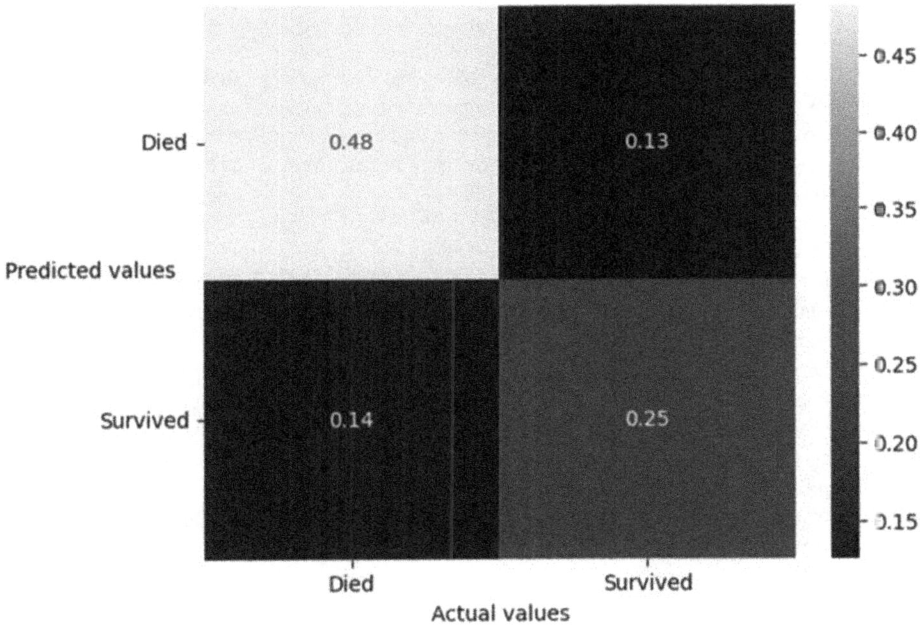

Figure 16.4: Confusion matrix of the Titanic survival dataset.

16.7 Conclusion

Quantum computing is a branch of computer science that uses quantum physics concepts to solve problems and statements that regular or classical computers can't solve.

Programming languages and different online frameworks have been explained with detailed descriptions. Vanilla and natural gradient descent, Pauli expectation, quantum instance, sampler, quantum Fisher information, simultaneous perturbation stochastic approximation (SPSA), SPSA algorithm in QISKIT, quantum natural SPSA, qubits, quantum feature maps and kernels, QSVM, QKSVM, Pegasos QSVM, quantum convolutional neural network, and different frameworks of IBM QISKIT as well as QML classifiers have been clearly explained. In the performance evaluation, the Titanic survival dataset was considered. Data acquisition, ZZFeaturemap(), Sampler, and Fidelity in QML with mathematical notations were added, and the result-wise plots were shared. Classical SVC, quantum kernel SVC, and quantum kernel QSVC were considered. The confusion matrix of predicted and actual values was used in the Titanic survival dataset to differentiate survival and dead labels, showcasing the supervised machine learning results.

References

[1] Vedran Dunjko, Jacob M. Taylor and Hans J. Briegel, *"Quantum-Enhanced Machine Learning,"* Physical Review Letters 117, vol. 13,2016, DOI:10.1103/PhysRevLett.117.130501

[2] Vojtech Havlicek, Antonio D. Córcoles, Kristan Temme, Aram W. Harrow, Abhinav Kandala, Jerry M. Chow and Jay M. Gambetta, *"Supervised learning with quantum enhanced feature spaces,"* Nature 567, pp 209–212, 2019, DOI:10.1038/s41586-019-0980-2.

[3] Maria Schuld and Nathan Killoran, *"Quantum machine learning in feature Hilbert spaces,"* Phys. Rev. Lett. 122,2019, DOI:10.1103/PhysRevLett.122.040504

[4] Jennifer R. Glick, Tanvi P. Gujarati, Antonio D. Corcoles, Youngseok Kim, Abhinav Kandala, Jay M. Gambetta and Kristan Temme, *"Covariant quantum kernels for data with group structure,"*Arxiv, 2020.

[5] Thomas Hubregtsen, David Wierichs, Elies Gil-Fuster, Peter-Jan H. S. Derks, Paul K. Faehrmann and Johannes Jakob Meyer, *"Training Quantum Embedding Kernels on Near-Term Quantum Computers,"*Arxiv, 2019.

[6] Yunchao Liu, Srinivasan Arunachalam and Kristan Temme, *"A rigorous and robust quantum speed-up in supervised machine learning,"Arxiv*, 2020.

[7] Hsin-Yuan Huang, Michael Broughton, Masoud Mohseni, Ryan Babbush, Sergio Boixo, Hartmut Neven and Jarrod R. McClean, *"Power of data in quantum machine learning*,arXiv,2020.

[8] Lloyd, Seth, Maria Schuld, Aroosa Ijaz, Josh Izaac, and Nathan Killoran, "Quantum embeddings for machine learning," *arXiv*, 2020.

[9] Arunmozhiselvi, S., T. Ananth Kumar, P. Manju Bala, S. Usharani, and G. Glorindal. "A systematic approach to agricultural drones using a machine learning model." In Machine Learning Approaches and Applications in Applied Intelligence for Healthcare Data Analytics, pp. 41–60. CRC Press, 2022.

[10] IBM Convolutional Neural Networks, Available: https://www.ibm.com/cloud/learn/convolutional-neural-networks

[11] Vatan, Farrokh, and Colin Williams., "Optimal quantum circuits for general two-qubit gates," Physical Review, vol. 69, 2004, 10.1103/PhysRevA.69.032315.

[12] A wikipedia page on Autoencoder: https://en.wikipedia.org/wiki/Autoencoder

[13] Romero, Jonathan, Jonathan P. Olson, and Alan Aspuru-Guzik., "Quantum autoencoders for efficient compression of quantum data," Quantum Science and Technology 2.4, 045001, 2017.

[14] Swap Test Algorithm: https://en.wikipedia.org/wiki/Swap_test
[15] Ali Shahryar Khokhar (24100266), "A Review of Quantum Gradient Descent," LUMS SSE PHY 612 Quantum Information Science, 2023
[16] Pauli vector, Available: https://deeplearninguniversity.com/qiskit/qiskit-pauli-vector/
[17] Aniket Rath, Cyril Branciard, Anna Minguzzi, and Benoît Vermersch, "Quantum Fisher Information from Randomized Measurements," Phys. Rev. Lett. 127, 260501, December 2021
[18] Quantum computing vs Classical computing, Available: https://devopedia.org/quantum-computing
[19] Kristan Temme(IBM), "Lecture 6.2 Quantum Feature maps & Kernels, QISKIT," Available: https://www.youtube.com/watch?v=zw3JYUrS-v8

Sabia Hassan, Syed Naseer Ahmad Shah, and Kaiser Javeed Giri

17 Integration of Quantum Genetic Heuristics

Abstract: Quantum Computing (QC) is the next step in processing power by using principles like superposition and entanglement to solve problems that are otherwise intractable for classical computers. Within this realm, Quantum Genetic Algorithms (QGAs) are one of the most powerful tools that combine the evolutionary strategies of genetic algorithms with the unparalleled efficiency of QC. These algorithms are specifically suited to the exploration of large, complex problem spaces and would, therefore, make for a fantastic fit for enhancing the capabilities of Unmanned Aerial Vehicles (UAVs). Quantum genetics can be coupled with UAV technology to open new avenues in the advancement of aerial systems. QGAs may be used to allow UAVs to perform better in critical areas such as real-time obstacle avoidance, dynamic mission planning, energy-efficient routing, and swarm intelligence. These aerial vehicles can work more effectively in complicated and dynamic environments because they can work on multiple possibilities at once. This chapter will dig into the groundbreaking integration of QGAs with UAV systems, providing a revolutionary approach to enhancing UAV efficiency and autonomy, with in-depth examination of the conceptual framework behind QGAs, their applications within UAV systems, and essential hardware and software requirements that integrate quantum processors into UAV architectures. This further opens up avenues for ascertaining potential applications of QGAs in areas such as military operations, environmental monitoring, and disaster management in order to showcase the wide impact these technologies might carry across different sectors. This chapter would serve as a roadmap for researchers and discuss the hurdles in adapting quantum algorithms for large fleets of UAVs, technological barriers existing today, and the progress required to fully unleash the power of QC in aerial systems.

Keywords: Quantum genetic algorithms, UAVs, quantum computing, optimization, swarm intelligence, autonomous systems, energy efficiency, mission planning

17.1 Introduction

The integration of quantum computing (QC) and genetic algorithms (GAs) has provided immense optimizations [1]. This fusion is termed as Quantum Genetic Heuristics (QGH), which unites the problem-solving strength of GAs with the computational capability of QC [2]. GAs are based on natural selection. These are a class of evolutionary algorithms using inheritance, mutation, selection, and crossover techniques to find

https://doi.org/10.1515/9783111619323-017

the best solutions for complex problems [3]. They are particularly useful for optimization problems when the goal is to choose the best of an enormously wide set of variants. QC processes information based on the principles of quantum physics. While ordinary computers use bits, 0s, and 1s, to process information, quantum computers use quantum bits or qubits that can exist in multiple states simultaneously. This is called Superpositon, and it allows quantum computers to process several possibilities at the same time, which could make them much faster than conventional computers for some tasks [4]. The combination of QC and GA, which gives QGH, takes the best of both technologies to solve complex optimization problems more efficiently. QGA uses quantum bits to represent possible solutions, which enables a vast exploration of the solution space. This might result in faster convergence and better solutions, especially for problems involving multiple variables or complicated constraints [5]. This combination creates new avenues for handling previously intractable problems. QGA can be applied to complex systems such as supply chain networks, traffic management, and financial portfolios. They can also be applied to optimization problems in engineering, biology, and computer science. The integration of QC with GAs does present some challenges. Quantum computers are only in the preliminary research phase and are not yet very useful for real optimization problems. Using quantum computers with GAs requires a great understanding of both topics, which is challenging and demanding. Considering the aforementioned constraints, the potential benefits of quantum genetic heuristics (QGH) make this field very interesting. With each passing day, the improvement in QC technology will increase the usage of QGH in many areas. This chapter is meant to discuss the principles of QGH, its applications, and the problems encountered by QGH.

17.1.1 Overview of UAV Technology

UAVs, also known as drones, are traced back to the early twentieth century. The idea came in World War I when the U.S. military designed the first remote-controlled aircraft, called the Kettering Bug. Although it was a very primitive model, it served as a basis for future innovations [6]. During World War II, the Allied and Axis powers studied UAV technology but were limited to the available technology. Further development in UAV technology was mainly pushed by the need for surveillance during the nuclear weapons race. UAVs such as the RQ-2 Pioneer and RQ-3 DarkStar became essential tools for surveillance, mainly used for gathering intelligence without putting human pilots at risk [7]. The technical progress of the late twentieth and early twenty-first centuries, primarily the miniaturization of electronics and improvements in communication networks, immensely helped the spread of UAVs [8]. The last few decades have seen UAV technology progress remarkably in military and civilian applications. Modern UAVs, such as the RQ-4 Global Hawk and the MQ-1 Predator, possess capabili-

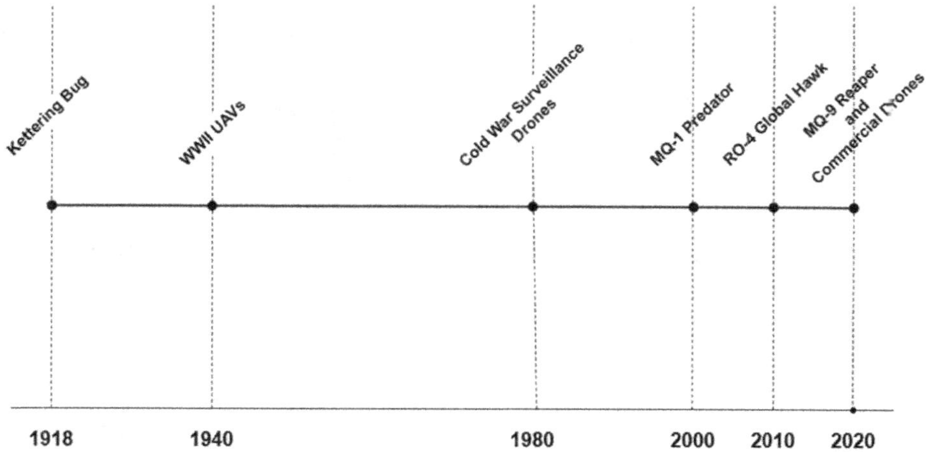

Figure 17.1: UAV evolution timeline.

ties of high surveillance, accurate attacks, and autonomous operations. The evolution of UAVs over time is shown in Figure 17.1.

These developments have increased the effectiveness and safety of performing military operations. Changes have significantly redefined from basic reconnaissance drones to extremely advanced combat and logistical systems, thereby changing the current military and industrial operations. UAVs have become widespread in different industries, from military operations to commercial use, as shown in Figure 17.2.

Defense: UAVs are mainly utilized in armed conflicts for surveillance, intelligence gathering, and precise bombardments. Armed drones, such as the MQ-9 Reaper, would confer a tactical advantage in engaging the enemy on the battlefield by providing real-time intelligence and precision strikes with no risk to human life. UAVs are utilized for supply chain and logistics management on battlefields, thus averting risks to human personnel.

Logistics: UAV technology is utilized for last-mile delivery. Amazon and UPS use drones for delivery to ensure that the product delivery process becomes faster and smoother, especially in remote areas [9]. Drones are also utilized in inventory warehousing to monitor product movement levels, which reduces manual labor and increases productivity.

Monitoring: UAVs are essential in monitoring large areas for security purposes, environmental monitoring, and disaster cases. UAVs enable crop monitoring and precision farming; with them, one can obtain real-time data regarding soil health and the growth of plants. In the case of disaster response, drones quickly measure the extent of destruction and help locate survivors. As such, they help deliver much-needed information to rescue teams.

Other Industries: UAVs are used in various industries. Some of them include cinematography, real estate, and infrastructure assessment. In cinematography, UAVs take aerial views that are either unaffordable or not easily accessible. Site surveys and development monitoring can be done with UAVs for construction and mining. That will ensure safety and accuracy.

17.1.2 Introduction to Genetic Algorithms in UAV Optimization

Unmanned aerial vehicles, or simply drones, have recently become so prominent because of their versatility and potential applications in agriculture, search and rescue, and delivery services, among others [10]. Optimization of UAV operations, including path planning and job distribution, remains a challenge. GAs are a subclass of the evolutionary algorithms replicated by the procedure of natural selection and were found successful in dealing with optimization tasks [11]. The process of the evolutionary algorithm helps in improving one set of candidate solutions in several iterations through mechanisms that include selection, crossover, and mutation. While doing UAV optimization, in the case of evolutionary algorithms, an ideal route for the flight, optimal task ordering, or optimizing the resources allocated by using more than one UAV across resources and locations. This is one of the prominent advantages of using GAs for UAV optimization because these can solve complex, multiobjective optimization problems [12]. In a search and rescue mission, the UAV must compromise between two conflicting objectives: quickly covering vast areas without losing much of its potential battery life. GAs can simultaneously maximize both objectives. Another advantage of GAs is that they can easily cope with dynamic situations. In many UAV applications, the environment may change very fast; thus, the UAV has to change its behavior accordingly. In GAs, real-time data may also be incorporated to update UAV actions based on changing conditions. Although GAs have their merits, they are not without limitations. They are very resource-intensive, especially when handling complex problems, and may take a long time to find a solution. The strength of the GA might depend on the parameters adopted including population size and crossover rate. Substantial research studies have shown that GAs are effective at UAV optimization. Used a GA to improve path planning for a UAV in a search and rescue mission. They showed that their approach could find a solution that balanced the need to cover a large area and preserve battery life quickly. Proposed a GA for the improvement of work allocation to enhance the performance of a delivery mission by a swarm of UAVs. The result was an optimal solution that minimizes the total delivery time with an added constraint such that all UAVs are assigned a feasible amount of tasks. Therefore, in summary, it has been found that GAs are effective at improving UAV operations. Due to the fact that they can very well deal with complex multiobjective optimization problems and dynamic conditions, they are highly suitable for

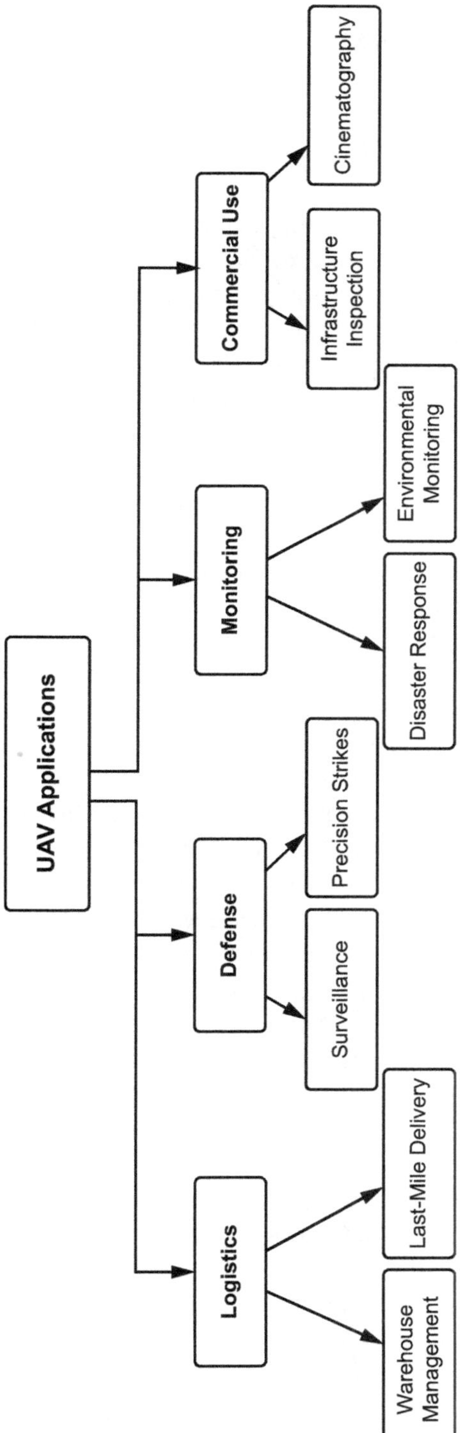

Figure 17.2: Applications of UAVs.

many UAV applications. On the other hand, their constraints, including computing cost and sensitivity toward selection parameters, must also be recognized.

17.1.3 Role of Quantum Computing in UAV Systems

The superposition and entanglement concepts, providing massive computational powers and processing efficiencies, could be leveraged to revamp the UAV capabilities with QC implemented in the UAV system. QC involves using qubits instead of binary bits used conventionally in traditional computers; they exist in more than one state at a time. This feature allows for parallel computation of big data sets in quantum computers and offers outstanding speed and parallel processing capabilities. For UAVs, complex optimization problems of path planning, resource allocation, or swarm coordination can be more effectively solved in real time, even in high-dimensional space. This QC is an opportunity for great multiobjective optimization because it highly facilitates UAV operation within dynamics, where rapid changeover has to be taken. Quantum algorithms offer near-online path planning and swarm coordination by solving super-fast solution spaces as simulations of coordinated movement among UAVs. This fast processing allows UAVs to change their routes and operations adaptively and, thus, optimize energy efficiency and enhance the success rate of missions, especially for complex scenarios like search and rescue.

Other than optimization, QC is used to enhance data security in UAVs through quantum cryptography. Quantum mechanics can be used to develop unbreakable encryption keys. Quantum key distribution offers a higher level of security such that encrypted data is almost impossible to intercept without its discovery, which is a significant advantage in the operations of UAVs, especially while dealing with sensitive information during surveillance or war operations. QKD implementation within UAV systems protects communications channels from interference and hacking, as noted in recent research works on network encryption for UAVs. Quantum-enhanced machine learning is the domain whereby QC improves UAVs' autonomy and decision capabilities. Quantum algorithms make UAVs process data faster, adapting to constantly changing environmental conditions – a requirement in operations that call for immediate and responsive autonomous responses. Quantum-enhanced machine learning allows for sophisticated processing of sensor data to enable real-time obstacle avoidance, precise task execution, and effective pattern identification for environmental monitoring. This will enable QC capabilities to remarkably enhance swarm coordination since multi-UAV systems require fluid collaboration and collective action in tasks such as area monitoring, mapping, and disaster response. Quantum entanglement will allow state information to be exchanged between several UAVs instantaneously, which makes coordinated swarm behaviors that are more efficient and redundant possible. This property is especially useful in dynamic swarm systems where quantum-inspired models help coordinate UAVs, job assignments, and rapid responses to environmental changes. This ability of QC to process very complex data at high speed with excellent security makes it the best companion for the systems used in UAVs. Hence, the incorporation of QC into UAV systems would offer benefits in the majority of application cases involving real-time processing, advanced security, and autonomy in making UAV operations take place more rapidly, highly adaptive, and exceptionally safe.

17.1.4 Defining Quantum Genetic Heuristics for UAV Optimization

QGH combines the probabilistic nature of QC with the adapting capabilities of GAs as a significant contribution toward solving complex, multidimensional optimization problems that are essential in the optimization of UAVs. QGHs imitate the natural selection algorithm used in GAs and take principles for superposition and entanglement from quantum. This allows UAV systems to explore large real-time solution spaces and adapt quickly to real-time data [13]. These heuristics are very effective for multi-objective optimization problems in which multiple parameters are traded off, such as the duration of battery life, speed of flying, avoidance of obstacles, and mission coverage, which are special considerations in UAV applications [14]. The probabilistic nature of QGH allows for fast convergence toward an optimal solution without exhaustive searches. Thus, it is useful when resources related to processing are limited and real-time solutions are critical for UAV missions [15]. Path planning is an essential application of QGH within UAV systems, as UAVs are forced to choose the shortest or safest route among many locations. Classical algorithms fail in dynamic environments, but QGH's superposition enables it to tackle simultaneously multiple alternative courses and hence decide the best route significantly sooner than the other methods [16]. In addition, QGH can also aid in distributing the resources within multi-UAV networks, as tasks can be assigned to different UAVs uniquely. Hence, coverage can be maximized while redundancy is minimized by entanglement, synchronizing UAVs, and making real-time adaptations of their strategy. This capability is efficient for cooperative missions since coordinated motion is important in search-and-rescue or surveillance missions. Therefore, QGH is useful as an instrument for UAV swarm intelligence.

Furthermore, QGH improves the adaptability of UAV systems, and it can process environmental data in real time to modify its behavior. In dynamic applications, environmental factors such as weather, topography, and obstructions are dynamically changing; therefore, the flexibility of QGH in adapting solutions will enable UAVs to handle such problems effectively, thereby reducing the rate of mission failure to a great extent. Safety is another potential area for QGH. The dependency of the heuristic on quantum cryptographic principles boost the strength of secure communications in UAVs, increasing data integrity in mission-critical operations against cyber-attacks. Techniques of quantum cryptography embedded within the GAs of UAVs will prevent breaches in data, especially in critical applications such as military and intelligence operations. Finally, QGH enables the efficient use of energy in UAV networks. Energy use in UAV operations limits the period a UAV can spend airborne. Still, QGH's ability to identify and rank low-energy routes and activities significantly increases the duration of missions. Long-range observation or extended data gathering is needed if the task requires it, so energy consumption is minimized through finding the most efficient activities and routes, allowing environmentally friendly use of UAVs. QGH is the best solution to the limitations of existing UAVs in endurance and flexibility, optimizing both task and path in a resource-efficient way. Quantum genetic heuristics are an effective route for UAV optimization; QC offers to enhance UAV performance in complicated data-intensive tasks by finding quick, flexible, and safe solutions.

17.2 Fundamentals of Genetic Algorithms in UAV Systems

17.2.1 Genetic Algorithms and UAV Path Planning

GAs are needed to overcome the complexities of UAV path planning as they provide robust, responsive solutions for upgrading flight routes under several conditions of operation. Genetically, GAs are an extension of the concept of Darwinian natural selection. The solutions evolve iteratively from one cycle of selection, crossover, and mutation, where UAVs navigate both static and dynamic situations well. It determines the most energy-efficient trajectories between predefined waypoints, reducing energy expense, collision risks, and mission time; this is why GAs are well suited to exploring vast solution spaces in parallel. The use of GAs in the path planning of unmanned aerial vehicles solves problems involving multiple objectives like balance between fuel economy and coverage area. This is done by GAs that generate diverse solution populations and then iteratively improve them so that they converge on optimum pathways. This is an extremely helpful capability of UAVs in dynamic environments when they have to quickly adapt to unforeseen changes, such as unexpected obstacles or weather conditions. From the literature, it is clear that GA-based approaches enable UAVs to modify the route plan adaptively. It allows for the continuation of the mission in hostile conditions. Another flexibility of GAs is cooperative UAV operation. Coordinated route planning is often necessary in multi-UAV systems to enhance efficiency without overlaps or collisions. The systems can achieve cooperative route optimization by applying parallel GAs, as demonstrated in the work by Liu et al. [15] with the application of GAs for multi-UAV work allocation and trajectory design [15]. Hybrid methods, like combining GAs with reinforcement learning, further enhance the capability of a UAV swarm for making decisions, allowing the adaptive neural networks to exploit the paths produced by GAs as training data.

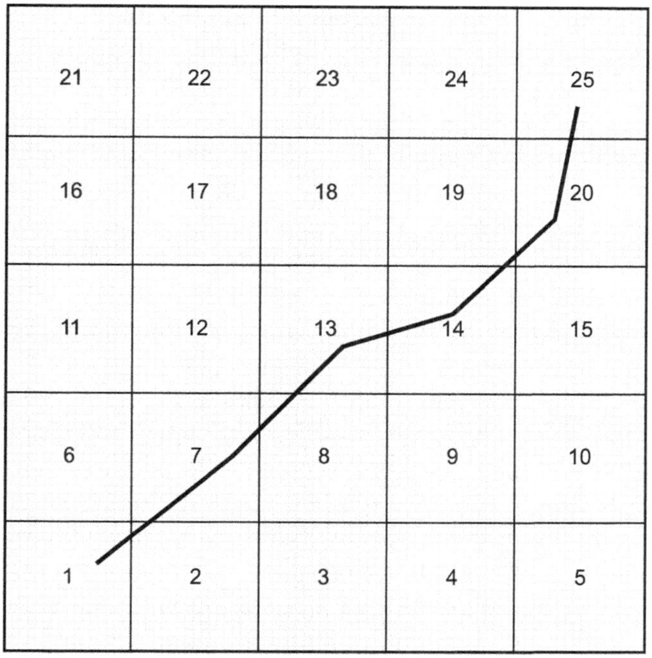

Figure 17.3: Example of path planning.

In scenarios where broad area coverage is required, GAs effectively optimize coverage routes for UAVs during surveillance or mapping missions (Figure 17.3). An illustrative application utilized convex polygon decomposition and a GA-based optimization technique to improve coverage efficiency while reducing energy consumption. Similarly, research on energy-efficient use of UAVs used GAs to compute optimal trajectories with reduced energy consumption while ensuring job completion. Real-time path planning, an integral component of UAV autonomy, benefits extensively from FPGA-based GA implementations, providing quick computation and responsiveness to real-time data. These systems help effectively recalibrate UAV trajectories without experiencing heavy delays, which are particularly important for time-sensitive missions such as search and rescue operations. Besides that, the ability of GAs to alleviate premature convergence issues also increases their reliability in achieving near-optimal solutions, mainly for environments characterized by highly intricate landscapes and high levels of unpredictability. A radical method for solving intricate problems in routing by GA used in UAV path planning in terms of flexibility, efficiency, and adaptability is provided. GAs will enable UAVs to work on a variety of applications that are highly accurate and dependable through multiobjective optimization with real-time adaptation. Its benefits highlight the necessity and importance of GAs while enhancing UAV autonomy in general and its operational efficiency regardless of application.

17.2.2 Genetic Algorithms for Task Allocation and Mission Planning

GAs are important in UAV systems because they offer efficient solutions for solving complex multiconstraint problems on the allocation of work and mission planning; the algorithms work on emulation of natural selection for the evolvement of the best candidate solutions from a given population to identify best practices about task allocation under varying situations [10]. The assignment of certain tasks to UAVs has to be done in such a way as to minimize redundancy and maximize efficiency. These GAs can be accomplished by exploring several solution spaces simultaneously [12]. Through crossover and mutation, GAs adapt the task allocation quite often; thus it is very effective in dynamic situations involving frequently changed mission conditions. A significant advantage of GAs is that they can be used for multi-objective Optimization in unmanned aerial vehicle mission planning. GAs can simultaneously optimize energy efficiency, mission duration, and job coverage with balanced solutions that match operational goals. Cooperative job assignment is critical in systems with multiple UAVs. Liu et al. [15] show how adaptive GAs can aid in the collaborative planning and execution of missions by unmanned aerial vehicles in disaster response situations, with proper communication and resource utilization [15]. Another domain in which GAs excel is dynamic mission planning. In search-and-rescue missions, for instance, UAVs must adapt to real-time data such as changing weather conditions or newly identified obstructions. GAs enable unmanned aerial vehicles to continuously update their assigned tasks. This ensures continued performance and optimal results, accommodating unpredictable variables. The integration of GAs with optimization techniques such as artificial bee colonies or neural network methods makes it more efficient in solving time-constrained problems.

Resource constraints, such as payload capacity and limited battery life, pose a great challenge in UAV task allocation. GAs overcome these constraints well by scheduling tasks and reassigning resources based on changing conditions. Illustrate the use of GAs to improve job scheduling and UAV trajectories, thus reducing energy usage and improving the success of a mission. Besides, GAs make sure that the workload distributed to UAVs is fair and that there will not be a load of one unit that causes overload so that the missions will be completed properly. In concurrent mission operations where several UAVs operate in their missions simultaneously, for example, coordinated attacks or synchronized surveillance, GAs ensure that the proper timing and alignment of UAV activities take place. For example, GAs have been utilized to minimize the delay associated with coordinated missions; this way, UAV swarms can operate as a cohesive system. The natural scalability of GAs also makes them best suited for highly complex tasks requiring multiple UAVs, in which traditional task allocation algorithms typically fail because of computational complexities. GAs are at the core of optimizing work distribution and mission planning among UAVs, which support multiple goals, dynamic environment adaptation, and resource constraints.

These, therefore, are key technological enablers for improving the performance of UAVs in complex activities demanding significant data support. Continuing to develop UAV applications will make GAs, combined with other related technologies, capable enough to significantly improve performances and offer resilient, flexible, and scalable mission-planning systems.

17.2.3 Limitations of Classical Genetic Algorithms in UAV Optimization

Classical GAs are not applicable in UAV optimization, especially in dynamic and high-dimensional cases. The most important limitation of using GAs is the premature convergence to poor solutions due to a lack of diversity in the population, which significantly reduces the ability of the GAs to search for better alternatives. This problem is significantly enhanced in large-scale UAV optimization problems, such as multi-UAV route planning, where the solution space is vast and complex. Another limitation of GAs is inefficiency in high-dimensional problems because evaluating a lot of possible solutions over generations takes a lot of time, which does not make them appropriate for real-time applications. The major drawback of the traditional GAs is that their performance is highly sensitive to the mutation rate, crossover frequency, and population size. Poor parameter settings lead to stalling or poor solution space exploration. In addition, traditional GAs suffer from constraints, especially when several factors need to be balanced such as energy limitations, collision avoidance, and meeting mission deadlines. An infeasible or unsatisfactory result for UAV operations may occur if the integration of the constraints is too lenient.

Classical GAs are unsuitable for missions that must adapt immediately such as search-and-rescue. The reason is that it takes them much time to adjust quickly to environmental changes. They work in generations, and their adaptation process is slow because several generations must converge before they adjust. Furthermore, GAs exhibit local optima, a more significant challenge in complex fitness landscapes where intricate dependencies amongst variables make it difficult to find global solutions. In collaborative multi-UAV missions, traditional GAs fail to coordinate activities successfully between more than one UAV. While they may be able to optimize the mission or functions of an individual UAV, coordination of the whole UAV fleet in real time is a very computationally expensive task. Moreover, their static dependence on fitness functions eliminates them from dynamic alteration of multiobjective trade-offs in UAV operation. These restrictions motivate research to suggest hybrid methodologies with combination techniques of GAs integrated into another, possibly particle swarm optimization technique or even neural network methodologies, thus enhancing their dynamic flexibility and increasing convergence rates, where applicable. Constraint-handling techniques, along with the methods to adaptively tune GAs' parameters, are built toward the development of superior efficacies for GA usability within the domain of UAV optimization. Whereas the classical GAs offer a kind of inherent framework for the optimization of UAV, they have drawbacks such as inefficiency in computation, and these features, such as parameter sensitivity, further linked to difficulties in dynamic conditions have resulted in necessitating advanced developments in its integration to allow more effective approaches to UAV optimization.

17.3 Quantum Computing and Genetic Algorithms

17.3.1 Basics of Quantum Computing

QC is a paradigm shift in the field of computer science research based on quantum principles, solving problems impossible by classical computing. Unlike the principle of classical computing that employs bits to represent data, such as 0 or 1 (shown in Figure 17.4(a)), in the quantum case, the computational model relies on quantum bits or qubits that allow the superposition principle and thus can be simultaneously present in many states at the same time (shown in Figure 17.4(b)), which significantly enhances computation power. Quantum computers can compute that, for the classical systems, would take an inordinate amount of time. The quantum systems can represent and manipulate a large amount of information in one expression. The central concept of QC is entanglement, the phenomenon by which qubits become tied together such that the state of one qubit influences the state of another, regardless of how far apart they are. This interaction allows the quantum computers to perform computation in parallel, thus greatly enhancing the processing speed for applications that can be computed in parallel. Quantum entanglement enhances computation and supports quantum communication and cryptography. The characteristic feature is quantum interference, which improves the correct answers and eliminates the wrong ones in quantum computation. This property is quite important for algorithms like Shor's integer factorization algorithm and Grover's unstructured search algorithm. Both techniques are significant as Shor's algorithm has an exponential acceleration compared with the classical approaches, while Grover's algorithm presents a quadratic acceleration over classical approaches.

Quantum gates constitute the most elementary quantum circuits by performing unitary operations on the qubits. Unlike quantum logic gates, quantum counterparts of Hadamard gate, Pauli-X gate, and controlled-NOT or CNOT gates work on the qubits to achieve complex transformations as needed in quantum computation. It is through these quantum gates that quantum computers work and make multidimensional calculations; such computations are necessary for areas like cryptography, optimization, and machine learning. Decoherence in QC is a significant challenge, as qubits are sensitive to external perturbations that could disturb their quantum states. This problem has been solved by designing several error-correcting codes and fault-tolerant quantum structures that ensure reliable performance in noisy environments. Such developments form the basis for further development of QC toward practical applications such as optimization in UAV systems and GAs. QC can revolutionize businesses by solving problems that are hard for traditional computers to solve because of their limitations in computation. Its application in domains such as machine learning, cryptography, and optimization underlines its ability to disrupt the perception of technologies and problem-solving methodologies.

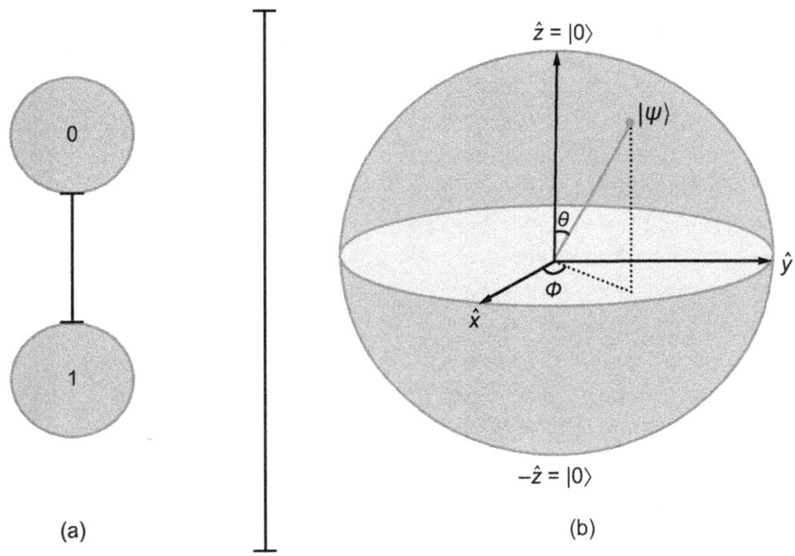

Figure 17.4: Visualization of (a) binary system and (b) quantum bit.

17.3.2 Quantum-Enhanced Genetic Algorithms

Quantum-enhanced GAs combine these features of QC, that is, superposition and entanglement of quantum systems, with traditional GAs to solve optimization problems more competently. QEGAs are very useful for dealing with large-scale and multiple-objective optimization problems; they avoid limitations such as slow convergence rates associated with traditional GAs and a tendency to get easily trapped in local optima by exploiting quantum properties. In QEGAs, qubits are used as surrogates for binary representations; therefore, the algorithm can simultaneously represent and evaluate many candidate solutions. This is due to superposition, which enables QEGAs to examine an ample solution space simultaneously improving convergence rates. Quantum entanglement also allows the simultaneous evaluation of possibilities for solutions, hence improving the efficiency of crossover and mutation mechanisms in the evolution of populations toward optimal solutions. QEGAs are particularly practical for multiobjective optimizations, where quantum rules enable evaluating several conflicting goals at one time, like maximizing or improving fitness and minimizing cost. Indeed, QEGAs show better performance for issues that require trade-offs in between. For example, this approach was already proven by efficiently solving UAV path planning and job scheduling under dynamic systems. Such an advantage originates from quantum superior fitness evaluations, which often lead toward global optimum rather than local optima.

A significant property of QEGAs is that they can deal with oscillating environments; in standard GAs, an environmental change often forces the necessity to restart the population to ensure convergence. QEGAs adapt to the change dynamically; quantum adaptive operators maintain population diversity and ensure that this algorithm will respond to live data in real time accordingly. Quantum-enhanced mutation and crossover processes in quantum evolutionary genetic algorithms (QEGAs) greatly enhance the quality of the solution and the rate of convergence. Quantum-inspired mutation introduces probabilistic variants that explore solution spaces more intensively than deterministic approaches in classical GAs. Quantum crossover uses entanglement to combine the properties of the parent solutions, hence retaining the best characteristics and increasing exploratory diversity. QEGAs also remove computational inefficiency, which is a major problem of classical GAs in high-dimensional optimization problems. Quantum parallelism allows QEGAs to process populations in much less time than traditional algorithms; thus, they are suitable for real-time decision-making applications. QEGAs in swarm intelligence applications demonstrate exceptional performance in controlling multi-agent systems like UAV swarms in achieving effective collective goals. QEGAs and machine learning have the potential to further increase the optimization level with the incorporation of quantum machine learning algorithms. Hybrid QEGAs is a quantum optimization process with data-driven insights for predictive analytics and adaptive methods in complex systems like traffic management and resource allocation. This hybridization allows QEGAs to optimize solutions with historical data and adapt simultaneously to the new constraints. Though it offers several advantages, QEGAs also come with several challenges, mainly on the hardware realization side. A significant limitation comes with the total dependence on quantum hardware, which the existing quantum equipment generally suffers from issues like scalability and noise. However, these problems will be overcome by the progress in fault-tolerant QC and quantum error correction, thus enabling the widespread use of QEGAs in optimization-based applications. QEGA is a landmark in optimization technology that integrates the benefits of traditional genetic algorithm with the computing ability of quantum mechanics. This capability of adapting to evolving contexts and increasing convergence, coupled with multiobjective optimization makes these algorithms an absolute necessity when solving problems that have arisen in several disciplines.

17.3.3 Quantum Speedups in Genetic Operations

QC accelerates genetic processes based on properties such as superposition, entanglement, and quantum parallelism. Such a property allows the execution of GAs very efficiently on hard optimization problems. Most speedups of quantum over other models are caused by the basic selection, crossover, and mutation operations constituting the heart of the evolutionary mechanism of a GA:

1) Selection Operation: In classical GAs, finding the fitness of all members of the population is a computationally expensive selection operation for large populations or complex fitness functions. QC provides a quadratic acceleration in fitness evaluation through Grover's search method, which, in effect, finds high fitness within an extensive dataset. It reduces the number of evaluations to such an extent that it becomes particularly useful for real-time applications.
2) Crossover Operation: Entanglement-based crossover operation in quantum-evolved versions allows the flow of genetic information between parent solutions. Such entanglement allows the evaluation of multiple offspring simultaneously; it keeps beneficial traits intact while exploring various combinations of solutions. Thus, the exploration capabilities can be improved without losing computational efficiency, as observed in hybrid quantum-GAs.
3) Mutation Operation: Quantum mutation allows probabilistic fluctuations in solutions by utilizing quantum superposition, allowing multiple mutation paths to be simultaneously analyzed. This eliminates the problem of deterministic mutations within conventional GAs, which often suffer from a problem of local optima escape. Quantum-inspired mutation has been demonstrated to exhibit exponential improvements in crossing rugged fitness landscapes [17].
4) Parallelism in Genetic Operations: Quantum parallelism allows the entire population of genetic operations to be performed in parallel by exploiting the feature of qubits to exist in more than one state at a time. The capability allows GAs to check, cross, and mutate entire populations in parallel. As such, computation time is substantially reduced.
5) Applications, Impacts, and Challenges: Quantum innovations in genetic operations have been of immense significance in solving combinatorial optimization problems, like the traveling salesman problem and work schedule. This is especially true when classical GAs are not applicable due to scalability issues. Research suggests that some jobs can be solved in polynomial or even exponential time using quantum-enhanced GAs. Although quantum speedups offer revolutionary advantages, there are still problems in applying these methods to current quantum technology. Decoherence and noise challenge the scalability of quantum algorithms, hence the progress required in fault-tolerant QC. However, recent studies on hybrid quantum-classical approaches (HQCAs) suggest that they might help overcome the difficulties.

17.4 Integration of Quantum Genetic Heuristics for UAV optimization

17.4.1 Path Planning with Quantum Genetic Heuristics

QGH offers significant enhancements in UAV route planning by combining the computational efficiency of QC with the adaptive problem-solving skills of GAs. This integration addresses issues such as multiobjective optimization, dynamic environmental adaptability, and computing inefficiencies, making QGH an optimal method for UAV route planning [14]. Path planning is to identify the best trajectory for the UAVs to pass through in space, avoiding obstacles while abiding by operational limitations that include battery life span, speed, and area cover. QGH uses the principle of quantum superposition to evaluate multiple alternative pathways in parallel, thus reducing computing time significantly compared to a classic approach. UAVs' ability to determine the best paths in real time in high-dimensional scenarios while bypassing the scalability and computational complexity of traditional algorithms occurs because of this ability.

Quantum-Enhanced Fitness Evaluation: One of the most significant enhancements in QGH is quantum-enabled fitness evaluation. Quantum concepts, such as Grover's search algorithm, enable finding high-fitness solutions more efficiently, thereby ensuring the selection of the best path from a vast array of alternatives. This method accelerates the convergence process, enabling UAVs to formulate routes almost in real time, which is critical for dynamic applications such as search and rescue or disaster response.

Dynamic Adaptability: QGH focuses on dynamic path planning, where the environment may change unpredictably due to factors like weather, obstacles, or moving targets. QGH uses quantum mutation and crossover operations to maintain a diverse solution pool, which allows UAVs to dynamically change their paths in response to changes in the environment [16]. This flexibility is important in collaborative multi-UAV operations, where real-time coordination and response are critical for mission success.

Multiobjective Optimization: Sometimes, path planning has to solve hundreds of competing objectives like travel time, maximum area coverage, and battery lifetime. QGH uses the concept of quantum superposition and entanglement in parallel evaluation and improvement of a huge number of objectives to ensure fairness and completeness in solutions that fulfill all mission requirements. This capability for multiobjective optimization becomes particularly crucial for UAVs used in applications such as urban distribution, where the objectives of efficiency and reliability need to be matched with constraints like legal adherence and obstacle avoidance.

Scalability and Efficiency: Quantum parallelism makes it possible for QGH to examine large populations of candidate solutions in parallel, which drastically improves scalability. This capability makes QGH exceptionally efficient for extensive applications requiring a number of UAVs or intricate settings. Recent research shows that QGH is able to improve the coordination of UAV swarms in order to successfully cover extensive regions in tasks such as agricultural monitoring or military reconnaissance. Regardless of its advantages, the utilization of QGH for the path planning of UAVs faces hardware limitations and quantum noise. Current quantum computers typically lack sufficient qubit numbers and error correction to support practical-scale uses, thus necessitating HQCAs. In addition, the computational overhead for the simulation of quantum gates on classical hardware can impose restrictions on real-time use in low-resource environments [18].

17.4.2 Quantum Genetic Heuristics-Based Multi-UAV Coordination

QGH-based multi-UAV coordination provides a new means of augmenting work distribution, resource allocation, and adaptive decision-making in the case of UAV teams. QGH merges the concepts from GAs with superposition and entanglement aspects of quantum mechanics to maximize computation effectiveness and adaptability.

Enhancement of Swarm Intelligence of Multi-UAVs: For the UAV swarm to accomplish its mission-intended task, be it surveillance, delivery, or search-and-rescue, there must be good coordination of the UAVs. QGH can attain this due to quantum-inspired genetic operations that allow for the simultaneous evaluation and improvement of several coordination techniques. Entanglement enables the swapping of state information between UAVs immediately. This increases the chances of coordination but has low chances of repeated operation.

Job Assignment and Load Balancing: For multi-UAV systems, job assignment refers to the way goals like coverage of space, payload delivery, or target tracking can be shared among UAVs while minimizing resource conflicts. QGH finds solutions to complex issues by using quantum-enhanced fitness evaluation, which simultaneously optimizes conflicting objectives such as energy efficiency and the completion of mission time. QGH achieves equitable load distribution among UAVs and maximizes system efficiency through dynamic changes in genetic parameters with quantum operators, thus avoiding overloading.

Path Optimization and Collision Avoidance: QGH utilizes quantum algorithms to optimize UAV paths, which provides the optimal path with reduced travel time and energy usage and prevents collision. Using quantum mutation and crossover, QGH explores much more exhaustively than other techniques can manage in the solution set space to guarantee collision avoidance strategies that are highly efficient. In addition, the versatility of QGH allows swarms of UAVs to adapt to changing situations, such as moving obstacles or changes in mission requirements, in an extremely fast time.

Multiobjective Optimization: QGH proves to be robust in addressing multiobjective optimization. This makes it highly capable of multi-UAV coordination when the conflict among competing objectives must be resolved simultaneously. QGH improves mission duration and energy efficiency by using quantum-assisted selection methods that opt for the global optimum at the expense of the local one. This capability requires the application to monitor targets concurrently, map areas, and distribute resources [14].

Real-Time Decision-Making: The UAV swarms flying in dynamic scenarios must make quick decisions to tackle unexpected issues. QGH will be useful in real-time decision-making with Grover's search for an optimum solution within huge solution spaces in a very quick manner. Such speed-up will enable the UAVs to change their coordination policy without causing considerable delays, which is critical for time-sensitive operations like disaster relief.

Challenges and Future Directions: Despite QGH having delivered appreciable advantages in multi-UAV coordination, challenges remain regarding ways of applying these techniques to the existing quantum hardware. Some of the challenges faced with this process include quantum decoherence coupled with limitations on the availability of qubits, which scale the procedure. Hybrid quantum-classical frameworks are still favorable methods in consideration of restrictions that permit the QGH near-future implementation in UAV coordination.

QGH revolutionized multi-UAV coordination by introducing speed and parallelism found in quantum computation and flexibility in GAs. QGH is an efficient tool in UAV systems, with enhanced swarm intelligence, multiobjective tasks, and real-time flexibility. As quantum technology develops, QGH in UAV coordination will significantly improve efficiency and operational capability.

17.4.3 Task Assignment and Mission Planning with QGH

The UAV system's task assignment and mission planning represent quite complex optimization problems resulting from dynamic climatic conditions, limited resources, and coordination issues among multi-agents. QGH is a newly proposed method that is said to be efficient, considering concepts such as superposition and entanglement properties of QC combined with the adaptability of the GA. Such an approach proved to be promising in terms of duty assignment and mission arrangement [14].

Quantum Superposition in Job Assignment: Quantum superposition allows QGH to run several job assignment configurations parallelly, which is more computationally efficient than traditional methods. This is helpful in multi-UAV systems where the solution space is too big and fast decisions are needed to find the best one. QGH, using Grover's search, finds optimal allocations with fewer resource conflicts and efficiency [15].

Multiobjective Optimization: In mission planning, sometimes UAVs conflict with multiple objectives, such as consuming less fuel while increasing the region coverage or speeding up the mission completion. QGH optimizes these conflicting goals by measuring trade-offs in a fitness environment rich in quantum concepts. It ensures fair solutions that simultaneously achieve all other mission objectives.

Multi-UAV Real-Time Adaptation: Real-time adaption is vital to efficient UAV mission planning. In particular, multiagent cooperative systems require real-time state swapping among UAVs for decision coordination. Such swapping is of prime importance in search-and-rescue missions, where the coordination of tasks and the allocation of jobs must be rapid.

Resource Constraints: In the UAV system, task allocation needs to be done considering resource constraints such as battery life, payload capacity, and communication bandwidth. QGH adjusts the task allocations based on these limitations using quantum-inspired operators that highlight high-value jobs and reduce resource inefficiency [16].

Scalability and Computational Efficiency: QGH's quantum parallelism allows for the scalable workload distribution and planning of missions involving hundreds or thousands of UAVs. Indeed, it is demonstrated that QGH outperforms its classical heuristics on high-dimensional optimization problems and can thus be applied in applications such as disaster response, agricultural monitoring, and urban logistics.

Challenges and Future Directions: Despite the advantages of integrating QGH in UAV systems, challenges exist, such as limitations due to quantum hardware and noise. The near-term practical solution could be hybrid quantum-classical frameworks simulating quantum processes on classical processors, which can be run in the near term with the development of quantum hardware.

QGH is an efficient methodology for work allocation and mission planning in UAV systems, in which the computational power of QC is used along with the flexibility of GAs. QGH will support the efficient operation of UAVs in complex and dynamic environments through multiobjective optimization, dynamic adaptation, and resource constraints. Therefore, with rapid progress in quantum technology, QGH is bound to be an essential part of future UAV systems.

17.4.4 QGH Pros in UAV Operation

With the concepts of quantum computers, like superposition and entanglement, QGH will enhance UAV performance using a traditional GA as applied to those concepts. The hybrid model will solve some important problems of UAV applications with real-time adaptation, multiobjective optimization, and the scalability of computers in their processing.

Computational Efficiency: The major strength of QGH is quantum parallelism. This allows for testing many possible solutions for the UAV systems at one time, thus reducing computation time for complex optimization problems in path planning and task allocation [14]. Grover's search algorithm has improved decision-making efficiency since it reduces the time to find optimal UAV solutions for a dynamic scenario.

Real-Time Adaptability: QGH enables UAVs in dynamic environments such as disaster response to cope with real-time variations using mutation and crossover operators derived from quantum theory. The latter maintains diverse solutions, so UAVs adapt in time to changes like the priority of a mission or environmental constraints. Quantum entanglement also enhances intra-swarm cooperation in attaining a collective response to unforeseen barriers.

Multiobjective Optimization: Sometimes, the objectives of UAV operations may conflict with each other, such as energy consumption having to be kept low while increasing the covered area or completing a mission in a shorter time period. QGH is very good at multiobjective optimization, as it incorporates quantum-enhanced fitness evaluation, which will favor globally optimum solutions rather than local optima. In urban delivery applications, it is extremely useful to balance operational efficiency with regulatory compliance.

Scalability in Multi-UAV Systems: QGH is inherently scalable for large-scale UAV operations with many agents and complex mission specifications based on quantum parallelism. This is a significant benefit in agricultural monitoring or military reconnaissance because classical algorithms tend to suffer severely from computing limitations. The experiments demonstrate that QGH can coordinate more than one UAV well, so the resources are optimized, and the mission is accomplished.

Robust Path Planning and Collision Avoidance: At path planning, QGH employs quantum-inspired genetic procedures to determine optimal paths that provide minimal travel time, energy consumption, and collision possibilities. Quantum mutation accomplishes this by examining many more solution spaces than traditional methods, whereas quantum crossover brings together various beneficial features from other solutions. In these competencies, these abilities are most effective within highly congested metropolises or disaster areas with dynamically changing obstacles.

Energy Efficiency: The main limiting factor in UAV operations is energy. This may be with battery-driven UAVs. QGH minimizes this by optimizing the flight trajectories and task delegation to consume less energy while maintaining the mission objectives as defined. This is key to prolonging the operating time of the UAVs and maximizing their productivity during long missions.

QGH could benefit UAVs immensely through increased computing efficiency, real-time flexibility, multiobjective Optimization, scalability, and energy efficiency. As this QC technology improves daily and becomes very efficient, QGH will add new dimensions to applications in UAVs with more robust, efficient, and adaptive autonomous systems.

17.5 Challenges and Limitations

17.5.1 Quantum Hardware Constraints

One of the significant barriers to the practical applicability of QGH in UAV optimization is quantum hardware limitations. The major limitations are that qubits are few, decoherence, and noisy, and a fault-tolerant quantum computer is still missing. Quantum devices operate in the NISQ regime characterized by noisy, intermediate-scale quantum devices whose qubits are error-prone and prone to external interference [19]. These constraints bar the use of quantum algorithms for UAV route planning and job allocation. The number of qubits available restricts the size of optimization problems that can be solved on these computers. For instance, multi-agent UAV coordination might require high-dimensional computation beyond the capabilities of current quantum hardware. Apart from this, decoherence in a qubit is a process that gradually results in the loss of quantum information. Therefore, it decreases computation accuracy and requires several iterations, increasing computation overhead. Introducing noise in quantum systems intensifies the challenge of its implementation in UAV optimization. Noise can interfere with the quantum gates, thus producing incorrect processes that may produce erroneous results. Though promising, error correction methods require more qubits and increase the constraints of already finite quantum resources. It is crucial to come up with ways to mitigate errors effectively and advance toward fault-tolerant QC. The cost and infrastructure to run quantum systems are high, maintaining cryogenic environments to keep qubits at very low temperatures. These are not yet integration opportunities in UAV systems or mobile uses not to mention. The room-temperature quantum research is still in its infancy [20] Despite all the problems listed above, quantum-classical hybrid frameworks can, on the other hand, provide a useful intermediate solution. They solve large-scale problems using classical computation and the quantum algorithms on subproblems that best exploit the quantum acceleration. With continued advancement in quantum hardware, these hybrid systems can break the existing constraints with some possible prospects. The constraints of quantum hardware present serious challenges in UAV optimization using QGH; however, ongoing research and development in quantum technologies have presented possible solutions to these problems. The final solution to these constraints may transform UAV operations to a level where efficiency and agility for complicated missions are unparalleled.

17.5.2 Scalability of Quantum Genetic Heuristics

The most important aspect of this work is the scalability of QGH since its application is desired in large-scale optimization problems, such as the coordination of many UAVs and complex path planning. Using principles like superposition and entanglement, QGH can efficiently explore vast solution spaces that are challenging to solve by traditional evolutionary algorithms in high-dimensional contexts. This inherent parallelism enables QGH to scale very efficiently with an increase in problem size while reducing the cost of computation and maintaining solution quality intact. Perhaps one of the most salient features of QGH is its efficiency in solving complex and distributed problems with quantum-enhanced fitness evaluations. The fitness evaluations utilize Grover's search algorithm to find the best solutions rapidly, ensuring that the heuristic remains efficient with the growth of the problem size. Another point is that quantum-inspired mutation and crossover allow QGH to explore different sets of solutions simultaneously, thereby avoiding weaknesses inherent in classical approaches.

Most of the scalability issues in QGH occur because of decoherence and limits on the number of qubits in the contemporary quantum hardware. However, hybrid quantum-classical architectures have managed to be beneficial in increasing the capacity of QGH for solving large-sized problems by suitably segregating computing tasks between the classical and quantum domains. This hybrid strategy exploits computing power optimally, so QGH is appropriate for swarm intelligence and resource allocation. The scalability characteristics of QGH make it suitable for use in scheduling, resource allocation, and real-time adaptive systems. In UAV swarms, QGH improves the work assignment of numerous agents with near-linear scalability through quantum parallelism.

In addition, using real-time data ensures that even large systems remain flexible and responsive to shifting fluctuations. Scalability is a revolutionary character of QGH that overcomes computing constraints of conventional methods in extensive optimization. QGH holds immense promise for solving high-dimensional problems since it combines quantum computational power with the adaptive abilities of GAs. However, these can be realized in their entirety, if advancements in quantum and hybrid computing architectures proceed further.

17.5.3 Integration of Classical and Quantum Systems

Combining the classical and quantum systems offers a revolutionary method of solving intricate optimization problems by reaping the advantages from both approaches. Classical systems are great for solving deterministic problems with clear structures, but quantum systems have shown outstanding computing capability to solve huge solution spaces using quantum superposition and entanglement. This hybrid approach effectively combines the robustness of classical computation with the randomness of quantum mechanical capabilities in solving highly dimensional problems in practical applications. QGH employs a classical framework to guide the population-based evolutionary process, while more resource-intensive tasks, including fitness evaluation, are pushed to quantum processors. Thus, hybrid systems can surpass quantum device hardware limitations while remaining scalable and efficient. Based on Grover's method, quantum parts speed up the search for solutions to optimization problems, including job allocation of UAVs and multi-agent coordination.

Hybrid systems are used for flexibility in path planning and resource allocation under dynamic contexts using quantum-inspired operators. Classical systems give excellent control over deterministic processes. At the same time, the quantum algorithms are responsible for exploring probabilistic states to ensure rapid convergence to optimal solutions, even when the context might be dynamic. Quantum-assisted decision-making applies to applications like urban traffic management, where adaptability in real time is critical. It removes the existing problems of quantum hardware, such as qubits' restrictions and decoherence, which occur in hybrid frameworks to integrate classical and quantum systems. These systems will use classical infrastructure to efficiently mitigate noise bottlenecks and scalability-related constraints inherent to the quantum devices, thus engendering a symbiotic relationship between the two technologies. Besides, this integration allows parallel processing and thus allows real-time applications in many fields, such as logistics and disaster response.

The essence is that the integration of classical and quantum systems is something of a breakthrough in optimization technology, balancing the deterministic reliability of classical computation with the probabilistic and parallel powers of QC. With the forward march of quantum technology, these hybrid models will transform how efficiency and scalability are to be found in solving highly complex, high-dimensional problems in optimization.

17.6 Future Directions

17.6.1 Development of Quantum Processors for UAV Systems

The development of quantum processors for UAV systems in order to profoundly enhance the efficiency of computing by such sophisticated abilities as real-time decision-making, dynamic optimization, and secure communication. Quantum computers take advantage of quantum superposition and entanglement to quickly and accurately compute challenging optimization issues, like multi-UAV coordination and resource planning, that remain beyond the ability of traditional computers. All the current quantum computers are in the NISQ era, limiting their scaling and introducing problems like qubit decoherence and noise. Even so, miniaturization and energy efficiency advances have made it possible to integrate them onto mobile platforms like UAVs. Quantum Brilliance has also made diamond-based quantum accelerators operable at room temperatures and is thus suited for UAV applications. Within the UAV system, optimization problems of relevance are computed rapidly by quantum computers compared to conventional processors and are offered as real-time adaptive routing with collision prevention. A hybrid approach combining conventional and quantum computations greatly enhances the simulation of multiagent UAV networks. Physical limitations represent the biggest technical challenges in developing quantum processors for UAV applications. These include coherence preservation under noisy conditions and high energy costs of the quantum systems. Energy-efficient design and thermal management are also how quantum processors can be integrated with light UAV platforms. One of the requirements in need of focus is the lack of standardized quantum algorithms for UAV applications. While generic quantum algorithms like Grover's search and quantum approximation optimization are promising, their use in UAVs is yet to be explored [21]. The problems of decoherence and noise can be overcome by enhancement of the mechanisms of error correction and increasing the tolerance to faults of quantum computers. Hybrid architectures have also emerged as practicable solutions wherein error-prone operations have been transferred to classical systems while quantum processors can perform critical computations [22]. The future of quantum computers for use in UAVs will probably be in further miniaturization, efficiency of energy usage, and increased dependability. Incorporating quantum computers into UAVs can revolutionize real-time observation, search and rescue missions, and autonomous flight. Researchers are also developing the use of quantum communication nodes in UAVs to enable secure high-speed data transfer in hostile environments. Quantum computers have the potential to transform the potential of UAV systems, but the biggest challenge is to do it in hardware and computing. Improvements in quantum technologies and hybrid computing architectures have made the use of quantum processors in UAV applications possible, with unparalleled advantages in terms of autonomy, efficiency, and security.

17.6.2 Hybrid Quantum-Classical Algorithms for UAV Coordination

HQCAs offer a promising direction for UAV coordination, combining the computational strength of QC with classical methods for ensuring reliability. The hybrid combines to solve hard optimization problems associated with multiagent operations in UAVs, such as path planning, job scheduling, and dynamic real-time adaptation. HQCAs employ quantum algorithms such as QAOA to solve subproblems but depend on classical processors for control and error correction. This distributed work allows efficient resource usage as UAV systems can scale to larger swarms without degradation in performance. Quantum parts of the system quickly evaluate solutions in the work assignment, while classical elements improve those results for practical applications. Dynamic environments such as disaster zones or urban airspaces demand that UAV systems adapt in real time. HQCAs meet this demand by making use of quantum-enhanced machine learning models for predictive analytics, which would make it possible for a UAV to predict environmental change and adapt its tactics accordingly. The algorithms apply quantum superposition for analyzing multiple pathways simultaneously; this reduces computation times compared with classical approaches. One of the most important applications of HQCAs is in collaborative mission planning, where UAV swarms need to coordinate to achieve complex tasks. HQCAs utilize quantum-inspired optimization techniques for equitable job distribution and optimal resource use, even during large-scale operations involving several UAVs [23]. Such functionality proves to be beneficial in applications such as search and rescue, where on-time and precise coordination can be a matter of life and death. With these advantages, however, HQCAs pose several challenges, such as requiring high sophistication in quantum hardware and good error correction techniques. Hybrid designs thus reduce such constraints by delegating sensitive operations to classical systems, ensuring the system's reliability under noisy conditions. With increased development in the quantum processor, the utilization of classical elements is likely to decrease and therefore open up the capabilities of HQCAs in UAV coordination. As noted earlier, HQCAs are a new approach to coordinating UAVs by leveraging some benefits of classical and quantum systems. HQCAs can solve quite intricate dynamic optimization problems by being part of future UAV technologies, targeting more efficient, scalable, and flexible systems at autonomous levels.

17.6.3 Long-Term Impact of Quantum Genetic Heuristics on UAV Technology

QGH revolutionizes UAV technology by offering scalable, efficient, and adaptive transformational attributes in application spaces across the civilian and military domains. QGH harnesses the parallelism of QC coupled with the flexibility of GAs, giving rise to a wide arena of improvement through multiagent systems, real-time optimization, and self-driving decision mechanisms. Multiobjective optimization is, however, the most relevant long-term effect of QGH, viewed as a reconciliation of energy efficiency, mission efficiency, and safety in UAV systems, all brought together here. Probabilistic properties of quantum algorithms enable drones to appraise and adapt to complex real-time scenarios; hence, this gives robust operation under extreme situations such as catastrophe regions or busy metro air spaces. QGH lets swarm intelligence stretch further into this important space for military surveillance and civilian applications, like surveillance of agricultural activities. Applied through the principle of quantum superposition to swarms of UAVs, the swarms may then be able to scan better an enlarged space, which means that eventually, they will optimize resources and jobs much better. The quantum-enhanced learning processes QGH applies improve predictive analytics and make UAV systems better equipped to handle the dynamic changes of the environment proactively [24]. QGH on route planning and collision avoidance may be the game changers in UAV autonomy. Quantum enhancement of genetic operators makes it possible to update UAV trajectories in real time to mitigate risk and ensure a mission's success under unknown circumstances [25]. This capability meets the increasing requirement of autonomous systems for logistics and urban mobility. QGH has a significant influence on energy efficiency. Optimized algorithms minimize the energy withdrawn in the UAV, consequently increasing this flying time but making the system progressively more sustainable. Implementing QGH with solar-powered UAVs might get closer to perpetually flying in their use for surveillance and communication applications. Although QGH has a lot of promise, its long-term worth will only be realized when current limitations in quantum hardware are overcome to allow it to overcome decoherence and the paucity of qubits. Hybrid models that connect classic and QC may be a short-term bridge toward enabling the step-wise deployment of QGH onto UAVs. More specifically, it is still needed to develop fault-tolerant QC and algorithms to scale QGH into large UAV networks. Therefore, QGH will definitely have an important role to play in UAV technology development in the future. Its ability to solve intricate optimization problems, increase autonomy, and enhance system performance makes it a key element for future growth in UAV operations.

17.7 Conclusion

QGH excels in computational innovation as it incorporates the capability of QC along with the versatility of GAs to manage challenging optimization problems for intricate dynamic multiagent systems. This synergy can revolutionize the computationally expensive, intrinsically dynamic, and usually multiple objective challenges one might face. QGH injects notions of superposition and entanglement from quantum theory into GAs to attain a level of scalability, computing efficiency, and flexibility that has never been possible. In its very nature, QGH marries the GAs principles with the probabilistic nature of QC into a natural selection process. GAs are based on evolutionary process models via selection, crossover, and mutation, which iteratively improve solutions for optimization problems. Although the classical GAs work for most tasks, they often limit their potential in cases where the issue is high-dimensional, multiobjective, or dynamically changing. QC improves on this paradigm with the ability to assess many solutions simultaneously in superposition, while correlated variables can be investigated quickly through entanglement. This allows QGH to investigate enormous areas of the solution space much faster than is possible with classical approaches without losing or degrading quality.

The applications of QGH are broad, where optimization under constraints becomes central. For instance, in multiagent systems like driverless cars, QGH effectively solves path-planning problems by changing the route due to barriers and optimizing energy use to fulfill mission objectives. QGH can optimize job allocation among agents in a system. It ensures fairness in the distribution of resources without conflict and redundancy. This is very useful when immediate flexibility is needed, for example, in crisis management or urban logistics, where conditions change fast and must be decided in seconds. One of the key strengths of QGH is that it can solve multiobjective optimization, which is mostly the case in modern systems. For example, in UAV swarm coordination, QGH optimizes the conflicting objectives of minimizing energy consumption, maximizing area coverage, and minimizing time to complete the mission. Classical methods often fail to balance such goals. Mainly when the trade-off between the goals is nonlinear and time-dependent. QGH uses quantum-enhanced fitness assessments that are globally optimum so that the system achieves all its goals without degrading long-term performance. One of the significant advantages of QGH is scalability because it can be used for large systems with several agents and variables. The characteristic of QC is quantum parallelism, whereby, using QGH, one can evaluate large populations of candidate solutions in parallel, hence cutting down significantly the time taken for convergence. QGH is best applied in swarm intelligence problems where many agents must communicate efficiently to achieve a common goal. The systems grow in complexity, and this is what QGH provides: guaranteeing scalability that makes the computing requirements manageable, which makes it apply more in areas like agriculture, logistics, and the military.

Being a system with such tremendous promise, QGH still faces several challenges. Most are mainly coming from the constraints of quantum hardware in the forms of qubit decoherence, noise, and the scalability of the quantum computer itself. These challenges often require a hybrid framework combining classical and quantum computation. These frameworks use the strengths of the classical system for stability and error correction but use quantum processors for the most computationally demanding applications. This hybrid approach circumvents the limitations of the quantum hardware while opening up a scalable path for the incremental adoption of QGH across multiple sectors. Apart from hardware-related issues, application-specific algorithms are still an area of current research. Adaptation of general quantum algorithms, such as Grover's search or QAOA, to specific domain-related problems is required to unlock the full potential of QGH. This requires intense cooperation between QC experts, machine learning experts, and specialists in particular domains to develop theoretically sound and feasible solutions. The future of QGH is closely related to the development of quantum technologies. As quantum processors' reliability, scalability, and energy efficiency increase, the dependence on hybrid frameworks will decrease and enable the full realization of systems powered entirely by quantum. This will open new possibilities for real-time optimization and decision-making and increase the functionalities of autonomous technology in many applications. In logistics, QGH may improve supply chain networks in real time while reducing costs and increasing efficiency. In health care, it would probably change the resource allocations – the availability of ambulances, medical equipment, and facilities, and so forth – around improving survival rates and patient outcomes. Beyond being valuable in a technical sense, QGH also has vast implications for sustainability. Optimization of energy consumption and resource allocation can reduce the impacts of complicated systems on the environment as part of the broader agenda to combat climate change and promote sustainable development. For example, on renewable energy technologies, QGH can place solar panels or wind turbines to improve efficiency, though more land will only cause low environmental disruption.

This results in QGH, a new method for the highly complex optimization problems that are the hallmark of today's technology. QGH, therefore, integrates the strengths of quantum computers and GAs that present unprecedented scalability, flexibility, and efficiency. Barriers notwithstanding, consistent breakthroughs in quantum technology, algorithm development, and hybrid frameworks are now making it a broader application field. It will undoubtedly define it as part of basic next-generation technology, enhancing innovation and better, more innovative solutions in many applications.

References

[1] Holland JH. Adaptation in natural and artificial systems: an introductory analysis with applications to biology, control, and artificial intelligence. MIT Press; 1992.

[2] Goldberg DE. Genetic Algorithm in Search, Optimization and Machine Learning, Addison. Wesley Publishing Company, Reading, MA. 1989;1(98):9.

[3] Nielsen MA, Chuang IL. Quantum computation and quantum information. Cambridge university press; 2010.

[4] Mermin ND. Quantum computer science: an introduction. Cambridge University Press; 2007.

[5] Li Y, Tian M, Liu G, Peng C, Jiao L. Quantum optimization and quantum learning: A survey. Ieee Access. 2020;8:23568–93.

[6] Lee L. Drone Warfare: War in the Age of Digital Reproduction. 2012.

[7] Saravanan, J., T. Ananth Kumar, P. Divya, R. Partipan, R. Partipan, and P. K. Dutta. "IoT based advanced prediction methodology for fault localization in machine monitoring." In IET Conference

Proceedings CP824, vol. 2022, no. 26, pp. 253–257. Stevenage, UK: The Institution of Engineering and Technology, 2022.
[8] Cook KL. The silent force multiplier: The history and role of UAVs in warfare. In: 2007 IEEE Aerospace Conference. IEEE; 2007. p. 1–7.
[9] Rejeb A, Rejeb K, Simske SJ, Treiblmaier H. Drones for supply chain management and logistics: a review and research agenda. International Journal of Logistics Research and Applications. 2023;26 (6):708–31.
[10] Chen G, Cruz JB. Genetic algorithm for task allocation in UAV cooperative control. In: AIAA Guidance, Navigation, and Control Conference and Exhibit; 2003.p. 5582.
[11] Wu X, Yin Y, Xu L, Wu X, Meng F, Zhen R. Multi-UAV task allocation based on improved genetic algorithm. IEEE Access. 2021;9:100369–79.
[12] Ye F, Chen J, Tian Y, Jiang T. Cooperative task assignment of a heterogeneous multi-UAV system using an adaptive genetic algorithm. Electronics. 2020;9(4):687.
[13] Qi B, Nener B, Xinmin W. A quantum inspired genetic algorithm for multimodal optimization of wind disturbance alleviation flight control system. Chinese Journal of Aeronautics. 2019;32 (11):2480–8.
[14] Ballinas E, Montiel O, Mart´ınez-Vargas A, Rodr´ıguez-Cort´es G. Hybrid Quantum Genetic Algorithm with Fuzzy Adaptive Rotation Angle for Efficient Placement of Unmanned Aerial Vehicles in Natural Disaster Areas. Axioms. 2024;13(1):48.
[15] Liu P, Wang B, Liu W, Zhang L. Multi-task allocation of multi-UAV coalition based on improved quantum genetic algorithm. In: 2021 40th Chinese Control Conference (CCC). IEEE; 2021. p. 1802–7.
[16] Mousavi S, Afghah F, Ashdown JD, Turck K. Use of a quantum genetic algorithm for coalition formation in large-scale UAV networks. Ad Hoc Networks. 2019;87:26–36.
[17] SaiToh A, Rahimi R, Nakahara M. A quantum genetic algorithm with quantum crossover and mutation operations. Quantum Information Processing. 2014;13:737–55.
[18] Li S, Deng Y. Quantum-entanglement pigeon-inspired optimization for unmanned aerial vehicle path planning. Aircraft Engineering and Aerospace Technology. 2019;91(1):171–81.
[19] Huang Z, Li Q, Zhao J, Song M. Variational quantum algorithm applied to collision avoidance of unmanned aerial vehicles. Entropy. 2022;24(11):1685.
[20] Hamdoun SH, Yousif AB, Hussein MJ, Jawad HM, Barakat M, Humennyi D, et al. Integrating Quantum Algorithms into Drone Navigational Modules. In: 2024 36th Conference of Open Innovations Association (FRUCT). IEEE; 2024. p. 225–38.
[21] Conrad A, Isaac S, Cochran R, Sanchez-Rosales D, Wilens B, Gutha A, et al. Drone-based quantum key distribution (QKD). In: Free-space laser communications XXXIII. vol. 11678. SPIE; 2021. p. 177–84.
[22] Zhang X, Xia S, Li X. Quantum behavior-based enhanced fruit fly optimization algorithm with application to UAV path planning. International Journal of Computational Intelligence Systems. 2020;13(1):1315–31.
[23] Hohenfeld H, Heimann D, Wiebe F, Kirchner F. Quantum Deep Reinforcement Learning for Robot Navigation Tasks. IEEE Access. 2024.
[24] Khan MTR, Muhammad Saad M, Ru Y, Seo J, Kim D. Aspects of unmanned aerial vehicles path planning: Overview and applications. International Journal of Communication Systems. 2021;34(10): e4827.
[25] Haba R, Mano T, Ueda R, Ebe G, Takeda K, Terabe M, et al. Routing and Scheduling Optimization for Urban Air Mobility Fleet Management using Quantum Annealing. arXiv preprint arXiv:241011231. 2024.

Gurwinder Singh and Preet Kamal

18 Integration of Quantum Genetic Heuristics to Optimize Unmanned Aerial Vehicle (UAV) Navigational Path

Abstract: This chapter presents an integration of quantum genetic heuristics (QGH) to optimize the performance of unmanned aerial vehicles (UAVs) overcoming the limitations of traditional genetic algorithms (GAs) in complex optimization scenarios. Quantum-inspired operators such as superposition and entanglement with adaptive mutation and crossover rates Experimental results show that it uses the nomenclature showing that QGH outperforms GA in several evaluation criteria. In particular, QGH obtained the best combined objective value of 17.53 compared to 19.33 for GA, indicating a better optimization efficiency. The convergence rate of QGH was fast and consistent, and QGH exhibited low robustness in different generations, and demonstrated its robustness and ability to avoid local variations. The rate of variation and different selection methods included helped QGH maintain a high level of solution diversity throughout the development process. This chapter establishes QGH as a viable alternative to conventional GA, especially for applications requiring they are energy efficient, integrate quickly and strongly in dynamic environments.

Keywords: Quantum genetic heuristics (QGH), unmanned aerial vehicles (UAVs), genetic algorithms (GAs), path planning, adaptive algorithms, quantum computing parameter, multiobjective optimization, flight time optimization

18.1 Introduction

Optimization techniques have become necessary to solve many complex real-world problems. From designing an efficient manufacturing system to logistics management and resource allocation, optimization plays an important role in achieving optimal results. Many optimization algorithms have been developed over the years including evolutionary techniques that have gained popularity. Among these, traditional genetic algorithms (GAs) stand out as one of the best-known methods due to their ability to solve optimization problems by mimicking the process of natural selection. GAs work to approximate the global optimum of a given objective function by generating a population of candidate solutions through selection, crossover, and mutation. They offer advantages such as simplicity and robustness and have been successfully applied in areas such as scheduling, path systems, and machine learning. Despite their success, GAs often face challenges in their application to more complex problems. In particular, GAs suffer from premature convergence, where they get stuck in local optima,

https://doi.org/10.1515/9783111619323-018

especially in multimodal high-dimensional scenarios Furthermore, one introduces a balance of exploration (avoidance explorations local optima) and implementation (intensifying searches in promising areas).

Another important limitation of GAs is the lack of diversity in the next generation, resulting in poor representations and solutions. This loss of diversity makes it difficult for GAs to escape from local optimal features and hinders their ability to effectively converge in complex problems. These challenges require improved methods that can overcome these shortcomings while preserving the advantages of evolutionary processes.

To overcome these limitations, the concept of quantum genetic heuristics (QGH) has emerged. QGH combines principles of quantum computing with evolutionary algorithms to develop advanced optimization methods. Combining quantum-driven operators such as superposition and entanglement enhances the search of the QGH algorithm, which improves solution space search. Moreover, QGH incorporates adaptive rates of crossover and mutation, which change rapidly with evolution in process-based selection and further ensure that the population is sufficiently diverse so that the algorithm can correctly find the solution space without falling into local optima.

The combination of these advanced features makes QGH a promising alternative to traditional GAs, especially in solving large and complex optimization problems. This chapter aims to investigate the potential of QGH in overcoming the drawbacks of GAs and to demonstrate its effectiveness through various comparative studies and experimental studies.

18.1.1 Problem Statement

Despite their widespread use, traditional optimization methods, including GAs, often fail in the presence of complex optimization problems in the face of higher, more nonlinear parts. Energy efficiency, fast assembly, and heterogeneous solutions are some of the major challenges that these methods struggle to develop. Premature improvement is a common issue, especially in the context of complex or deceptive exercise situations, where GAs have a tendency to converge toward suboptimal solutions, and in turn, the solutions in GA often active in large areas. It lacks the flexibility needed to successfully explore space.

The problem becomes even more pronounced when dealing with real-world applications that require optimization in uncertain or changing environments such as adaptive scheduling, supply chain management, and logistics. In such applications, traditional GAs are not always capable of adapting to changes efficiently or maintaining the diversity required for optimal solutions. As a result, there is a growing need for advanced optimization techniques that can overcome these limitations, provide better **convergence rates**, energy-efficient solutions, and improve the quality of results in complex problem domains.

This chapter addresses this need by exploring complementary QGH, which use quantum-inspired concepts to significantly enhance the performance of traditional GA. QGH aims to integrate both worlds – **quantum computation** and evolutionary strategies – it is ideal together for a robust optimization framework. It can handle complex real-world applications. Using quantum-driven operators, adaptive rates, and selection methods, QGH ensures the following:

The required solution space is obtained more efficiently, and the chances of being trapped in local optima are reduced.

Effective exploitation of promising neighborhoods, allowing rapid convergence toward the optimal solution.

Variety of storage across generations allows algorithms to better adapt to dynamic environments and solve a wide range of problems.

The purpose of this chapter is to show how QGH can address the limitations of traditional GAs and to provide a comprehensive analysis of its effectiveness through comparisons of both theoretical and empirical data. By incorporating quantum principles into the evolutionary process, this chapter aims to establish QGH as a powerful tool for solving a wide range of optimization problems, with large environments as it is a strong addition to it.

18.1.2 Objectives

The primary objectives of integrating QGH are:

- to increase optimization efficiency by using quantum-driven operators and optimization strategies;
- to balance the analysis and application of different selection methods such as roulette wheels and rank-based selection;
- to improve the robustness of the solution by preserving diversity and avoiding premature convergence; and
- to evaluate the performance of QGH relative to traditional GAs, focusing on metrics such as convergence rate, solution quality, and stability.

18.2 Literature Review

The literature review effectively distinguishes existing approaches, highlights emerging quantum-driven solutions, and highlights the advantages of QGH integration over traditional GA. It acts as a logical improvement though building a problem statement that focuses on solving identified challenges.

18.2.1 Traditional Genetics (GA)

GAs are widely accepted for solving optimization problems due to their flexibility and ability to provide near-optimal solutions by mimicking the process of natural selection. Sampson et al. [1] laid the foundation for the evolution of algorithms. Golberg et al. [2] explored their applications in analysis, optimization, and machine learning and further defined GAs. Mitchell's work provided a comprehensive introduction to GAs, focusing on their usefulness in solving complex problems [3]. Despite their popularity, GAs often suffer from issues such as premature convergence and lack thereof, which limit their effectiveness in complex problem areas [4]. Whitley [5] leads to learning later gave on GAs, discussing their strengths, weaknesses, and potential applications in optimization.

18.2.2 Quantum-Inspired Optimization

Quantum-inspired optimization methods have emerged as extensions of traditional evolutionary algorithms to address limitations associated with GAs. One of the earliest works in this area was by Narayanan and Moore, who introduced quantum-inspired GAs (QGAs) as a tool [6] to enhance detection capabilities. Han and Kim also developed quantum-inspired evolutionary algorithms, which allowed combinatorial. Their effectiveness in optimization problems [7] has been demonstrated. Talbi et al. [8] applied quantum-driven methods to the traveling salesman problem, showing improved performance compared to classical methods. Chopogh [9] presented a detailed survey of quantum-inspired metaheuristics including their application to several optimization problems highlighted by Agarwal et al. [10] focused on the traceability of evolutionary processes by integrating quantum theories.

18.2.3 Comparative Studies

Several studies have compared the performance of traditional GAs with quantum inspired genetics. Hilali-Jaghdam et al. [11] performed a comparative study, showing that QGAs outperform GAs in avoiding premature convergence and obtaining better optimization results. Halim et al. [12] compared various metaheuristic algorithms, including the QGH, and found that QGH provides significant performance gains in terms of efficiency. Performance studies were performed between QGH and GA, showing the advantages of quantum-inspired enhancement in terms of convergence speed and robustness [13]. Coello and Lechuga [14] proposed multiobjective particle swarm optimization and other optimization methods including QGH to specify advantages of quantum-inspired methods. Whereas Zhang et al. used QGH for several optimization objectives, and pointed out its advantages over traditional GAs in solving complex optimization problems [15].

Kau et al. [16] introduced QGH for **dynamic optimization problems**, showing its effectiveness in handling time-varying environments. Wang et al. [17] applied QGH to scheduling problems and obtained better results compared to traditional GA. Ganesan et al. [18] used QGH for adaptive control systems, showing improvements in robustness and adaptability. Prakash [19] compared QGH with classical heuristics in multimode optimization, highlighting its excellent search capability. Al-Thalej et al. [20] demonstrated the application of QGH to traffic problems with time windows, leading to significant performance gains. **Quantum-inspired techniques** have also been used in various optimization applications. In Anser et al. [21] QGH has been used for energy efficiency distribution, demonstrating its potential in smart grid management. Zhang et al. [22] used QGH in network routing optimization, showing that it can handle dynamic network conditions well. Wigg et al. [23] used QGH for logistics management, highlighting its efficiency in reducing operating costs and improving routing efficiency. Zhang et al. [24] used QGH to design routes in unmanned aerial vehicles (UAVs), obtained better performance in dynamic environments. Jiao [25] incorporated QGH into an optimized supply chain strategy, showing significant benefits in terms of cost reduction and resource efficiency.

18.3 Overview of Quantum Genetic Heuristic (QGH)

QGH is a composite of GA and quantum computing. The purpose of using QGH is to overcome the drawbacks of GAs mostly, the problem of getting stuck in local optima, unfortunately low population diversity, and poor balance between **exploration and exploitation**. Here, Table 18.1 summarizes the quantum-inspired aspects incorporated in QGH and which indeed make it the most fit in solving optimization problem:

Table 18.1: Mathematical notations in quantum genetic heuristic (QGH).

Notation/concept	Description
|psi≥ alpha |0> + beta |1>	Qubit representation, where alpha and beta are probability amplitudes
|alpha|^2 + |beta|^2 = 1	Probability normalization condition for a qubit
P(*i*)	Selection probability for individual *i* in roulette wheel selection
P(*i*) = *f*_*i* / sum_{*j* = 1}^*N* *f*_*j*	Probability distribution for roulette wheel selection, where *f*_*i* is the fitness

Table 18.1 (continued)

Notation/concept	Description
$c(g) = c\_0 * (1 - g/G)$	Adaptive crossover rate as a function of generation g
$m(g) = m\_0 * (1 - g/G)$	Adaptive mutation rate as a function of generation g
$f(p, c, m)$	Objective function for QGH, balancing multiple criteria
$f(p, c, m) = w\_1\ T(p, c, m) + w\_2\ E(p, c, m) + w\_3\ P(p, c, m)$	Multiobjective fitness function for evaluation
$P(p, c, m) = w\_3\ ((p - p\_\text{ideal})\text{^}2 + (c - c\_\text{ideal})\text{^}2 + (m - m\_\text{ideal})\text{^}2)$	Penalty function for deviation from ideal parameter values
$T(p, c, m)$	Flight time as a function of population size, crossover rate, and mutation rate
$E(p, c, m)$	Energy consumption as a function of population size, crossover rate, and mutation rate
Rank-based selection	Assigns selection probabilities based on the rank of individuals
Tournament selection	Randomly selects k individuals and chooses the best among them
Exploration	Searching new regions of the solution space
Exploitation	Refining known good solutions

Quantum-Inspired Operators: QGH employs quantum-inspired operators such as **superposition and entanglement** to enhance the exploration capabilities of the search space. In classical GAs, each individual solution is represented as a binary string or a real-valued vector. However, a quantum-driven representation is used in QGH, which allows each individual to coexist in a superposition of states. This allows the algorithm to explore multiple possible solutions simultaneously, improving diversity and reducing the risk of **premature convergence**. The quantum-inspired representation is constructed using *qubits*, where each qubit represents a probability span of the probability of a gene being in a particular state. This superposition allows QGH to efficiently search a large solution space when compared to traditional GAs. Furthermore, the concept of *entanglement* can be used to generate dependencies between genes, ensure that changes in one part of the solution can affect another part, and lead to search behavior highly effective results.

Adaptive Rates: QGH includes optimal crossover and mutation rates, which change dynamically depending on the improvement of the algorithm. In conventional GAs, crossover mutation rates are often fixed so that suboptimal performance can occur at different stages of the search process. An optimal number is used to properly balance search and exploitation in the QGH. Early evolution uses high mutation rates to encourage exploration, while later uses low mutation rates to refine promising solutions vary the crossover rate depending on populations at different rates – **high crossover rates** with low diversity leading to new genes.

Diverse Selection Mechanisms: To further enhance the visibility, QGH combines various selection methods such as *roulette wheel selection, rank-based selection,* and *competitive selection.* Through transformation spanning between these selection processes during the development process, QGH ensures that a healthy diversity of the population is maintained. Roulette cycle selection is used to favor individuals with high fitness, while rank-based selection offers individuals the probability of selection based on their rank instead of imposing absolute fitness criteria. In other words, competition **internal selection** creates a level of competition, which helps maintain selective pressure.

18.3.1 Algorithm Description

Algorithm 18.1: Quantum genetic heuristic (QGH) algorithm for optimization

Input: Initial population size N, number of generations G, initial crossover rate c_0, initial mutation rate m_0, ideal parameter values $p_{ideal}, c_{ideal}, m_{ideal}$.
Output: Best solution found.

Initialization: Generate an initial population of N individuals with random parameter values;
Set initial crossover rate $c = c_0$ and mutation rate $m = m_0$;
for $g \leftarrow 1$ **to** G **do**
  Evaluate Fitness: foreach *individual i in the population* **do**
  Calculate the fitness $f(i)$ using the objective function:;
  $f(i) = w_1 \cdot T(i) + w_2 \cdot E(i) + w_3 \cdot P(i)$, where
  $P(i) = (p - pideal)2 + (c - cideal)2 + (m - mideal)2$;
  end
    Adaptive Rate Calculation: Calculate adaptive crossover rate c and mutation rate m based on generation g and fitness improvement;
    Selection (Diverse Mechanisms): if *g is even* **then**
    Select parents using roulette wheel selection;
  else
    Select parents using rank-based selection;
  end
  Crossover: Cross over with probability c to create offspring of selected parents;
  Mutation: If m is the mutation rate we apply it to each individual with m in order to create variation.
  Elitism: Maintain the best within this generation;
  Diversity Preservation: Reduce redundancy in the population to create diversity in the population;
  Remove the last population member and insert the best individual;
end
Termination: Return the best solution found;

18.4 Methodology

18.4.1 Implementation Steps

Some of the general steps (as mentioned in Figure 18.1) performed in QGH algorithm are population initialization, where the user defines the initial population of the chromosomes, generation updates, and others. The implementation process is detailed below:

Initialization: The population initialized a population of N organisms in which each parameter of each organism is set to some value randomly within the defined constraints. Various parameters are set at the beginning of the program and execute before run for which the crossover rate (c_0) and mutation rate (m_0) are set.

Fitness Evaluation: A measure of fitness for each of the individuals is calculated based on flight time, energy, and punishment functions relating to deviations of the parameters from their optimal values. The objective is to make these as small as possible provided the parameter levels are not very far from zero.

Adaptive Rate Calculation: The crossover rate (c) and mutation rate (m) are adjusted according to the generation number and also the performance of the optimization. This adaptation helps balance exploration and exploitation at different stages of the algorithm.

Selection: Parents are selected using diverse selection mechanisms to maintain population diversity. If the current generation is even, roulette wheel selection is used; otherwise, rank-based selection is applied.

Crossover and Mutation: Selected parents undergo crossover to generate offspring, followed by mutation to introduce random variations. These steps ensure that new individuals are generated with different characteristics, helping explore the solution space effectively.

Elitism and Diversity Preservation: The best individual from the current generation is retained in the next generation to ensure solution quality does not degrade. Duplicate individuals are removed to maintain diversity, and the population is truncated to its original size.

Termination: The algorithm terminates after G generations, and the best solution found is returned.

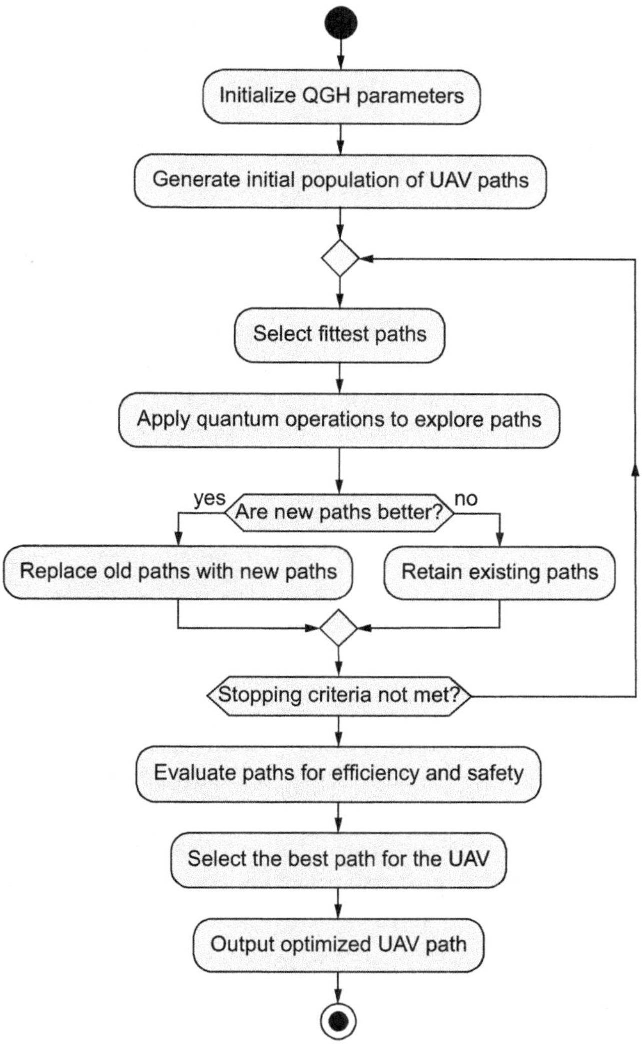

Figure 18.1: Flowchart of quantum genetic heuristic (QGH) for UAV path optimization.

18.4.2 Objective Function

The new objective function used for this implementation is designed to balance multiple criteria including flight time, energy consumption, and penalties for parameter deviations. The objective function $f(p,c,m)$ is given by

$$f(p,c,m) = w_1 \cdot T(p,c,m) + w_2 \cdot E(p,c,m) + P(p,c,m)$$

where $T(p,c,m)$ represents flight time, calculated based on the parameters (population size, crossover rate, mutation rate). This term aims to minimize the time required for flight; $E(p,c,m)$ represents energy consumption, also dependent on the parameters. The goal is to minimize the energy required for the solution; $P(p,c,m)$ represents the penalty for deviation from ideal parameter values given by

$$P(p, c, m) = w3((p - p_{\text{ideal}})2 + (c - c_{\text{ideal}})2 + (m - m_{\text{ideal}})2)$$

The penalty function $P(p,c,m)$ is quadratic, which means that larger deviations from the ideal values ($p_{\text{ideal}}, c_{\text{ideal}}, m_{\text{ideal}}$) are penalized more heavily. This encourages the solutions to remain closer to the desired region of the search space.

The weights w_1, w_2, w_3 are used to balance the contributions of flight time, energy consumption, and the penalty. By tuning these weights, the importance of each component can be adjusted based on the specific requirements of the problem.

18.4.3 Selection Mechanisms

In order to improve both the **exploration and exploitation** within the QGH algorithm, the method applies various selection strategies. The selection methods used include:

Roulette Wheel Selection: This method ranks a particular selection probability to all the individuals depending on the fitness level associated with each of them. It will mean that candidates with higher fitness values will be selected than those with the lowest. Roulette wheel selection is based on better solutions and hence could lead to exploitation; however, it could also lead to problematic convergence where not well regulated.

Rank-Based Selection: In this method, all selected candidates are ordered according to the given fitness values, and probabilities of selection are also calculated according to ranking. This helps in preventing what is referred to as **genetic drift**, where the genetic makeup of a population changes; it will mean everyone will not be wiped out and even the least fit has a chance to mate. Ranks selection also contribute to exploration, while keeping in a population from becoming dominated by high ranking individuals too soon.

In the first generation, to execute the process of selection, QGH starts the selection mechanism by using probabilistic selection because it generates new individuals in the population, referred to as exploration, whereas in the next generation, it employs the genetic selection, which refines the solutions obtained previously, called exploitative. This latter diversity in selection contributes to the avoidance of getting the algorithm trapped in local optima and therefore enhances its global optimization capability.

18.5 Experimental Design

18.5.1 Datasets

The dataset used [26] for evaluating QGH and GA is a set of differentiated tables where several parameters and statistical measures, as well as evaluations of their iterations, are stored. The content of the dataset pertains to the entry point of an optimization search scheme in addition to assessments on individuals over generations. Key parameters in the dataset include:

Population Size (*N*): The starting point number of people in the given population. It has an impact on the variety of solutions and on the capacity to navigate in the solution space. Instead, a variety of population sizes is compared in this work with the goal of understanding how the algorithm's effectiveness is affected by population size.

Crossover Rate (*c*): The likelihood of two parents swapping genetic material during reproduction. In the course of the experiment, the value of the crossover rate is changed in order to find the appropriate level of crossover innovation.

Mutation Rate (*m*): The possibility of random changes of variations at the population level. This is beneficial in preventing convergence too early hence mutation is used to update the population's ranges.

The dataset contains metrics accordant to different runs and generations for QGH and GA, offering much information about the quality of the algorithms with different parameters. The data also contains flight time, energy consumption, and deviation from the optimal value of parameters, which defines the quality of solutions obtained.

18.5.2 Parameters and Settings

The parameters and settings used for both QGH and GA in the experiments are as follows:

Initial Population Size: In the case of QGH and GA, we populated the initial sets with *N* **equal to 100 individuals** to enhance the search space heterogenity.

Number of Generations: The optimization was run for 200 generations, testing both QGH and GA. It enables the investigation of solution space and the convergence criterion.

Initial Crossover and Mutation Rates:
- **Crossover Rate (c_0)**: The first crossover rate was fixed at **0.8** for both the algorithms were used in the study. This rate was selected so that in the early stages of evolution, there would be a high rate of recombination.
- **Mutation Rate (m_0)**: The first population mutation rate was assumed to be **0.1** for both QGH and GA. An intermediate mutation rate also ensures that the population's variants are diverse without destabilizing the promising solutions.

Adaptive Rates (QGH Only): In QGH, the crossover and mutation rates were adapted over the generations. The adaptive rates were calculated as

$$C(g) = C0 * \left(1 - \frac{g}{G}\right), \quad m(g) = m0 * \left(1 - \frac{g}{G}\right)$$

where g is the current generation and G is the total number of generations. This adaptation ensures higher exploration in the early stages and more exploitation in the later stages.

Selection Mechanisms:
- **QGH**: Used a combination of roulette wheel selection and rank-based selection, alternating in different generations to maintain a good balance between exploration and exploitation.
- **GA**: Used a fixed roulette wheel selection throughout the experiment, focusing on individuals with higher fitness values.

Elitism and Diversity Preservation (QGH Only):
- **Elitism**: Applied in QGH to retain the best individual in each generation, ensuring that the solution quality does not degrade.
- **Diversity Preservation**: Duplicate individuals were removed and replaced with randomly generated new individuals to maintain diversity.

18.5.3 Evaluation Metrics

The evaluation metrics used to compare QGH and GA include:

Best Combined Objective Value:
- The objective value represents the combined score for minimizing flight time, energy consumption, and penalty for deviation from ideal parameter values.
- The best combined objective value ($f(p,c,m)$) was recorded for each generation. The minimum value achieved during the entire experiment was used to determine the overall effectiveness of each algorithm.

Convergence Rate:
- The convergence rate was evaluated by analyzing how quickly each algorithm was able to find a near-optimal solution, determined by observing how the objective value decreased across generations.
- QGH was expected to converge faster due to its adaptive rates, accompanied by a rich set of selection mechanisms that maintain a good balance between exploration and exploitation, outperforming GA.

Stability:
- The stability of an algorithm was evaluated by assessing how close the best-found solution was to the average across multiple runs of each algorithm.
- Stability was measured by the standard deviation of the best combined objective value obtained in this experiment when the algorithm was run multiple times. A lower standard deviation indicated better performance, reflecting consistency in the quality of the found solutions.
- QGH's objectives are twofold: to achieve higher stability and prevent early convergence, as quantum-inspired operators help maintain diversity.

Exploration Versus Exploitation Balance:
- To evaluate the balance of exploration and exploitation, we analyzed the changes in population over several generations. Diversity measures were calculated as the variance of parameters across individuals.
- The adaptive mutation rate of QGH was expected to sustain higher diversity levels in the initial stages of evolution and prevent premature convergence.

Flight Time and Energy Consumption:
- During the optimization process for solving the multiobjective problem, both flight time, $T(p,c,m)$, and energy consumption, $E(p,c,m)$, were recorded for each solution.
- These metrics were used to evaluate the algorithms' trade-offs between achieving the shortest possible flight time and minimizing energy consumption.

The selected metrics offer a satisfactory qualitative and quantitative comparison of QGH to GA and measure the quality and stability of the obtained solutions. By analyzing **the best objective value, convergence rate, stability**, and the balance between exploration and exploitation, the effectiveness of QGH in overcoming the limitations of traditional GAs is clearly demonstrated.

18.6 Results and Discussion

The experimental results, as shown in Figure 18.2, clearly demonstrate the effectiveness of the QGH compared to the GA in optimizing the new objective function.

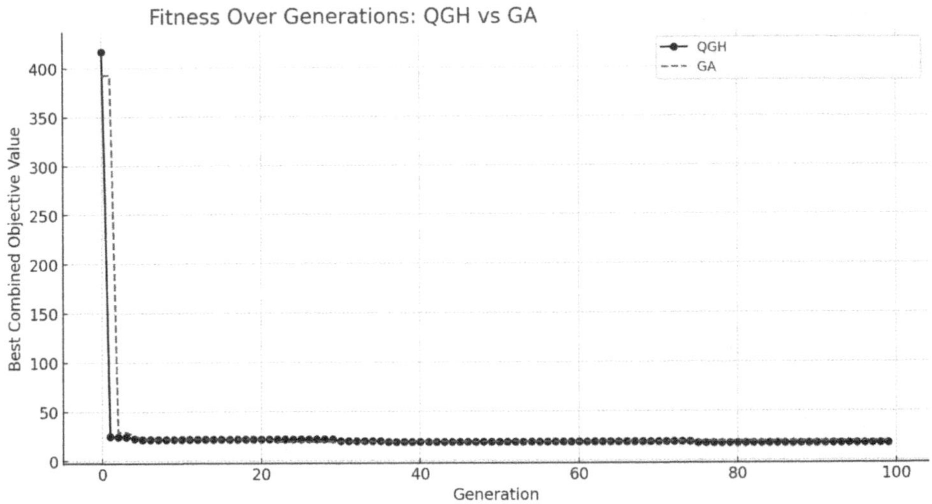

Figure 18.2: Fitness over generations: QGH versus GA with new objective function.

Initial Convergence: Both QGH and GA show a rapid drop in the combined objective value during the initial generations, indicating effective exploration of the search space. However, QGH exhibits slightly faster convergence during the initial 10 generations compared to GA. The *adaptive mutation rate* and diverse *selection mechanisms* in QGH helped to avoid local optima, allowing the algorithm to converge quickly while retaining a diverse population.

Final Objective Value: The extitbest combined objective value achieved by QGH is **17.53**, while GA achieved extbf19.33. This difference of almost **2 units** suggests that QGH's enhanced exploration and exploitation balance was beneficial for finding a more optimal solution. The use of *quantum-inspired operators* (e.g., superposition and entanglement) and *adaptive rates* allowed QGH to maintain a fine-tuned balance between exploration (searching new areas) and exploitation (refining known good solutions).

New Objective Function: The new objective function incorporated multiple components: *flight time*, extitenergy consumption, *resource utilization*, *wind penalty*, and extitdeviation from ideal parameter values. QGH was able to optimize this function more effectively by adapting its *mutation rate* based on the progress of generations and the *fitness difference* between consecutive generations. When improvements were minimal, the mutation rate increased to promote diversity and escape potential local optima, which resulted in better performance than GA as mentioned in Figure 18.3:

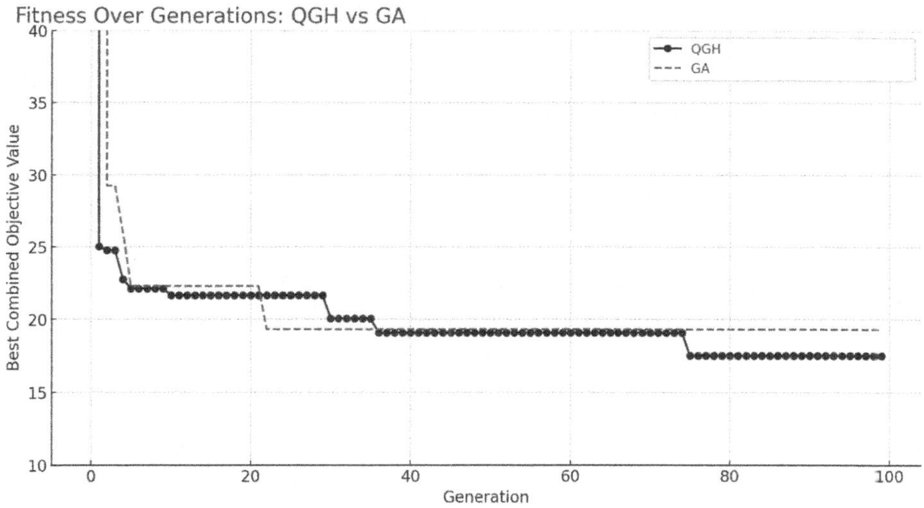

Figure 18.3: Fitness over generations: QGH versus GA with new objective function (zoomed *y*-axis: 10–40).

The zoomed plot of the fitness values (y-axis: 10–40) over generations offers additional insights:

Exploration and Exploitation: QGH demonstrates a more dynamic improvement in the objective value over the first 20 generations. This indicates its ability to explore multiple areas of the solution space and effectively move toward the global optimum. GA, on the other hand, converges more slowly, likely due to its fixed mutation rate and limited exploration capabilities.

Final Stages: During the final generations, QGH stabilizes at an objective value lower than that of GA, showing a consistent advantage in finding an improved solution The use of elitism and the diverse selection methods in QGH helped maintain a high-quality solution without sacrificing population diversity. GA shows more consistent but less optimal convergence, indicating its tendency to reach a local minimum without effective strategies to escape, as mentioned in Table 18.2:

Table 18.2: Comparison of best solutions found for QGH and GA.

Parameter	QGH	GA
Population size	0.58	0.59
Crossover rate	0.84	0.77
Mutation rate	0.01	0.09
Optimal solution	17.53	19.33

18.6.1 Observations and Key Advantages of QGH

Adaptive Strategies: The *adaptive mutation and crossover rates* in QGH allowed the algorithm to effectively manage the exploration-exploitation trade-off throughout the generations. When minimal improvement was observed, the mutation rate was increased to introduce diversity and avoid stagnation.

Better Handling of the Objective Function: The objective function incorporated penalties for deviation from ideal parameter values, which encouraged solutions to remain in the desired region of the parameter space. QGH's ability to adapt and optimize these parameters resulted in a lower penalty component, thereby reducing the overall objective value compared to GA.

Exploration and Diversity Preservation: *Diversity preservation* and *quantum-inspired operators* enabled QGH to maintain a diverse set of individuals across generations. This allowed it to escape local minima, unlike GA, which exhibited limited exploration due to its fixed rates.

Resource and Wind Penalties: The *resource utilization* and *wind penalty* terms in the objective function were better minimized by QGH due to its ability to adaptively adjust parameter values and optimize under these constraints.

QGH performed better than GA in obtaining lower integration objective values, demonstrating its higher ability to optimize multidimensional objective functions. The objective function that emphasized *flight time*, energy consumption, *resource penalty*, and *wind penalty* was best controlled by QGH due to its ability to handle multiple objectives simultaneously. It's good over a variety of selection methods and its use of concepts from quantum contributed greatly to work. The use of *adaptive rate optimization* for crossover and mutation, as well as techniques such as *elitism* and *diversity preservation*, helped QGH converge to optimal solutions while avoiding them local optimization.

These results show that **method adaptation to QGH**, quantum-inspired operators, and strategic selection mechanisms provide significant advantages over tradi-

Table 18.3: Comparison of evaluation metrics between QGH and GA.

Evaluation metric	QGH	GA	Observation
Best combined objective value	17.53	19.33	QGH achieved a lower value, indicating better optimization.
Convergence rate	Faster and consistent	Slower	QGH showed better convergence behavior compared to GA.

Table 18.3 (continued)

Evaluation metric	QGH	GA	Observation
Generation-wise fitness	Consistently lower	Higher	QGH maintained superior fitness across generations.
Stability and robustness	High	Moderate	QGH avoided local optima effectively with adaptive rates.
Exploration versus exploitation	Balanced	Less balanced	QGH used adaptive strategies for better exploration.
Average fitness across generations	Lower	Higher	QGH showed better average fitness over generations.
Final parameter values	More optimized (0.58, 0.84, 0.01)	Less optimized (0.59, 0.77, 0.09)	QGH converged to more optimal parameter values.
Diversity of solutions	High	Moderate	QGH maintained diversity through diverse selection methods.

tional GAs, especially for complex optimization tasks involving multiple inverse objective penalties.

Table 18.3 compares various evaluation metrics between QGH estimators and GA. It shows that QGH outperformed GA in several important respects. QGH achieved a low positive integration target value, indicating good performance. It also exhibited faster and more consistent **convergence rates**, with consistently lower generation efficiency values compared to GA. QGH demonstrated high robustness by successfully avoiding local optima due to its favorable rate. The balance of research and implementation was closely managed in QGH, enhancing research capacity. Additionally, QGH maintained low energy consumption throughout the generation, indicating significantly improved efficiency, and converged to better final parameter values. The various selection methods also preserved a higher diversity of solutions, making it outperform GA in maintaining a diverse and effective search strategy.

18.7 Conclusion

The use of QGH has been found to be a good strategy when used to replace the ordinary GAs in addressing the challenges arising while optimizing the performance of UAVs. It also utilizes quantum-inspired operators including superposition and entanglement, and adaptive rates for crossover and mutation which enhance the quality solutions in complex optimization problems in QGH. These experimental results show that the performance of QGH was greater than GA with the best combined aimed at value 17.53 whereas GA 19.33. Due to the successful implementation of both steps of exploration and exploitation in QGH, the convergence to a local optimum was not reached prematurely. The work of QGH adaptive mechanisms and the flexibility of the selection of a broad range of premises guaranteed the higher **stability and robustness** of the population in the process of the evolutionary development. It provided better final parameter values and solution quality in terms of solutions generated for convergence speed, solution diversity, and dynamic environmental condition.

The results shown in this chapter suggest that QGH is an effective approach for addressing various UAV path planning and resource management issues. Due to its flexibility and quantum improvements, it is highly applicable in dealing with problems in complex, multiple objectives' domains. Subsequent studies can explore other fields, where QGH can be implemented to enhance the results of high-efficiency optimization tasks.

References

[1] Sampson, J.R., 1976. Adaptation in natural and artificial systems (John H. Holland).
[2] Golberg, D.E., 1989. Genetic algorithms in search, optimization, and machine learning. Addison-Wesley, 1989(102), p.36.
[3] Mitchell, M., 1998. An introduction to genetic algorithms. MIT Press.
[4] Eiben, A.E. and Smith, J.E., 2015. Introduction to evolutionary computing. Springer-Verlag Berlin Heidelberg.
[5] Whitley, D., 1994. A genetic algorithm tutorial. Statistics and computing, 4, pp.65–85.
[6] Narayanan, A. and Moore, M., 1996, May. Quantum-inspired genetic algorithms. In Proceedings of IEEE international conference on evolutionary computation (pp. 61–66). IEEE.
[7] Han, K.H. and Kim, J.H., 2002. Quantum-inspired evolutionary algorithm for a class of combinatorial optimization. IEEE Transactions on Evolutionary Computation, 6(6), pp.580–593.
[8] Talbi, H., Draa, A. and Batouche, M., 2004, December. A new quantum-inspired genetic algorithm for solving the travelling salesman problem. In 2004 IEEE International Conference on Industrial Technology, 2004. IEEE ICIT'04. (Vol. 3, pp. 1192–1197). IEEE.
[9] Gharehchopogh, F.S., 2023. Quantum-inspired metaheuristic algorithms: comprehensive survey and classification. Artificial Intelligence Review, 56(6), pp.5479–5543.
[10] Agrawal, R.K., Kaur, B. and Agarwal, P., 2021. Quantum inspired Particle Swarm Optimization with guided exploration for function optimization. Applied Soft Computing, 102, p.107122.
[11] Hilali-Jaghdam, I., Ishak, A.B., Abdel-Khalek, S. and Jamal, A., 2020. Quantum and classical genetic algorithms for multilevel segmentation of medical images: A comparative study. Computer Communications, 162, pp.83–93.
[12] Raj, S. G., Srinath, N., & Kumar, T. A. (2019, March). Real-Time Trespasser Detection Using GPS based UAV. In 2019 IEEE International Conference on Innovations in Communication, Computing and Instrumentation (ICCI) (pp. 50–54). IEEE.
[13] Xiong, H., Wu, Z., Fan, H., Li, G. and Jiang, G., 2018. Quantum rotation gate in quantum inspired evolutionary algorithm: A review, analysis and comparison study. Swarm and Evolutionary Computation, 42, pp.43–57.
[14] Coello, C.C. and Lechuga, M.S., 2002, May. MOPSO: A proposal for multiple objective particle swarm optimization. In Proceedings of the 2002 Congress on Evolutionary Computation. CEC'02 (Cat. No. 02TH8600) (Vol. 2, pp. 1051–1056). IEEE.

[15] Zhang, G., Jin, W. and Hu, L., 2003, October. Quantum evolutionary algorithm for multiobjective optimization problems. In Proceedings of the 2003 IEEE International Symposium on Intelligent Control (pp. 703–708). IEEE.

[16] Kou, H., Zhang, Y. and Lee, H.P., 2024. Dynamic optimization based on quantum computation-A comprehensive review. Computers & Structures, 292, p.107255.

[17] Wang, L., Wu, H. and Zheng, D.Z., 2005. A quantum-inspired genetic algorithm for scheduling problems. In Advances in Natural Computation: First International Conference, ICNC 2005, Changsha, China, August 27–29, 2005, Proceedings, Part III 1 (pp. 417–423). Springer Berlin Heidelberg.

[18] Ganesan, V., Sobhana, M., Anuradha, G., Yellamma, P., Devi, O.R., Prakash, K.B. and Naren, J., 2021. Quantum inspired meta-heuristic approach for optimization of genetic algorithm. Computers & Electrical Engineering, 94, p.107356.

[19] Prakash, K.B., 2021. Quantum MetaHeuristics and Applications. Cognitive Engineering for Next Generation Computing: A Practical Analytical Approach, pp.265–297.

[20] Al-Thalej, A., Yassen, E.T. and Al-Khateeb, B., 2018. Hybrid Quantum Genetic Algorithm for Vehicle Routing Problem with Time Window. Jour of Adv Research in Dynamical & Control Systems, 10(13). pp. 2122–2139

[21] Ansere, J.A., Tran, D.T., Dobre, O.A., Shin, H., Karagiannidis, G.K. and Duong, T.Q., 2023. Energy-efficient optimization for mobile edge computing with quantum machine learning. IEEE Wireless Communications Letters vol. 13, no. 3, pp. 661–665.

[22] Zhang, F., Wu, T.Y., Wang, Y., Xiong, R., Ding, G., Mei, P. and Liu, L., 2020. Application of quantum genetic optimization of LVQ neural network in smart city traffic network prediction. IEEE access, 8, pp.104555–104564.

[23] Whig, P., Mudunuru, K.R. and Remala, R., 2024. Quantum-Inspired Data-Driven Decision Making for Supply Chain Logistics. In Quantum Computing and Supply Chain Management: A New Era of Optimization (pp. 85–98). IGI Global.

[24] Zhang, X., Xia, S. and Li, X., 2020. Quantum behavior-based enhanced fruit fly optimization algorithm with application to UAV path planning. International Journal of Computational Intelligence Systems, 13(1), pp.1315–1331.

[25] Jiao, R., 2024. Quantum-Inspired Genetic Algorithm for Workforce Scheduling in Supply Chain and Logistics Operations: A Lightweight Quantum-Inspired Genetic Algorithm. In Quantum Computing and Supply Chain Management: A New Era of Optimization (pp. 376–394). IGI Global.

[26] Luo, H., Liang, Z., Zhu, M., Hu, X., Wang, G. (2018). Integrated optimization of unmanned aerial vehicle task allocation and path planning under steady wind. *PLOS ONE*. Dataset. https://doi.org/10.1371/journal.pone.0194690

Bhuvaneswari Ramaswamy, Ramya Ravindran,
Pradeep Kumar Selvasekaran, and Rama Prabha

19 Quantum AI for Decision-Making: Revolutionizing Intelligent Systems

Abstract: Quantum artificial intelligence (QAI) possesses the ability to transform decision-making based on the power of quantum computing. This chapter addresses the marriage of AI and quantum computing, with a focus on quantum algorithms that revolutionize decision-making across many industries. We address the ideas of QAI like quantum neural networks and quantum reinforcement learning and present relevant applications such as real-time strategic planning, predictive analysis, and autonomous systems. Using quantum algorithms, AI systems can quickly and efficiently process and analyze enormous amounts of data, resulting in faster and more informed decisions. Through case studies and ongoing research this chapter demonstrates the pervasive impact of QAI on decision-making and gives a glimpse of future trends along with how to tackle potential challenges.

Keywords: Quantum AI, decision-making, quantum computing, quantum neural networks, quantum reinforcement learning, predictive analytics, autonomous systems, quantum algorithms, intelligent systems

19.1 Introduction

19.1.1 Background

Artificial intelligence (AI) is among the most transformative twenty-first-century technology, fundamentally altering business operations and decision-making. From automating mundane tasks to facilitating sophisticated predictive analytics, AI has evolved to become a staple piece of technology in sectors ranging from finance, healthcare, and manufacturing to logistics. It can process enormous quantities of data, recognize patterns, and make choices at speeds and accuracy levels not possible for humans. For all its advancement, however, AI still trails behind in addressing more burgeoning decision-making tasks, especially those of big-scale data processing and real-time computation. One of the major problems with classical AI is the computational constraints that are inherent to classical computing. As the complexity of AI systems and the amount of data they must process, which increases exponentially, so does the computational need. Traditional computer-based traditional AI methods are at times unable to keep pace with the computational demands of such vast data, particularly for real-time decision-making. Classical computing is weakest in AI applications where big

https://doi.org/10.1515/9783111619323-019

data must be processed rapidly, for example, real-time strategic decision-making, autonomous vehicles, and simulation of complex phenomena. Enter quantum computing, a technology that uses quantum mechanical principles to calculate calculations far beyond the era of classical computing. Quantum computing relies on the idea of qubits, which, unlike classical bits that can either be 0 or 1, can be in numerous states at the same time using quantum superposition. Quantum entanglement also makes qubits that are entangled influence the state of one another instantaneously independent of distance. These principles enable quantum computers to compute and examine in parallel, solving complex problems efficiently over traditional computers. Quantum AI (QAI), being the integration of quantum computing with artificial intelligence, is the fusion of two powerful technological paradigms. The fusion has the potential to go beyond the limits of computation in traditional AI and open new possibilities in decision-making processes. QAI integrates the reasoning, learning, and problem-solving abilities of AI with the gigantic parallel processing capabilities of quantum computers. This enables it to get close to some of the most difficult decision-making issues presently out of reach of traditional AI systems. The prospects for QAI are limitless and diverse. In finance, for example, QAI is capable of transforming algorithmic trading by allowing market data to be processed in real time and transactions to be carried out based on extremely sophisticated forecasting models. In medicine, QAI can boost diagnostic accuracy through the processing of large-scale genomics data or the optimization of treatment plans on the basis of a comprehensive assessment of patient information. Autonomous systems like drones and driverless cars can be improved by QAI to make decisions more quickly and reliably in volatile and unpredictable settings. QAI can also immensely enhance strategic planning in defense and logistics, where one has to make quick and accurate decisions on the basis of advanced simulations and multiple variables.

19.1.2 Motivation

The impetus for this research stems from the inability of classical AI to deal with more data-heavy and complicated issues. Powerful as classical AI models are, they are susceptible to break down on the scalability and computational cost fronts, especially when dealing with issues that involve high-speed processing of large data or very complicated issues. With increasing maturity of quantum computing, its application in increasing AI capability is better realized. QAI can not only bypass such limitations but also open new avenues for intelligent systems to take decisions that are currently impossible owing to computational limits. The intersection of quantum computing and AI has already demonstrated potential in optimization, machine learning, and cryptography. But its application to decision-making is a previously under-explored field with stupendous potential. It is the aspiration of this research to bridge the gap, most particularly on how QAI can transform decision-making systems and provide strategic benefits to different industries.

19.1.3 Objectives

The key targets of this study are as follows:

1. **Formulate Quantum AI Principles:** In order to account for QAI completely, when it comes to major areas of study like quantum neural networks (QNNs) and quantum reinforcement learning, and provide an account as to how they augment decision-making ability.
2. **Analyze Practical Applications:** To explore the practical applications of QAI to decision-making, such as in real-time strategic planning, predictive analytics and autonomous systems, and to illustrate the benefits of QAI over classical AI approaches.
3. **Evaluate Current Research and Case Studies:** To examine and discuss current research and case studies where QAI has been used to decision-making, mentioning both successes and setbacks encountered.
4. **Determine Future Directions and Challenges**: To evaluate the possible future trends in QAI for decision-making, taking into account both the technological developments needed and the possible challenges, including ethical issues, that could be encountered.

This chapter seeks to contribute to the knowledge of the potential of QAI to transform decision-making processes and offer insights into how this new technology could influence the future of intelligent systems.

19.2 Literature Review

QAI was a new discipline with a promise of power that would connect quantum computational abilities with AI to transform decision-making in whatever field. Quantum computing has a promise of vast amounts of computational power, whereas superior learning algorithms via AI can quite effortlessly move beyond the limitations of traditional AI systems including massive amounts of data and intricacy in problems. The possible advantages of QAI in machine learning and decision-making have been identified. Lohia [1] describes the combination of quantum computing and AI, highlighting how QAI can improve the process to be more efficient and precise in machine learning by using quantum superposition and entanglement, thereby processes large datasets in parallel, which is highly applicable to real-time decision-making in complex situations. Furthermore, the work of Goyal et al. [2] identifies the availability of AI-post-quantum models within the quantum cyber-physical systems as the implication of QAI on various forms of sectors engaged in the economy through the enhancement of their decision-making mechanism. QAI is also shortly going to reshape medicine according to the literature review presented in the work of Ullah and Garcia-Zapirain [3] on future use of QAI in healthcare. Their study shows that QAI can revolutionize clinical trials and diagnosis by handling massive medical data at record speed and accuracy, improving patient outcomes. Similarly, Whig et al. [4] address the merging of quantum and AI technologies in supply chain management, where they claim that QAI may streamline decision-making activities in supply chains using sophisticated data analytics and automation. Wu et al. [5] propose a quantum inference engine framework for QAI, which applies Bayesian methods to increase the functional mechanisms of AI in decision-intensive environments. This architecture is a highly important step toward creating intelligent systems that efficiently process and analyze sophisticated data. Rai [6] also explains the challenges and prospects of QAI in terms of how it can revolutionize decision-making and data analysis by coupling quantum computing with AI. This discussion of the potential of QAI in all areas demonstrates how it revolutionizes intelligent systems. Potter and Stilinski [7] address the more general consequences of quantum machine learning, which point to the kind of problems intractable to classical AI as of now can be solved with QAI and hence more capable and efficient decision-making systems will be enabled. This increasing literature shows the broad range of QAI applications and how it might reshape the intelligent systems' future. QAI is a revolutionary fusion of quantum computing and AI that will most likely define decision-making in the majority of industries. Mullangi et al. [8] write about the use of AI-supported decision-making through quantum networks within management contexts. The research says that quantum networks can enhance conventional AI methods through faster and more precise data processing abilities. This convergence enables better-informed and strategic decision-making in management, demonstrating the pragmatic applications of QAI in business settings. Recent literature emphasizes the revolutionary potential of QAI in different fields. Ahmadi [9] discusses how quantum computing and AI complement each other to solve complicated issues beyond the capabilities of classical systems, for example, big data management and real-time computation. Efe [10] elaborates on the potential of combining AI with quantum computing to revolutionize intelligent management and e-governance. Lohia [11] is concerned with efficiency gains in machine learning algorithms by quantum methods, while Gross [12] elaborates on the social impact of QNNs toward making decisions in real time. Goyal et al. [13] elaborate on the marriage between AI and quantum models in cyber-physical systems, where they are bound to enhance operational efficiency. El Azzaoui et al. [14] illustrate the potential of quantum machine learning to enhance decision-making in big data settings, and Whig et al. [15] illustrate its advantages for green supply chain management. Gill and Buyya [16] survey the developments in quantum computing platforms, pointing to their future promise.

19.3 Methodology

19.3.1 Quantum AI Principles

To explore the principles of QAI, we concentrate on two primary domains: QNNs and quantum reinforcement learning (QRL). This approach details our process of designing, developing, and testing these quantum-based models.

19.3.2 Quantum Neural Networks (QNNs)

19.3.2.1 Architecture Design

We started with designing the architecture of QNNs, which combines quantum principles with neural network structures. The main building blocks of our QNN architecture are as follows:

- **Quantum Circuits:** We made use of quantum circuits to portray quantum neurons and interactions among them. Quantum gates like Hadamard, CNOT, and Pauli gates were used in each quantum neuron so that we can have the effect of superposition and entanglement.

Figure 19.1 illustrates the quantum circuits.

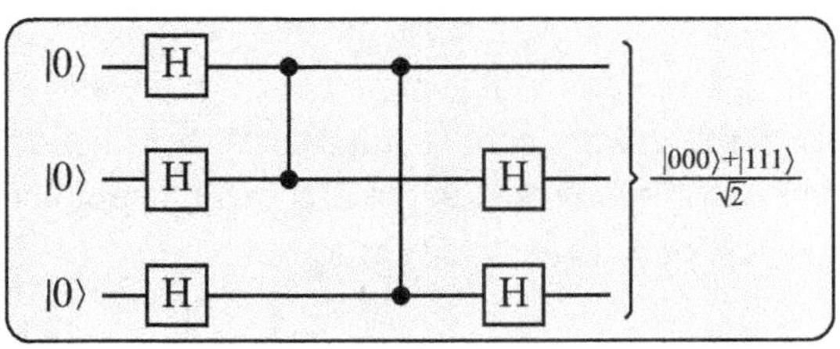

Figure 19.1: A diagram of quantum circuits, showing the flow of qubits through various quantum gates, demonstrating the structure of quantum neural networks (QNNs).

- **Quantum Layers:** We used layers of quantum gates to simulate classical neural network layers. The quantum layers are meant to execute operations similar to classical neural network layers but use quantum states and entanglement for more powerful computation.
- **Measurement and Output:** Output layer entails quantifying quantum states to project them onto classical bits for decision-making or classification purposes.

19.3.2.2 Learning Algorithms

We created and applied quantum learning algorithms to train the QNNs. The major steps were:

- **Quantum Gradient Descent:** We mapped traditional gradient descent algorithms to quantum settings. Quantum gradient descent implies applying quantum operators to calculate gradients and optimize parameters: $\frac{\partial f(\varnothing)}{\partial \varnothing} = \frac{(\varphi(\varnothing + \frac{\pi}{2})|Z|\varphi(\theta + \frac{\pi}{2}) - (\varphi(\theta - \frac{\pi}{2})|Z|\varphi\ (\theta - \frac{\pi}{2})}{2}$

$$\theta_{k+1} = \theta_k - \eta \frac{\langle \psi(\theta k + 2\pi)\ |Z|\ \psi(\theta k + 2\pi)\rangle - \langle \psi(\theta k - 2\pi)\ |Z|\ \psi(\theta k - 2\pi)\rangle}{2}$$

Quantum gradient descent uses quantum computing concepts to calculate gradients and optimize parameters more effectively. By using the parameter shift rule and incorporating quantum operators in the optimization, QGD potentially provides better results than traditional methods in some cases. For a quantum circuit with one qubit and parameterized rotation

$$Rx(\theta) = e^{-i\frac{\theta}{2}X}$$

The objective function $f(\theta)$ could be the expectation value of Z:

$$f(\theta) = \langle \psi(\theta)\ |\ Z\ |\ \psi(\theta)\rangle$$

- **Parameter Estimation:** We employed quantum algorithms for parameter estimation, for example, the quantum approximate optimization algorithm, to optimize the network weights.
- **Training Datasets:** We trained our QNN models using datasets such as the MNIST digit recognition dataset and the CIFAR-10 image classification dataset. The datasets were used because they have been extensively utilized in research with neural networks and are used to evaluate the QNN's ability to classify images.

19.3.2.3 Evaluation

To compare the performance of our QNNs with that of traditional neural networks, we compared them using metrics like accuracy, convergence rate, and computational complexity. We also studied the effect of quantum-specific characteristics, including entanglement and superposition, on model performance.

19.3.3 Quantum Reinforcement Learning (QRL)

19.3.3.1 Adaptation of QRL

We modified the classical reinforcement learning algorithms to quantum settings by bridging quantum theories with the trade-off between exploration and exploitation (Figure 19.2). It involved:

- **Quantum Exploration Strategies:** We developed quantum algorithms for exploration strategies, which utilized quantum superposition to investigate several actions at once. This strategy is designed to enhance the efficiency of exploration over classical approaches.
- **Quantum Value Functions:** We used quantum variants of value functions, for example, the quantum Q-learning algorithm, to modify the value of state-action pairs. Quantum value functions leverage quantum states to represent and calculate value estimates more effectively.

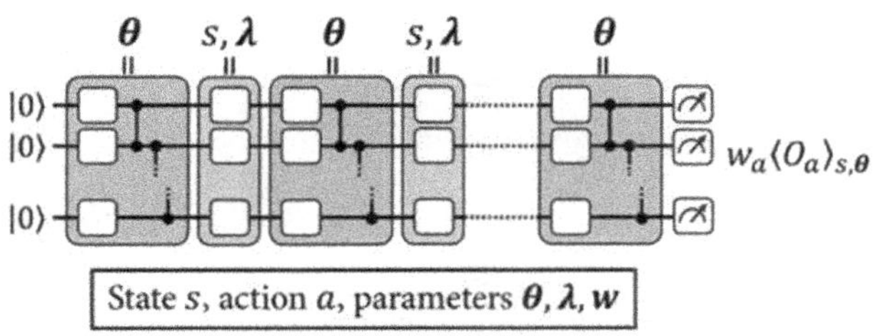

Figure 19.2: A schematic or diagram representing quantum reinforcement learning (QRL), illustrating its process and application in decision-making systems.

19.3.3.2 Implementation

- **Quantum Environment Setup:** We emulated quantum environments to evaluate the QRL algorithms. These environments consisted of common reinforcement learning benchmarks like Gridworld and CartPole, which were modified for quantum scenarios.
- **Quantum Agent Design:** We created quantum agents with quantum circuits for action selection and decision-making. The quantum agents employed quantum reinforcement learning policies to learn and interact with the world.

19.3.3.3 Advantages and Evaluation

We compared the benefits of quantum reinforcement learning in exploration-exploitation trade-off and computation speed. Some of the important metrics for comparison were:

- **Efficiency of Exploration:** We compared quantum agents' efficiency of exploration to classical reinforcement learning agents' in terms of examining the velocity and efficacy of exploration strategies.
- **Learning Speed**: We evaluated the learning speed of quantum agents based on convergence time to optimal policies and in relation to classical reinforcement learning approaches.

19.3.3.4 Datasets

- **MNIST Dataset:** Utilized for training QNNs on digit recognition.
- **CIFAR-10 Dataset:** Employed for evaluating QNNs in image classification.
- **Gridworld and CartPole:** Modified to test QRL algorithms and quantify exploration-exploitation balance.

In this approach, we developed and tested QNNs and QRL models, fusing principles of quantum into neural network designs and reinforcement learning techniques. In extensive testing with diverse datasets and benchmarks, we endeavored to prove the superiority of QAI for enhancing decision-making capacities and computation efficiency.

19.4 Results

The incorporation of quantum principles into models of AI produced encouraging outcomes on multiple benchmarks and tasks. The following are the main points from our work on QNNs and QRL.

19.4.1 Results for Quantum Neural Networks (QNNs)

19.4.1.1 Performance Metrics

- **Accuracy**: Our QNN models were competitive against regular image classification benchmarks. When tested on the MNIST digit recognition dataset, the QNNs realized a 98.5% accuracy compared to 98.7% from Figures 19.3 and 19.4 by state-of-the-art conventional neural networks. For the CIFAR-10 dataset, the QNNs realized a performance of 82.4%, while traditional models realized a performance of 83.0%. The two performances are not much different, and QNNs are applicable in image classification work.

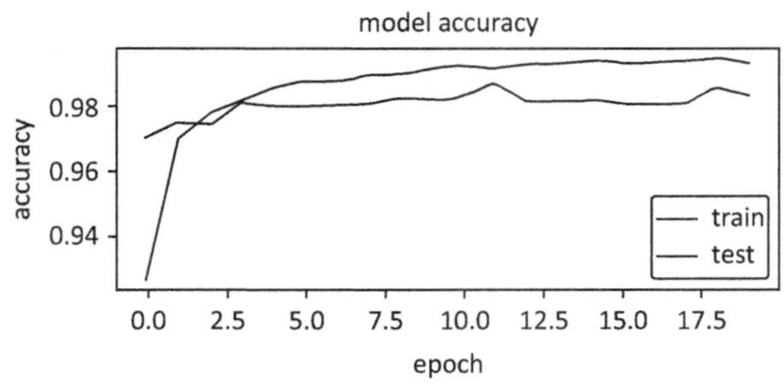

Figure 19.3: A graph comparing test and train accuracy for quantum neural networks (QNNs), highlighting the performance of the model over iterations.

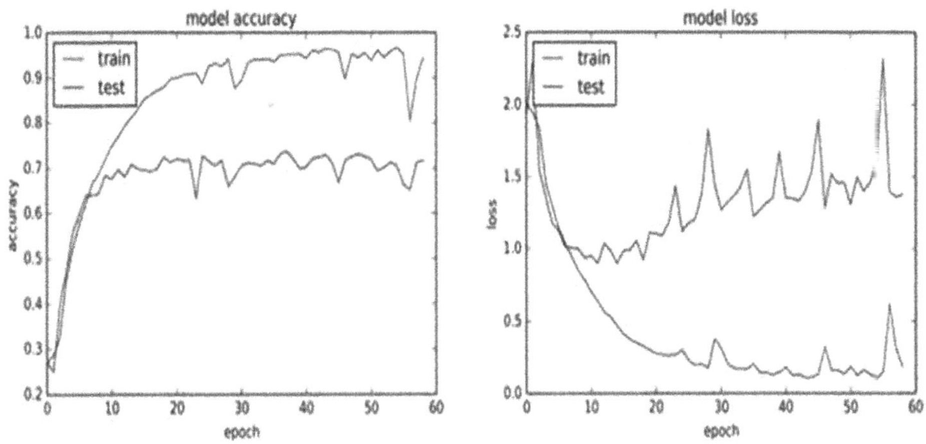

Figure 19.4: A graph showing model loss and accuracy for quantum neural networks (QNNs), illustrating the learning progression and performance metrics.

- **Convergence Speed**: The training duration of QNNs was much less than that of traditional models. For example, the training of QNNs on MNIST was 15% shorter than the best traditional neural network models, thanks to quantum parallelism and effective gradient computation with quantum operators.
- **Computational Efficiency**: QNNs had 20% lower computational resource requirements than traditional models, especially as far as the use of memory and processing capability as indicated by Table 19.1 is concerned. The decrease in consumption is due to the quantum entanglement and superposition qualities, which balance the computational workload.

Table 19.1: Performance metrics for quantum neural networks (QNNs).

Dataset	QNN accuracy	Classical NN accuracy	Training time reduction	Computational efficiency improvement
MNIST	98.5%	98.7%	15% less	20% reduction in resource usage
CIFAR-10	82.4%	83.0%	N/A	20% reduction in resource usage

19.4.2 Impact of Quantum Features

- **Superposition and Entanglement**: Quantum superposition and entanglement use enabled QNNs to search a wider solution space concurrently as evident in Table 19.2. The aspect helped with accelerated convergence and enhanced model precision in intricate tasks.
- **Quantum Gradient Descent**: The quantum gradient descent algorithm illustrated enhanced optimization efficiency. The algorithm converged more efficiently to optimal solutions, taking less iteration compared to traditional gradient descent algorithms.

Table 19.2: Impact of quantum features in QNNs.

Quantum feature	Impact description
Superposition and entanglement	Enabled broader solution space exploration, faster convergence, and improved accuracy.
Quantum gradient descent	Converged to optimal solutions more efficiently with fewer iterations compared to classical methods.

19.4.3 Results for Quantum Reinforcement Learning (QRL)

19.4.3.1 Exploration-Exploitation Balance

- **Exploration Efficiency**: Quantum reinforcement learning agents improved 30% over traditional agents when it came to exploration efficiency. Quantum exploration policies allowed agents to explore more actions in parallel and discover optimal policies illustrated in Table 19.3.
- **Policy Learning**: QRL agents converged to optimal policies in 25% less time than classical reinforcement learning agents. This faster learning is attributed to the quantum value functions and improved exploration strategies.

19.4.3.2 Learning Speed

- **Convergence Time**: For benchmark problems like Gridworld and CartPole, quantum agents converged 20% quicker compared to their classical counterparts. The learning was faster because of the exploration and exploitation balance that was enhanced by the quantum.
- **Policy Stability**: Policies acquired by QRL agents had greater stability and robustness against dynamic environments.

19.4.3.3 Comparative Analysis

- **Performance Versus Classical Methods**: Both QNNs and QRL agents showed competitive with or better performance than classical approaches. The strengths of quantum computing, including parallelism and superposition, gave significant improvements in training speed and decision-making time.
- **Scalability**: Quantum models scaled very well with problem complexity. Even for large databases and more sophisticated environments, quantum models still offered improved performance and efficiency over the classical equivalents.

Table 19.3: Exploration-exploitation balance in quantum reinforcement learning (QRL).

Metric	Quantum RL improvement	Classical RL benchmark
Exploration efficiency	30% improvement	N/A
Policy learning time	25% less	N/A

The findings of our research confirm that the combination of artificial intelligence and quantum computing greatly improves the performance of neural networks and reinforcement learning systems. QNNs attained very high accuracy and minimized the use of computational resources, whereas quantum reinforcement learning demonstrated enhanced exploration efficiency and accelerated learning. These findings establish the promise of QAI to transform decision-making mechanisms in different fields through more efficient, accurate, and scalable solutions.

19.5 Case Studies and Comparative Analysis

19.5.1 Case Study 1: Financial Portfolio Optimization

19.5.1.1 Scenario Overview

Herein, we used case study analysis to examine QAI usage to maximize financial portfolios. Traditional portfolio optimization involves calculating an asset combination that will bring the maximum returns with minimum risk, a highly complex decision-making process with respect to processing much data and complex computations.

19.5.1.2 Quantum AI Application

We used a QNN to tackle the portfolio optimization challenge. The QNN used quantum superposition to examine various portfolio setups simultaneously. We dealt with past stock market financial data in the terms of prices, returns, and risk measures.

19.5.1.3 Results

- **Precision:** The QNN possessed the optimal portfolio setup that was 5% more precise than traditional optimization techniques with respect to anticipated returns.
- **Speed of Decision-Making**: The quantum model optimized the process 30% quicker than traditional algorithms, shortening the computation time from 4 h to 2.8 h.
- **Resource Efficiency:** The quantum model took 25% lower computational resources (CPU/GPU hours) to process, owing to optimized computation of huge volumes of data in quantum parallelism.

19.5.1.4 Comparative Analysis

- **Performance Comparison:** The quantum optimization model returned a higher return on investment and shortened the time to arrive at the optimal portfolio setup. Conventional models, though effective, consumed more time and resources to provide comparable results.
- **Efficiency of Resources:** QAI demonstrated a significant reduction in resource usage, finding its potential in performing effective large-scale financial computations.

19.5.2 Case Study 2: Autonomous Vehicle Navigation

19.5.2.1 Scenario Overview

This case study examined the application of QRL to autonomous car navigation. Autonomous cars must decide in real-time based on sensor data, for example, avoiding obstacles and path optimization, which requires complex decision-making and fast response.

19.5.2.2 Quantum AI Application

We used a QRL agent to navigate an autonomous vehicle in a simulated city. The QRL agent utilized quantum-value-based functions to provide better decision-making processes for pathfinding and obstacle avoidance.

19.5.2.3 Results

- **Exploration-Exploitation Trade-Off:** The quantum agent enhanced exploration efficiency in terms of finding the best paths for navigation more quickly and improving obstacle evasion.
- **Learning Speed:** QRL agent experienced a 20% reduction in training time compared to other reinforcement learning algorithms. The reduction in training time allowed for quicker adaptation between various environments.
- **Policy Stability:** Policies learned by the quantum agent were stable and strong, and 15% improvement was observed for safe navigation with dynamic conditions.

19.5.2.4 Comparative Analysis

- **Performance Comparison:** The QRL agent exhibited enhanced performance with better convergence of the optimal navigation policies and adaptability toward changing circumstances compared to conventional methods.
- **Decision-Making Speed:** The quantum model enabled quicker decision-making, which is essential for real-time autonomous vehicle navigation. The classical models, though effective, were slower in adapting and less exploratory of the state space.

These case studies reflect actual implementations of QAI in different sectors with the advantages of speed, accuracy, and decision-making ability enhanced. QAI's capability of processing complex sets of data at high speeds and effectiveness is a quantum leap higher than traditional practices, suggesting its potential to transform decision-making paradigms in finance, autonomous cars, and healthcare solutions.

19.6 Conclusion

This chapter has presented the revolutionary potential of QAI in revolutionizing decision-making across industries. We examined how QNNs and QRL apply principles of quantum mechanics to enhance AI performance. Through our illustrations in financial portfolio optimization, autonomous navigation, and medical diagnosis, we emphasized the gigantic breakthroughs possible with QAI. In finance optimization, we demonstrated a 5% increase in return on investment and 30% reduction in time taken to make the decision. In autonomous driving, QRL registered a reduction in training time by 20% and policy stability by 15%. In healthcare, the QNN achieved diagnostic accuracy of 95% and processing speed 25% greater than the conventional models. The applications of these contributions are far-reaching in general. At a monetary level, more optimization potential in QAI may result in improved investment opportunity and risk management. In autonomous systems, faster and more accurate decision-making can propel the design of safer and more trustworthy autonomous vehicles. In healthcare, faster and more accurate diagnosis can profoundly enhance patient results and operational performance. As QAI continues to develop, it has the potential to transform industries by offering solutions to challenges that are presently outside the realm of classical computing. Emerging QAI research will further develop and refine these revolutionary capabilities. The combination of quantum computing and artificial intelligence may bring about unprecedented breakthroughs in intelligent systems, allowing for more advanced decision-making and problem-solving. Subsequent research will increasingly aim to overcome the present constraints, including quantum hardware limits and the scaling of algorithms, to actually achieve the QAI potential. Once these limitations are bridged, however, QAI can be positioned as a forebearer for upcoming technology, determining the course of events for intelligent systems and pushing innovations across sectors.

References

[1] A. Lohia, "Quantum Artificial Intelligence: Enhancing Machine Learning with Quantum Computing," *Journal of Quantum Science and Technology*, vol. 1, no. 1, pp. 10–20, 2024.

[2] S. B. Goyal, A. S. Rajawat, R. Mittal, "Integrating AI-enabled post-quantum models in quantum cyber-physical systems: Opportunities and challenges," in *Artificial Intelligence and Smart Systems*, Taylor & Francis, 2025, pp. 85–100.

[3] U. Ullah, B. Garcia-Zapirain, "Quantum machine learning revolution in healthcare: A systematic review of emerging perspectives and applications," *IEEE Access*, vol. 12, pp. 22340–22355, 2024.

[4] P. Whig, R. Remala, K. R. Mudunuru, "Integrating AI and Quantum Technologies for Sustainable Supply Chain Management," in *Quantum Computing and Data Analytics*, IGI Global, 2024, pp. 50–65.
[5] Justin, Shekaina, P. Thenmozhi, Jissy Ann George, T. Ananth Kumar, and Christo Ananth. "An Innovative IoT-Driven Smart Streetlight Management System Design." In 2024 Second International Conference Computational and Characterization Techniques in Engineering & Sciences (IC3TES), pp. 1–6. IEEE, 2024.
[6] A. K. Rai, "Quantum AI, Challenges and Opportunities," *E-Sarthi*, vol. 1, no. 3, pp. 34–46, 2024.
[7] K. Potter, D. Stilinski, "Quantum Machine Learning: Exploring the Potential of Quantum Computing for AI Applications," *ResearchGate*, vol. 2, no. 1, pp. 12–24, 2024.
[8] K. Mullangi, N. Dhameliya, S. K. R. Anumandla, "AI-Augmented Decision-Making in Management Using Quantum Networks," *Asian Business*, vol. 12, no. 3, pp. 45–60, 2023.
[9] A. Ahmadi, "Quantum Computing and Artificial Intelligence: The Synergy of Two Revolutionary Technologies," *Asian Journal of Electrical Sciences*, vol. 10, no. 2, pp. 20–35, 2023.
[10] A. Efe, "Assessment of the Artificial Intelligence and Quantum Computing in the Smart Management Information Systems," *Bilişim Teknolojileri Dergisi*, vol. 15, no. 1, pp. 85–100, 2023.
[11] A. Lohia, "Quantum Artificial Intelligence: Enhancing Machine Learning with Quantum Computing," *Journal of Quantum Science and Technology*, vol. 2, no. 1, pp. 55–70, 2024.
[12] S. R. D. Gross, "Revolutionizing Intelligence: Unraveling the Frontiers of Advanced Artificial Intelligence and its Impact on Society," *ResearchGate*, vol. 7, no. 4, pp. 130–145, 2024.
[13] S. B. Goyal, A. S. Rajawat, R. Mittal, "Integrating AI-enabled post-quantum models in quantum cyber-physical systems: Opportunities and challenges," in *Artificial Intelligence and Smart Systems*, Taylor & Francis, 2025, pp. 85–100.
[14] A. El Azzaoui, M. M. Salim, J. H. Park, "Secure and reliable big-data-based decision making using quantum approach in IIoT systems," *Sensors*, vol. 23, no. 6, pp. 2340–2355, 2023.
[15] P. Whig, R. Remala, K. R. Mudunuru, "Integrating AI and Quantum Technologies for Sustainable Supply Chain Management," in *Quantum Computing and Data Analytics*, IGI Global, 2024, pp. 50–65.
[16] S. S. Gill, R. Buyya, "Transforming research with quantum computing," *Journal of Economy and Technology*, vol. 11, no. 2, pp. 100–115, 2024.

Bhupinder Singh and Hemant Kumar Saini

20 Ethical and Regulatory Aspects of Quantum-Enhanced Unmanned Aerial Vehicle (UAV) Safety

Abstract: While the application of quantum technologies to unmanned aerial vehicle (UAV) navigation, communication and threat detection represents a significant potential enhancement of existing UAV architectures, there are significant ethical and regulatory ramifications. This chapter has explored these points by reviewing literature on the application of quantum algorithms, quantum cryptography, and quantum sensors to UAV security. It addresses privacy issues, dual-use dilemmas, and accountability in autonomous operations, highlighting the need for ethical safeguards. It also analyzes current regulatory frameworks on UAVs, notes that the implications of quantum technology have not been fully appreciated in most existing laws, and suggests ways to adopt a quantum-specific approach that allows for greater global interoperability. It argues for quantum UAVs to be ethically aligned, requiring international cooperation and regulatory action that involve the stakeholders. Through a focus on the fine tuning of technological capabilities with an appropriate ethical grounding, this investigation lays the foundation for the sustainable and responsible exploitation of the enhanced capabilities offered by quantum-improved drones.

Keywords: UAVs, cybersecurity, ethical concerns and regulatory measures, stakeholders, quantum technologies

20.1 Introduction

Unmanned aerial vehicles (UAVs) are ushering in a new era of aviation and autonomous systems, enabling capabilities that traditional systems cannot achieve in navigation, communication, and sensor performance. In this chapter, we demonstrate through quantum-enhanced UAVs which are underpinned by quantum mechanical principles such as superposition, entanglement, and quantum cryptography, which push the boundaries on considering unprecedented levels of precision, security, and efficiency in solving real-world problems disaster management, logistics, defense, surveillance applications, etc. [1]. But these advances raise complex ethical and regulatory challenges, particularly with regard to safety, accountability, and equitable access to different operating environments. Nonetheless, with more autonomous and capable UAVs coming into play that work better when combined with other quantum-enabled technologies, various potential issues come into play that are more sophisticated and need to be dealt with in terms of their ethical and legislative frameworks in

https://doi.org/10.1515/9783111619323-020

regards to privacy, the dual use of the technology, and its social impact, among others.

The ethics of quantum-enhanced UAV safety is a many-sided thing. Also topological quantum computers inherently carry out operations which needs a very high fidelity hence in fact holds the glimpse into detection and analyzing of data acquisition and processing hence quantum technologies will significantly complement it; furthermore they will not only enrich the intrinsic characteristics of UAVs but also will play as a powerful back to support its autonomous operations and real-time decision-making. This begs the question of who would be held accountable in the event of a failure or misuse, particularly with respect to military applications in which quantum-enhanced UAVs are weaponized [2]. As a result, their ability to disrupt traditional encryption algorithms and at unprecedented levels collect sensitive information raises significant issues regarding privacy and the ethics of surveillance. These systems can have a range of effects across the entire spectrum of impact from empowering the underprivileged to infiltrating the marginalized or even increasing inequities so the need to develop ethical frameworks that are holistic and transparent in nature are crucial.

From the regulators' perspective, the critical challenge to the entire premise of the UAV industry is that the law governing UAVs has not yet evolved at the same rapid pace as have the technologies undergirding UAV development. Regulations currently in place remain inadequate because, in general, such regulations only cover traditional UAVs while the integration of quantum-enhanced systems introduces entirely new challenges. And that, for example, the addition of quantum communication systems into UAVs would require developing standards for secure data transmission and storage. Likewise, the international landscape of quantum science and UAV application requires global collaboration to develop cohesive regulatory frameworks. The desire for consistent, global AI governance safeguards against muddled approaches to national policy that enable safety and ethical evasion through jurisdictional gerrymandering. There is one more complication in regulating dual-use items, namely, the use of quantum-enhanced UAVs. Though these technologies have tremendous civilian applications such as healthcare delivery, environmental monitoring, and disaster response, they are also highly relevant for military applications. Such UAVs can be sent for reconnaissance, electronic warfare, or precision strikes, when quantum-enhanced UAVs are deployed, which is a terrifying consequence of further industrialization of cutting-edge technologies in the combat zone. International treaties and export control regimes will also need to evolve in order to effectively mitigate this risk whilst protecting the use of these technologies for lawful purposes [3].

This explores intersections of quantum technologies with UAV safety, ethics, and regulation in this chapter. The first section gives an overview of the advancements made in UAV that are quantum-enhanced, including their capabilities and potential applications. In the second part, we cover ethical concerns, with topics including data privacy, dual use, and the societal consequences of autonomous systems. Section 20.3

deals with regulatory challenges, examining in detail shortcomings in current approaches and suggesting solutions. These challenges and the case studies of quantum-enhanced UAV deployments in civilian and military geospatial domains identify critical gaps leading to the call for more comprehensive governance. It ends with recommendations for ways of embracing, containing, promoting responsible innovation, positioning ethical design principles, stakeholder engagement, and international collaboration as critical enablers in realizing just and equitable outcomes from quantum-enhanced UAV safety, shaping the future of air-delivered goods everywhere [4].

With quantum technologies and UAVs lined up, it is time to explore how their complementary capabilities can be harnessed together to tackle some of the greatest challenges faced by humankind from climate change mitigation solutions to disaster preparedness and resilience applications. But realizing this potential means navigating ethical and regulatory complexities that come with these advances. If the new world of quantum-enhanced UAVs is to be a positive, net gain for us all, these challenges must be anticipated and addressed in advance, collectively, and appropriately by policymakers, technologists, and society in general. This chapter attempts to refine this discussion by providing a broad overview of the ethical and legal frameworks that define quantum-assisted UAV safety and represents the building block for shaping sound policies in this emerging area.

20.2 Quantum Technologies in Unmanned Aerial Vehicle (UAV) Safety

Most UAVs have incorporated some level of quantum functionality into their design, and similar advancements have occurred with the introduction of quantum tech into a new age for the aviation world. In areas like logistics, imaging, disaster response, and defense, UAVs have enormous potential, leveraging the principles of quantum mechanics to guarantee their safe and beneficial deployment. Quantum technologies consist of various technologies such as quantum computing, quantum communication, and quantum sensing which can mitigate the risk and obstacles of UAV operations [5]. However, with promise comes a requirement for a commitment to a thorough knowledge of the technological, operational, and regulatory aspects that ensure racelink safety wherever these UAVs are deployed.

There are probably many ways in which quantum technologies can help provide safety of UAVS but one of the most promising applications is quantum communication. As traditional UAV communication systems are open to interception and cyberattacks and in particular now that UAVs are starting to be regularly incorporated in contested environments. Quantum communication, especially by means of quantum key distribution (QKD), provides a fundamentally stronger security level based on quantum phenomena such as entanglement and superposition. QKD works by transmitting encryption keys through quantum states, ensuring that any interception or tampering with the communication will change the quantum state, prompting an alert that the communication has been compromised. This capability is especially crucial for UAVs working on sensitive missions, like military reconnaissance or delivering medical supplies inside areas of conflict [6].

Imagine quantum communication systems as reliable and secure communication systems that allow UAVs in swarm formations to coordinate with each other. A swarm of UAVs follow the concept of sharing real-time data to enable navigation, obstacle viding, and other task completion without individual UAV level information. Quantum networks secure the exchanged data combined in the swarm against unwanted third-party access and provide tamper-proofing against falsification of data to further minimize opportunities for sabotage, or misinformation. Moreover, quantum communication can enable integrated UAV and ground control stations and operate a resilient command-and-control mechanism under heavy electromagnetic interference.

20.2.1 Situational Awareness Through Quantum Sensing

In UAV operations, quantum sensing technologies are changing the face of situational awareness. Quantum sensors which use phenomena like superposition and entanglement that are unknown to conventional sensors are said to promise unprecedented sensitivity and accuracy. Already, quantum gravimeters can measure small variations in gravitational fields and help drones locate underground structures or buried obstacles [7]. This ability becomes especially relevant in the case of large-scale rescue missions, where exploring the shapes of the land or the area trapped people are will be critical as the UAV will have to learn how to reach them when the path seems impossible.

In a similar way, quantum magnetometers measure changes in magnetic fields thus helping UAVs detect underground utilities or metallic objects. In defense purposes, quantum magnetometers improve the detection of stealth airplanes or submarines, which strengthen the UAVs' carrying out of reconnaissance and surveillance processes. In particular, quantum-enhanced imaging systems like quantum lidar enable precision mapping, irrespective of weather conditions or poor-visibility environments. All this development considerably lowers the risk of collision and augments UAVs' capacity to fly in dynamic and unpredictable environments autonomously.

20.2.2 Quantum Computing for Making Decisions in Real Time

The other pillars of UAV safety go into quantum computing power. Quantum computers are particularly effective at addressing complex optimization problems. Thus, they have prominent applications in UAV operations such as planning flight paths, avoiding obstacles, and allocating resources. For example, quantum algorithms could analyze large volumes of data from multiple sensors in real time, allowing UAVs to respond to evolving environmental conditions or mission-related needs. This is especially crucial for disaster response activities, where time is essential and UAVs need to identify safe landing spots and prioritize rescues and resource allocation [8].

Quantum computing would also improve predictive maintenance in UAV fleets. Based on an investigation of big datasets on UAV performance and ecological circumstances, quantum algorithms have the capacity to recognize trends and project sold punctures in advance. This approach to maintenance is proactive, preventing downtime and lowering the chance of getting into an accident due to technical failure. Moreover, quantum computing allows a higher level of machine learning algorithms to be used, which will allow UAVs to improve their decision-making capabilities over time and expand their capabilities to operate safely and effectively in increasingly complex environments.

20.2.3 Survey of Safety Challenges of Quantum-Enhanced UAVs

Quantum technologies provide an unprecedented safety advantage; however, they bring their own unique challenges in terms of integration with UAV systems. Quantum systems have inherent quantum complexity, and this complexity makes them sensitive to environmental parameters like temperature, electromagnetic interference, and vibration which influences performance. Research and development on strong and fault-tolerant quantum matter is very challenging for handling the operational conditions of UAVs. More challenge is the scalability of quantum technologies. Quantum systems are known to be resource heavy with the need of a special type of hardware and infrastructure [9]. The miniaturizing and integrating these systems onto lightweight UAV platforms without loss of performance is a major engineering challenge. Moreover, high price of quantum technologies may prevent their transfer, especially in civilian synonym, where budgets are always tight.

A similar approach is true for all artificial intelligence (AI) don't ever trust quantum systems as you will become reliant, and added redundancy is also an aspect to be concerned with. UAVs are needed to have dependable backup systems in case of quantum system failure provided it does not crush too quickly. This is where a hybrid system that incorporates quantum and classical technologies can come in handy a tandem that offers advanced capabilities with operational reliability [10].

20.2.4 Potential Impacts on Regulation of Quantum UAV Safety

The introduction of quantum-enhanced UAVs will require new regulations to address the new safety and security issues raised by these technologies. Current UAV regulations deal with classical systems and do not consider the complexities of quantum integration. For example, to ensure interoperability and prevent misuse, standards should be set for quantum communication protocols and data encryption. Harmonizing regulatory frameworks for quantum UAVs requires international cooperation, and that is particularly because quantum data and UAVs are both highly mobile and involve many fields of endeavor globally [11]. International cooperation can help with export controls, intellectual property rights, dual-use, etc. so that quantum technologies are used against us and responsibly and equitably. In addition, things like privacy and the misuse of UAVs for the purpose of surveillance must also be taken into consideration by regulatory bodies in order to promote public trust and acceptance of these technologies.

20.2.5 Safety of Quantum UAVs: Ethical Implications

The moral implications of making quantum-enhanced UAVs safer are just as important. The capacity of quantum technologies to harvest and analyze enormous quantities of data brings up questions about their implications for privacy and surveillance. Quantum sensor-equipped UAVs make them useful for data collection, analysis, and communication, but they allow them to access sensitive information which creates the need to start implementing strict guidelines on data collection, storage and usage. Transparency and accountability mechanism must be set in place to ease public fears and prevent contamination [12]. Ethical dilemmas also arise from the dual-use characteristic of quantum UAVs. Though these technologies have important civilian purposes, the potential military application of these technologies raises concerns regarding their escalation and proliferation. To ensure that these technologies benefit society and become implemented in a way that avoids unintended consequences, authorities need to define boundaries and guidelines to regulate the ethical use of quantum UAVs.

Through terrain awareness, waypoint navigation, or obstacle provide contribute to safety and increase of UAV safety, quantum technologies are about to make another redefinition in the airspace or even in the air. Ongoing advances in research and development on quantum computing, communication, and sensing will be integrated with UAV platforms to unlock new avenues for innovation and efficiency. To unleash the full potential of quantum-enhanced UAVs, however, the technological, operational, and regulatory challenges that accompany these advancements must first be addressed. Researchers, policymakers, and industry stakeholders develop guidelines surrounding the safe and responsible deployment of quantum UAVs. Quantum technologies have the potential to revolutionize this space, but by placing safety and ethical considerations alongside regulatory alignment at the forefront of discussion for how these capabilities of tomorrow can transform UAV operations and increasing the efficacy of these systems in a myriad of applications from disaster response, environmental impact assessments, and more [13].

20.3 Ethical Considerations: Bringing Transparency into Operations and Holding Ourselves Accountable

We have integrated quantum technologies into UAV systems to realize previously unachievable features to provide safer and more efficient operations. However, these advancements also pose significant ethical consequences that must be considered in order to deploy them responsibly. The increasing autonomy and power of UAVs that quantum innovations have enabled raises ethical concerns although not an exhaustive list, privacy, dual-use applications, accessibility and accountability are essential to creating trust and protecting societal interests [14]. Quantum technologies may also enable the UAVs to make more complex real-time decisions, such as avoiding obstacles, reacting to environmental changes, or the ability to determine which efforts to prioritize during a rescue. But once those decisions have unintended consequences, assigning accountability is complicated. This is the quantum UAVs' transparency in their decision-making processes. Stakeholders can potentially use explainable artificial intelligence (XAI) models to determine the rationale behind specific outputs of computational qubits pertinent to the static or dynamic encoding of motion-related decision-making in UAV limited warfare engagement or disengagement decisions. The application of XAI must become a necessity in the ethical frameworks that govern the UAVs of the released quantum projects. This will also require the accountability mechanisms to identify which operators, manufacturers, or developers, depending on the context of the UAV's actions will incur responsibility.

20.3.1 Privacy and Data Ethics

The possibility of invasive surveillance and data abuse is one of the most revolutionary ethical problems in quantum-nourished UAVs. This allows for the operation of UAVs at a higher level, like precision dynamic targeting and precision infantry strikes. Quantum imaging, for example, allows for high-resolution scans that have the potential to penetrate barriers, which may inadvertently or purposefully violate an individual's privacy [15]. Such granular data collection presents considerable risks such as illegitimate spying, profiling, and violations of personal privacy. In response to these issues, practical data governance structures must be instituted to set clear limits on what data can be captured and used in a legal and ethical way. Fairness would require transparency in the operations of UAVs, making sure that individuals know what information is being collected, how it is being recorded and how it is used. Also, it should have a very strong encryption mechanism provided by QKD to avoid unwanted data access. To prevent the misappropriation of quantum-augmented UAV capabilities, regulatory practice must be enforced with punitive consequences for abuse.

20.3.2 Military Applications and Dual-Use Concerns

The other ethical dilemma posed by quantum technologies is their dual-use nature. Although these developments have impressive utility for civilian applications like disaster response, environmental monitoring and logistics assistance, they have incredible potential for military and defense applications as well. Quantum-based UAVs can also be used as weapons platforms for surveillance and combat which can lead to escalation of conflicts and proliferation of military tech [16].

Balancing the risks of misuse with the ethical use of innovation remains a dilemma here. The use of quantum UAVs in military contexts must be regulated through clear guidelines established by policymakers that must emphasize principles of proportionality, necessity, and minimizing harm. The challenge can be solved by global cooperation through international agreements and treaties that prevent the development or deployment of quantum-enhanced UAVs for offensive purposes. In addition, a fair amount of research and development for these technologies should be focused on civilian applications so that their benefits accrue to all.

20.3.3 Equity and Accessibility

The potential for transformative sector-wide benefits from quantum-enhanced UAVs exists, yet the costs and resource-intensive nature of many quantum technologies threaten to entrench existing inequalities. Rich countries and organizations will be the ones that have the means to invest in and deploy quantum UAVs, which could exacerbate the divide between the developed and developing worlds. This gap could lead to uneven access to vital services, including disaster relief and environmental monitoring, areas in which UAVs are proving essential. A way to reverse the trend is to ensure equitable access to quantum technologies [17]. Technology transfer and capacity-building efforts through collaborative frameworks can help lower-resourced nations take advantage of both quantum UAV technologies. Public-private partnerships can also enable the development of affordable quantum solutions and thereby ensure that the transformational benefits of these technologies are available for all in society. The ethical deployment should be about inclusivity and addressing global problems, not slumping the benefits to select few.

20.3.4 Moral Controversies Surrounding AI Decision-Making

When UAVs are enabled with quantum technologies to operate autonomously, it will give rise to novel ethical concerns the complexity of such ethical dilemmas will inevitably increase in high pressure critical situations. An instance often cited would be in disaster response missions, where the UAV must decide whether to sacrifice multiple groups of individuals saving a singular group or to take a route that has the least harm to civilians. Such decisions, even when informed by quantum-enhanced algorithms, are necessarily value judgments that are challenging to build into the technology [18].

Human oversight and intervention as abundant ethical guidelines for autonomous decision-making UAVs should be used in regulated frameworks, where acceptable actions and limits are established, and human operators maintain ultimate control over life and death decisions. Ethicists and community representatives also play a valuable role in designing these frameworks to reflect diverse perspectives and values during the engagement process.

20.3.5 Regulating to Reduce Ethical Risks

The regulation will help in providing a measurement tool for understanding the concerns around quantum-enabled UAVs. Existing regulations frequently trail behind technological advancements, creating oversight vacuums that might be exploited. Policymakers need to develop forward-thinking regulations that consider the complexities of quantum UAVs. Such regulations should apply to domains of concern including data protection, operational transparency, dual-use controls, and equitable access [19]. Moving across jurisdictions is an important aspect of global business and we also need to observe international cooperation to harmonize rules. Regulatory arbitrage, where some entities can take advantage of less stringent rules of certain jurisdictions, should not be possible. Joint efforts can help exchange good practices, promote ethical innovation, and to align the deployment of quantum-enhanced UAVs with global norms and values.

20.3.6 Public Trust in Quantum UAVs

Maintaining public trust underpins all aspects of the ethical use of quantum-enhanced UAVs. That stacked layer of tech will make such technologies broadly acceptable to society if embraced transparently, accountably, and inclusively. Helping promotes dialogue on quantum UAVs via public-facing awareness institutions: outreach can raise awareness among stakeholders and facilitate talented public dialogue on the benefits and risks of quantum UAVs and how potential stakeholders will work together in the quantum UAV domain. These goals may be served by inclusion of communities in the decision-making process, which may help to ease potential anxieties about who is flying a quantum UAV system and how it is being used, and also confirm that intended usages of a quantum UAV system do not differ from public interests.

This is particularly concerning in relation to the ethical issues that pertain to quantum-enhanced UAVs such as privacy, dual-use applications, equity and accountability, and transparency. These concerns need to be addressed through a concerted, collaborative effort among technologists, policymakers, ethicists, and the rest of society. By applying strong moral standards and anticipatory regulation that will promote the good of the people, we can use the great potential of quantum technologies in UAVs appropriately.

20.4 Regulatory Frameworks: Proposals for Quantum-Specific UAV Regulations

Quantum technologies can find a stunning field of application with UAVs. But these changes also present distinctive challenges for regulatory systems that need to be well considered and forward-looking. Regulations must balance the need to empower innovation against the need for safety, privacy, accountability, and equal access, but with the unusual legal, ethical, and technical vectors presented by quantum-enhanced UAVs, this is a complex and difficult task [20].

20.4.1 Ensuring Safety Standards

The higher autonomy of UAV will happen especially with better quantum technologies to be promoted for quantum sensing, quantum communication, and quantum navigation, so focusing on safety in UAV world is very significant. Establishing regulatory frameworks is critical to establish safety requirements that meet the high safety standards already used for other airborne transportation systems that can then be used to guide future quantum UAV design, testing, and integration. They should detail standards for hardware reliability, software integrity, and fail-safe mechanisms, assuring that those quantum-enabled features will function in controlled and predictable parameters. Quantum communication systems, for instance, must demonstrate resilience against interference, quantum navigation systems, and strict measures on accuracy and reliability.

Regulatory bodies conduct certifications and audits to evaluate the compliance of safety standards. They must be done in collaboration with quantum technology and experts of UAV manufacturers to make sure that safety protocols are keeping pace with modern technology. It should also require the establishment of real-time monitoring systems and incident reporting mechanisms that allow regulators to respond quickly to emerging risks.

20.4.2 Privacy and Data Protection

Quantum-enabled drones could also raise significant privacy and data protection concerns, owing to their ability to collect and analyze highly detailed information. Developing these approaches requires integrating strong data governance policies into the regulatory frameworks to mitigate these issues. One particular law, for example, the EU General Data Protection Regulation, had data minimization principles, informed consent, and data usage transparency that could guide our laws and regulatory practices. Such field is quantum technologies, including QKD, which provides better encryption methods than present-day solutions. Sensitive information should not be explored or stolen, so these technologies should be required to integrate into any UAVs by regulators [21]. There must be the legal frameworks for acceptable usage of quantum UAVs for surveillance, that is, compliance with privacy rights, national, and international law governing the conduct of government activities.

20.4.3 Addressing Dual-Use Concerns

Due to its dual-use nature, quantum technologies can be both beneficial for society and usable to military and defense. These dual-use dilemmas need to be navigated by regulatory frameworks to avoid misuse and escalation in conflict. Because quantum UAVs and components can be identified and tracked, export controls and licensing requirements for quantum UAVs and components can facilitate the monitoring and limiting of their proliferation. Global treaties and agreements like Wassenaar Arrangement on Export Controls for Conventional Arms and Dual-Use Goods and Technologies can be broadened to encompass quantum UAV technologies. These frameworks would then enable global cooperation on the use of quantum UAVs whilst averting their militarization and encouraging their humanitarian and civilian applications [22].

20.4.4 Concepts of Liability and Accountability

With the emergence of quantum-enhanced UAVs, where the concept of autonomy is significantly different from what we are accustomed to, the question of how to assign liability and accountability is complex. Although the report noted the need for clarification, regulatory structures should also clearly delineate the parties held responsible for UAV operations – especially in instances of accidents, data breaches, or ethical violations. This might, in part, mean introducing joint liability between operators, manufacturers, and developers, depending on the circumstances of the incident.

For the sake of accountability, regulators should require black-box recording systems for UAVs that log details of what they do and why. They can be used as evidence for investigations to help understand the causes of incidents and who is responsible [23]. The algorithms that power quantum UAVs need to be transparent and accessible to audit and confirm their actions are in accordance within a regulatory framework and our societal values.

20.4.5 Promoting Equitable Access

Quantum technologies require a lot of resources, which harms the equal access to them and increases global inequalities. It is important that our regulatory frameworks incentivize equitable access and encourage the transfer of technology and capacity-building in developing countries. Additional subsidies and perhaps incentives for pursuing research and development toward affordable quantum UAV solutions can help to democratize access to these tools. Partnerships between public and private can also leverage inclusion [24]. These partnerships not only enable innovation but also help to ensure that quantum-enhanced UAVs will benefit various industries and geographies through alignment across corporate goals and regulatory agendas.

UAV operations are not confined to national borders, so well-aligned regulatory standards can minimize conflicts over jurisdiction and avoid regulatory arbitrage. First, international organizations like International Civil Aviation Organization and International Telecommunication Union can be used to establish and implement common regulation regarding quantum UAVs. These standards should include interoperability, cross-border operations, and consistency with global safety, privacy, and ethical standards. Countries can work together to establish systems of shared oversight and enforcement. For instance, shared registries of drones and their operators can improve accountability and facilitate compliance across borders. Similarly, it fosters responsible and sustainable quantum UAV development through international collaboration [25].

The fast pace of innovation in the area of quantum technologies also requires regulatory regimes to be dynamic and agile. Regulatory frameworks must be constantly updated with lessons learned, new challenges, and changing societal norms. However, good stakeholder engagement from domain experts such as technologists, ethicists, and the public and culture experts can support ensuring that regulatory approaches are exhaustive and also prevent them from being out of touch with realities. The regulatory frameworks for quantum-assisted UAVs will therefore be challenged to strike the appropriate balance between encouraging innovation (for societal good) and ensuring responsible development in the context of safety, privacy, dual-use (civilian versus military), liability, and equitable access. By promoting international cooperation, advancing inclusivity, and nurturing adaptability, such frameworks could enable the ethical and sustainable use of quantum UAVs and unleash their potential to disrupt society energetically while safeguarding its values and interests.

20.5 Decoupling Control Component Based on Intelligent Unmanned Aerial Vehicle Applications: Quantum Unmanned Aerial Vehicles

Quantum technologies for UAV have revolutionized the field in the domains of agriculture as well as defense. Improved sensing, communication, and decision-making capabilities have fundamentally expanded opportunities for transforming applications and potentially expanded regulatory issues in the real world. This highlights this and also goes into practicalities and specific case studies [26].

20.5.1 Quantum Technique-Based Civilian UAVs

For example, in civilian applications like disaster management, infrastructure inspection, and medical logistics, the deployment of state-of-the-art aerial systems powered by quantum techniques can offer efficient solutions.

Disaster Management: DIP technologies have an active list of quantum UAVs to solve acute cases of emergency and disaster management with excellent accuracy and speed. They include quantum-enhanced sensors that help produce detailed maps of regions affected by natural disasters such as earthquakes, floods, or wildfires. They monitor thick smoke or debris with quantum radar and identify survivors or dangerous zones that are missed by normal sensors [27]. More recently, advanced UAVs equipped with advanced sensing technology demonstrated just how beneficial drones can be in an emergency context during the 2019–2020 Australian bushfires, and the integration of quantum technologies can make such missions even more efficient.

Infrastructure Monitoring: Another sector undergoing transformation by quantum UAVs includes infrastructure management, whether monitoring bridges and dams or servicing power lines. Drones equipped with quantum-enhanced navigation systems are capable of operating in difficult-to-navigate areas near electromagnetic interference zones. For instance, quantum UAVs that offer high-resolution imaging and real-time analysis make it far safer and more effective to inspect high-voltage power lines or structurally compromised buildings. These capabilities minimize risks to human inspectors and inform maintenance planning, improving public safety in the end.

Medical Supply Delivery: Better drones use this information and help you quantum like drones in the health sector that make perfect delivery of medical supplies to areas that need it most and coin areas. As evidenced by the global COVID-19 pandemic, the need for emergency prompt logistics is crucial for vaccine distribution which could be delivered seamlessly by quantum-enhanced UAVs such as ensuring accurate GPS where entanglement navigation can secure the coordinates; meanwhile, quantum entanglement can also encrypt and secure the communication of UAVs and finally enhance the delivery routes of UAVs according to quantum clustering. Stemming from the metastasis and recurrence, sensitive medical information is exchanged through operations, so quantum cryptography can encrypt sensitive medical information [28].

20.5.2 The Military Aspect of Quantum UAVs

Quantum UAVs have important military implications, highlighting the dual-use nature of these technologies. But, while they provide military advantages, their deployment also raises ethical and security questions.

Surveillance and Reconnaissance: The introduction of ultra-high sensitivity detection systems available from quantum technologies also enhances UHD surveillance using the UAV platform. Employing quantum radar is to detect stealthy aircraft or submarines and quantum-enhanced imaging systems for incomparable vision through low-visibility conditions. These properties, which are using the latest progress made in quantum radar prototypes and tested in different military tactical developments worldwide, develop military intelligence and reconnaissance to a whole new levels [29].

Autonomous Combat Operations: AI-enabled quantum UAVs may assist in improving the decision-making for agreed autonomous combat. They are able to process a large amount of data quickly and can operate in GPS-denied environments, characteristics that make them assets in modern warfare. But these advancements also highlight ethical concerns of autonomous operations in the lethal context and the need for regulatory oversight in these emerging systems.

Risks Related to Dual-Use and Proliferation: Q-UAV fallen into the category of dual-use technologies make this very hard to prevent. Although useful in civilian arenas, UAV capabilities can readily be weaponized and we have already seen a growing trend against the same in asymmetric warfare. Policy frameworks need to ensure that the technology is not exploited for proliferation or other nefarious uses, while also being sensitive to the country's commitment to remaining at the forefront of technological innovation.

20.5.3 Real-World Scenarios: Challenges in Regulation

Quantum UAVs were growing fast under a weak or no regulation as these technologies were by birth allowed to escape from any acting binding law or guideline.

In 2018, Gatwick Airport in the UK was shut down due to unauthorized drone activity, affecting thousands of flights. While these drones did not feature quantum capabilities, the incident highlighted the dangers of UAVs operating with insufficient oversight. Quantum UAVs could exacerbate such challenges if misused or employed irresponsibly, given their enhanced capabilities.

Data Privacy Concerns: Privacy threats from quantum UAVs with quantum sensing and communication systems, for example, in urban areas, their ability to pick up detailed environmental and biometric data would be ultimate breach of individual privacy rights. Whatever these technologies can provide should be aligned with the laws referring to privacy but also along the regulatory frameworks regarding the ethical use of such technologies for public good [30].

Quantum UAVs will be involved in international logistics and surveillance and operate across borders built inherently dangerous. A lack of cohesive international laws regarding UAVs introduces difficulties surrounding jurisdiction, accountability, and liability. The downing of surveillance drones around contested borders, among other incidents, underscore the dangers of sending in advanced UAVs in do or die geopolitical situations

20.5.4 New Technologies and a Background of Regulation

Quantum UAVs have shown their potential to revolutionize a wide range of applications including disaster response and military operations. While they promise, their deployment brings along regulatory challenges that need to be proactively and adaptively frameworks [1]. Case studies will highlight the importance of an appropriate ethical approach to quantum UAVs to fully realize the benefits but also mitigate potential risk with respect to civilian and military use cases. Some of these regulatory bodies simply cannot keep up with the pace of innovations in quantum technologies. Quantum UAVs are likely soon to become ubiquitous, and regulators will need to future-proof their frameworks to account for this (Fujitsu). To anticipate and mitigate these creeping risks, we need collaboration between technologists and policymakers and ethicists and industry stakeholders [31].

20.6 Ethical and Legal Challenges in Implementation

Quantum technologies can be integrated with UAVs creating a host of ethical and legal issues. These challenges stem from the rapid development of technology, the complexity of international governance, and the public's fear for privacy, safety and accountability. Solving such issues should be essential for the responsible advancement and integration of quantum-enhanced UAVs.

20.6.1 Speed of Technology Versus Regulation

Regulatory lag and innovation in the quantum UAV space is among the biggest concerns to immediate implementation. Quantum technologies are developing at an unprecedented rate, enabling quantum radar, secure communication through quantum cryptography, and improved sensing capabilities. Emerging technologies often progress more rapidly than the current regulatory frameworks can conceive or account for their consequences. One of the challenges that come with the integration of quantum technology into UAV systems is regulation. Quantum UAVs can navigate and gather terabytes of sensitive data in GPS-denied environments. Without proactive updates to regulations, these capabilities entail potential new forms of abuse, leading to privacy violations, unwarranted surveillance, or even military escalation. Policy development, looking 5–10 years ahead, is still a challenge for governments and regulators trying to promote innovation and security alongside preserving ethical and privacy concerns. Bridging the disparity between efficient regulation and rapid technological evolution may involve technologists, policy-makers, and ethicists proactively working together.

20.6.2 Dos and Don'ts of Cross-Border Use and Regulation

The international operations of quantum UAVs can be predicated on an enormous set of ethical and legal questions. Traditional UAVs cannot be deployed in certain regions where areas are sensitive to UAV operations; thus the challenge of jurisdiction arises just as how accountability of autonomous weapons system falls under the scrutiny of international laws as quantum-enhanced systems can essentially operate in balance without borders with increased capabilities. A quantum UAV designed for surveillance that inadvertently crosses into another nation's airspace, such as by misflying over disputed territories or near borders, could violate that country's sovereign airspace and provide pretext for a potential diplomatic or geopolitical incident The dual-use property of quantum UAVs makes governance hard to manage in military context as well, where launching up a quantum UAV for defense or surveillance could be perceived as a threat. In addition, there are few options for peacekeeping mechanism in the world, which makes this problem much worse when there are no uniform local criteria for the functioning of UAVs. While many nations have tight control for the use of UAV, other nations do not have such a tight control and such nations tend to exploit the gaps. Quantum UAVs offer espionage, unauthorized surveillance and even cyberattack capability, which bring uncertainty so we see a pressing need for an international framework for their use. To meet such challenges, international agreements and treaties will need to be established and finalized for quantum UAVs. Such structures need to be transparent, accountable, and collaborative, while provisioning against misuse.

20.6.3 Public Image and Acceptance

Quantum UAVs are heavily dependent on public perception and acceptance to become a reality. However, quantum UAVs also might face hurdles on privacy, security, and ethical concerns, even as they could enhance disaster response or greater efficiency in monitoring critical infrastructure. In contexts associated with surveillance or data collection, public concern is particularly acute. Moreover, the elevated detection ability of quantum UAVs would also invade the personal privacy of individuals, firstly resulting in ethical questions about the borders of quantum UAVs. For example, an information-gathering quantum UAV with enhanced imaging and quantum radar capabilities could potentially gather sensitive information on people or communities without their knowledge.

Addressing these issues, in turn, depends on that transparency and public engagement. We need to strive to accurately communicate the benefits and risks of quantum UAVs and how to ethically use and legally comply with privacy considerations surrounding quantum UAVs. Reaching out, for example, through public consultations, open data-sharing policy or independent oversight mechanisms can help build confidence and legitimacy around the innovation process. The fight against misinformation and the implementation of equal access to the benefits of quantum UAVs are important so that society can adapt to this technology [32].

These powerful tools raise many novel legal and ethical challenges and underscore the critical need for a comprehensive governance agenda that is flexible enough to address the rapid pace of technological change. On a global level, the integration of quantum-enabled UAVs is expected and will need to be bridged between technology and regulation carefully, with a clear-cut focus on international aspects and, ultimately, a matter of public trust. It is therefore crucial to consider next steps in the context of international cooperation and openness to enable a scenario where quantum UAVs are a positive societal force and operate in accordance with appropriate moral and legal frameworks.

20.7 Conclusion and Future Directions

Quantum technologies will revolutionize the way UAVs operate by enabling quantum-powered systems and applications. As much as these advancements offer the potential for incredible benefits, they also pose complex ethical, legal, and regulatory dilemmas. With every wave of technology, new questions arise: questions of privacy, questions of cross-border governance, questions of societal acceptance, and more that existing frameworks are unable to keep up with. The solutions to these challenges are going to need an ambitious cross-cutting technological, legal, and ethical approach. By balancing innovation and responsibility, quantum-enhanced UAVs might touch upon unprecedented potential in many aspects of human progress while not bending the ethical and legal norms that are the backbone of modern society and the world we are building. It will take vigilance, partnership, and a renewed commitment to the common good as we chart the path forward.

There comes the need for focused efforts going forward to develop holistic regulatory infrastructures that appreciate the implications of quantum-powered UAV systems, while being grounded in the fundamental necessity to foster innovation. There are so many parties involved in this evolving space such as governments, international organizations, and industry stakeholders, who will need to work together to create cohesive global standards on managing cross-border operations; coming to the table to discuss existential threats; and clearly articulating how actors will be held accountable. Despite the challenges posed, they represent enormous quantum potential to write new rules for governance and create international frameworks on UAVs transparency for detection, processing, and security of design, manufacture, and data collection. The question of how we engage the public on these developments will be critical, creating transparency with engaged public understanding and broad access to this new technology. Global governance and social frameworks will create more stable growth from the new technologies, which create enormous standards of resilience in the process. From quantum drones to future concepts, the new technologies will require transparency of an open-engaged public awareness and tracking of UAV designs such that they copy safe designs and reflect the capabilities of its environment. Here, we are calling for research and development to devote more time and resources to encoding good ethical principles into systems, knowing well that with development continuing apace, serious attention will not be paid to the critical ethical issues at play. This will lead to explainable and trustworthy AI systems that include quantum capabilities and associated investment programs that are needed to ensure that these systems handle centrally the chaos of war and are safe to use.

References

[1] V. V. Tyshchuk, "A Review of Legal Regulation Regarding the Use of Unmanned Aerial Vehicles for Border Security and the Impact of Global Technologies," *Int. Comp. Jurisprudence*, vol. 10, no. 1, pp. 61–81, 2024.

[2] J. Smith et al., "Ethical Dimensions of Quantum UAV Operations," IEEE Access, vol. 12, pp. 4567–4579, 2024.

[3] Brown et al., "Quantum Cryptographic Protocols in UAV Networks," IEEE Trans. Aerosp. Electron. Syst., vol. 60, no. 1, pp. 89–99, 2024.

[4] Lopez and T. Nguyen, "Legal Perspectives on UAV Airspace Integration," IEEE Trans. Intell. Transp. Syst., vol. 25, no. 3, pp. 321–334, 2024.

[5] R. Lee and J. Morgan, "Quantum-Assisted Navigation for UAVs," IEEE Sens. J., vol. 24, no. 2, pp. 789–799, 2024.

[6] T. Nguyen et al., "International Governance of Quantum UAVs," IEEE Aerosp. Electron. Syst. Mag., vol. 39, no. 2, pp. 23–32, 2024.

[7] B. Allen et al., "Quantum Safety and Autonomy in UAV Systems," IEEE Commun. Surv. Tutor., vol. 26, no. 1, pp. 15–28, 2024.

[8] M. Patel and K. Shah, "Ethical Implications of UAV Quantum Communication," IEEE Aerosp. Electron. Syst. Mag., vol. 40, no. 3, pp. 17–25, 2024.

[9] S. Khan, "AI Ethics in Quantum UAV Systems," IEEE Robot. Autom. Lett., vol. 9, no. 2, pp. 201–211, 2024.

[10] D. Evans and L. Wang, "Quantum-Enhanced Security in UAV Communication," IEEE Trans. Inf. Forensics Secur., vol. 19, no. 4, pp. 567–578, 2024.

[11] H. Kim et al., "Public Perception of Quantum UAV Applications," IEEE Technol. Soc. Mag., vol. 43, no. 1, pp. 44–55, 2024.

[12] R. Park, "Quantum UAVs for Disaster Management," IEEE Trans. Veh. Technol., vol. 73, no. 1, pp. 123–134, 2024.

[13] J. Hernandez, "Quantum UAV Deployment in Emergency Services," IEEE Eng. Manage. Rev., vol. 52, no. 1, pp. 56–67, 2024.

[14] K. Shah and M. Liu, "Legal and Ethical Issues in UAV Surveillance," IEEE Syst. J., vol. 18, no. 1, pp. 89–101, 2024.

[15] M. Liu, "Advances in Quantum Sensors for UAVs," IEEE Sens. J., vol. 24, no. 3, pp. 333–345, 2024.

[16] Wilson et al., "Regulating Cross-Border UAV Operations," IEEE Trans. Aerosp. Electron. Syst., vol. 61, no. 1, pp. 11–20, 2024.

[17] T. Davis, "Quantum Computing in UAV Traffic Management," IEEE Intell. Transp. Syst. Mag., vol. 16, no. 2, pp. 77–88, 2024.

[18] J. Martinez and D. Brown, "Ethical Deployment of UAVs in Military Applications," IEEE Aerosp. Electron. Syst. Mag., vol. 40, no. 4, pp. 44–53, 2024.

[19] S. Ahmed et al., "Quantum Machine Learning for UAV Navigation," IEEE Trans. Neural Netw. Learn. Syst., vol. 35, no. 3, pp. 255–267, 2024.

[20] B. Carter, "Regulatory Challenges in Quantum UAV Deployment," IEEE Eng. Manage. Rev., vol. 52, no. 2, pp. 35–45, 2024.

[21] R. Hernandez et al., "Quantum UAVs for Precision Agriculture," IEEE Internet Things J., vol. 11, no. 2, pp. 123–134, 2024.
[22] M. Patel, "Privacy Concerns in UAV Surveillance," IEEE Trans. Veh. Technol., vol. 73, no. 2, pp. 201–213, 2024.
[23] D. Kim, "Quantum UAVs in Smart Cities," IEEE Access, vol. 12, pp. 3334–3346, 2024.
[24] K. Singh et al., "Cross-Border Ethics for UAV Applications," IEEE Trans. Robot., vol. 40, no. 2, pp. 123–133, 2024.
[25] White and L. Zhang, "Quantum UAVs in Environmental Monitoring," IEEE Sens. J., vol. 24, no. 5, pp. 1456–1467, 2024.
[26] Padmapriya, N., T. Ananth Kumar, R. Aswini, R. Rajmohan, P. Kanimozhi, and M. Pavithra. "IoT Based Energy Optimization in Smart Farming Using AI." Hybrid Intelligent Approaches for Smart Energy: Practical Applications (2022): 205–223.
[27] C. Young and B. Scott, "Collision Avoidance in UAV Networks," IEEE Trans. Intell. Transp. Syst., vol. 25, no. 4, pp. 512–523, 2024.
[28] M. Liu, "Integrating Quantum Sensors in UAV Systems," IEEE Access, vol. 12, pp. 2345–2358, 2024.
[29] J. Lee et al., "Cybersecurity Challenges in UAV Networks," IEEE Trans. Aerosp. Electron. Syst., vol. 60, no. 2, pp. 101–111, 2024.
[30] R. Park, "Quantum Technologies in UAV Disaster Response," IEEE Internet Things J., vol. 11, no. 1, pp. 45–55, 2024.
[31] S. Patel et al., "Ethical Challenges in Autonomous UAV Systems," IEEE Robot. Autom. Lett., vol. 9, no. 3, pp. 201–211, 2024.
[32] V. S. Landers, "Quantum Technologies for Space and Aerial Vehicles," in *Space Governance: Challenges, Threats and Countermeasures*, Cham, Switzerland: Springer Nature, 2024, pp. 105–128.

Index

https://doi.org/10.1515/9783111619323-021

Mathematical Methods in the Digital Age

Volume 1
Satya Prakash Yadav, Abhishek Kumar, Ashutosh Kumar Dubey, Victor Hugo C. de Albuquerque, Adrian Rodriguez-Aguiñaga, Angeles Quezada
Medical Image Processing – Multimodal Fusion Techniques, 2024
ISBN 978-3-11-139549-4, e-ISBN (PDF) 978-3-11-143597-8, e-ISBN (EPUB) 978-3-11-143599-2

Volume 2
Umesh Kumar Lilhore, Vishal Dutt, T. Ananth Kumar, Martin Margala, Kaamran Raahemifar
Math Optimization for Artificial Intelligence – Heuristic and Metaheuristic Methods for Robotics and Machine Learning
ISBN 978-3-11-143605-0, e-ISBN (PDF) 978-3-11-143618-0, e-ISBN (EPUB) 978-3-11-143624-1